Calculus
the language of change

SECOND EDITION

Calculus
the language of change

SECOND EDITION

Keith D. Stroyan

Mathematics Department
University of Iowa
Iowa City, Iowa

ACADEMIC PRESS

San Diego   London   Boston
New York   Sydney   Tokyo   Toronto

Mathematica is a registered trademark of Wolfram Research, Inc.
Maple is a registered trademark of Waterloo Maple, Inc.

ACADEMIC PRESS
525 B Street, Suite 1900, San Diego, CA 92101-4495, USA
1300 Boylston Street, Chestnut Hill, MA 02167, USA
http://www.apnet.com

ACADEMIC PRESS LIMITED
24–28 Oval Road, London NW1 7DX, UK
http://www.hbuk.co.uk/ap/

Library of Congress Cataloging-in-Publication Data

Stroyan, K. D.
    Calculus: the language of change/Keith D. Stroyan—2nd ed.
        p.   cm.
    Rev. ed. of: Calculus using Mathematica. c 1993.
    Includes bibliographical references and index.
    ISBN 0-12-673030-X (acid-free paper)
    1. Calculus—Computer-assisted instruction.   I. Title.
QA303.5.C65S77   1998
515'.078'553—dc21   97-49617
                                                          CIP

Printed in the United States of America
97 98 99 00 01 IP 9 8 7 6 5 4 3 2 1

Preface to the second edition: *Calculus: The Language of Change*

In the preface to the first edition I wrote: "The primary goal of this new calculus curriculum is to show students first hand how calculus acts as the language of change and to answer their question, 'What good is it?' " This has not changed. The wide variety of student projects are prototypes for our student's future professional lives. Students may choose projects *they* find interesting on topics ranging from the price of breakfast rolls to the fate of a bungee diver. As professionals they will use computers for technical details like the integral of secant cubed. (Rather than spending a week reviewing a nineteenth century calculus text.) They still need to know when a quantity of interest is given by an integral.

The ideas of calculus are profoundly useful in many subjects and the projects show the students why. The hard bound text is the backbone of ideas needed to use calculus, but I believe it is important to let students *use* calculus on projects while they are learning these basics. The paperback book of projects is one source for this valuable practice. I hope to support a web site which will have others as they are developed.

The CD included with the book contains computer programs that accompany the text. These programs help students learn basic calculus as well as how to use the computer. This edition uses the much improved *Maple* V and *Mathematica* 3.0 with marvelous graphs, animations, and typeset formulas. The section-by-section programs enable students to work projects a few times per year as a professional would, combining math and computing in a clear presentation of the problem and their solution.

The mathematical background theory appears as an electronic text on the CD. It is not printed and will hopefully also be available on a web site. Some sections of the text also appear only on the CD. Most courses will still want to study some of these CD sections.

Many instructors and students used the first edition of these materials and I tried to benefit from their experience in making the second edition. All of us who taught from the materials tried to assess student progress in many ways. The most dramatic example was a follow-up study of over two thousand students at BYU. They compared the performance of students using three approaches to calculus in 7 subsequent (traditional) technical courses. The students who used *Calculus Using Mathematica* out-scored the other 2 groups by about 10% in 6 courses and averaged 2.93 vs. 2.95 in the 7th. Statistics don't prove anything, but I think it shows we did no harm by our alteration of the traditional syllabus. All the evidence suggests that the term-paper length projects are the source of our student's improved performance.

Modern computing has changed what we should teach our students. The calculus reform opponents who say, 'We still have to teach students all the old drills.' are wrong. Modern computing can free students from many un-interesting old details, and more important, help make real problems accessible to beginners. These are the ways we use computing in this approach to calculus.

I made many changes in response to difficulties or complaints. The projects are quite open-ended and students overwhelmingly appreciate *their* projects. At the same time, they frequently complain about the sometimes vague *Problems*. The difference is that they work long enough on projects to take intellectual ownership of them. Vagueness is my fault and the clear solutions are their doing. It is important to have students formulate the complete question sometimes, but I have tried to make it clearer when this is part of the question and when an *Exercise* is more like high school template work.

I have removed all technical discussion of Robinson's modern infinitesimal numbers (or hyperreal numbers) from the text. The book still has lots of intuitive approximations. The intuitive ideas are explained in detail in the Math Background text on the CD, both in terms of limits and in terms of infinitesimals.

The most peculiar change in the second edition is that the title has changed. First, we think the new title is more descriptive. Calculus is the language that will allow our students to successfully study many technical disciplines from engineering to finance. Second, this edition also has *Maple* Worksheets that are functionally similar to the *Mathematica* materials that accompany it. *Maple* can now do numerics, symbolics, and graphics on the level we need in this new approach to calculus. Both *Maple* and *Mathematica* continue to improve both in being easier to use and in performing more sophisticated computations. Take your pick in software, they're both wonderful.

We don't know all the best answers to questions of teaching calculus with computers. Which skills and details are most important? How can the computer help students learn them? When should we use a computer and when should we work by hand? How much do you have to learn about software? The many instructors who have used *Calculus Using Mathematica* have a good start at answering these questions. We know our students do better in later work and have a more professional attitude about the course itself. I hope you will use *Calculus: The Language of Change*, improve on it, and have fun bringing calculus into the 21st century. Most of all, I hope "C:TLC" contributes substantially to students' professional and intellectual lives. It's hard work, but it's worth it. Calculus is one of the greatest creations of the human intellect and it is also profoundly useful.

Preface to the first edition, *Calculus using Mathematica*

Calculus is primarily important because it is the language of science. It is profound mathematics and a key to understanding physical science and engineering, but calculus also has an expanding role in economics, ecology, and some of the most quantitative parts of business and social science. Beginning college students should learn calculus, because without calculus they close the door to many scientific and technical careers. This book is intended for beginning students who want to become users of calculus in any one of these areas.

A vast majority of our students are convinced of the importance of calculus by working on problems that they find interesting. The primary goal of this new calculus curriculum is to show students first hand how calculus acts as the language of change and to answer their question, 'What good is it?' We 'show' students this by developing their basic skills and then having them apply those skills to their choice of topics from a wide variety of scientific and mathematical projects. Students themselves answer such questions as,

Why did we eradicate polio by vaccination, but not measles?
They present their solution in the form of term papers. (Three large papers and several small ones per year.) This core text does NOT stand on its own; rather it is one of four parts of our materials:

Core text:	*Calculus using Mathematica*
Science projects:	*Scientific Projects for Calculus using Mathematica*
Computing:	*Mathematica NoteBooks for Calculus using Mathematica*
Math projects:	*Mathematical Background for Calculus using Mathematica*

Computing with *Mathematica* has changed both the topics we treat and the way we present old topics. It allows us to achieve our primary goal of having students work real scientific and mathematical projects within the first semester. A number of topics formerly considered too advanced are a major part of our course. *Mathematica* can numerically solve basic differential equations and create a movie animating its 'flow.' This allows us to treat deep and important applications in a wide variety of areas (ecology, epidemiology, mechanics), while only developing basic skills about traditional exponential functions. The ingredients in studying nonlinear 2-D systems are high school math for describing the law of change and exponentials for local analysis.

Mathematica has accessible 3-D graphics, which it can also animate, so we can study problems in more than one variable in the first year. Most problems in science have more than one variable and several parameters.

Mathematica has a convenient front end editor (called NoteBooks) that helps us keep the 'intellectual overhead' to a minimum. Our intention is to use computing to study deep mathematics and applications, not to let the tail wag the dog. We weave *Mathematica* into the fabric of the course and introduce the technical features gradually.

Mathematica also helps students learn the central mathematics of calculus. Our students learn the basic skills of differentiation and integration, but also learn how to use *Mathematica* to perform very elaborate symbolic and numerical computations. We don't labor some of the esoteric 'techniques of integration,' or complicated differentiations. If students' basic skills are backed up with modern computing, it is not necessary to drill them ad nauseam in order to make them proficient mathematical thinkers and users of calculus. Our students demonstrate this in several major term papers on large projects. In addition, their performance on traditional style tests is very good (though the tests only comprise half of their grade.) Good understanding of the main computations and knowledge of how to use them with help from modern graphic, numeric and symbolic computation focuses our students' efforts on the important issues.

How Much Does it Count?

Our course grades were first derived in the following way, though we are shifting credit away from exams as we go.

Traditional Skill Exams (computers not available)

     Exam #1 -15%

     Exam #2 -15%

     Final Exam - 20%

Daily Homework (about 3/4 traditional drill and word problems) - 15%

Weekly electronic homework (submitted electronically) - 10%

Term paper length projects (almost all use computing) - 25%

Suggested Course Syllabi

Two suggested course syllabi that we have used are included in the *Mathematica* Note-Books. The accelerated syllabus is for students who are very well prepared in high school math. We have used it at Iowa and BYU. The regular syllabus is for students with ordinary preparation for mainstream calculus in college. We have used it at Iowa, UNC and UW-L.

Acknowledgements

Many people contributed ideas and effort toward developing these materials. The National Science Foundation made the first edition possible. I am grateful to them all.

Undergraduate students at the University of Iowa, The University of Northern Colorado, The University of Wisconsin - LaCrosse, and Brigham Young University have worked hard in response to our taking them seriously. They showed us which applications of calculus they find interesting. Almost all the projects were developed in close collaboration with students in the course or with undergraduate assistants who took the course. The course is old enough that now some of the assistants are physicians, engineers, computer scientists, actuaries, mathematicians, and medical researchers themselves.

Faculty and graduate students at these schools also contributed a great deal. The first edition faculty leaders were Walter Seaman at Iowa, Steve Leth at UNC, John Unbehaun at UW-L and Gerald Armstrong at BYU. We have been blessed at Iowa with many marvelous graduate assistant teachers. Francisco Alarcon, Randy Wills, Asuman Oktac, Robert Dittmar, Monica Meissen, Srinivas Kavuri, Kathy Radloff, Robert Doucette, Oki Neswan and others helped make the course a success from the start. Recently, I worked with Eric Smith, Dan Coroian, Steve Benson, Chris Cartwright, and Matt Scheutte. Colleagues at Iowa who used the materials and made suggestions for the second edition include Paul Muhly, Ken Atkinson, Norm Johnson, Oguz Durumeric, Dennis Roseman, and Walter Seaman, as well as the graduate assistants who worked with them. Gerald Armstrong and John Unbehaun continued to offer valuable advice.

Ideas from other new calculus projects also contributed to our developments. Frank Wattenberg at the University of Massachusetts was especially helpful. The philosophy of the Five College Project, which Wattenberg's materials spin off, contributed a lot to our course. Lang Moore and David Smith at Duke, Elgin Johnston and Jerry Mathews at Iowa State traded ideas for student projects with us. Deborah Hughes-Hallett, Andrew Gleason and David Lomen of the Harvard project shared rough ideas and offered encouragement at many meetings. Jerry Uhl, Horacio Porta and William Davis of the *Calculus & Mathematica* project did likewise. Our calculus projects share a common goal to show students that calculus is important and thus help train the 21st century's scientists, engineers, mathematicians and technical managers. We are all doing this in different ways and at different levels. The successful feature common to our new courses is: more actively involved students.

Contents

Part 3
3-D Geometry

Part 6
Infinite Series

Part 7
Appendices on CD

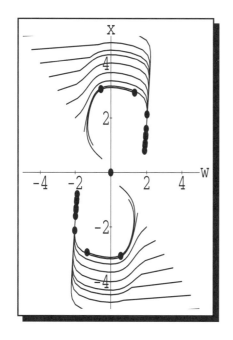

Part 1

Differentiation in One Variable

CHAPTER 1

Calculus with Computers

This book, with its software called "Mathematica NoteBooks" and "Maple Worksheets," is a new approach to an old subject: calculus.

Calculus means "stone" or "pebble" in Latin. Roman numerals were so awkward that ancient Romans carried pebbles to compute their grocery bills (and gamble) and, as a result, "calculus" also came to be known as a method for computing things. The specific calculus of this book is "differential and integral calculus," formerly called "infinitesimal calculus." The fundamental things this calculus computes are rates of change of continuous variables through differentials and the inverse computation: accumulation of known rates of change through integrals. Derivatives and integrals are illustrated in Figures 1.1 and 1.2.

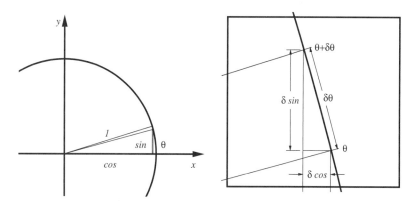

Figure 1.1: Derivatives of sine and cosine

Calculus is one of the great achievements of the human intellect. It has served as the language of change in the development of scientific thought for more than three centuries. During that time it became a coherent polished subject, growing hand in hand with physical science and technology. The contemporary importance of calculus is actually expanding into economics and the social sciences, as well as continuing to play a key role in its traditional areas of application. Chapter 2 previews its use in the study of epidemic diseases, where the models can show us such things as the differences in successful vaccination strategies for polio and measles. Similar models have helped make national health policy changes in screening for gonorrhea.

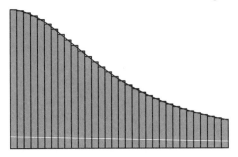

Figure 1.2: Arctangent as an integral

The model in Chapter 2 is a system of differential equations, so it may seem technically a little ahead of the story. The basic ideas are not difficult and require only high school math. There is a straightforward description of the daily rates of change in numbers of sick people that depends on how many people are sick and how many are susceptible. The objective of Chapter 2 is to show that it is relatively easy to describe the spread of a disease such as measles by writing formulas that say how it changes. One new thing in our approach to calculus is that *you* will be expected to *use* calculus to describe applications that interest you. (We offer many choices, including mathematical ones, ranging from bungee jumping to drug absorption to price adjustment. See the accompanying book of projects.) This will show you the role of calculus as a language – you will "speak" it.

Computers make it possible for us to begin the course with a real-life problem such as the study of epidemics. By describing epidemic changes in computer notation, we will immediately get approximate answers to many questions about the spread of diseases. Throughout the course, we will use scientific computing, either *Mathematica* or *Maple*, that is both powerful and relatively easy to use. Computing will help us understand and explore mathematics and to expand the power and applicability of the mathematics we learn. You will leave this course with a good start on learning 21st-century scientific computing.

You will also learn all the old things from traditional calculus in this course. In particular, you will learn to compute derivatives and integrals. Good high school algebra preparation and effort during the semester are all you need. The rote computations that often dominate traditional calculus take practice but not very much if you are proficient at high school math. If you make this effort, you will be well prepared for your later science and math courses. *Mathematica* or *Maple* may even help you check some of your rote skills. You will be far better prepared than traditional calculus students in understanding the role of calculus in science and mathematics.

1.1   Previews of Coming Attractions

Besides basic computing, this chapter shows you some simple examples of the traditional limit that defines the derivative:

$$\lim_{\Delta x \to 0} \frac{f[x + \Delta x] - f[x]}{\Delta x} = f'[x]$$

Examples of the relationship between a function $f[x]$ and its derivative $f'[x]$ are

if $y = f[x] = x^2$,   then $\dfrac{dy}{dx} = f'[x] = 2x^1$

if $y = f[x] = x^3$,   then $\dfrac{dy}{dx} = f'[x] = 3x^2$

if $y = f[x] = x^3 - x^2 - 5x$,   then $\dfrac{dy}{dx} = f'[x] = 3x^2 - 2x - 5$

if $y = f[x] = \text{Sin}[x]$,   then $\dfrac{dy}{dx} = f'[x] = \text{Cos}[x]$

The various derived functions $f'[x]$ go with the original functions, $f[x]$. No doubt you can see a pattern in the first few results above, but do not worry if you have not seen this before. The rules for finding $f'[x]$ from formulas for $f[x]$ are the "little pebbles" of calculus. We return to a systematic development of rules of differentiation in Chapters 5 and 6.

Rules are important but are not the whole story. In this chapter, you will see how the computer can help with the "approximation" part of the rest of the story. Each of the expressions above means that the rate of change of the function $f[x]$ is approximately $f'[x]$ or that "small" changes in the function are approximately linear at rate $f'[x]$. (Linear functions change at a constant rate.) Where is the "approximation" in the foregoing examples? Where is the linearity? Use of rules alone skips the approximation and only gives the answer. The computer together with some basic high school math can help us see the approximation.

The idea of a limit is based on an approximation. The meaning of

$$\lim_{\Delta x \to 0} \frac{f[x + \Delta x] - f[x]}{\Delta x} = f'[x]$$

is that the expression $\dfrac{f[x + \Delta x] - f[x]}{\Delta x}$ gets close to $f'[x]$ as Δx gets small. Or,

$$\frac{f[x + \delta x] - f[x]}{\delta x} \approx f'[x], \quad \text{when } \delta x \approx 0$$

How close and how small are the technical details of the approximation that we treat later in the course.

In this chapter we will see two graphical approximations and animate them with the computer. We return to another view of the main approximation of differential calculus in Chapters 4 and 5. At that point, you will learn the rules and approximations more systematically.

1.2   An Introduction to Computing

One goal of this chapter is to have you run your first Mathematica or Maple program. (They are similar, and you will use whichever one you or your school owns.)

Both *Mathematica* NoteBooks and *Maple* Worksheets include instructions to you, have space for instructions to the computer, and have places for you to type your own interpretations of results. These programs can do numeric, symbolic, and graphical computations in the same program.

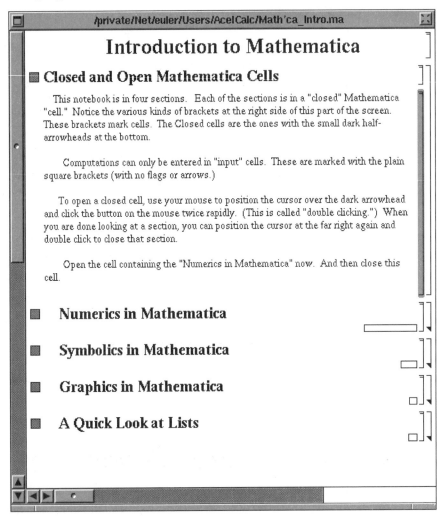

Figure 1.3: A *Mathematica* NoteBook window

The folder **aComputerIntro** uses features of the program to introduce you to *Mathematica* or *Maple*. When you open the programs from that folder, you will see a window that looks something like the one above, though the control buttons may be a little different in your version.

You may be using your own computer or using a network of various kinds of computers that belong to your school. Except for the location of the control buttons, *Mathematica* or *Maple* programs run the same way on all computers. The appropriate programs should be installed on your hard drive or your school's network from the CD that comes with the book.

There is a *Mathematica* version of the **aComputerIntro**, and there is a *Maple* version. They are different in detail but contain the same information. You need to run only one version of each program depending on your system.

If your machines are networked, you need to learn about the details from your instructor. These vary from place to place, even with the same kind of machine.

Exercise set 1.2

1. *If your computers are networked, log in, change your password, and log out.*

You need to learn to do "mouse" editing. Of course, it is a little frustrating sometimes to use the computer to get an introduction to the computer. Stick with it; once you learn a little "mouse editing," you will find that almost all the computer information you need is contained in the *Mathematica* or *Maple* programs. You could also ask a friend or lab monitor to help.

2. *Log on to your account with your new password, run the programs in **aComputerIntro**.*
Read the instructions in the programs and do the exercises in it. When you are done, save it.

Your instructor will tell you how you should save your electronic homework. This depends on the kind of network and disk drives that are installed with your system.

1.3   Linearity in Local Coordinates

Since we want to understand how calculus approximates nonlinear functions by linear ones, we will first reformulate linear functions in terms of their changes. This is describing a line by its "differential equation."

We will look at this formulation symbolically (or algebraically), graphically (or geometrically), and numerically, since it is often helpful in mathematics to view a topic from all three of the computational points of view described in the programs of **aComputerIntro**.

The change in a variable is a difference, the new value minus the old value, so we abbreviate change by Greek capital delta, Δ (for difference, small delta is written δ). Using these symbols, we have the definition

Definition 1.1. *y Varies Linearly with x*
Let y be a real valued function of a real variable x, y = f[x]. We say the dependent variable, y, varies linearly with the independent variable, x, if the ratio of the change in y over a corresponding change in x always equals the same constant, m,

$$\frac{\Delta y}{\Delta x} = m$$

for any pair of (x, y) points satisfying $y = f[x]$. For example, if $y_1 = f[x_1]$, $y_2 = f[x_2]$ and $y_3 = f[x_3]$, then the following ratios of differences are all equal,

$$\frac{y_2 - y_1}{x_2 - x_1} = \frac{y_2 - y_3}{x_2 - x_3} = \frac{y_3 - y_1}{x_3 - x_1} = m$$

We use this definition to get equations for a line three ways: with "local coordinates" directly, and with ordinary coordinates to give the point-slope formula, and the two-point form of a line. Algebraic simplification of these forms leads to the familiar slope-intercept form of a line:

$$y = f[x] = m\,x + b$$

This algebraically simplified form is often *not* simplest when we are interested in the change of a nonlinear function near a certain point. Local coordinates put it in the form "$y = m\,x$," but since (x, y)-coordinates are already used, we introduce new variables (dx, dy) to do this.

Example 1.1. *Geometric Linearity*

Two pairs of points on the graph of a linear function are shown in Figure 1.4 along with horizontal and vertical segments corresponding to the changes in x and y. Notice that the two triangles are similar.

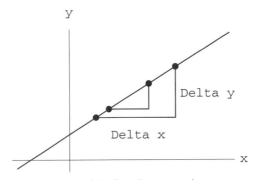

Figure 1.4: Similar triangles

Definition 1.1 is equivalent to the statement that any two triangles along the curve are similar. This is because similarity of the geometric figures means the ratios of corresponding sides are equal. Geometrically, the constant m in Definition 1.1 is the slope, so "a line has constant slope" is the equivalent geometric statement that y varies linearly with x.

Example 1.2. *Numerical Linearity*

Data from an accurate experiment might look something like the following (where the two place accuracy removes measurement errors).

Linear (x, y) Data			
x	1.0	2.0	5.0	6.3
y	3.0	1.0	-5.0	-7.6

Table 1.1: Linear data

These data are linear because the ratios of all changes equal the same constant.

The change in x from the first point to the second is $2.0 - 1.0 = 1.0$, the change in y from first to second is $1.0 - 3.0 = -2.0$ (taking the difference in the same order), and

$$\frac{\Delta y}{\Delta x} = \frac{1.0 - 3.0}{2.0 - 1.0} = -2.0$$

Similarly, the ratio of changes from second to third is

$$\frac{\Delta y}{\Delta x} = \frac{-5.0 - 1.0}{5.0 - 2.0} = \frac{-6.0}{3.0} = -2.0$$

The ratio of changes from third to fourth is

$$\frac{\Delta y}{\Delta x} = \frac{-7.6 - (-5.0)}{6.3 - 5.0} = \frac{-2.6}{1.3} = -2.0$$

The ratio of changes from second to fourth is

$$\frac{\Delta y}{\Delta x} = \frac{-7.6 - 1.0}{6.3 - 2.0} = \frac{-8.6}{4.3} = -2.0$$

The numerical point of Definition 1.1 is that any way we compute the ratio of changes, we get the same constant.

There is no reason to separate the numerical, graphical, and symbolic views of a topic; in particular, it is perfectly natural to graph the data in Table 1.1 as well as to find a formula that describes the graph and data. Both the graph and formula are more powerful representations in the sense that they condense information about all the points between the data points in convenient forms (although the data may be all that are scientifically known.)

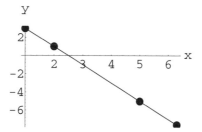

Figure 1.5: Graph of Table 1.1

The formula derived from any two (x, y) points of Table 1.1 also provides a simpler way to demonstrate linearity. The remaining ones can just be checked by substitution into the formula.

Notice that the computer chose axes starting at the first point of the table, rather than at $(0, 0)$. In calculus, we often need a formula for the line with a given slope through a certain point. We will see this as the tangent line problem in the second half of this section.

For this reason, we will develop special "local" coordinates for this problem. We start with a fresh example and leave a unified approach to the tabular data to one of your exercises.

Example 1.3.   *Local Coordinates for the Point-Slope Form of a Line*

Suppose we know that a line goes through the point $(-4, 3)$ and has slope -2. The easy way to sketch this line would be to start at the point $(-4, 3)$ and move $+1$ in the x direction, then -2 in the y direction, put our ruler down, and draw. We want a symbolic version of this procedure of changing from a base point.

We define new coordinates (new x, new y) $= (dx, dy)$ called *local coordinates through* $(-4, 3)$ (Figure 1.6). These have their origin at $(-4, 3)$ and are parallel to the (x, y) coordinates. Notice that an amount of dx represents a *change* in x from the point -4 and an amount of dy represents a change in y from 3. [The quantities Δx and Δy of Definition 1.1 represent any differences, not necessarily based at $(-4, 3)$.] In local coordinates, Definition 1.1 says the equation of our line is

$$\frac{dy}{dx} = -2 \qquad \text{or} \qquad dy = -2\ dx$$

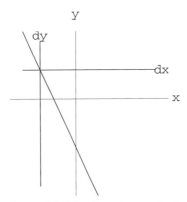

Figure 1.6: The line $dy = -2\ dx$

Example 1.4.   *The (x, y) Point-Slope Formula from the Local Formula*

How can we change the local equation $dy = -2\ dx$ to our original (x, y)–coordinates? The (dx, dy) origin is the same as the (x, y) point $(-4, 3)$. Local coordinates measure changes in x and y from this point. This means we could have defined the local coordinates symbolically by

$$dx = x - (-4) = x + 4$$
$$dy = y - 3$$

so the line can be written in either of the forms

$$\frac{dy}{dx} = -2 \qquad \Leftrightarrow \qquad \frac{y - 3}{x + 4} = -2$$

Here is another explanation. The change in y moving from 3 to the generic value y is the difference in y, $\Delta y = y - 3$, and the change in x in moving from -4 to the generic value x is the difference in x, $\Delta x = x - (-4) = x + 4$, as shown in Figure 1.7.

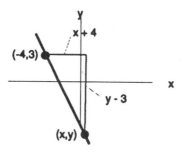

Figure 1.7: Changes from $(-4, 3)$ to (x, y)

The ratio of any pair of corresponding differences equals the constant slope m, $\frac{\Delta y}{\Delta x} = m$, and our particular changes are dx and dy, so the following are equivalent:

$$\frac{dy}{dx} = m \qquad \Leftrightarrow \qquad \frac{\Delta y}{\Delta x} = m \qquad \Leftrightarrow \qquad \frac{y - 3}{x + 4} = -2$$

We can simplify this equation algebraically, but it is often best left in this form when we are interested in the changes in the variable near $(-4, 3)$. Local coordinates build this feature in by moving the origin to this point. The algebraically simplified formula has the more compact form $y = m\,x + b$, but the difference form (above at the right) shows the point and the slope. Local coordinates are already simplified at their own origin.

In general, the local coordinates can be defined by

Definition 1.2. *(dx, dy) Coordinates*
The local coordinates through the fixed (x, y) point (x_0, y_0) are given by

$$dx = x - x_0$$
$$dy = y - y_0$$

Local coordinates are nothing mysterious, but they are very useful in calculus.

Procedure 1.3. The General (x, y) Point-Slope Formula

Find the (x, y) equation of a line of slope m through the fixed point (x_0, y_0).
(a) Write the local equation:
$$\frac{dy}{dx} = m$$
(b) Replace dy and dx by their definition in terms of x and y:
$$\frac{y - y_0}{x - x_0} = m$$

You should verify that this is the same as the point-slope formula if you memorized that in high school.

Example 1.5. *Two-Point Formula from the Change Formula*

Here is another example of using change to describe a line. Suppose our line goes through the points $(x, y) = (-2, 3)$ and $(x, y) = (1, 4)$.

Before we can use Procedure 1.3, we need to compute the slope. The two given points tell us the slope:

$$\frac{\Delta y}{\Delta x} = \text{constant}$$

$$\frac{4 - 3}{1 - (-2)} = \frac{4 - 3}{1 + 2} = \frac{1}{3}$$

The differences needed to compute the slope are shown on Figure 1.8.

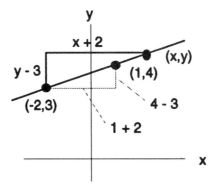

Figure 1.8: Two points determine the line $y = \frac{1}{3}x + \frac{11}{3}$

Now step (1) of Procedure 1.3 says the equation in local coordinates at $(x, y) = (-2, 3)$ is:

$$\frac{\Delta y}{\Delta x} = \text{constant}$$

$$\frac{dy}{dx} = \frac{1}{3}$$

Step (2) replaces local coordinates at $(x, y) = (-2, 3)$ by differences in the original coordinates:

$$\frac{dy}{dx} = \frac{y - 3}{x - (-2)} = \frac{y - 3}{x + 2} = \frac{1}{3}$$

The general differences $y - 3$ and $x + 2$ are shown on Figure 1.8.

We can also solve this problem using the change in moving from $(1, 4)$ to the general (x, y) point:

$$\frac{dy}{dx} = \frac{1}{3}$$

$$\frac{y - 4}{x - 1} = \frac{1}{3}$$

In local coordinates the equation is $\dfrac{dy}{dx} = \dfrac{1}{3}$, but the point at which we localize affects the $x - y-$ form. The equations we got from the different base points are equivalent:

$$\frac{y-3}{x+2} = \frac{1}{3} \qquad \Leftrightarrow \qquad \frac{y-4}{x-1} = \frac{1}{3}$$

The easiest way to see this is to simplify both algebraically, putting them in the form $y = m\,x + b$:

$$\frac{y-3}{x+2} = \frac{1}{3} \qquad\qquad\qquad \frac{y-4}{x-1} = \frac{1}{3}$$

$$y - 3 = \frac{1}{3}(x+2) \qquad\qquad\qquad y - 4 = \frac{1}{3}(x-1)$$

$$y - 3 = \frac{1}{3}x + \frac{2}{3} \qquad\qquad\qquad y - 4 = \frac{1}{3}x - \frac{1}{3}$$

$$y = \frac{1}{3}x + \frac{2}{3} + 3 \qquad\qquad\qquad y = \frac{1}{3}x - \frac{1}{3} + 4$$

$$y = \frac{1}{3}x + \frac{11}{3} \qquad\qquad\qquad y = \frac{1}{3}x + \frac{11}{3}$$

In high school, you learned various symbolic forms of the equation of a line: the point-slope formula, the two-point formula, as well as slope-intercept formula. You do not need to remember the high school formulas as long as you can use Definition 1.1. In any case, start to think in terms of changes rather than only the fixed form $y = m\,x + b$.

Example 1.6. *Linearity and Constant Rates*

When we view the independent variable as time, Definition 1.1 of linearity is equivalent to the statement that the dependent variable changes at a constant rate. Time rates of change are important in many applications of calculus, and we will see them often later in the course. Once we understand an idea in x and y, it is often advantageous to understand it in terms of other variables, such as time, t.

Exercise set 1.3

1. Local Coordinates for a Line

(a) *Suppose a line goes through the point $(5, -4)$ and has slope -3. Define local coordinates (dx, dy) with origin at the point $(5, -4)$, giving formulas for dx and dy in terms of x and y.*
What is the (dx, dy) equation for this line?
What is the (x, y) equation of the line?
What is the (x, y) slope-intercept form $(y = m\,x + b)$ of this line?

(b) *Suppose a line goes through the fixed point (x_0, y_0) and has slope m.*
What are the local (dx, dy) coordinates through (x_0, y_0)?
What is the (dx, dy) equation of the line?
What is the (x, y) equation of the line?
What is the (x, y) slope-intercept form $(y = m\,x + b)$ of this line $(b = ?)$?

2. Point-Slope Form to Slope-Intercept Form

(a) *Algebraically simplify the expression* $\dfrac{y-3}{x+4} = -2$ *showing that it is equivalent to the expression* $y = -2\,x - 5$ *(when $x \neq -4$).*

i)   *This line has slope -2. Use both formulas to show this.*

ii)   *This line contains the point $(3,-4)$. Use both formulas to show this.*

iii)   *This line crosses the y-axis at $y = -5$. Use both formulas to show this.*

(b) *Simplify the following expressions and give the slope and y-intercept:*

i) $\dfrac{y-2}{x-3} = -4$ *ii)* $\dfrac{y-2}{x+3} = 4$ *iii)* $\dfrac{y+2}{x-3} = -4$

iv) $\dfrac{y}{x-3} = 1$ *v)* $\dfrac{y-3}{x} = -2$ *vi)* $\dfrac{y}{x} = 3$

(c) *Simplify the expression* $\dfrac{y-y_0}{x-x_0} = m$, *showing that it is equivalent to $y = m\,x + b$ (when $x \neq x_0$), where $b = y_0 - m\,x_0$. What is the slope of this line? Use both formulas to show that the line contains the point (x_0, y_0).*

3. Two Points Determine a Line

(a) *Suppose a line goes through the points $(3,-2)$ and $(-4,5)$. Find its slope.*

i)   *Use the slope and the point $(3,-2)$ to find its (x,y) equation.*

ii)   *Use the slope and the point $(-4,5)$ to find its (x,y) equation.*

iii)   *Compare the two equations by putting both in slope-intercept form.*

(b) *Find the slope-intercept equation of the line that passes through:*

i) $(1,2)$ *and* $(3,4)$ *ii)* $(2,-1)$ *and* $(-3,4)$ *iii)* $(-2,3)$ *and* $(5,-7)$

4. Graphics to Symbolics
Find an equation for each of the graphs shown below.

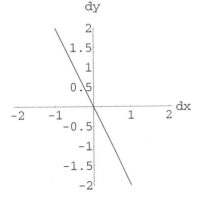 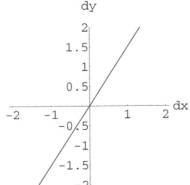

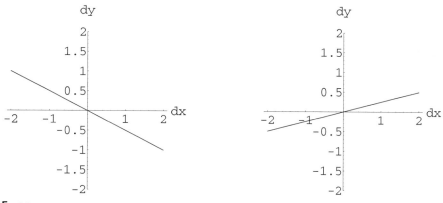

5. Numerics to Symbolics

   (a) *Find an equation satisfied by all the points on Table 1.1.*
   (b) *(Optional) Solve this with the computer using the program **LinearIntro**.*

6. The **LinearIntro** Program

   *Run the program **LinearIntro** in the Chapter 1 folder of the course software and answer the questions in it. Save the results in your computer account (or homework disk). **LinearIntro** shows you how to solve Exercise 1.3.4 with the computer. Comparing your paper-and-pencil work with your electronic work should help you begin the course computing.*

7. *Newton's Law of Cooling says that the rate of cooling of an object is proportional to the difference between the temperature of the object and the temperature of its surroundings. Let F denote the Farenheit temperature of the object and t denote the time in minutes beginning with $t = 0$ at our first observation. (See Chapters 4 and 8.)*
   (a) *Suppose an object is 75° when we first observe it and it cools to 60° ten minutes later. What is its (average) rate of cooling over this time period?*
   (b) *Sketch the line segment passing through the points $(t, F) = (0, 75)$ and $(10, 60)$.*
   (c) *What is the slope of this segment?*
   (d) *What is the equation of the line through these two points?*
   Newton's Law of Cooling does not give the temperature as a function of the time, but rather says, $R = k(F - F_0)$, where R is the rate of cooling, F_0 is the temperature of the surroundings, and k is a constant. The line through the points above is not a good long-term prediction of temperature because its slope is constant and that slope is the rate referred to in Newton's Law. Calculus will give us a simple exponential formula.

8. *The downward velocity v of an object released in vacuum near the surface of the earth satisfies a linear equation, $v = gt + v_0$, where g is Galileo's universal gravitational constant, $g = 9.8$ m/sec² or $g = 32$ ft/sec². (See Chapter 10.)*
   (a) *Suppose we first observe a falling object traveling 13 ft/sec downward. How fast will it be falling 2 seconds later?*
   (b) *Suppose we first observe an object thrown 13 ft/sec upward. How fast will it be falling 2 seconds later?*

1.4   Derivatives for Explicit Formulas

This section begins with more data. The data are NOT linear because the ratios of all changes do NOT equal the same constant.

Nonlinear (x, y) Data				
x	-1.000	0.000	1.000	1.500	2.000
y	-1.000	0.000	1.000	3.375	8.000

Table 1.2: Nonlinear Data

Four differences from first to second, second to third, etc. are:

$$\frac{0.0 - (-1.0)}{0.0 - (-1.0)} = 1, \quad \frac{1.0 - 0.0}{1.0 - 0.0} = 1, \quad \frac{3.375 - 1.0}{1.5 - 1.0} = 4.75, \quad \frac{8.0 - 3.375}{2.0 - 1.5} = 9.25$$

A graph of the data is shown in Figure 1.9

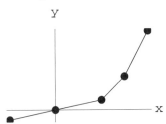

Figure 1.9: The data of Table 1.2 connected by lines

The slopes of segments connecting points on this curve do NOT remain the same. These points lie on the curve $y = f[x] = x^3$ as shown in Figure 1.10

Figure 1.10: The data on $y = f[x] = x^3$

The slope of the curve varies "continuously" as a point moves along it. The derivative measures this changing slope.

Example 1.7.   *The Symbolic Derivative of $y = x^3$*

One way to approach the derivative is to choose a fixed (x, y) point and look at the slope of the segment that differs by a small amount Δx in x. The corresponding change in y now must be computed with the function $f[x] = x^3$ at the new x point $x + \Delta x$. This is

$$f[x + \Delta x] = (x + \Delta x)^3 = x^3 + 3x^2 \, \Delta x + 3x(\Delta x)^2 + (\Delta x)^3$$

The change in y is the new value minus the old value:

$$
\begin{aligned}
\Delta y = f[x + \Delta x] - f[x] &= (x + \Delta x)^3 - x^3 \\
&= x^3 + 3x^2 \Delta x + 3x(\Delta x)^2 + (\Delta x)^3 - x^3 \\
&= 3x^2 \, \Delta x + 3x(\Delta x)^2 + (\Delta x)^3 \\
&= 3x^2 \cdot \Delta x + (3x + \Delta x) \cdot (\Delta x)^2
\end{aligned}
$$

The ratio of these changes is

$$\frac{\Delta y}{\Delta x} = 3x^2 + (3x + \Delta x) \cdot (\Delta x)$$

In particular, if $x = 1.0$ and $\Delta x = 1.0$, we have

$$
\begin{aligned}
\frac{\Delta y}{\Delta x} &= 3 \, x^2 + (3 \, x + \Delta x) \cdot (\Delta x) \\
&= 3 \cdot 1^2 + (3 \cdot 1 + 1) \cdot 1 \\
&= 7
\end{aligned}
$$

We could have seen this particular result more easily from

$$(f[x + \Delta x] - f[x])/(\Delta x) = (2^3 - 1)/(1 - 0) = (8 - 1)/1 = 7$$

but the symbolic expression is more powerful. It lets us see what happens as Δx gets small:

$$
\begin{aligned}
\lim_{\Delta x \to 0} \frac{\Delta y}{\Delta x} &= \lim_{\Delta x \to 0} 3x^2 + (3x + \Delta x) \cdot (\Delta x) \\
&= 3x^2 + \lim_{\Delta x \to 0} (3x + \Delta x) \cdot (\Delta x) \\
&= 3x^2 + 0
\end{aligned}
$$

The limit is $3 \, x^2$ because $3 \, x^2$ does not change as Δx gets smaller and smaller, whereas the remaining expression,

$$(3x + \Delta x) \cdot (\Delta x)$$

tends to zero. It gets small as Δx gets small since it is multiplied by the small number Δx. For example, suppose we want the error $(3x + \Delta x) \cdot (\Delta x)$ to be less than $1/1000$ and we know that $|x| < 1,000,000$. The term $|3x + \Delta x| < 3,000,001$, so if we take any $|\Delta x| < 1/(4 \times 10^9)$, the error is less than a thousandth.

To summarize the example so far, we have shown

$$\lim_{\Delta x \to 0} \frac{f[x + \Delta x] - f[x]}{\Delta x} = f'[x]$$

$$\lim_{\Delta x \to 0} \frac{(x + \Delta x)^3 - x^3}{\Delta x} = 3x^2$$

What does this computation mean? Graphically, we have found that the slope of the segment from the point $(x, f[x])$ to the point $(x + \Delta x, f[x + \Delta x])$ gets closer and closer to $3 \, x^2$ as Δx gets smaller and smaller. In other words, the slope of the line through closer and closer points on the curve tends to $3\,x^2$. You can see this dynamically on the computer program **DerivLimit** in the course software. Figure 1.11 shows three successively closer points.

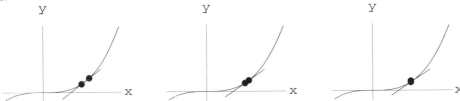

Figure 1.11: Successively closer points on $y = x^3$

The successive lines tend toward the graph of the line $dy = m\,dx$ in local coordinates through $(x, f[x])$ where $m = 3\,x^2$ [x is considered fixed for the (dx, dy)-plot]. The equation

$$dy = 3x^2\,dx$$

gives the local coordinates of the tangent line to the curve at the point shown in Figure 1.12:

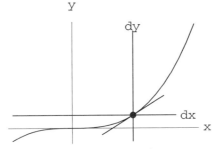

Figure 1.12: $y = x^3$ and $dy = 3\,x^2\,dx$ on one graph

We summarize the limit computation by the following statement:

$$\text{If}\quad y = f[x] = x^3, \quad \text{then} \quad \frac{dy}{dx} = f'[x] = 3x^2$$

The new function $f'[x] = 3x^2$ is called the derivative of $f[x] = x^3$, and we say that the slope of the curve at the single point x is $f'[x]$ since its tangent line at this point has this slope.

We can also see the convergence of the limit

$$\lim_{\Delta x \to 0} \frac{f[x + \Delta x] - f[x]}{\Delta x} = f'[x]$$

by plotting both $(f[x + \Delta x] - f[x])/\Delta x$ and the function $f'[x]$ on the same (x, y) graph for smaller and smaller Δx. The program **DerivLimit** also shows the convergence of the approximate formulas to the limiting formula. It can even do this dynamically as an animation or computer generated movie.

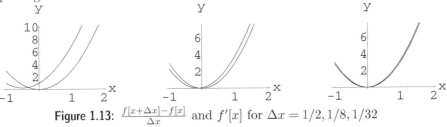

Figure 1.13: $\frac{f[x + \Delta x] - f[x]}{\Delta x}$ and $f'[x]$ for $\Delta x = 1/2, 1/8, 1/32$

━━━━━━━━━━━━━━━ **Exercise set 1.4** ━━━━━━━━━━━━━━━

1. *For the function $f[x] = x^3$, verify that when $x = 1.0$ and $\Delta x = 0.5$ the expression*

$$\frac{\Delta y}{\Delta x} = \frac{f[x + \Delta x] - f[x]}{\Delta x} = 4.75$$

but that when $x = 1.5$ and $\Delta x = 0.5$ the expression

$$\frac{\Delta y}{\Delta x} = \frac{f[x + \Delta x] - f[x]}{\Delta x} = 9.25$$

2. *Let $f[x] = x^2$. This exercise has you show step by step that $f'[x] = 2x$.*

Expand the expression $f[x + \Delta x] = (x + \Delta x)^2$.
Use your expansion to show that $f[x + \Delta x] - f[x] = 2x \cdot \Delta x + (\Delta x)^2$.
*Show that the $\lim_{\Delta x \to 0} \frac{f[x+\Delta x]-f[x]}{\Delta x} = 2x$. Check your work with the **Symbolics** program in aComputerIntro.*

3. *Let $f[x] = x^4$. This exercise has you show step by step that $f'[x] = 4x^3$.*

Expand the expression $f[x + \Delta x] = (x + \Delta x)^4$.
Use your expansion to show that $f[x + \Delta x] - f[x] = 4x^3 \cdot \Delta x + (\Delta x)^2 \cdot (stuff)$.
*Show that the $\lim_{\Delta x \to 0} \frac{f[x+\Delta x]-f[x]}{\Delta x} = 4x^3$. Check your work with the **Symbolics** program in aComputerIntro.*

4. *Run the computer program **DerivLimit**, and verify convergence behind the rules*

$$if \quad y = f[x] = x^2, \quad then \quad \frac{dy}{dx} = f'[x] = 2x^1$$

$$if \quad y = f[x] = x^4, \quad then \quad \frac{dy}{dx} = f'[x] = 4x^3$$

$$if \quad y = f[x] = x^3 - x^2 - 5x, \quad then \quad \frac{dy}{dx} = f'[x] = 3x^2 - 2x - 5$$

$$if \quad y = f[x] = \mathrm{Sin}[x], \quad then \quad \frac{dy}{dx} = f'[x] = \mathrm{Cos}[x]$$

1.5 Review of High School Math

Chapter 28 on the CD reviews some of the most important high school math topics needed in calculus. Take a brief look at it now, and return to it as needed throughout the course. Here are a few topics from the CD review chapter.

1.5.1 Function Notation and Substitution of Expressions in Functions

The CD Chapter 28 has a careful explanation of function notation and substitution of expressions in functions such as the one used in our preview of derivatives:

$$\frac{f[x + \Delta x] - f[x]}{\Delta x} = \frac{(x + \Delta x)^3 - x^3}{\Delta x}$$

$$= \frac{(x^3 + 3x^2\,\Delta x + 3x(\Delta x)^2 + (\Delta x)^3) - x^3}{\Delta x}$$

$$= 3x^2 + 3x(\Delta x) + (\Delta x)^2$$

1.5.2 Types of Explicit Functions

The basic high school functions defined by explicit formulas are

Linear Functions	$y = mx + b$ or	$\dfrac{\Delta y}{\Delta x} = m$
Polynomials	$y = a_0 + a_1x + \ldots + a_nx^n,$	a_j constants
Trigonometric Functions	$y = \mathrm{Sin}[\theta], \quad y = \mathrm{Cos}[\theta], \quad y = \mathrm{Tan}[\theta]$	
Inverse Trig Functions	$\theta = \mathrm{ArcTan}[y]$	
Exponential Functions	$y = a \cdot b^x,$	$a, b > 0$ constants
The Natural Logarithm	$x = \mathrm{Log}[y]$	

Note: Natural log is sometimes denoted $\ln(y)$ in high school texts, where $\log(y)$ denotes the base 10 log. We do not need base 10 logs.

1.5.3 Format for Homework

Chapter 28 includes a general approach to "word problems." In homework problems involving applications

Procedure 1.4. Homework Format
(a) Explicitly list your variables with units and sketch a figure if appropriate.
(b) Translate the information stated in the problem into formulas in your variables. (If this translation is difficult, you may not have chosen the best variables.) Often it is helpful to balance units on both sides of an equation.
(c) Formulate the question in terms of your variables and solve.
(d) Explicitly interpret your solution.

There are practice word problem examples in the CD Chapter 28.

1.5.4 Review as You Go

We will review several important high school topics in the main text at the beginning of sections where they are needed. At those places, we will also reference the relevant parts of the CD Chapter 28. We urge you to forge ahead with calculus and brush up on high school material as needed when you need it.

1.6 Free Advice

We know that the first few computer exercises take a long time and may be frustrating for people who are not used to "windows" and "mouse editing." These initial frustrations will seem very simple in a week or so if you confront them now. DO NOT GET BEHIND IN YOUR HOMEWORK.

We encourage you to work with a friend on the computing in this course. Often two heads are better than one because one of you can concentrate on the machine details while the other thinks about the main mathematical task you are trying to accomplish. As long as you change roles occasionally, you will both learn faster.

There are several kinds of problems for you to work in this course.

1.6.1 Drill Exercises

Routine algorithms that just need practice and not much thinking are called "drill exercises." Some of these are expanded on in the Mathematical Background book on the CD that accompanies this core text. Some instructors like to give drill tests as "barrier exams." This means that you have several chances to pass a skill exam but that you must get 90% of the problems completely correct. You do not get a grade for the skill test but cannot pass the course unless you hurdle it. These are necessary skills, but the skills alone do not mean that you understand the material.

1.6.2 Exercises

Regular "exercises" may require a little more thought about the current material than drill exercises. Calculus is important because the basic skill algorithms have important meanings. The skills themselves are not the whole story, and exercises sometimes take a step in the direction of their meaning. Exercises are sometimes "electronic"; that is, they require you to use a *Mathematica* NoteBook or *Maple* Worksheet. Most instructors assign one large electronic homework assignment per week. These assignments show you how to do basic calculations on the computer. Once you have done these, you can use the computer to solve regular homework exercises and work projects.

1.6.3 Problems

"Problems" are larger exercises that are meant to help you organize your thinking. They go beyond exercises in that they usually have several parts and ask you to write summary explanations of the combined meaning of the parts. Your written explanations are very important. You will find that it is sometimes difficult to put your ideas into words. You "understand" but find it hard to explain the problem. Wrestling with this difficulty improves your understanding and helps you to combine the parts of calculus that you have learned into coherent understanding.

1.6.4 Projects

"Projects" are larger problems - serious applications of calculus, not fragmented parts of applications or text exercises contrived to come out in whole numbers. Scientific Projects comprise a paperback book accompanying this text. You should look through the table of contents of the Scientific Projects to get an idea of how widely YOU can apply calculus. There are projects on topics ranging from bungee jumping to extinction of whales.

CHAPTER 2

Using Calculus to Model Epidemics

This chapter shows you how the description of changes in the number of sick people can be used to build an effective model of an epidemic. Calculus allows us to study change in significant ways.

In the United States, we have eradicated polio and smallpox, yet, despite vigorous vaccination campaigns, measles remains a persistent pest. Why were we able to eradicate polio but not measles? You can work out the answer to this question in the Herd Immunity project after you have studied this chapter. Calculus as the language of change really does give us deep insights. Plain English is not as powerful in understanding certain consequences of change. We want you to see an example immediately because the primary goal of our course is to show you that calculus has important things to contribute to many real problems.

The computer makes this study possible at such an early stage. The change model involves simple but tedious arithmetic. The formulas for this arithmetic look awfully messy, but you will be able to make computations easily in the short tables contained in the exercises. The computer will carry out much longer computations based on the same idea; the electronic exercises show you how to program the appropriate computations. The computer conveniently provides numerical and graphical results, so we can analyze our results and begin to develop insight into model epidemics.

Differential equations are ahead of our story only in a technical, not conceptual sense. You will see that the equations say what the various rates of change are, but you will not "integrate" them symbolically. You don't need to know anything about symbolic derivatives or integrals (not even the ones in Chapter 1). Concentrate on the idea of how infectious individuals change the number of sick people over time, and let the computer do the arithmetic. The moral of the chapter is that these simple rates of change give us important information with the help of the computer. Seeing this will also raise new questions but at a higher level. Later, calculus and the computer together will answer these questions.

2.1   The First Model

An epidemic is a large short-term outbreak of a disease. This section develops a simple model of the spread of a disease.

Human epidemics are often spread by contact with infectious people, although sometimes there are "vectors," such as mosquitos, rats and fleas, or mice and ticks involved in disease transmission. There are many kinds of contagious diseases, such as smallpox, polio, measles, and rubella, that are easily spread through casual contact. Other diseases, such as gonorrhea, require more intimate contact.

Another important difference between the first group of diseases and gonorrhea is that measles "confers immunity" to someone who recovers from it, whereas gonorrhea does not. In other words, once you recover from rubella, you cannot catch it again. This feature offers the possibility of control through vaccination. In this section we will formulate our first model of the spread of a disease such as rubella to which people are susceptible, then infectious, and finally, recovered and immune. (Rubella is commonly called "German measles." It is usually mild, but may produce severe birth defects if contracted by a pregnant woman during the first trimester.)

A scientific model is necessarily a simplification. We cannot include contacts between particular individuals or even other diseases of the population, diet, the amount of travel, the weather and other factors that have "small" effects on the spread of the disease, because we would not be able to make computations and predictions with such a complicated model. By identifying the "main" effects, we can formulate a model that is simple enough to use to compute and predict the overall spread of a disease. That is the good news. The bad news is that our model cannot predict specifics, such as when you will contract rubella. Some exercises will ask you to use the model to make predictions, but others will ask you to think about and discuss limitations of the model. The usefulness of a model requires us to make correct scientific judgments about which effects are "small" and which are not. We need to test our models against known data both to measure important parameters and to verify that we have neglected only small effects. Once this is done, we can use the model to make predictions in new situations.

2.1.1 Basic Assumptions

We will make the following assumptions in formulating our model:

(a) **SIR:** All individuals fit into one of the following categories:
    Susceptible: those who can catch the disease
    Infectious: those who can spread the disease
    Removed: those who are immune and cannot spread the disease

(b) **The population is large** but fixed in size and confined to a well-defined region. You might imagine the population to be a large university during the semester, when relatively little outside travel takes place.

(c) **The population is well mixed;** ideally, everyone comes in contact with the same fraction of people in each category every day. Again, imagine the multitude of contacts a student makes daily at a large university.

When formulating models, it is very important to make your assumptions clear. We have listed them explicitly to help make them clear. The exercises will ask you about the limitations of these assumptions.

The categories of people and the way we have assumed that they move between the categories can be summarized in the graphic "compartments" shown in Figure 2.1.

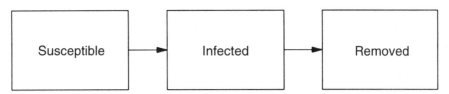

Figure 2.1: Disease compartments

We want our mathematical model to be able to compute the number of people in each of the compartments at any given time. In mathematical models it is also important to keep a list of variables. Variables are letters that stand for quantities that can change. We must express the main properties of the epidemic in terms of these variables.

2.1.2   Variables for Model 1

$$
\begin{aligned}
t &= \text{the time in days with t=0 at the start of observation} \\
S &= \text{the number of susceptible individuals} \\
I &= \text{the number of infectious individuals} \\
R &= \text{the number of removed individuals}
\end{aligned}
$$

These variables keep track of the number of individuals in each of the compartments as shown in Figure 2.2. Time t is the independent variable; and S, I, and R are dependent on time. In other words, they are functions of time, but they are not functions given by explicit formulas like the ones in Chapter 28 on high school review (such as $y = x^3$. Even though we will be able to compute $S = S[t]$, there is no explicit formula for S in terms of t.)

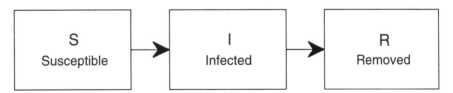

Figure 2.2: Compartment sizes at time t

2.1.3   Derivation of the Equations of Change

Now we want to add a mathematical description of the way individuals move among the susceptible, infectious, and removed compartments. This will add two parameters to our model, a and b, that are specific to a particular S-I-R disease. Parameters are letters that have a fixed value during the entire problem but can change from one problem to the next. We begin with the simplest change - how one recovers.

A person with rubella is infectious for about 11 days. The infectious compartment will contain people who have had the disease for 1 day, 2 days, and so on up to 11 days. At 12 days an individual moves from the infectious to the removed compartment. Of course, not everyone gets the disease at exactly 8:00 am. Also, as the epidemic spreads, more people may be in the 1 day group than in the 11 day group. Nevertheless, we will assume that the different groups in the infectious compartment are roughly the same size so that we can make a simple statement about recovery: For rubella, the number of people added to the removed compartment tomorrow is $\frac{1}{11}$th of the infectious compartment today,

$$\text{new people in } R\text{-compartment tomorrow} = \frac{1}{11}I$$

A general disease, from which one recovers and gains immunity, has the day's change in the recovered compartment given by,

$$\text{new people in } R\text{-compartment tomorrow} = b\,I$$

where the parameter b controls how quickly people recover, or $1/b$ is the number of days one stays in the infectious compartment. For rubella, $b = 1/11$.

Notice that the units of the right side of the above equation for the day's change in R are "number of people per day." We will keep careful account of units.

The next change we describe is how one spreads the disease. If you are infectious with rubella and you sit through a whole class next to a person who is susceptible, then you will probably spread the disease to that person. If you sit through class next to an immune person, you will not spread the disease. We are supposing that our population is well mixed, so people contact others from each compartment every day. The disease-causing contacts by a single infectious person is given by the number of "close contacts" made with susceptible people during the period of infectiousness. We will assume that we can find an average number of such contacts for everyone and denote this number by a parameter c

$$c = \text{the average number of contacts per infective during the whole infectious period}$$

The parameter c is the average total number of "close" contacts per infective. Some diseases are more contagious than others, so we want a parameter that describes the contagiousness. The parameter c is a combined measure of the level of mixing in the population and the contagiousness of the disease. The problem is: How do we measure c? Later we will show that c can be measured indirectly with the help of calculus.

The number of contacts per infective each day is given by $a = c \cdot b$, because b equals $1/(\text{the number of days infectious})$. This means that if I people are infectious on a particular day,

$$a \cdot I = \text{the total number of adequate contacts per day}$$

Only the contacts with susceptible people result in a new case of the disease, so we multiply this total times the fraction of susceptible people. Our final parameter, n, is the fixed total population. This makes the number of new cases of the disease,

$$\text{new cases tomorrow} = a\frac{S}{n}I$$

where

$$n = \text{the fixed size of the population}$$

2.1.4   Parameters of the S-I-R Model

$b =$ 1/(the average number of days infectious)

$c =$ the average total number of contacts per infectious person

$a = b \cdot c$ the daily rate of contacts per infective

$n =$ the total number of people in the population

Notice the units of a, b, and c:

c in average number of close contacts per infective

b in $\dfrac{1}{\text{days}}$ or $\dfrac{1}{b} = $ number of days infective

a in $\dfrac{\text{number (of adequate contacts)}}{\text{number (infectives)} \times \text{days (infectious)}}$

New cases cause the size of S to decrease and the size of I to increase by the same amount. Recoveries cause the size of I to decrease and the size of R to increase by that amount. It is best to write all our equations in terms of increases in the variables S, I, R. We simply take a "negative increase" to mean a decrease. The variable S has a negative increase $-a\frac{S}{n}I$, the variable R has a positive increase bI, and I has an input from the susceptible compartment and an output to the removed compartment, $a\frac{S}{n}I - bI$. The direction of change is shown in Figure 2.3.

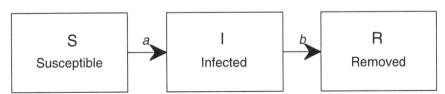

Figure 2.3: Compartments with Parameters

2.1.5   The First Equations of Change

$$\text{tomorrow's increase in } S \;\; = \;\; -a\frac{S}{n}I$$

$$\text{tomorrow's increase in } I \;\; = \;\; a\frac{S}{n}I - bI$$

$$\text{tomorrow's increase in } R \;\; = \;\; bI$$

Mathematical models need to be expressed in terms of variables. The "equations" above are not written in terms of the variables we listed. In fact, the left-hand side is a phrase. The phrase is clear, and we want you to make some computations and think about formulating the equation in terms of the variables.

Exercise set 2.1

1. *Use the daily changes above to compute 4 days of a rubella epidemic. You need the specific parameters, a, b, and c.*

   (a) *For rubella, we know that $\frac{1}{b} = 11$ and the average contact number is $c = 6.8$. Use these values to compute the parameter a.*

   (b) *Suppose a population initially (at $t = 0$) has*

$$
\begin{aligned}
S &= 10,000 \\
I &= 1,000 \\
R &= 19,000
\end{aligned}
$$

   Use the equations of change above to complete a table showing the first four days of the epidemic.

t	0	1	2	3	4
S	10,000				
I	1,000				
R	19,000				

Figure 2.4: *Tabular epidemic*

   (c) *Use your completed table to graph S versus t, I versus t, and R versus t.*

The table you have just filled out is one way to record S, I, and R as function of t. To find I at time 3, you simply look in the I row below the 3 in the t row. The mathematical notation for a function is

$$S[t], \qquad I[t], \qquad R[t]$$

The value of $I[3]$ is recorded in the I row below the 3 column of the t row.

You computed $S[1]$, $I[1]$, and $R[1]$ from the change equations and the known values $S[0]$, $I[0]$, and $R[0]$. Next, you used $S[1]$, $I[1]$, and $R[1]$ to compute $S[2]$, $I[2]$, $R[2]$, and so on. This is a recursive calculation: step 3 requires steps 1 and 2.

2. *Write formulas for (an unknown) $S[t+1]$, $I[t+1]$, and $R[t+1]$ in terms of a (known) $S[t]$, $I[t]$, and $R[t]$. Write your formulas so that $S[t+1]$, $I[t+1]$, and $R[t+1]$ are on the left-hand side of the equalities and expressions in $S[t]$, $I[t]$, and $R[t]$ are on the right.*

The answer to Exercise 2.1.2 is our first formal mathematical epidemic model. It allows us to recursively compute the course of an S-I-R epidemic in steps of 1 day.

3. *The equations of change are not needed to compute all three compartment sizes. If $S[t]$ and $I[t]$ are known, express $R[t]$ in terms of n, $S[t]$, and $I[t]$. (Hint: What is the total population in terms of these categories?)*

4. *What are the units of the expressions:*

   a) $a \cdot I$ b) $a \cdot I \cdot \frac{S}{n}$ c) $b \cdot I$

   What do they each measure?

C

5. *Log onto your computer account, and run the program **FirstSIR** in the Chapter 2 folder. It will help you check the values in your table and learn how to compute with the computer.*

6. Conjectures
We want you to speculate on the course of different kinds of diseases. Suppose disease "A" is very virulent; that is, it is very easily spread. Sketch the graph of I versus t for such a disease. Now do the same for disease "B," which is not so virulent. Suppose both diseases have an infectious period of 14 days. What is the value of b for both models of these diseases? How does the parameter a for disease A, a_A compare to the parameter a for disease B, a_B? Later in this chapter, you should compare your speculations with computed results.

7. *Which parts of our model are invalid for describing a flu epidemic on our campus that begins just before Thanksgiving break? Is our model valid for a flu epidemic that begins just after Thanksgiving break?*

Problem 2.1. ───────────────────────────────▼

Build a model for a disease that confers a period of immunity for a limited time after recovery but for which immunity is lost after this period. What other basic assumptions do you make? Write a detailed explanation.
▲

2.2   Shortening the Time Steps

> *A whole day of exposure may not be needed to transmit disease, so people may move from the susceptible compartment to the infectious compartment in a shorter period of time. This section adds that feature to our model.*

Time units of "days" are natural for rubella because $b = 1/11$ days comes directly from the period of infection. We want to keep these units. Consider the new infections with daily rate

$$\text{number of new infectious per day} = a\frac{S}{n}I$$

In 12 hours, $\frac{1}{2}$ day, there will be only half of the day's contacts, so

$$\text{new infections in } \frac{1}{2} \text{ day} = a\frac{S}{n}I \cdot \frac{1}{2}$$

In 4 hours, there will be $4/24 = \frac{1}{6}$th of the day's contacts, so

$$\text{new infections in 4 hours} = a\frac{S}{n}I \cdot \frac{1}{6}$$

In 1 hour, there will be $\frac{1}{24}$ th of the day's contacts, so

$$\text{new infections in 1 hour} = a\frac{S}{n}I \cdot \frac{1}{24}$$

In 1 minute, there will be $1/(24 \times 60) = 1/1440$ th of the day's contacts, so

$$\text{new infections in 1 minute} = a\frac{S}{n}I \cdot \frac{1}{1440}$$

In general in Δt days, the number of new infections will be

$$\text{new infections in } \Delta t \text{ days} = a\frac{S}{n}I \cdot \Delta t$$

Notice the units,

$$\left(a\frac{S}{n}I\right) \times \Delta t \quad \text{in (successful contacts per day) times (fractional days)}$$
$$= \text{successful contacts during the } \Delta t \text{ period}$$

If we calculate cases in half-day steps, the new cases are as above and the number of susceptibles in a half day equals the old susceptibles less the new cases,

$$S\left[\frac{1}{2}\right] = S[0] - \left(a\frac{S[0]}{n}I[0]\right) \cdot \frac{1}{2}$$

This quantity could then be used to compute $S[1] = S[1/2]-$ new cases during the second half day. We would need to compute the infectious population at time $= 1/2$ using

$$I\left[\frac{1}{2}\right] = I[0] + \left(a\frac{S[0]}{n}I[0] - bI[0]\right) \cdot \frac{1}{2}$$

because $I[1/2]$ is needed to find the rate of change of S during the next half day. Once this is known,

$$S[1] = S\left[\frac{1}{2}\right] - \left(a\frac{S\left[\frac{1}{2}\right]}{n}I\left[\frac{1}{2}\right]\right) \cdot \frac{1}{2}$$

and

$$I[1] = I\left[\frac{1}{2}\right] + \left(a\frac{S\left[\frac{1}{2}\right]}{n}I\left[\frac{1}{2}\right] - bI\left[\frac{1}{2}\right]\right) \cdot \frac{1}{2}$$

This looks more complicated than it really is. All it amounts to is using the present values of S and I to compute the next value. To see why, just complete the table in the next exercise.

Exercise set 2.2

1. (a) *Complete the table below where* $t = 0, \frac{1}{2}, 1, \frac{3}{2}, 2, \frac{5}{2}, 3, \frac{7}{2}, 4$:

t	0	$\dfrac{1}{2}$	1	$\dfrac{3}{2}$	2	$\dfrac{5}{2}$	3	$\dfrac{7}{2}$	4
S	10,000								
I	1,000								
R	19,000								

Figure 2.5: *Half-day table*

(b) *Graph S, I, and R vs. t.*

2.

(a) *What does the following quantity measure and in what units?*

$$\left(a\frac{S[0]}{n} I[0] - bI[0] \right) \cdot \frac{1}{2}$$

(b) *Explain why the computation you did to complete the third column of the table in the previous exercise is represented functionally by*

$$S\left[\frac{3}{2}\right] = S[1] - \left(a\frac{S[1]}{n} I[1] \right) \cdot \frac{1}{2}$$

$$I\left[\frac{3}{2}\right] = I[1] + \left(a\frac{S[1]}{n} I[1] - bI[1] \right) \cdot \frac{1}{2}$$

(c) *How many new infections will there be in 1 hour? Write formulas for* $S[t + \frac{1}{24}]$ *and* $I[t + \frac{1}{24}]$ *in terms of* $S[t]$ *and* $I[t]$, *which we assume are known, and where t has one of the values* $t = \frac{k}{24}$, $k = 0, 1, 2, 3, \ldots, 95$.

(d) *How many total computations would you need to do to compute* $S[4]$, $I[4]$, *and* $R[4]$ *in one hour steps? (Count all additions and multiplications.* $t = 4$ *means 4 days.)*

(e) *Explain why the general recursive procedure for computation of* $S[t]$ *and* $I[t]$ *is*

$$\textit{Start with} \quad S[0]$$
$$I[0]$$

For each successive $t = \Delta t, 2\Delta t, 3\Delta t, \cdots$, *compute the new function values,*

$$S[t + \Delta t] = S[t] - \left(a\frac{S[t]}{n} I[t] \right) \cdot \Delta t$$

$$I[t + \Delta t] = I[t] + \left(a\frac{S[t]}{n} I[t] - bI[t] \right) \cdot \Delta t$$

The computer does not mind doing all the computations required even to find $S[400]$, $I[400]$, and $R[400]$ in one hour steps. You will use a "Do loop" to write a computer program for this computation.

3. Programming the Recursion

*Run the program **SecondSIR** from the courseware. The program will not only do all the calculations necessary to solve the S-I-R model for small time steps but also will produce graphs of the results. The details are explained (within the program itself) first step by step, and then as a complete computation.*

*First, use **SecondSIR** to verify your hand computations from Exercise 2.2.1 above. Second, use **SecondSIR** to calculate and graph a 3-month rubella epidemic.*

2.3   The Continuous Variable Model

This section extends our disease model so that time varies continuously.

We will summarize what you should have learned from the text and exercises so far. The summary involves the awfully messy formulas mentioned in the introduction, but these lead to the simpler differential equations and, as you already know, they correspond to much simpler computer statements in **SecondSIR**.

First, it is fairly easy to give formulas for the average daily changes in S, I, and R, such as

$$\text{tomorrow's increase in } R = b\,I$$

Second, it is easy to use these equations to recursively tabulate the course of an epidemic as you did in the first exercise. Third, the *change* in a function of time such as $S[t]$ as time changes from t to $t + \Delta t$ is given by the difference

$$S[t + \Delta t] - S[t]$$

A general form of the solution to Exercise 2.2.2

$$S[t + \Delta t] = S[t] - \left(a\frac{S[t]}{n}I[t] \right)\Delta t$$

$$I[t + \Delta t] = I[t] + \left(a\frac{S[t]}{n}I[t] - b\,I[t] \right)\Delta t$$

for $t = 0, \Delta t, 2\Delta t, 3\Delta t, \ldots, k\Delta t, \ldots$ may be written in the form

$$\frac{S[t + \Delta t] - S[t]}{\Delta t} = -a\frac{S[t]}{n}I[t]$$

$$\frac{I[t + \Delta t] - I[t]}{\Delta t} = a\frac{S[t]}{n}I[t] - b\,I[t]$$

so both sides of the equations are rates of change. (Compute $R[t] = n - S[t] - I[t]$.) The first form is most useful in computing and looks simpler in **SecondSIR**. The second form is written as a rate of change in units of numbers of people per day,

$$\frac{S[t + \Delta t] - S[t]}{\Delta t} \quad \text{in numbers/day}$$

Recall that we verified that the units of the right-hand sides are numbers per day.

The time step Δt (or difference in time, abbreviated by Greek capital delta-t) is arbitrary, so it is not difficult to imagine t and Δt taking arbitrary continuous real values. We also want to let our basic variables take continuous real values so that we can formulate a calculus model.

2.3.1   The Continuous S-I-R Variables

For our continuous epidemic model, we choose

$$t \quad = \quad \text{time measured in days continuously from } t{=}0 \text{ at the start of the epidemic}$$

$$s \quad = \quad \text{the fraction of the population that is susceptible} \quad = \frac{S}{n}$$

$$i \quad = \quad \text{the fraction of the population that is infectious} \quad = \frac{I}{n}$$

$$r \quad = \quad \text{the fraction of the population that is removed} \quad = \frac{R}{n}$$

where n is the (fixed) size of the total population. Dividing both sides of our previous equations by n, we obtain

$$\left(\frac{S[t + \Delta t]}{n} - \frac{S[t]}{n} \right) \frac{1}{\Delta t} = -a \frac{S[t]}{n} \frac{I[t]}{n}, \text{ etc.,}$$

so that, in terms of the fraction variables, we obtain rate equations

$$\frac{\Delta s}{\Delta t} \quad = \quad \frac{s[t + \Delta t] - s[t]}{\Delta t} \quad = \quad -a\, s[t]\, i[t]$$

$$\frac{\Delta i}{\Delta t} \quad = \quad \frac{i[t + \Delta t] - i[t]}{\Delta t} \quad = \quad a\, s[t]\, i[t] - b\, i[t]$$

where $\Delta s = s[t + \Delta t] - s[t]$ and $\Delta i = i[t + \Delta t] - i[t]$ represent the differences in the susceptible and infectious fractions corresponding to the change in time Δt. (We can compute $r[t] = 1 - s[t] - i[t]$.) Briefly, this is the expression

$$\frac{\Delta s}{\Delta t} \quad = \quad -a\, s\, i$$

$$\frac{\Delta i}{\Delta t} \quad = \quad a\, s\, i - b\, i$$

Both sides of the above equations are average rates of change from time t to time $t + \Delta t$.

In calculus, the limit of the rate of change of a continuous function as the time step tends to zero is called the derivative,

$$\frac{ds}{dt} = \lim_{\Delta t \to 0} \frac{\Delta s}{\Delta t} = \lim_{\Delta t \to 0} \frac{s[t + \Delta t] - s[t]}{\Delta t}$$

$$\frac{di}{dt} = \lim_{\Delta t \to 0} \frac{\Delta i}{\Delta t} = \lim_{\Delta t \to 0} \frac{i[t + \Delta t] - i[t]}{\Delta t}$$

Since the right-hand sides of the above rate equations do not depend on Δt, we can interpret the limiting equations as the system of S-I-R differential equations.

2.3.2   S-I-R Differential Equations

(S-I-R DE's)

$$\frac{ds}{dt} = -a\,s\,i$$

$$\frac{di}{dt} = a\,s\,i - b\,i$$

Methods of calculus will let us compute interesting things about this continuous model, but for now we know that if we let Δt be "small," then we can find $s[t]$ and $i[t]$ in recursive steps of Δt by specifying a, b, $s[0]$, $i[0]$ and using a computer program like **SecondSIR** corresponding to the formulas

$$s[t + \Delta t] = s[t] - (a\,s[t]\,i[t])\Delta t$$

$$i[t + \Delta t] = i[t] + (a\,s[t]\,i[t] - b\,i[t])\Delta t$$

The discrete recursive form is closer to the "differential" form of the equations

$$ds = -a\,s\,i\,dt$$

$$di = (a\,s\,i - b\,i)\,dt$$

where

$$ds \approx s[t + \Delta t] - s[t] = -a\,s[t]\,i[t]\,\Delta t$$

$$di \approx i[t + \Delta t] - i[t] = (a\,s[t]\,i[t] - b\,i[t])\Delta t$$

The interpretation of the approximation or "limit" is more difficult for the differential form of the equations, because both sides tend to zero as Δt tends to zero. Actually, the functions $s[t]$ and $i[t]$ computed with the recursive formulas depend on the step size, Δt. You already may have seen that this process "converges" when you worked with **SecondSIR**. This convergence is called "Euler's Method" of approximate solution of differential equations. It is an important "limit" based on a very simple idea and will be a recurring theme in this course.

───────────── (**Exercise set 2.3**) ─────────────

1. *What is the meaning of fractional values of S, I, and R in Exercise 2.1.1 of Section 2.1? Are the fractional variables s, i, and r meaningful in a small population?*

2. *The total population is part of a model with S, I, and R, but not with s, i, and r. Explain how we might use the history of a rubella epidemic with one size population to make predictions about a new and different size one.*

C

3. Convergence and a Fast Solver

*Are the recursive formulas for $s[t + \Delta t]$ exact solutions to the S-I-R differential equations? Run the **ThirdSIR** program and compare solutions with step sizes of $1, \frac{1}{2}, \frac{1}{4}, \frac{1}{8}$ over the time interval $0 \le t \le 4$.*

The point of Euler's method is that you can start at a value and move a little in the direction of change given by a differential equation to find a nearby value. Once a new value is known, you can repeat the process with the known value. If the steps are small enough, this is approximates the solution. Try this in the next exercise, and remember that you need only a rough sketch, rather than high numerical accuracy.

4. The Geometry of Euler's Method

Let x and y be real variables. You are given that when $x = 0$, then $y = 1$,

$$y[0] = 1 \quad \text{and that } y \text{ changes with respect to } x \text{ by the rule} \quad \frac{dy}{dx} = -y$$

Sketch the graph of $y = y[x]$ by graphically using Euler's Method.
HINTS: What is the slope of the graph at $(x, y) = (0, 1)$? What value of y do you get if you move along a graph of slope -1 for 0.1 x-units? What is the slope of the graph if you are at the point $(x, y) = (0.1, 0.9)$? What is the slope of the graph at a point $(x, y) = (?, 0.2)$? What is the slope of the graph at a point $(x, y) = (?, 0.0)$?

Now we are ready to start using the model we have developed.

Problem 2.2. DIFFERENCES IN PARAMETERS
*Run the program **SIRsolver** from the courseware for the following cases of the parameters b and c.*
  (a) *A rubella epidemic, where $1/b = 11$, $c = 6.8$ and $a = b \cdot c$.*
  (b) *Measles, with a very high contact number $c = 15$ and $b = 1/8$.*
  (c) *Influenza, with a low contact number of $c = 1.4$ and $b = 1/3$.*
  The most important part of this problem is to analyze your results. What are the main differences among these three diseases? Which one is highly contagious, which is moderate and which is least contagious? Which one allows a larger portion of the population to escape infection? How many escape in each case? Write your summary analysis of the comparison among these diseases in a few brief paragraphs. (You may use a text cell in your copy of the program.)
  If you made some conjectures in Exercise 2.1.6, test them by running this program with the appropriate parameters.

2.4   Analysis of Change

Now that we have a mathematically simple way to describe change in an epidemic and an effective way to experiment with our model using the computer, we want to return to some scientific questions about epidemics.

The point of our work so far in this chapter is this:

If we know **initial values** and **instantaneous rates of change**,

$$s[0] = s_0 \qquad \frac{ds}{dt} = -a\,s\,i$$

$$i[0] = i_0 \qquad \frac{di}{dt} = a\,s\,i - b\,i$$

then the computer program **SIRsolver** can give us the functions $(s[t], i[t])$ satisfying these two properties.

The computer solution can be used either numerically or graphically, but the functions $s[t]$ and $i[t]$ cannot be expressed as formulas (like high school functions such as $y = x^3$); we can only compute them by hand recursively (as in Exercise 2.2.1). (Explicit formulas are also important, and we will study their calculus later.) We want you to use the graphed computer solution together with the description of change given by the differential equations to analyse when an epidemic is increasing or declining (see Problem 2.3.)

Recall from Chapter 1 that $\frac{di}{dt} = \lim_{\Delta t \to 0} \frac{i[t+\Delta t] - i[t]}{\Delta t}$ represents the slope of the graph of i vs. t (although we used x and y there). This is because $i[t + \Delta t] - i[t]$ is the change in i during the time interval from t to $t + \Delta t$. If the slope is positive, the graph is rising, or i is increasing as t increases. Since the change law is $\frac{di}{dt} = a\,s\,i - b\,i$, this means that i is increasing when $a\,s\,i - b\,i$ is positive. The combination of graphics and this formula is a powerful tool.

Problem 2.3. EXPANDING AND DECLINING EPIDEMICS ⸺⸺⸺⸺▼

We say an epidemic is expanding as long as the fraction of infectious people increases. It is declining when i decreases. See Figure 2.6 for the typical appearance of the graph of i vs. t in an epidemic. Where is i increasing on the graph? Where is it decreasing? When are the most people sick at one time? How many are sick at the maximum? At the point where i changes from increasing to decreasing, what can you say about the time rate of change of i, that is, the quantity $\frac{di}{dt}$? How is this related to the size of s, a, and b? (Hint: What is the formula for $\frac{di}{dt}$?)

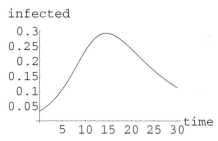

infected

Figure 2.6: Infectious vs. time

2.5   Long-Term Change

Will anyone escape infection in our epidemic?

If some people never get sick, then as t grows larger and larger, $s[t]$ will NOT get closer and closer to zero, but rather will stay above the fraction of the population that does not get sick. This is because we are studying an S-I-R disease from which you can never become susceptible once you are infected.

Since $s[t]$ can only decrease, mathematically, we are asking if

$$\lim_{t\to\infty} s[t] = 0 \quad\text{or}\quad \lim_{t\to\infty} s[t] > 0$$

The limit as t tends toward infinity simply means the $s[t]$ approaches a value as t gets larger and larger. We denote this limiting value by

$$s[\infty] = \lim_{t\to\infty} s[t]$$

This is a difficult limit to compute analytically (see Exercise 2.6.3), but it turns out to be a key to a complete understanding of the S-I-R model.

Problem 2.4. HOW MANY ESCAPE? ━━━━━━━━━━━━━━━━━━━━━━━━▼

1) As an epidemic fades out, the fraction of infectious people tends to zero,

$$\lim_{t\to\infty} i[t] = 0$$

*Why is this correct? Can you give scientific reasons, if not mathematical ones? Verify that our model of rubella predicts this by running the **SIRsolver** program to large values of t.*
2) Does the susceptible fraction tend to zero as an epidemic runs its course,

$$s[\infty] = \lim_{t\to\infty} s[t] = 0?$$

*What sorts of intuitive arguments can you make pro or con? Run the **SIRsolver** program for large values of time and estimate $s[\infty]$ from the computer solution. (The graph of $s[t]$ for a 4 month rubella epidemic is shown in the Figure 2.7. Does $s[t] \to 0$ on that graph?)*

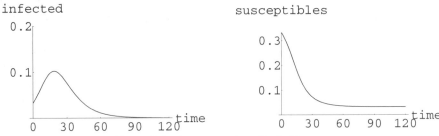

Figure 2.7: Four months of rubella

Problem 2.5. LIMITING SUSCEPTIBLES BY DISEASE ────────────────▼
*Experiment with $\lim_{t\to\infty} s[t]$ for rubella, measles, and influenza using the **SIRsolver** program. The necessary parameters, c and b, are given in Problem 2.2 above. The long-term graphs raise many questions such as, "How long does the epidemic last?" "Is there a formula for $s[\infty]$?" These are very difficult questions, but the computer will give you practical answers, if not formulas.*

Write a summary of your experiments on how $s[\infty]$ depends on c, b, and $s[0]$?
▲

2.6   Calculus and the S-I-R Invariant

Calculus gives us a key insight into the epidemic model by finding an "invariant formula." This formula gives us a way to measure the contact parameter c.

Up to this point, you should have followed all the mathematical derivations by working the exercises. With the help of the computer, you could even answer questions like

$$\text{Does } \lim_{t\to\infty} s[t] \text{ equal zero?}$$

At least approximately.

You might wonder, "Why bother with calculus if you have a computer?" The answer is that you can do more powerful calculations with both calculus and the computer than you can do with either one separately. We want to show you one example of calculus combined with computing, but we do NOT expect you to follow the details of the integral derivation at the end of this section. The derivation shows how explicit of high school functions ARE helpful, despite the fact that $s[t]$ cannot be given by a high school formula.

The derivation at the end of the section uses calculus to show that the S-I-R differential equations

$$\frac{ds}{dt} = -a\,s\,i$$
$$\frac{di}{dt} = a\,s\,i - b\,i$$

imply that

$$i + s - \frac{1}{c} \text{Log}[s] = k, \qquad \text{for the same constant } k \text{ and all } t$$

In other words, the combined quantity $i[t] + s[t] - \frac{1}{c} \text{Log}[s[t]]$ always has the same constant value, no matter what time t we take. (This is an example of a "first order invariant," something like the energy in physics.) This is the key to measuring the parameter c in the model. (If we could not measure c by some experiment, the model would only be "descriptive" or have a "fudge factor.")

Example 2.1. *Calculus Measures the Contact Number*

The invariant (in time) equation for s and i can be used to measure c from epidemic data. We have

$$s[\infty] - \frac{1}{c} \text{Log}\left[s[\infty]\right] = i[0] + s[0] - \frac{1}{c} \text{Log}\left[s[0]\right]$$

since $i[\infty] = 0$ and both sides equal the same constant k. At the start of an epidemic, $i[0]$ is small, so

$$s[\infty] - \frac{1}{c} \text{Log}\left[s[\infty]\right] \approx s[0] - \frac{1}{c} \text{Log}\left[s[0]\right]$$

and therefore approximately,

$$s[\infty] - s[0] = \frac{1}{c} \left(\text{Log}\left[s[\infty]\right] - \text{Log}\left[s[0]\right]\right)$$

so

$$c \approx \frac{\text{Log}\left[s[\infty]\right] - \text{Log}\left[s[0]\right]}{s[\infty] - s[0]}$$

By measuring the susceptible fraction at the beginning and the end of an epidemic, we can compute c and then we can find a from the computation $a = bc$. Calculus gave us the exact invariant equation that s and i satisfy at all times. No substitute in recursive computer computations provides this equation. However, once we have this information from calculus, the computer again adds to our knowledge.

Example 2.2. *How Calculus Finds the Time-Invariant Equation (Optional)*

Without justifying why this works, we compute the ratio

$$\left(\frac{di}{dt}\right) \bigg/ \left(\frac{ds}{dt}\right) = \frac{a\,s\,i - b\,i}{-a\,s\,i}$$

cancel dt, and do some algebra

$$\frac{di}{ds} = \frac{b}{a}\frac{1}{s} - 1 = \frac{1}{c\,s} - 1$$

$$di = \frac{1}{c}\frac{1}{s}\,ds - 1\,ds$$

We will learn later that if $y = s + k$ for any constant, then $\frac{dy}{ds} = 1$ and that if $\frac{dg}{ds} = \frac{1}{c} \cdot \frac{1}{s}$ for a constant c, then $g = \frac{1}{c} \text{Log}[s]$, the natural logarithm. These facts of calculus say that the indefinite integrals are as follows:

$$\int di = \frac{1}{c} \int \frac{1}{s} \, ds - \int ds$$

$$i = \frac{1}{c} \text{Log}[s] - s + k$$

for some constant k. This is the computation of the important time invariant equation for the epidemic model in five short steps.

Exercise set 2.6

1. Math Nerd Fever

   At the beginning of the semester last year, 96% of the students were not interested in calculus, but word of this chapter spread around campus infecting the population with a mythical disease Math Nerd Fever. Victims of the disease find that they become interested in calculus - and stay that way, though they only are successful in spreading the infection for a brief period when they are first excited about the chapter. At the end of the last academic year, 44% of the population was interested in calculus. What is the contact number for Math Nerd Fever? On the average, how many new math nerds does each infected student create?

2. *Suppose we measure a and b in a college population for a new strain of flu. What limitations would there be in using these parameters in predicting an epidemic in a nursing home?*

3. Computation of $\lim_{t \to \infty} s[t]$

   (a) *In the rubella epidemic with $c = 6.8$, $s[0] = 1/3$, $i[0] = 1/30$, find k.*

   (b) *We know from the previous section that $i[t]$ tends to zero as t tends to infinity, so the limiting susceptible fraction also satisfies*

   $$0 + s[\infty] - \frac{1}{c} \text{Log}[s[\infty]] = k$$

   with the same value of k as you computed in part (1) of the exercise. We could view this as the following question: Find all solutions in s of the equation

   $$s - \frac{1}{c} \text{Log}[s] - k = 0$$

   *Use the computer program **EpidemicRoots** to do this. Notice from the graph that there are two roots. The smaller one is $s[\infty]$.*

Later in the course, we will see analytical and graphical ways to understand the invariant equation more clearly. (See the phase diagrams of Chapter 22.)

2.7 Chapter Summary

Problem 2.6. THE BIG PICTURE ⸻▼

Write a paragraph explaining how the language of calculus is used to describe the changes in the susceptible, infectious, and removed fractions of a population. In other words, explain what the S-I-R differential equations "say" in English. Notice how cryptic the calculus is compared to English.

Write a paragraph summarizing the way the computer recursively finds the course of an epidemic by using the change rules given by the differential equations and an initial value of s, i, and r. (In other words, describe Euler's Method and the way you completed the tables in Exercises 2.1.1 and 2.2.1.) Be brief and give only a sample calculation and the main idea, rather than a lot of details. How does the computer interact with calculus to find the whole graph of the epidemic from the starting values and the formulas describing change due to a particular disease?

Finally, explain why the same model applies to a number of diseases by using different parameters a, b, and c. Do these parameters vary as time changes?

▲

2.8 Projects

Once or twice a semester we want you to explore a topic in detail. There is a separate book of detailed Scientific Projects that go with this text. The following subsections give a brief description of the projects most closely related to this chapter. You can work those projects now or come back to them later.

2.8.1 The New York Flu Epidemic

This project compares the S-I-R model with actual data of the Hong Kong Flu epidemic in 1968-69. The data we have to work with are "observed excess pneumonia-influenza deaths." In this case, our "removed" class includes people who have died. This is a little gruesome, but it is difficult to find data on actual epidemics unless they are extreme.

2.8.2 Vaccination Strategies for Herd Immunity

This project uses the mathematical model of an S-I-R disease to find a prediction of how many people in a population must be vaccinated in order to prevent the spread of an epidemic. In this project, you use data from around the world to make predictions on successful vaccination strategies. Dreaded diseases like polio and smallpox have virtually been eliminated in the last two generations, yet measles is a persistent pest right here on campus. The mathematical model can shed important light on the differences among these diseases. The mathematics is actually rather easy once you understand the basic concept of decreasing infectives.

2.8.3 Endemic S-I-S Diseases

Some infectious diseases do not confer immunity, such as strep throat, meningitis, or gonorrhea. In these diseases, there is no removed class, only susceptibles and infectives. As in the S-I-R diseases, we can make a mathematical model for S-I-S diseases.

An S-I-S disease has the potential to become endemic, that is, to approach a non-zero limit in the fractions of susceptible and infectious people. Once you have done some experiments, you will be able to show what happens mathematically and explore health policy implications.

CHAPTER 3
Linearity vs. Local Linearity

This chapter helps you understand the main approximation of differential calculus: Small changes in smooth functions are approximately linear. Calculus lets us use this approximation to replace nonlinear problems by linear ones. We view this first graphically by magnification, then numerically, and finally symbolically.

It is much easier to answer the question: "For which x is $f[x]$ equal to $g[x]$?" if $f[x]$ and $g[x]$ are of the form $f[x] = 1 + 0.07x$ and $g[x] = 0.3 + 0.27x$ than if $f[x] = \text{ArcTan}[x]$ and $g[x] = \text{Log}[x]$. Graphically, this is a question of finding a point of intersection (cf. Figures 3.1 and 3.2).

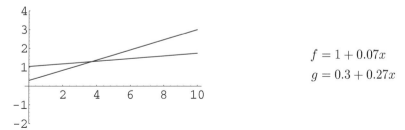

$$f = 1 + 0.07x$$
$$g = 0.3 + 0.27x$$

Figure 3.1: A common point on two linear graphs

Symbolically, the linear problem is simple, and the nonlinear one is not. Calculus lets us approximate smooth functions linearly on a small scale. (If you look very closely at the linear and nonlinear graphs near the point of intersection, they look the same.)

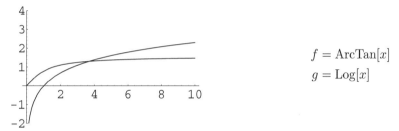

$$f = \text{ArcTan}[x]$$
$$g = \text{Log}[x]$$

Figure 3.2: A Common point on two nonlinear graphs

A linear function has the form $dy = m \cdot dx$ in local (dx, dy)-coordinates. The parameter m is the slope of the line that graphs the function. This linear function is increasing if $m > 0$, decreasing if $m < 0$, and horizontal (or constant) if $m = 0$.

Calculus tells when a nonlinear function $y = f[x]$ is increasing by computing an approximating linear function, the differential, $dy = m \cdot dx$. The slope m of the differential is the derivative, $m = f'[x]$. It is easy to tell if the linear function is increasing; we simply ask if the slope m is positive. A nonlinear function also increases where its linear approximation increases and the approximation is valid. The "uniform derivatives" in this book make the approximation valid at least in a small interval.

The "local approximation" of differential calculus means that a microscopic view of a smooth function appears to be linear as in Figure 3.3. In other words, calculus can be used to compute what we would see in a powerful microscope focused on the graph of a nonlinear function. If we "see" an increasing linear function, then the nonlinear function is increasing in the range of the microscope.

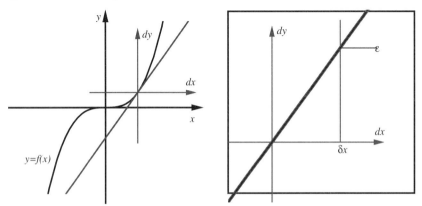

Figure 3.3: A highly magnified smooth graph

"Approximation" can mean many things. We take up a perfectly natural kind of approximation in the CO_2 Project in the accompanying separate book on Scientific Projects. Long-range approximation turns out not to be the kind that is so useful in calculus, but it is worth thinking about a little just for comparison. The next section gives you a clever case for comparison. "Local" microscopic approximation is probably not the first kind of approximation you would think to use, but it is the kind of approximation that makes calculus so successful.

3.1   Linear Approximation of Ox-bows

This section shows you one dramatic example of a nonlocal linear "approximation." The local linearity of calculus is only good for small steps; we hope Mark Twain's wit will help you remember this.

The chapter *Cut-offs and Stephen* of Mark Twain's *Life on the Mississippi* contains the following excerpt about ox-bows on the lower Mississippi. In a flood, the river can jump its banks and cut off one of its meandering loops, thereby shortening the river and creating an ox-bow lake.

In the space of one hundred and seventy-six years the Lower Mississippi has shortened itself two hundred and forty-two miles. That is an average of a trifle over a mile and a third per year. Therefore, any calm person, who is not blind or idiotic, can see that in the Old Oolitic Silurian Period, just a million years ago next November, the Lower Mississippi River was upwards of one million three hundred thousand miles long, and stuck out over the Gulf of Mexico like a fishing-rod. And by the same token any person can see that seven hundred and forty-two years from now the Lower Mississippi will be only a mile and three-quarters long, Cairo and New Orleans will have joined their streets together, and be plodding comfortably along under a single mayor and mutual board of aldermen. There is something fascinating about science. One gets such wholesome returns of conjecture out of such a trifling investment of fact.

Exercise set 3.1

1. Linear Ox-bows

How wholesome are Twain's returns? Express the length of the river (in miles) as a function of time in years using the implicit mathematical assumption of Twain's statement. What is this assumption? We are not given the length of the Lower Mississippi at the time of Twain's statement, but we can find it from the information. What is it? What is the moral of this whole exercise in terms of long-range prediction?

3.1.1 Chapter Plan

Here is the plan for the rest of the chapter. First, in Section 3.2, we use the computer to observe that the graphs of "typical" functions look "smooth." Specifically, when we "zoom in" at high magnification to any point on the graph $y = f[x]$, we see a straight line. We will identify this line as $dy = f'[x]dx$ in the local coordinates described in Chapter 1.

Of course, there are exceptional "non-typical" functions, and we look briefly at them at the end of the next section.

Section 3.3 is our first real effort to measure the deviation from straightness. A smooth graph becomes indistinguishable from the straight differential line because the gap between the curve and the tangent decreases as we increase the magnification. This section measures the gap in some specific cases and concludes with a general summary problem.

Section 3.4 uses the general formula first to verify your results from the previous special cases and then to extend the idea. Chapter 5 takes up the symbolic approach to the gap computations in general, whereas Chapter 4 shows how the gap formula looks in an approximate solution of a differential equation.

Chapter 6 gives the rules that allow us to find the derivative function $f'[x]$ by a "calculus." These rules are actually theorems that guarantee that the gap in the microscope is small at high enough magnification. If you want to take that on faith, you could skip Sections 3.3, 3.4, and Chapter 5, where we compute the gap graphically, numerically, and symbolically.

3.2   Graphical Increments

This section uses computer "microscopes" to give one of the most important interpretations of differentiability or "local linearity."

At high enough magnification, the graphs $y = x^2$, $y = x^3$, $y = x^{27}$, $y = 1/x$, and $y = \sqrt{x}$ appear straight (except at $x = 0$ on $1/x$ and $x \leq 0$ on $\sqrt{x}$.) The computer programs **Microscope** and **Zoom** will let you experiment with these and other functions yourself. The sequence of graphs in Figure 3.4 shows a "zoom" at $x = 1$ on the graph of $y = x^2$.

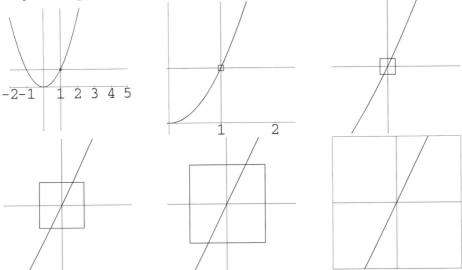

Figure 3.4: Successive magnifications of $y = x^2$ at $x = 1$

The computer programs **SecantGapZ** and **Zoom** help give you the geometric idea of the Increment Principle. These programs contain "animations" - a computer generated "movie" of graphs being magnified.

─────────────(**Exercise set 3.2**)─────────────

1. Computer Zooming

    (a) *Run the computer program **Zoom** with its built-in example function, $y = f[x]$. Read the instructions in the program on making the computer run an animation. You should see a "movie" of an expanding graph. Each "frame" of the movie is expanded 1.5 times. Once the magnification is high enough, the graph appears to be linear.*

    (b) *Focus the microscope at the point $x = 0$ instead of the built-in $x = 1$ and reenter the two computation cells.*

    (c) *Use the computer program **Zoom** and redefine the function $f[x]$ to make animations of microscopic views of some of the functions:*

i) $f[x] = x^2$ ii) $f[x] = x^3$ iii) $f[x] = \frac{1}{x}$

iv) $f[x] = \sqrt{x}$ v) $f[x] = \mathrm{ArcTan}[x]$ vi) $f[x] = \mathrm{Log}[x]$

vii) $f[x] = \mathrm{Exp}[x] = e^x$ viii) $f[x] = \mathrm{Cos}[x]$ ix) $f[x] = \mathrm{Sin}[x]$

Try focusing the microscope over the points $x = 1.5$, $x = 0$, and $x = -1$ when all these are possible. (Say why they if they cannot be done.)

2. Measuring the Derivative

Views of a function $y = f[x]$ are shown in the three figures below magnified by 50 and focused on the graph above three different x's. Measure the derivative at each point by finding the slope of the approximating straight line.

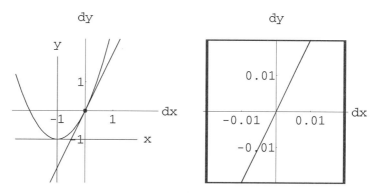

Figure 3.5: $y = f[x]$ magnified at $x = 1$, $f'[1] = ?$

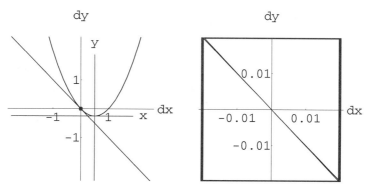

Figure 3.6: $y = f[x]$ magnified at $x = -1/2$, $f'[-1/2] = ?$

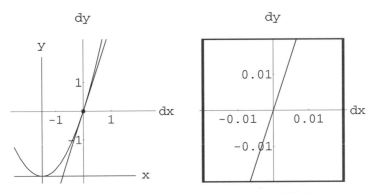

Figure 3.7: $y = f[x]$ *magnified at* $x = 3/2$, $f'[3/2] = ?$

3.2.1 Functions with Kinks and Jumps

Not all functions are smooth or locally linear. This means that when we magnify some functions, their graphs do not become more and more like straight lines. Which functions do, and which do not? Symbolic rules of calculus will answer this question easily, and it turns out that "most" functions are smooth so that microscopic views of their graphs do appear linear. Two exercises below experiment with some exceptional cases.

3. Jump Functions

*Run the computer program **Zoom** on the functions*

$$j_1[x] = \frac{\sqrt{x^2 + 2x + 1}}{x + 1} \qquad and \qquad j_2[x] = \frac{\sqrt{x^2 - x^4/2}}{x}$$

Focus the microscope at the points $x = +1$, $x = 0$, *and* $x = -1$. *Where are the functions smooth and locally linear? Where is the "trouble spot"? What happens to the functions at* $x =$ *the trouble spot? Why?*

4. Kink Functions

*Run the computer program **Zoom** on the functions*

$$k_1[x] = \sqrt{x^2 + 2x + 1} \qquad and \qquad k_2[x] = \sqrt{x^2 - x^4/2}$$

Focus the microscope at the points $x = +1$, $x = 0$, *and* $x = -1$. *Where are the functions smooth and locally linear? Where is the "trouble spot"? What happens to the functions at* $x =$ *the trouble spot? Why?*

3.2.2 Continuous, But Not Smooth

The main ideas of differential calculus are based on approximating small changes in the output of a function when a small change is made in its input. We need some notation to indicate a small change. If x_1 and x_2 are nearly equal, we will write

$$x_1 \approx x_2$$

For now, this will just be an intuitive notion; we will not say exactly how close they have to be in order to write $x_1 \approx x_2$.

> **Definition 3.1.** *Informal Definition of Continuity*
> *We say that a function $f[x]$ is continuous when small changes in x only produce small changes in the value of the function,*
>
> $$x_1 \approx x_2 \Rightarrow f[x_1] \approx f[x_2]$$

This is an intuitive formulation of the expression

$$\lim_{x_1 \to x_2} f[x_1] = f[x_2]$$

and it is important for you to have an idea of what this means before we try to formalize it technically (in Chapter 5.) The definition is a kind of approximation: $f[x_2]$ is approximately $f[x_1]$ when x_2 is approximately x_1, but it is NOT an approximation by a linear function. Local linear approximation graphically says the graph looks like a straight line at a small scale. (Naturally, a linear function is continuous because the function changes by a multiple of the change in x. We formulate linear approximation symbolically in the Section 3.4.)

Problem 3.1. CONTINUITY WITHOUT LOCAL LINEARITY ⎯⎯⎯⎯⎯⎯⎯▼
Explain in terms of the definition above why the functions $j_i[x]$ in Exercise 3.2.3 are not continuous (at certain points) but the functions $k_i[x]$ in Exercise 3.2.4 are continuous (at least where they are defined).

▲

The jump functions $j_i[x]$ are discontinuous at certain points, and the functions $k_i[x]$ are continuous but not locally linear (or smooth or differentiable) at certain points. The function (shown in Figure 3.8)

$$W[x] = \text{Cos}[x] + \frac{\text{Cos}[3\,x]}{2} + \frac{\text{Cos}[3^2\,x]}{2^2} + \frac{\text{Cos}[3^3\,x]}{2^3} + \cdots + \frac{\text{Cos}[3^k\,x]}{2^k} + \cdots$$

is continuous but is not locally linear at *any* point. It has a kink at every point on the graph! Weierstrass discovered this function, and its graph looks like lots of different size "Ws" all hooked together. It is an old example of a "fractal." We want you to graph it and look at it on several scales. When we study series later in the course, we will return to Weierstrass' function. (See Problem 25.2.2.)

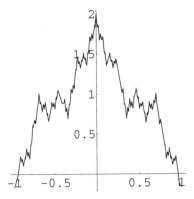

Figure 3.8: Weierstrass' nowhere differentiable function

Problem 3.2. WEIERSTRASS' WILD WIGGLES
Graph Weierstrass' function using the program **Weierstrass**. *Try several scales, "delta". The program keeps the width of the x-axis and the width of the y-axis on your graph the same. This prevents distortion of slopes.*

3.3   Algebra of Microscopes - CD

The goal of this section is to numerically and symbolically calculate the error of deviation from straightness in microscopic views of graphs.

Remainder of this section only on CD

3.4   Symbolic Increments - CD

The following formula for the change in a general function

$$f[x + \Delta x] - f[x] = f'[x] \cdot \Delta x + \varepsilon \cdot \Delta x$$

gives the gap ε one would measure at magnification $1/\Delta x$ between a straight line of slope $f'[x]$ and the curve as we move from x to $x + \Delta x$.

Remainder of this section only on CD

3.5   Projects

Projects for this chapter are:

3.5.1   CO2 Data

The Scientific Project on CO2 Data fits a linear function to data and compares it with long range prediction.

3.5.2   A Project on Functional Linearity

Linearity in function notation has a peculiar appearance. This is not difficult, just different. This project shows you the linear case of the main formula underlying differential calculus.

3.5.3   A Project on Functional Identities

The Project on Functional Identities includes further study of the additive identity

$$f[x + y] = f[x] + f[y]$$

This is related to a famous problem posed by Cauchy and solved by Hamel. If $f[0] = 0$ and $f[x]$ satisfies the identity $f[x + y] = f[x] + g[y]$,

$$f[x + \Delta x] - f[x] = g[\Delta x]$$
$$f[0 + \Delta x] - f[0] = g[\Delta x]$$
$$f[\Delta x] = g[\Delta x]$$

so that

$$f[x + \Delta x] - f[x] = f[\Delta x]$$
$$f[x + \Delta x] = f[x] + f[\Delta x]$$

Changing variables, we get, $f[x + y] = f[x] + f[y]$. In particular, if $f[x] = m\,x$, then $f[x]$ satisfies $f[x + y] = f[x] + f[y]$. Cauchy's question was: If an unknown function satisfies $f[x + y] = f[x] + f[y]$, does the unknown function have to be $f[x] = m\,x$ for some constant m? You answer this question in the exercises for the important case where $f[x]$ is smooth.

Notice that the identity

$$f[x + \Delta x] - f[x] = m \cdot \Delta x$$

is the microscope approximation

$$f[x + \Delta x] - f[x] = m \cdot \Delta x + \varepsilon \cdot \Delta x$$

with zero error, $\varepsilon = 0$.

Differential Equations and Derivatives

This chapter uses the "microscope" approximation to view derivatives as time rates of change and describe some practical changes.

The microscope approximation of the last chapter can be written

$$f[x + \Delta x] = f[x] + f'[x] \cdot \Delta x + \varepsilon \cdot \Delta x$$

If our independent variable is t rather than x and our function is $s[t]$ rather than $f[x]$, the equation becomes

$$s[t + \Delta t] = s[t] + s'[t] \cdot \Delta t + \varepsilon \cdot \Delta t$$

This is related to the differential equations in Chapter 2.

In Chapter 2 we approximated the susceptible and infectious fractions in a population by recursively computing

$$
\begin{aligned}
s[t + \Delta t] &= s[t] + s'[t]\Delta t \\
i[t + \Delta t] &= i[t] + i'[t]\Delta t
\end{aligned}
$$

where we computed the "prime" terms using the formulas:

$$s'[t] = -a\, s[t]\, i[t] \qquad \text{and} \qquad i'[t] = (a\, s[t] - b)i[t]$$

so

$$
\begin{aligned}
s[t + \Delta t] &= s[t] - (a\, s[t]\, i[t])\Delta t \\
i[t + \Delta t] &= i[t] + (a\, s[t]\, i[t] - b\, i[t])\Delta t
\end{aligned}
$$

The recursively computed functions $s[t]$ and $i[t]$ actually depended on the step size Δt, but we observed in practical terms that the approximate solutions converged with only a modestly small Δt. In any case, the exact solutions satisfy

$$
\begin{aligned}
s[t + \Delta t] &= s[t] + s'[t] \cdot \Delta t + \varepsilon_s \cdot \Delta t \\
i[t + \Delta t] &= i[t] + i'[t] \cdot \Delta t + \varepsilon_i \cdot \Delta t
\end{aligned}
$$

The recursive approximation discards the errors $\varepsilon_s \cdot \Delta t$ and $\varepsilon_i \cdot \Delta t$ at each successive step.

In this chapter, we formulate differential equations such as

$$\frac{dC}{dt} = -k\, C$$

and use the microscope approximation

$$C[t + \Delta t] = C[t] + C'[t] \cdot \Delta t + \varepsilon \cdot \Delta t$$

and the formula $C'[t] = -k \ C[t]$, so

$$C[t + \Delta t] = C[t] - k \ C[t] \cdot \Delta t + \varepsilon \cdot \Delta t$$

We approximate the exact solution $C[t]$ by discarding the error $\varepsilon \cdot \delta t$. The approximate solution is computed recursively: $C_a[0] = c_0$, and for later times,

$$C_a[t + \delta t] = C_a[t] - k \ C_a[t] \ \delta t$$

This satisfies $C_a[t] \approx C[t]$ when $\delta t \approx 0$ is small.

4.1 The Cool Canary

This section studies change and another kind of linearity - change that is linear in the dependent variable.

It was a cool January night in Iowa, about $0°$ F outside but cozy and around $75°$ F in our snug farm house. Jonnie was still mad at sister Sue for getting a better report card, so at 8:00 pm he put her covered canary cage outside. Ten minutes later the canary chirped as he always does at $60°$F.

The canary will die when the temperature in the cage reaches freezing ($32°$ F). How long do we have to rescue him?

This scenario is a little silly, but we are getting at some serious mathematics. The urgent need of the poor canary should hold your attention. We let C equal the temperature in degrees Fahrenheit and t equal the time in minutes, measured with $t = 0$ at the time Jonnie put the cage outside. Our data are $(t, C) = (0, 75)$ and $(t, C) = (10, 60)$. A linear model of the data looks like Figure 4.1, but is a linear model correct?

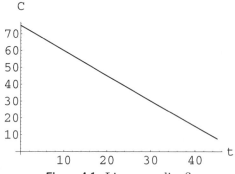

Figure 4.1: Linear cooling?

Will the temperature of the cage drop below zero? Of course not. Is the limiting value of the temperature zero? In other words, will the little canary's body temperature be zero when Susie finds him in the morning? Could the temperature decline linearly to zero and then abruptly stop declining and stay at zero?

A linear graph of temperature vs. time means that the rate of cooling remains the same for all temperatures. In other words, when the cage is much warmer than the outside air, it only cools as fast as when it has cooled down to almost zero. Does this seem plausible? Of course not. The graphs in Figure 4.1 and Figure 4.2 show a linear and a nonlinear fit to our same two data points and plot temperature for 50 minutes. Different nonlinear graphs are shown in Figure 4.3 for 600 minutes. Notice the "tapering off" of the nonlinear graph as the temperature gets near zero.

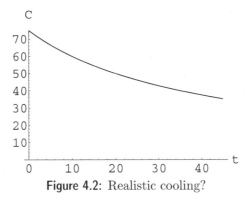

Figure 4.2: Realistic cooling?

There are many explicit formulas with the property that the rate of decrease itself decreases as the dependent variable tends toward zero. One student suggested a variation on $y = \frac{1}{x}$. If we sketch the graph of $y = \frac{1}{x}$ except make the variables C and t, we notice that there is a "singularity" at $t = 0$. We need to move the graph of $C = \frac{k}{t}$ to the left in order to make it at least a plausible model of the chilly canary. (Certainly the temperature does not tend to infinity as t tends to zero!) Why not also try a family of curves like $C = \frac{k}{(t+a)^2}$ or $C = \frac{k}{\sqrt{t+a}}$? This approach is explored in Exercise 4.1.3.

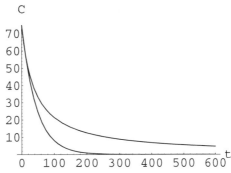

Figure 4.3: Two nonlinear cooling models

Guessing explicit formulas for temperature vs. time is a hard way to build a model of the cooling canary. It is better to go right to the underlying question of how the temperature *changes*. We want a mathematical relationship that says, "The rate of cooling decreases to zero as C decreases to zero." The simplest relationship is a linear equation between C and the rate of change of C with respect to t, C'. How do you "say" that C' is a linear function of C? This is Problem 4.1.

You should see that the ideas of the poor freezing canary's plight are simple - at least qualitatively. The rate of cooling decreases as the temperature decreases, but it is hard to express the simple ideas in terms of variables. This is because we do not have the proper mathematical *language*. A language for change needs symbols describing the change in other quantities.

Exercise set 4.1

1. A Practical Two-Point Formula
What is the equation of the line that passes through the points $(0, 75)$ and $(10, 60)$?
Use your equation to predict the temperature of the canary at 6:00 am, 10 hours after Jonnie does his vicious deed.

2. Tapering Off is a ?? in The Rate of Cooling
What happens to the rate of cooling on the nonlinear graphs above as the temperature decreases toward zero? Does the rate of cooling decrease or increase as the temperature decreases toward zero?
We expect that when C is big, the rate of cooling is big, and that when C is near zero, the rate of cooling is small. Why do we expect this? Which graphs in the Figures above have this feature?

3. Mindless Fits
Show that all of the functions below satisfy $C[0] = 75$ and $C[10] = 60$.
   (a) $C = \frac{a}{t+k}$, $a = 3000$, $k = 40$

   (b) $C = \frac{a}{\sqrt{t+k}}$, $a = 100\sqrt{10}$, $k = 160/9$

   (c) $C = \frac{a}{(t+k)^2}$, $a = 10000(27 + 12\sqrt{5}) \approx 538328$, $k = 40 + 20\sqrt{5} \approx 84.7214$

   (d) $C = 75 \, e^{-kt}$, $k = \text{Log}[5/4]/10 \approx 0.0223144$
Use the computer to graph each of these functions for 10 hours.

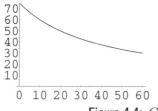

Figure 4.4: $C = \frac{a}{t+k}$

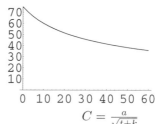

$C = \frac{a}{\sqrt{t+k}}$

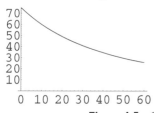

Figure 4.5: $C = \frac{a}{(t+k)^2}$

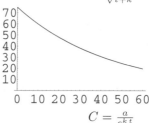

$C = \frac{a}{e^{kt}}$

Problem 4.1. LINEAR COOLING ▬▬▬▬▬▬▬▬▬▬▬▬▬▬▬▬▬▬▬▬▬▬▼

Let C' stand for the rate of cooling of the canary cage of temperature C at time t.

What is the relationship between rate of cooling, C', and the temperature, C? In other words, what does "rate of cooling" mean? For example, which units would you use to measure C'?

What feature of the graph of C vs. t represents C'?

What is the physical meaning of C'?

Write an equation that says, "The rate of cooling is proportional to the temperature."

▲

4.2 Instantaneous Rates of Change

Time rates of change are one important way to view derivatives. This section contains two examples. The reason that $-\frac{dC}{dt}$ is the instantaneous rate of decrease in temperature can be seen from the increment approximation.

Here is why $\frac{dC}{dt}$ is the rate of change of temperature - instantaneously. The change in temperature for a small change ($\delta t \approx 0$) in time, from time t to $t + \delta t$, is

$$C[t + \delta t] - C[t]$$

(in units of degrees). The rate at which it changes during this time interval of duration δt is the ratio

$$\frac{C[t + \delta t] - C[t]}{\delta t} = \frac{\delta C}{\delta t}$$

(in units of degrees per minute). The increment approximation for $C[t]$ is

$$C[t + \delta t] - C[t] = \frac{dC}{dt} \cdot \delta t + \varepsilon \cdot \delta t$$

with $\varepsilon \approx 0$, for $\delta t \approx 0$. Dividing results in the expression

$$\frac{C[t + \delta t] - C[t]}{\delta t} \approx \frac{dC}{dt}$$

In an "instant" δt becomes zero, the left hand side of the previous equation no longer makes sense mathematically but "tends" to the instantaneous rate of change as ε becomes zero. This means that

$$\text{the rate of change of } C \text{ in an instant of } t = \frac{dC}{dt}$$

Exactly what an instant is may not be clear, but the description of cooling makes it clear that it is useful.

4.2.1 The Dead Canary

The derivative $\frac{dC}{dt}$ of a function $C[t]$ is the instantaneous rate at which C changes as t changes. In the first section, we told you about Jonnie putting sister Susie's canary cage out into an Iowa winter night. The initial temperature was 75° F and 10 minutes later the temperature was 60° F. The rate of cooling would be fast when the temperature is high and gradually become slower as the temperature gets closer to zero. When the temperature reaches zero, the cage wouldn't cool any more. The simplest relationship with this property is that the rate of cooling is proportional to the temperature, or

$$\text{the rate of cooling} = kC$$

The poor canary has long since died, but that makes the mathematical model simpler anyway (since he no longer generates even a little heat.) We want to finish the story, mathematically speaking, before we bury the canary.

The simplest way to measure the rate of cooling of the canary cage would be to measure the rate at which the temperature *decreases* as time increases. In other words,

$$\text{the rate of cooling} = -\frac{dC}{dt}$$

Because the rate of cooling of the canary cage is proportional to the temperature, we have

$$\text{rate of drop in temperature} \propto \text{temperature}$$
$$\text{rate of drop in temperature} = k \cdot \text{temperature}$$
$$-\frac{dC}{dt} = kC$$
$$\frac{dC}{dt} = -kC$$

We also know that $C = 75$ when $t = 0$ and $C = 60$ when $t = 10$.

──────────────── **Exercise set 4.2** ────────────────

C

1. Euler Solution of the Canary Problem

   *Use the program **EulerApprox** to compute the temperature of Susie's canary. Run your program with various values of k until C = 60 when t = 10. (Hint: Start with k = 0.03 and 0.01.) Once you determine the value of k that makes C[10] = 60, change the final time to 10 hours = 600 and plot the whole canary's plight, so that you find the temperature at 6:00 am in particular.*

2. Graphical Euler's Method

   You only need to know where C starts (C[0] = 75) and your equation for change $\frac{dC}{dt} = -kC$ to sketch its curve. (This is not very accurate but gives the right shape.)

   (a) *Sketch the curve C = C[t] when k = 1/50, so $\frac{dC}{dt} = -\frac{C}{50}$. When C = 75, this makes the slope of the curve −3/2. Move a small distance along the line of slope −3/2 from the point (0, 75). If C = 60, how much is the slope of the curve? If C = 5, how much is the slope of the curve? If C = 0, what is the slope of the curve?*

(b) *Sketch the curve $C = C[t]$ on the same axes when $k = 0.01$, so $\frac{dC}{dt} = -C/100$.*

(c) *Explain the idea in the computer's recursive "Euler" computation of the temperature. Write a few sentences covering the main ideas without giving a lot of technical details but rather only giving a description and a few sample computations. In other words, convince us that you understand the computer program and its connection with sketching the curve $C = C[t]$ from a starting value and differential equation.*

The language of change in Problem 4.1 says, "the cage cools slower as temperature approaches zero." Specifically, the rate of cooling is a linear function of the temperature. This is called "Newton's Law of Cooling," and it really does work (for inanimate objects). But math is not magic, as the next problem shows.

4.2.2   The Fallen Tourist

You have a chance to visit Pisa, Italy, and are fascinated by the leaning tower. You decide to climb up and throw some things off to see if Galileo was right. Luckily, you have a camera and your glasses to help with the experiment. You drop your camera and glasses simultaneously and are amazed that they smash simultaneously on the sidewalk below. Unconvinced that a heavy object does not fall faster, you race down and retrieve a large piece and a small piece of the stuff that is left.

Dropping the pieces again produces the same simultaneous smash and you also notice, of course, that the objects speed up as they fall. Not only that, but both the heavy object and the light object speed up at the same rates. You know that a feather would be affected by air friction; but you theorize that in vacuum, there must be just one "law of gravity" to govern falling objects.

About then, the Italian police arrive and take you to a nice quiet room to let you think about the grander meaning of your experiments while an interpreter arrives from the American consulate.

Problem 4.2. GALILEO'S FIRST CONJECTURE ⎯⎯⎯⎯⎯⎯⎯⎯⎯⎯⎯▼
Since all objects speed up as they fall, and at the same rate neglecting air friction, you theorize that

> *the speed of a falling object is proportional to the distance it has fallen.*

Let D be the distance the object has fallen from the top of the tower at time t (in the units of your choice.)

(a) *What is the speed in terms of the rate of change of D with respect to t?*

(b) *Formulate your "law of gravity" as a differential equation about D.*

(c) *What is the initial value of D in your experiment?*

(d) *Use the computer or mathematics to decide if your "law of gravity" is correct.*
▲

We will follow up on the Tourist's progress later in the course when we study Bugs Bunny's Law of Gravity in Problem 8.5, Wiley Coyote's Law of Gravity in Problem 21.8, and Galileo's Law in Exercise 10.2.1.

4.3   Projects
4.3.1   The Canary Resurrected

The Scientific Project on the canary compares actual cooling data with the model of this chapter.

CHAPTER 5

Symbolic Increments

The main idea of differential calculus is that small changes in smooth functions are approximately linear. In Chapter 3, we saw that "most" microscopic views of graphs appear to be linear, but we want a symbolic way to predict the view in a powerful microscope without actually having to use graphical magnification. The computations in this chapter give us that prediction.

Direct computation of the microscopic "gap" is hard work, but the formulas show when the gap gets small. This guarantees that a magnified graph will appear linear before we draw the graph. The direct computations of this chapter are proofs of specific differentiation rules. The next chapter develops the rules systematically as a procedure or "calculus" of derivatives. You could just accept the specific results of this chapter and go on to Chapter 6 to learn the "rules," but differentiation rules are actually theorems that say a local linear approximation is guaranteed by certain systematic computations. You should understand that this means the "gap" tends to zero at all "good" points.

If the gap does not go to zero at a point, we will not see a straight line when we focus our microscope there. We know from Exercise 3.2.4 that the perfectly reasonable function

$$f[x] = \sqrt{x^2 + 2x + 1}$$

does not have the local linearity property at $x = -1$. (Its graph has a kink no matter how much you magnify it.) Some less reasonable ones like Weierstrass' function

$$W[x] = \text{Cos}[x] + \frac{\text{Cos}[3x]}{2} + \frac{\text{Cos}[3^2 x]}{2^2} + \cdots + \frac{\text{Cos}[3^n x]}{2^n} + \cdots$$

are continuous but not locally linear at any point (see Figure 3.8). Calculus gives a procedure to find out if a function given by a formula is locally linear.

Once we have the basic rules of this chapter and some functional rules from the next, we will differentiate the kink function

$$f[x] = \sqrt{x^2 + 2x + 1} \qquad \Rightarrow \qquad f'[x] = \frac{x + 1}{\sqrt{x^2 + 2x + 1}}$$

but then notice that $f'[-1]$ is not defined, so that we cannot make any conclusion about local linearity at this point. (General differentiation rules cannot say there is a kink, only that no general conclusion is possible when the rules do not apply.) At every other x, the rules do apply and this function is locally linear.

5.1   The Gap for Power Functions

The gap ε for $y = f[x] = x^p$ with $f'[x] = p\,x^{p-1}$

In Chapter 3, we formulated the deviation of $y = f[x]$ from a straight line geometrically. The observed error or gap between the curve and line at magnification $1/\Delta x$, denoted ε, satisfies

$$f[x + \Delta x] - f[x] = m \cdot \Delta x + \varepsilon \cdot \Delta x$$

The term $\varepsilon \cdot \Delta x$ is the actual unmagnified error that appears in the formula above. We observe ε because of magnification. The number m is the slope of the microscopic straight line $dy = m\,dx$ in (dx, dy)-coordinates focused over x. (See Chapter 1 for equations of lines in local coordinates.) "Approximate linearity" means that the error ε is too small to measure, $\varepsilon \approx 0$, when $\delta x \approx 0$ is "small enough."

Notice that m may depend on x, but not Δx, because the slope of the curve depends on the point but not the magnification. As we move the focus point of the microscope, the slope may change, but the graph should always appear straight under the microscope. Since m depends on x and not on Δx, it is customary to denote this slope by $f'[x]$ rather than m. The function $f'[x]$ is called the derivative of the function $f[x]$ and the local linear equation $dy = f'[x]\,dx$ is called the differential of $y = f[x]$. (It is the equation of the tangent line at a fixed value of x in local (dx, dy)-coordinates.)

We will compute the symbolic gap for all the examples in this section (and Exercise 5.1.1) by the steps

Procedure 5.1.   Computing the ε-Gap

1) Compute $\dfrac{f[x + \Delta x] - f[x]}{\Delta x}$

2) Simplify the expression from the first part and compute

$$f'[x] = \lim_{\Delta x \to 0} \frac{f[x + \Delta x] - f[x]}{\Delta x}$$

Give an intuitive justification of why your limit is correct.

3) Use your limit $f'[x]$ to solve for $\varepsilon = \dfrac{f[x + \Delta x] - f[x]}{\Delta x} - f'[x]$

4) Show that $\varepsilon \to 0$ as $\Delta x \to 0$, or $\varepsilon \approx 0$ is small when $\Delta x = \delta x \approx 0$ is small.

The error ε is the "gap" or amount of deviation from straightness we see above $x + \Delta x$ at power $1/\Delta x$. We want to let Δx get "small enough" so that ε is below the resolution of our microscope. This chapter shows symbolically that ε tends to zero for various functions at "good" points.

Example 5.1.   *Increments of $y = f[x] = x^3$*

Let $y = f[x] = x^3$ and calculate the increment corresponding to a change in x of Δx. First, we know from Example 1.7 and Example 28.4 that

$$\frac{f[x + \Delta x] - f[x]}{\Delta x} = 3x^2 + (3x + \Delta x)\Delta x$$

Second, the intuitive limit

$$\lim_{\Delta x \to 0} \frac{f[x + \Delta x] - f[x]}{\Delta x} = 3x^2$$

since $(\Delta x(3x+\Delta x) = 0$ when we plug $\Delta x = 0$ into the expression above. (This is technically correct, too, as long as x is bounded by some fixed amount.)

Third,

$$\varepsilon = (3x + \Delta x)\Delta x$$

comparing the lines above.

Fourth, $\boxed{\varepsilon \text{ tends to zero as } \Delta x \text{ tends to zero.}}$ Here is a rough argument that shows that the ε-gap error becomes small as Δx becomes small: If x is no more than a thousand, $|x| \leq 1000$, and we want the observed error to be less than one one millionth, $|\varepsilon| < 10^{-6}$, then Δx needs to be small enough so that $|\Delta x|$ times 3001 is less than a millionth, for example, $|\Delta x| < 10^{-10}$, because

$$|\Delta x(3x + \Delta x)| < 10^{-10}(3001) < 10^{-6}\frac{3001}{10,000} < \frac{1}{1,000,000}$$

As long as x is bounded, we can make the error as small as we please by choosing a small enough Δx. The exact formula for how small Δx needs to be is not so obvious, but it is clear that for any fixed bound on x, the error $\varepsilon \to 0$ as $\Delta x \to 0$ for all $|x| \leq b$.

Finally, we rewrite this in the form of the increment (or microscope) approximation:

> **Increment of $f[x] = x^3$:**
> $$(x + \Delta x)^3 - x^3 = 3x^2 \Delta x + (\Delta x(3x + \Delta x))\,\Delta x$$
> $$f[x + \Delta x] - f[x] = f'[x] \cdot \Delta x + \varepsilon \cdot \Delta x$$
> with $f'[x] = 3x^2$ and $\varepsilon = (\Delta x(3x + \Delta x))$.

We summarize the knowledge that ε can be made small by making Δx small by writing

$$y = x^3 \qquad \Rightarrow \qquad dy = 3x^2\,dx$$

This formula, called a "differential," omits the error term and is just the local (dx, dy)-equation at a fixed value of (x, y) for the tangent line to $y = x^3$. The complete calculation means that the nonlinear graph approaches that line as we magnify more and more; in other words, the "gap" in the microscope, ε, tends to zero for all x beneath some bound, $|x| \leq b$ (see Figure 5.1).

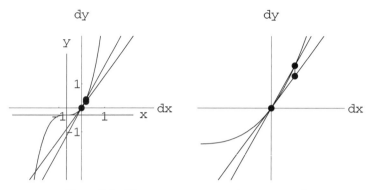

Figure 5.1: The gap near $x = 2/3$ on $y = x^3$

In the next example we calculate the gap for $y = \dfrac{1}{x}$ and show that there is a new complication in making ε tend to zero. There will sometimes have to be exceptional values of x. In these cases, we cannot even get close to the bad values of x. This is an annoying but necessary complication as you might suspect by looking at Figure 5.2 near zero.

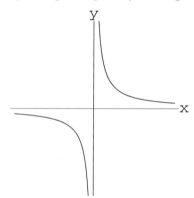

Figure 5.2: $y = 1/x$

Example 5.2. *Exceptional Numbers and the Derivative of $y = \dfrac{1}{x}$*

We follow the steps of Procedure 5.1 and Exercise 5.1.1. (See Exercise 28.7.3 for help with these computations.)

1) Compute $\dfrac{f[x + \Delta x] - f[x]}{\Delta x}$:

$$f[x + \Delta x] - f[x] = \frac{1}{x + \Delta x} - \frac{1}{x} = \frac{x - (x + \Delta x)}{x(x + \Delta x)}$$

$$= \frac{-1}{x(x + \Delta x)} \cdot \Delta x$$

$$\frac{f[x + \Delta x] - f[x]}{\Delta x} = \frac{-1}{x(x + \Delta x)}$$

2) Compute $\boxed{f'[x] = \lim_{\Delta x \to 0} \dfrac{f[x + \Delta x] - f[x]}{\Delta x}}$ intuitively.

It is "clear" that as $\Delta x \to 0$,

$$\lim_{\Delta x \to 0} \frac{-1}{x(x + \Delta x)} = \frac{-1}{x \cdot (x + 0)} = \frac{-1}{x^2}$$

(This is true unless $x \to 0$ at the same time. In particular, if $x = -\Delta x$, then $\frac{-1}{x(x+\Delta x)}$ is not even defined.) At least with x fixed, we should have

$$f'[x] = -\frac{1}{x^2}$$

3) We make the "gap" error explicit by the formula $\boxed{\varepsilon = \dfrac{f[x + \Delta x] - f[x]}{\Delta x} - f'[x]}$ and put

the expression on a common denominator

$$\varepsilon = \frac{-1}{x(x + \Delta x)} - \frac{-1}{x^2}$$
$$= \frac{1}{x^2} + \frac{-1}{x(x + \Delta x)}$$
$$= \frac{x + \Delta x}{x^2(x + \Delta x)} + \frac{-x}{x^2(x + \Delta x)}$$
$$= \frac{\Delta x}{x^2(x + \Delta x)}$$
$$= \Delta x \cdot \frac{1}{x^2(x + \Delta x)}$$

4) Show that $\varepsilon \to 0$ as $\Delta x \to 0$. It seems "clear" that

$$\lim_{\Delta x \to 0} \varepsilon = \lim_{\Delta x \to 0} \Delta x \cdot \frac{1}{x^2(x + \Delta x)} = 0 \cdot \frac{1}{x^2(x + 0)} = 0$$

but plugging in zero is really not quite enough because it misses the point to the *approximation* we want. We want the (x, y)-graph of the function

$$F_{\Delta x}[x] = \frac{f[x + \Delta x] - f[x]}{\Delta x}$$

to approximate the graph of $f'[x]$ when Δx is small. Another way to say this is that we want the whole function $\varepsilon[x, \Delta x]$ to be small independent of x provided Δx is sufficiently small. Then we can move the microscope over these values of x and continue to see a straight line approximating $y = f[x]$.

The factor Δx in $\Delta x \cdot \frac{1}{x^2(x+\Delta x)}$ does tend to zero; so, if we can bound the other factor, the whole function can be made small. The only way for $\frac{1}{x^2(x+\Delta x)}$ to get big is for x to be near zero. If we restrict x to stay away from the 'bad' point $x = 0$, then we can always make ε small. If $b > 0$ is fixed,

$$\varepsilon \approx 0 \quad \text{for all} \quad |x| \geq b > 0 \quad \text{when} \quad \delta x \approx 0$$

and the graphs of the functions $F_{\Delta x}[x]$ tend to the graph of $f'[x]$ for all $|x| \geq b$.

Increment of $f[x] = 1/x$:

$$\frac{1}{x + \Delta x} - \frac{1}{x} = \frac{-1}{x^2} \cdot \Delta x + \varepsilon \cdot \Delta x$$

$$f[x + \Delta x] - f[x] = f'[x] \cdot \Delta x + \varepsilon \cdot \Delta x$$

with $f'[x] = -\frac{1}{x^2}$ and $\varepsilon = \Delta x \cdot \dfrac{1}{x^2(x + \Delta x)}$.

We summarize the knowledge that ε can be made small by making Δx small by writing

$$y = \frac{1}{x} \qquad \Rightarrow \qquad dy = \frac{-1}{x^2}\, dx$$

This notation means that under sufficient magnification the gap between the curve and its tangent will be appear small as shown in Figure 5.3 for $x = 3/2$. The formulas are not valid if $x = 0$.

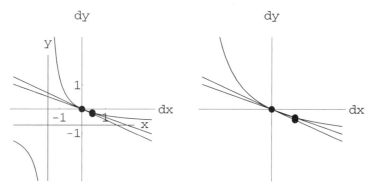

Figure 5.3: The gap near $x = 3/2$ on $y = 1/x$

Example 5.3.   *Exceptional Numbers and the Derivative of* $y = \sqrt{x}$

Here is another example of the algebraic part of computing increments. We follow the steps of Procedure 5.1 and Exercise 5.1.1. (See Exercise 28.7.3 for help with these computations.)

1) Compute $\dfrac{f[x + \Delta x] - f[x]}{\Delta x}$

$$f[x + \Delta x] - f[x] = \sqrt{x + \Delta x} - \sqrt{x}$$

$$= \frac{(\sqrt{x + \Delta x} - \sqrt{x})(\sqrt{x + \Delta x} + \sqrt{x})}{\sqrt{x + \Delta x} + \sqrt{x}}$$

$$= \frac{1}{\sqrt{x + \Delta x} + \sqrt{x}} \cdot \Delta x$$

$$\frac{f[x + \Delta x] - f[x]}{\Delta x} = \frac{1}{\sqrt{x + \Delta x} + \sqrt{x}}$$

2) Compute $\boxed{f'[x] = \lim\limits_{\Delta x \to 0} \dfrac{f[x + \Delta x] - f[x]}{\Delta x}}$

Although there certainly may be difficulties near $x = 0$ (see Figure 5.4), if x is fixed and positive,

$$\lim_{\Delta x \to 0} \frac{f[x + \Delta x] - f[x]}{\Delta x} = \lim_{\Delta x \to 0} \frac{1}{\sqrt{x + \Delta x} + \sqrt{x}} = \frac{1}{\sqrt{x + 0} + \sqrt{x}} = \frac{1}{2\sqrt{x}}$$

3) Calculate the error gap $\boxed{\varepsilon = \dfrac{f[x + \Delta x] - f[x]}{\Delta x} - f'[x]}$ (see Exercise 28.7.3):

$$\varepsilon = \frac{f[x + \Delta x] - f[x]}{\Delta x} - f'[x]$$

$$= \frac{1}{\sqrt{x + \Delta x} + \sqrt{x}} - \frac{1}{2\sqrt{x}}$$

$$= \frac{-1}{2\sqrt{x}(\sqrt{x + \Delta x} + \sqrt{x})^2} \cdot \Delta x$$

4) Show that $\boxed{\varepsilon \to 0}$ as $\Delta x \to 0$. It seems "clear" that

$$\lim_{\Delta x \to 0} \varepsilon = \lim_{\Delta x \to 0} \frac{-1}{2\sqrt{x}(\sqrt{x + \Delta x} + \sqrt{x})^2} \cdot \Delta x$$

$$= \frac{-1}{2\sqrt{x}(\sqrt{x + 0} + \sqrt{x})^2} \cdot 0$$

$$= \frac{-1}{8\,x\sqrt{x}} \cdot 0 = 0$$

and this shortcut computation (plugging in 0) is justified as a function approximation for all x as long as the term $\frac{-1}{8\,x\sqrt{x}}$ cannot get large. This is guaranteed by making $x \geq b$ for some fixed positive $b > 0$.

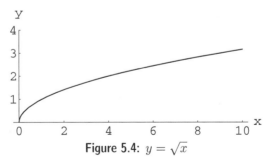

Figure 5.4: $y = \sqrt{x}$

Increment of $f[x] = \sqrt{x}$:

$$f[x + \Delta x] - f[x] = f'[x] \cdot \Delta x + \varepsilon \cdot \Delta x$$

$$\sqrt{x + \Delta x} - \sqrt{x} = \frac{1}{2\sqrt{x}} \cdot \Delta x + \varepsilon \cdot \Delta x$$

with $f'[x] = \frac{1}{2\sqrt{s}}$ and $\varepsilon = -\Delta x/(2\sqrt{x}(\sqrt{x + \Delta x} + \sqrt{x})^2)$.

We summarize the knowledge that ε can be made small by making Δx small by writing

$$y = \sqrt{x} \qquad \Rightarrow \qquad dy = \frac{1}{2\sqrt{x}}\, dx$$

This notation means that under sufficient magnification the gap between the curve and its tangent will be appear small as shown in Figure 5.5 for $x = 2/3$. These formulas are not valid if $x \le 0$.

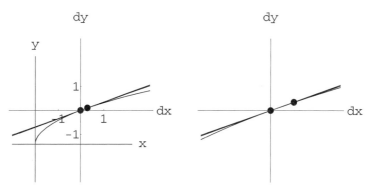

Figure 5.5: The gap near $x = 2/3$ on $y = \sqrt{x}$

Exercise set 5.1

1. $y = x^p \Rightarrow dy = p\, x^{p-1}\, dx,\ p = 1, 2, 3, \ldots$
For each $f[x] = x^p$ below:

(a) *Compute* $\dfrac{f[x + \Delta x] - f[x]}{\Delta x}$ *and simplify.*

(b) *Compute* $f'[x] = \lim_{\Delta x \to 0} \dfrac{f[x + \Delta x] - f[x]}{\Delta x}$ *Give an intuitive justification why your limit is correct. Does x need to be bounded, or can it vary arbitrarily as Δx tends to zero?*

(c) *Use your limit $f'[x]$ to solve for ε and write the increment equation:*

$$f[x + \Delta x] - f[x] = f'[x] \cdot \Delta x + \varepsilon \cdot \Delta x$$
$$= [\text{term in } x \text{ but not } \Delta x]\Delta x + [\text{observed microscopic error}]\Delta x$$

Notice that we can solve the increment equation for $\varepsilon = \dfrac{f[x + \Delta x] - f[x]}{\Delta x} - f'[x]$

(d) *Show that $\varepsilon \to 0$ as $\Delta x \to 0$. "Show" this in any way that you consider reasonable. Does x need to be bounded, or can it vary arbitrarily as Δx tends to zero?*

(1) *If $f[x] = x^1$, then $f'[x] = 1x^0 = 1$ and $\varepsilon = 0$.*
(2) *If $f[x] = x^2$, then $f'[x] = 2x$ and $\varepsilon = \Delta x$.*
(3) *If $f[x] = x^3$, then $f'[x] = 3x^2$ and $\varepsilon = (3x + \Delta x)\Delta x$.*
(4) *If $f[x] = x^4$, then $f'[x] = 4x^3$ and $\varepsilon = (6x^2 + 4x\Delta x + \Delta x^2)\Delta x$.*
(5) *If $f[x] = x^5$, then $f'[x] = 5x^4$ and $\varepsilon = (10x^3 + 10x^2\Delta x + 5x\Delta x^2 + \Delta x^3)\Delta x$.*

If you have difficulty with this exercise, see the high school review Exercise 28.7.1. Also see the program **SymbIncr** in this chapter's folder. If you want practice computing limits, see the Math Background chapter on computing limits on the CD accompanying this book.

Problem 5.1.

Use Procedure 5.1 to find $f'[x]$, and write the whole increment approximation

$$f[x + \Delta x] - f[x] = f'[x]\,\Delta x + \varepsilon\,\Delta x$$

showing that $\varepsilon \to 0$, when $\Delta x \to 0$.
 (a) $f[x] = x^n$,   *n   a positive integer*
 (b) $f[x] = \frac{1}{x^2}$
 (c) $f[x] = \sqrt[3]{x}$

*You can use the computer to help with your symbolic computations. See the program **SymbIncr** in this chapter's folder and verify your predictions using the program **Micro1D**.*

5.2   Moving the Microscope

This section uses interval notation to give a technical definition of local linearity. This "uniform" limit allows us to "move" the microscope.

A summary of the ε-gap computations so far is

$$y = f[x] = x^p \qquad \Rightarrow \qquad dy = f'[x]\,dx = p\,x^{p-1}\,dx, \quad p = 1, 2, 3, 4, 5$$

$$y = f[x] = \frac{1}{x} \qquad \Rightarrow \qquad dy = f'[x]\,dx = \frac{-1}{x^2}\,dx$$

$$y = f[x] = \frac{1}{x^2} \qquad \Rightarrow \qquad dy = f'[x]\,dx = \frac{-2}{x^3}\,dx$$

$$y = f[x] = \sqrt{x} \qquad \Rightarrow \qquad dy = f'[x]\,dx = \frac{1}{2\sqrt{x}}\,dx$$

$$y = f[x] = \sqrt[3]{x} \qquad \Rightarrow \qquad dy = f'[x]\,dx = \frac{1}{3\sqrt[3]{x^2}}\,dx$$

More than the summary, we know that the size of the gap (given by ε) viewed in a microscope of power $1/\Delta x$ goes to zero even as x varies, provided we avoid "bad" points. For the integer powers we need to have x bounded, $|x| \leq b$ for a fixed b as you saw in Exercise 5.1.1. ("$\pm\infty$" are the "bad" points in this case.) The other "bad" points are fairly obvious from the summary above. If $x = 0$, the function $1/x$ and its derivative $-1/x^2$ are undefined. We have to expect trouble there. If $x < 0$, the function $\sqrt{x}$ is undefined, but even if $x = 0$, where $\sqrt{x}$ is defined, the derived function $1/(2\sqrt{x})$ is undefined, so we expect trouble. All we need is a way to say where "good" approximations take place.

To give a general approach, we want to phrase the exceptions in terms of intervals.

Definition 5.2.   *Notation for Open and Compact Intervals*
   If a and b are numbers, we define open intervals as follows
   $(a, b) = \{x : a < x < b\}$
   $(-\infty, b) = \{x : x < b\}$
   $(a, \infty) = \{x : a < x\}$
   We define compact (or "closed and bounded") intervals by
   $[a, b] = \{x : a \leq x \leq b\}$

The condition of "tangency" is expressed by the microscopic error formula given next.

Informal Smoothness:
The function $f'[x]$ is the derivative of the function $f[x]$ if whenever we make a small change $\delta x \approx 0$ in input x in the interval of differentiability (a, b), then the change in output satisfies

$$f[x + \delta x] - f[x] = f'[x] \cdot \delta x + \varepsilon \cdot \delta x$$

with error $\varepsilon \approx 0$. This looks like

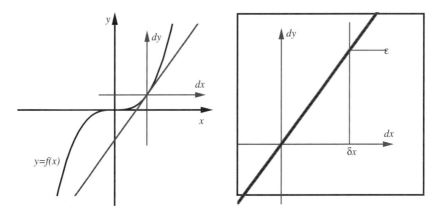

Figure 5.6: A symbolic microscope

Informally, this approximation remains valid if we move the microscope anywhere inside an interval of "good" points. The approximation means that a microscopic view of a tiny piece of the graph $y = f[x]$ looks the same as the linear graph $dy = f'[x] \cdot dx$. (The lower case (small) Greek delta δ, indicates intuitively that when the difference in x is a sufficiently small amount, $\delta x \approx 0$, then the error $\varepsilon \approx 0$ is small.) When we say $f'[x]$ is the derivative of $f[x]$ we mean that this local approximation is valid. We have shown this approximation directly for the functions summarized at the beginning of the section on appropriate compact intervals described in detail at the end of this section.

The rules of calculus are wonderful: They tell you where the trouble is going to occur.

Procedure 5.3. The Graph of the Linear Function Given by

$$dy = m \, dx$$

in local (dx, dy)-coordinates at $(x, f[x])$ is the tangent line to the explicit nonlinear graph

$$y = f[x]$$

provided $m = f'[x]$ and $f'[x]$ can be computed by the rules yielding a formula valid in an interval around x.

Specifically,

Theorem 5.4. *Successful Rules Imply Linear Approximation*
Suppose the derivative $\frac{dy}{dx} = f'[x]$ can be computed from an explicit formula $y = f[x]$ using the rules of Chapter 6. Also, suppose that both $f[x]$ and $f'[x]$ are defined on the compact interval $[\alpha, \beta]$. Then the size of the gap, ε in

$$f[x + \Delta x] - f[x] = f'[x] \cdot \Delta x + \varepsilon \cdot \Delta x$$

can be made small for all x in $[\alpha, \beta]$ by choosing a sufficiently small Δx. (For all x in $[\alpha, \beta]$ if $\Delta x = \delta x \approx 0$, then $\varepsilon \approx 0$.)

The complete technical definition of smoothness is given in the Mathematical Background materials on the CD accompanying this text. The background also gives a proof of Theorem 5.4. Here is the technical definition.

Definition 5.5. *Technical Smoothness (See Mathematical Background for Details)*
A real function $f[x]$ is called smooth (or differentiable or derivable) on the open interval (a, b) if there is a function $f'[x]$ such that for every compact subinterval $[\alpha, \beta]$ in (a, b), the function limit

$$\lim_{\Delta x \to 0} \frac{f[x + \Delta x] - f[x]}{\Delta x} = f'[x] \qquad \text{uniformly for all } x \text{ in } [\alpha, \beta]$$

In the Mathematical Background we show that the following are equivalent to this definition:

(a) A real function $f[x]$ is smooth on the real interval (a, b) if there is another real function $f'[x]$, called the derivative of $f[x]$, such that whenever $a < x < b$ and x is a bounded hyperreal number and not infinitely near a or b, then an infinitesimal increment of the dependent variable is approximately linear, that is,

$$f[x + \delta x] - f[x] = f'[x] \, \delta x + \varepsilon \, \delta x$$

where the error ε is infinitesimal, whenever δx is infinitesimal.

(b) A real function $f[x]$ is smooth on the open interval (a, b) if there is a function $f'[x]$ such that for every c in (a, b), the double limit converges,

$$\lim_{x \to c, \Delta x \to 0} \frac{f[x + \Delta x] - f[x]}{\Delta x} = f'[c]$$

The Mathematical Background also has a chapter on computing uniform limits of formulas like the gaps ε we have found so far.

We do not have to worry about all the technical details, but we do want to understand the role of points where either $f[x]$ or $f'[x]$ is an undefined formula. The following examples are all smooth functions. By Theorem 5.4, the "proof" that they are smooth just amounts to valid use of basic rules, in this case the Power Rule.

Example 5.4.   *Domains of Approximation for $y = x^p$, $p = 1, 2, 3, \ldots$*

The functions $y = x^n$ and their derivatives $\frac{dy}{dx} = n\,x^{n-1}$ for $n = 1, 2, 3, \ldots$ are defined for all real x in $(-\infty, \infty)$. Theorem 5.4 says they are differentiable on the open interval $(-\infty, \infty)$, and this means that for any pair of real numbers, $\alpha < \beta$, the gap ε can be made small over the compact interval $[\alpha, \beta]$ by choosing a single, small enough Δx. You cannot make ε small for the whole real line, $(-\infty, \infty)$.

Example 5.5.   *Domains of Approximation for $y = x^p$, $p = -1, -2, \ldots$*

The functions $y = \dfrac{1}{x^n}$ and their derivatives $\dfrac{dy}{dx} = -\dfrac{n}{x^{n+1}}$, for $n = 1, 2, 3, \ldots$ are defined for all nonzero real x in $(-\infty, 0)$ or $(0, \infty)$, but not at $x = 0$. Theorem 5.4 says they are differentiable on the open intervals $(-\infty, 0)$ and $(0, \infty)$. This means that for pairs of real numbers, $\alpha < \beta < 0$ or $0 < \alpha < \beta$, the gap ε can be made small over the whole compact interval $[\alpha, \beta]$ by choosing a single, small enough Δx. You cannot make ε small for the whole interval, $(-\infty, 0)$ or $(0, \infty)$ or for $[\alpha, \beta]$ if $\alpha < 0 < \beta$.

Example 5.6.   *Domains of Approximation for $y = x^p$, $p = 1/2, 1/3, \ldots$*

The function $y = \sqrt{x}$ and its derivative $\dfrac{dy}{dx} = \dfrac{1}{2\sqrt{x}}$ are defined for all positive real x in $(0, \infty)$. Both are NOT defined at $x = 0$. Theorem 5.4 says they are differentiable on the open interval $(0, \infty)$, so for pairs of real numbers $0 < \alpha < \beta$, the gap ε can be made small over the compact interval $[\alpha, \beta]$ by choosing a single, small enough Δx. You cannot make ε small for the whole interval $(0, \infty)$.

───────────────── (**Exercise set 5.2**) ─────────────────

You can see this function convergence for yourself on the computer.

1. ε on the Computer

*Run the program **DfctLimit** to show graphically that the gap errors of all the following functions tend to zero AS FUNCTIONS OF x away from the "bad" points.*

$$y = f[x] = x^p \qquad \Rightarrow \qquad dy = f'[x]\,dx = p\,x^{p-1}\,dx, \quad p = 1,2,3,4,5$$

$$y = f[x] = \frac{1}{x} \qquad \Rightarrow \qquad dy = f'[x]\,dx = \frac{-1}{x^2}\,dx$$

$$y = f[x] = \frac{1}{x^2} \qquad \Rightarrow \qquad dy = f'[x]\,dx = \frac{-2}{x^3}\,dx$$

$$y = f[x] = \sqrt{x} \qquad \Rightarrow \qquad dy = f'[x]\,dx = \frac{1}{2\sqrt{x}}\,dx$$

$$y = f[x] = \sqrt[3]{x} \qquad \Rightarrow \qquad dy = f'[x]\,dx = \frac{1}{3\sqrt[3]{x^2}}\,dx$$

2. A View in the Microscope

You are told that a certain function $y = f[x]$ has a derivative for all values of x, $f'[x]$. At the point $x = 1$, we know that $f'[1] = -2/3$. Sketch what you would see in a very powerful microscope focused on the graph $y = f[x]$ above the point $x = 1$.

Compare your work on the next problem about kinks with Exercise 3.2.4.

Problem 5.2. ε ON THE COMPUTER AND ANALYTICALLY FOR A KINK ⎯⎯⎯⎯⎯▼
*Run the program **DfctLimit** on the function*

$$y = f[x] = \sqrt{x^2 + 2x + 1}$$

which has derivative

$$\frac{dy}{dx} = f'[x] = \frac{x+1}{\sqrt{x^2 + 2x + 1}}$$

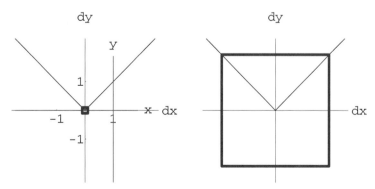

Figure 5.7: $y = \sqrt{x^2 + 2x + 1}$ *near* $x = -1$

Show that the function and its derivative are NOT BOTH defined at $x = -1$. Verify analytically that the gap you see on the computer does NOT tend to zero.

5.3 Trigonometric Derivatives - CD

The gaps ε for $y = f[x] = \mathrm{Sin}[x]$, $y = f[x] = \mathrm{Cos}[x]$, and $y = f[x] = \mathrm{Tan}[x]$ are calculated in this section by comparing the length of a segment of the unit circle with the vertical and horizontal projections from the ends of the segment.

Remainder of this section only on CD

5.4 Derivatives of Log and Exp - CD

The gaps ε for $y = e^x$ and $y = \mathrm{Log}[x]$ are discussed in this section.

Remainder of this section only on CD

5.5 Continuity and the Derivative - CD

This section shows that locally linear implies continuous and uniform derivatives are continuous.

Remainder of this section only on CD

5.6 Projects and Theory
5.6.1 Hubble's Law and the Increment Equation

$$R[t + \delta t] = R[t] + R'[t]\,\delta t + \varepsilon\,\delta t$$

Evidence of an expanding universe is one of the most important astronomical observations of this century. Light received from a distant galaxy is "old" light, generated millions of years ago at a time t_e when it was emitted. When this old light is compared to light generated at the time received t_r, it is found that the characteristic colors, or spectral lines, do not have the same wavelengths. All the current wavelengths are longer, $\lambda[t_e] < \lambda[t_r]$. This means that light is "redder" now; this is the famous red shift.

The Scientific Project on Hubble's Law shows you an explanation for the expanding universe that is based just on using the increment approximation. Recently, there has been some reexamination of Hubble's Law indicating that Hubble's "constant" may not be constant. This is still compatible with an increment derivation of the law, which relies on the tiny time increment of only a few human generations.

5.6.2   Numerical Approximation of Exponential Derivatives

The project on Exponential Derivatives has you calculate the constants

$$k_a = \lim_{\Delta x \to 0} \frac{a^{\Delta x} - 1}{\Delta x}$$

and then adjust a until you make $k_a = 1$. This is one way to compute the natural base $e \approx 2.71828\ldots$.

5.6.3   Small Enough Real Numbers or "Epsilons and Deltas"

The increment approximation used to estimate $\text{Sin}[\text{"29 degrees"}]$ was very close, but how do we know that the increment approximation gets close for real increments, not just close for small increments? The Mathematical Background chapter on "epsilons and deltas" answers this question. All the theorems are proved in detail in the Mathematical Background.

CHAPTER
6

Symbolic Differentiation

This chapter presents the "method of computing" or "calculus" of derivatives by giving symbolic rules for finding formulas for derivatives when we are given formulas for the functions.

When we compute a derivative, we want to know that the increment approximation is valid. You must use high school algebra and trig, but you do not have to establish the increment approximation directly as we did in Chapter 5. The graphical and symbolic theorems of 1-variable differentiation say the following:

If we can compute the derivative $f'[x]$ of a function $f[x]$ using the rules from this chapter, then a sufficiently magnified view of the graph $y = f[x]$ appears linear at each point of the interval where the formulas for $f[x]$ and $f'[x]$ are valid.

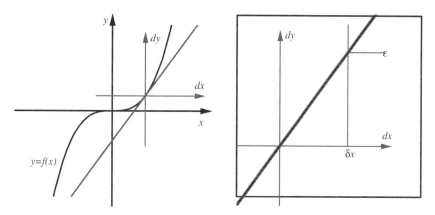

Figure 6.1: $y = f[x]$ and $dy = m \cdot dx$ through a powerful microscope

The line with local coordinates $dy = m \, dx$ "looks" the same as $y = f[x]$ under magnification when $m = f'[x]$. The slope $f'[x]$ depends on the center of magnification, x.

This condition of "tangency" is expressed symbolically by the approximation formula that says the nonlinear change is a linear term plus something small compared to the change:

$$f[x + \delta x] - f[x] = f'[x] \cdot \delta x + \varepsilon \cdot \delta x$$

with the magnified error ε small, $\varepsilon \approx 0$, whenever the input perturbation is small, $\delta x \approx 0$, and x lies in an interval $[\alpha, \beta]$ where both $f[x]$ and $f'[x]$ are defined.

The microscope equation above expresses the nonlinear change, $f[x + dx] - f[x]$, in terms of a change dx or "local variable" dx, with x fixed. The linear term in dx is called the differential,

$$dy = f'[x] \cdot dx \qquad \text{or} \qquad dy = m \cdot dx$$

in (dx, dy) coordinates (with x fixed), where dy represents the change from $f[x]$. When $dx = \delta x \approx 0$ is small, the difference between these terms is small compared to δx because the difference is a product of a small term ε and the small change δx. On magnification by $1/\delta x$, the term $\varepsilon \cdot \delta x$ appears to be the size of ε. If this is small enough (by virtue of large enough magnification), we do not see it and the graph appears linear.

The results of this chapter ensure that the error is small whenever x lies in an interval $[\alpha, \beta]$ where both $f[x]$ and $f'[x]$ are defined. (At a fixed high magnification the graph appears straight simultaneously for every microscope at an x-focus point in $[\alpha, \beta]$.) The rules of calculus are theorems which guarantee that this approximation is valid, provided the resulting formulas are defined on intervals. This is a powerful yet practical theory. Here is a brief example of how it is used.

Example 6.1. *$f'[x_0]$ is the Slope of $y = f[x]$*

The slope of the line (in local coordinates)

$$dy = m \cdot dx$$

is m and the line points upward if $m > 0$. Because a microscopic image of the graph $y = f[x]$ cannot be distinguished from the graph of the linear equation $dy = m \cdot dx$ when $m = f'[x_0]$, the graph $y = f[x]$ is increasing at the approximate rate $f'[x_0]$ near x_0.

Example 6.2. *Using the Theory*

The theory is easy to use once you learn the rules from this chapter. Here are two examples where the theory breaks down. The breakdown is easy to detect. By the end of the chapter, you will be able to apply rules and compute the following two derivatives for

$$f[x] = \sqrt{x^2 + 2x + 1} \qquad \text{and} \qquad y = x^{\frac{2}{3}}$$

obtaining

$$f'[x] = \frac{x + 1}{\sqrt{x^2 + 2x + 1}} \qquad \text{and} \qquad \frac{dy}{dx} = \frac{2}{3\sqrt[3]{x}}$$

After computing without fear, you need to check the formulas to see that

$$f'[-1] = \frac{-1 + 1}{\sqrt{(-1)^2 - 2 + 1}} = \frac{0}{0} \qquad \text{and} \qquad \frac{dy}{dx} = \frac{2}{3\sqrt[3]{0}} = \frac{2}{0}$$

are undefined. When the formulas are not valid, the theory does not predict anything; but, in this case, we have seen that there is a kink in the graph of $f[x]$ at $x = -1$ (see Exercise 3.2.4). There is a vertical cusp on the graph of $y = x^{\frac{2}{3}}$ at $x = 0$ (see Problem 6.1).

The rules of this chapter guarantee that the increment approximation for tangency holds when the resulting formulas are valid on intervals. Of course, your first task now is to learn:

Example 6.3. *All the Rules of Differentiation*

There are only eight rules in this chapter and you must memorize them:

$$y = x^p \quad \Rightarrow \quad \frac{dy}{dx} = p\,x^{p-1}, \quad p \text{ constant}$$

$$y = \mathrm{Sin}[\theta] \quad \Rightarrow \quad \frac{dy}{d\theta} = \mathrm{Cos}[\theta]$$

$$y = \mathrm{Cos}[\theta] \quad \Rightarrow \quad \frac{dy}{d\theta} = -\mathrm{Sin}[\theta]$$

$$y = e^x \quad \Rightarrow \quad \frac{dy}{dx} = y = e^x$$

$$x = \mathrm{Log}[y] \quad \Rightarrow \quad \frac{dx}{dy} = \frac{1}{y}$$

$$\frac{d\,(a\,f[x] + b\,g[x])}{dx} = a\,\frac{df[x]}{dx} + b\,\frac{dg[x]}{dx}$$

$$\frac{d(f[x]\,g[x])}{dx} = \frac{df[x]}{dx}\cdot g[x] + f[x]\cdot\frac{dg[x]}{dx}$$

$$\frac{dy}{dx} = \frac{dy}{du}\cdot\frac{du}{dx}, \quad \text{when } y = f[u] \quad \& \quad u = g[x]$$

The hard thing is to learn to combine these rules with high school algebra and trig.

6.1 Rules for Special Functions

This section gives the five specific differentiation rules for basic functions.

The algebra of exponents together with the derivatives we computed in Chapter 5 suggest a single rule that includes the examples of Exercise 5.1 and before. To understand why this rule covers the cases of roots and reciprocals, you must understand the laws of exponents in the review Chapter 28, especially Exercise 28.4.2.

Theorem 6.1. *The Power Rule*
For any constant p,

$$y = x^p \quad \Rightarrow \quad \frac{dy}{dx} = p\,x^{p-1}$$

In other words, functions that can be expressed as powers are locally linear with derivative as above – provided the formulas on both sides of the implication are defined on an interval.

We showed directly in the last chapter that if $y = \sqrt{x}$, then $dy = \frac{1}{2\sqrt{x}}\,dx$.

Example 6.4. $\frac{d(\sqrt{x})}{dx}$ *by Rules*

This is one special case of the Power Rule with $p = 1/2$, because

$$y = \sqrt{x} = x^{\frac{1}{2}}, \quad \text{so} \quad \frac{dy}{dx} = \frac{1}{2}\, x^{\frac{1}{2}-1} = \frac{1}{2}\, x^{-\frac{1}{2}}$$

$$= \frac{1}{2}\, \frac{1}{x^{\frac{1}{2}}} = \frac{1}{2}\, \frac{1}{\sqrt{x}} = \frac{1}{2\sqrt{x}}$$

Notice that our final formula is only valid on the open interval $(0, \infty) = \{x : 0 < x < \infty\}$. The open interval of validity is part of the Power Rule, but you compute first and then think. Do not forget the second step.

Example 6.5. $\frac{d(1/x^2)}{dx}$ *by Rules*

This is a special case of the Power Rule with $p = -2$, because

$$y = \frac{1}{x^2} = x^{-2}, \quad \text{so} \quad \frac{dy}{dx} = -2\, x^{-2-1} = -2\, x^{-3}$$

$$= \frac{-2}{x^3}$$

Notice that our final formula is only valid on the open interval $(0, \infty) = \{x : 0 < x < \infty\}$ or the interval $(-\infty, 0)$ but not on any interval of the form $[a, b]$ with $a < 0 < b$.

Example 6.6. $\frac{d(x\sqrt{x})}{dx}$ *by Rules*

This is a special case of the Power Rule with $p = 3/2$ because

$$y = x\sqrt{x} = x^1\, x^{1/2} = x^{1+\frac{1}{2}} = x^{\frac{3}{2}}$$

$$\frac{dy}{dx} = \frac{3}{2}\, x^{\frac{3}{2}-1} = \frac{3}{2}\, x^{\frac{1}{2}}$$

Notice that our final formula is valid only for $x \geq 0$. The largest open interval where the function and derivative are defined is $(0, \infty)$.

In the last chapter we directly proved derivative formulas for the sine and cosine using a microscopic view of the circle. The angles must be measured in radians in order to compare differences in sine and cosine with length along the unit circle. Here are the formulas:

Theorem 6.2. *The Sine and Cosine Rules*
   For θ in radians,

$$y = \text{Sin}[\theta] \qquad \Rightarrow \qquad \frac{dy}{d\theta} = \text{Cos}[\theta]$$

$$y = \text{Cos}[\theta] \qquad \Rightarrow \qquad \frac{dy}{d\theta} = -\text{Sin}[\theta]$$

The sine and cosine rules are valid for all real θ. This means that the increment approximation holds on $(-\infty, \infty)$.

We will postpone the proof of the exponential and log rules but include them here because they are the only other special function rules you need to learn.

Theorem 6.3. *The Log and Exponential Rules*

$$y = e^x \qquad \Rightarrow \qquad \frac{dy}{dx} = y = e^x$$

$$x = \text{Log}[y] \qquad \Rightarrow \qquad \frac{dx}{dy} = \frac{1}{y}$$

The exponential rule is valid for all real x or, in other words, the increment approximation holds on $(-\infty, \infty)$. The natural logarithm rule makes sense only if the log and the formula for the derivative are both defined, so the increment approximation for $\text{Log}[y]$ is valid on $(0, \infty)$.

Exercise set 6.1

1. *You cannot divide by zero or take even roots of negative numbers (as real functions). Show that the Power Rule does not apply at $x = 0$ for $p = \frac{1}{3}$ or at $x = -2$ for $p = \frac{1}{4}$.*

$$y = x^p = x^{1/3} \qquad \Rightarrow \qquad \frac{dy}{dx} = p\,x^{p-1} = ?$$

and

$$y = x^p = x^{1/4} \qquad \Rightarrow \qquad \frac{dy}{dx} = p\,x^{p-1} = ?$$

Use the computer to graph the two functions $y = x^{1/3} = \sqrt[3]{x}$ and $y = x^{1/4} = \sqrt[4]{x}$ for $-3 \leq x \leq 3$. Explain your "bad" analytical result above in terms of the graph.

We will not prove the general Power Rule Theorem now, but ask you to check it for the cases we already know in the next exercise.

2. (a) *Show that the Power Rule agrees with all the derivatives we computed directly in and before Exercise 5.1.1 as well as those that you computed in Problem 5.1. They are the following:*

$$y = x \Rightarrow \frac{dy}{dx} = 1 = 1x^{1-1} = x^0 \qquad y = \frac{1}{x} = x^{-1} \Rightarrow \frac{dy}{dx} = \frac{-1}{x^2} = -1x^{-2}$$

$$y = x^2 \Rightarrow \frac{dy}{dx} = 2x^1 = 2x^{2-1} \qquad y = \frac{1}{x^2} = x^{-2} \Rightarrow \frac{dy}{dx} = \frac{-2}{x^3} = -2x^{-3}$$

$$y = x^3 \Rightarrow \frac{dy}{dx} = 3x^2 = 3x^{3-1} \qquad y = \sqrt{x} = x^{\frac{1}{2}} \Rightarrow \frac{dy}{dx} = \frac{1}{2\sqrt{x}} = \frac{1}{2}x^{-\frac{1}{2}}$$

$$y = x^n \Rightarrow \frac{dy}{dx} = nx^{n-1} \qquad y = \sqrt[3]{x} = x^{\frac{1}{3}} \Rightarrow \frac{dy}{dx} = \frac{1}{3\sqrt[3]{x^2}} = \frac{1}{3}x^{-\frac{2}{3}}$$

(b) *Differentiate the following by first converting to power form and then applying the Power Rule. Convert your derivatives back to radical notation.*

a) $y = \frac{1}{x^5}$
b) $y = \sqrt[5]{x}$
c) $y = \frac{1}{\sqrt[3]{x^2}}$

d) $y = x^2 \sqrt{x}$

e) $y = x^4 \sqrt{x^3}$

f) $y = \dfrac{x^2}{\sqrt{x}}$

g) $y = x \sqrt[3]{x}$

h) $y = x^5 \sqrt[3]{x^2}$

i) $y = \dfrac{x^2}{\sqrt[3]{x^2}}$

(c) *The derivative of a constant function is zero. Why? The Power Rule also includes a case of this in the form,*

$$y = x^0 \quad \Rightarrow \quad \frac{dy}{dx} = 0 = 0\,x^{0-1}$$

What is the value of x^0? Is 0^0 defined?

It is important to be able to apply the differentiation rules to functions defined in terms of letters other than x. At first, it is simplest to learn the manipulations with one letter, that's true, but it is also important to move beyond that. Here is some practice:

3. Other Variables

a) $y = x^2 \quad \Rightarrow \quad \frac{dy}{dx} = ?$

b) $u = \frac{1}{v^2} \quad \Rightarrow \quad \frac{du}{dv} = ?$

c) $y = \sqrt{x} \quad \Rightarrow \quad \frac{dy}{dx} = ?$

d) $u = \frac{1}{\sqrt{v}} \quad \Rightarrow \quad \frac{du}{dv} = ?$

e) $y = x^3 \sqrt{x^3} \quad \Rightarrow \quad \frac{dy}{dx} = ?$

f) $u = \frac{v^2}{\sqrt[3]{v^2}} \quad \Rightarrow \quad \frac{du}{dv} = ?$

g) $y = \mathrm{Sin}[x] \quad \Rightarrow \quad \frac{dy}{dx} = ?$

h) $u = \mathrm{Cos}[v] \quad \Rightarrow \quad \frac{du}{dv} = ?$

i) $y = \mathrm{Log}[x] \quad \Rightarrow \quad \frac{dy}{dx} = ?$

j) $u = e^v \quad \Rightarrow \quad \frac{du}{dv} = ?$

Problem 6.1. A CUSP ⎯⎯⎯⎯⎯⎯⎯⎯⎯⎯⎯⎯⎯⎯⎯⎯⎯⎯⎯⎯⎯⎯▼

*When the graph of $y = x^{2/3}$ is magnified at $x = y = 0$, what do we see? The Power Rule does not apply at $x = 0$. Why? Still, we can either use small increments directly or look at microscopic views for smaller and smaller values of δx. The question is th following: In an tiny microscope, do we see a "VEE" or a vertical straight line segment? (HINT: Run the animation in the computer program **Zoom**, then explain what you see analytically.)*

⎯⎯⎯⎯⎯⎯⎯⎯⎯⎯⎯⎯⎯⎯⎯⎯⎯⎯⎯⎯⎯⎯⎯⎯⎯⎯⎯⎯⎯⎯⎯⎯⎯⎯▲

The Power Rule, the Sine and Cosine Rules, and the Log and Exponential Rules are the only particular function rules you need to learn. The other general rules of differentiation allow you to use these to build a host of formulas that you can differentiate. The general function combination rules take up the next three sections.

6.2   The Superposition Rule

This section shows that the sum of the derivatives is the derivative of the sum. The physical Superposition Principle says that the response to a sum of stimuli is the sum of the responses to the separate stimuli. These are closely related ideas.

Another way to express the physical Superposition Rule is

Output[stimulus 1 + stimulus 2] = Output[stimulus 1] + Output[stimulus 2]

This is a simple property that is often violated in real life. For example, the combined effect of a cup of coffee and an aspirin is not the same as the two separate effects. Systems that satisfy the Superposition Principle are often called "linear systems," because you can apply the "output" to a linear combination $a f[x] + b g[x]$ or form the same linear combination of the separate outputs,

$$\text{Out}[a f[x] + b g[x]] = a\text{Out}[f[x]] + b\text{Out}[g[x]]$$

Chapter 23 and the associated projects develop important applications where physical superposition does apply and "linearity" of the derivative is at the heart of the matter. In the case of differentiation, "Output" means derivative and "stimulus" means input function.

Theorem 6.4.   *The Superposition Rule (or Linearity of Differentiation)*
If $f[x]$ and $g[x]$ are smooth real functions for $\alpha < x < \beta$ and a and b are real constants, then the linear combination function $h[x] = a\,f[x] + b\,g[x]$ is also smooth for $\alpha < x < \beta$ and

$$\frac{d\,(a\,f[x] + b\,g[x])}{dx} = a\frac{df[x]}{dx} + b\frac{dg[x]}{dx}$$

In words the theorem says that the derivative of a linear combination of functions is the same linear combination of their derivatives.

PROOF OF SUPERPOSITION FOR DIFFERENTIATION:

The general proof of the Superposition Rule is little more than algebra. We have the Increment Formula for $f[x]$ and $g[x]$, so

$$
\begin{aligned}
h[x + \delta x] - h[x] &= (a\,f[x + \delta x] + b\,g[x + \delta x]) - (a\,f[x] + b\,g[x]) \\
&= a\,(f[x + \delta x] - f[x]) + b\,(g[x + \delta x] - g[x]) \\
&= a\,(f'[x]\delta x + \varepsilon_1 \delta x) + b\,(g'[x]\delta x + \varepsilon_2 \delta x) \\
&= (a\,f'[x] + b\,g'[x])\,\delta x + (a\varepsilon_1 + b\varepsilon_2)\delta x
\end{aligned}
$$

Because $\varepsilon_1 \approx 0$ and $\varepsilon_2 \approx 0$ (by the Increment Formula for f and g), we know $(a\varepsilon_1 + b\varepsilon_2) \approx 0$, thus $h'[x] = a\,f'[x] + b\,g'[x]$ and the theorem is proved.

Example 6.7. $\frac{d(5\cdot\sqrt{x}-\pi\cdot x^2)}{dx}$

Let

$$f[x] = \sqrt{x}, \qquad a = 5, \qquad g[x] = x^2, \qquad b = -\pi$$

then

$$a\,f[x] + b\,g[x] = 5\cdot\sqrt{x} - \pi\cdot x^2$$

and the derivatives of the pieces are $\frac{df}{dx} = \frac{1}{2\sqrt{x}}$ and $\frac{dg}{dx} = 2x$, so

$$\frac{d\,(a\,f[x] + b\,g[x])}{dx} = a\frac{df[x]}{dx} + b\frac{dg[x]}{dx}$$

$$\frac{d(5\cdot\sqrt{x} - \pi\cdot x^2)}{dx} = 5\cdot\frac{1}{2\sqrt{x}} - \pi\cdot 2x$$

A shortcut way to write this computation is

$$\frac{d(5\sqrt{x} - \pi\cdot x^2)}{dx} = 5\frac{d(\sqrt{x})}{dx} - \pi\frac{d(x^2)}{dx} = 5\frac{d(x^{1/2})}{dx} - \pi\frac{d(x^2)}{dx}$$

$$= 5\cdot\frac{1}{2}\cdot x^{\frac{1}{2}-1} - \pi\cdot 2\cdot x^{2-1} = \frac{5}{2}x^{-\frac{1}{2}} - 2\pi\,x$$

$$= \frac{5}{2}\frac{1}{x^{1/2}} - 2\pi\,x$$

$$= \frac{5}{2\sqrt{x}} - 2\pi\,x$$

Example 6.8. *The Constant Multiple Rule,* $\frac{d(a\,f[x])}{dx} = a\frac{df}{dx}[x]$

We may take $b = 0$ and $g[x] = 0$ in the Superposition Rule. If $y = \frac{e^\pi}{\sqrt[3]{x}}$, let $a = e^\pi$ and $f[x] = \frac{1}{x^{1/3}}$, $b = g[x] = 0$, so

$$\frac{d\,(a\,f[x])}{dx} = a\frac{df[x]}{dx}$$

$$\frac{d(\frac{e^\pi}{\sqrt[3]{x}})}{dx} = e^\pi\frac{d(\frac{1}{\sqrt[3]{x}})}{dx} = e^\pi\frac{d(\frac{1}{x^{1/3}})}{dx} = e^\pi\frac{d(x^{-1/3})}{dx}$$

$$= e^\pi\frac{-1}{3}x^{-\frac{1}{3}-1} = e^\pi\frac{-1}{3}x^{-\frac{1}{3}-\frac{3}{3}} = e^\pi\frac{-1}{3}x^{-\frac{4}{3}}$$

$$= -\frac{e^\pi}{3}\frac{1}{x^{\frac{4}{3}}} = -\frac{e^\pi}{3}\frac{1}{\sqrt[3]{x^4}}$$

Notice that e^π is a constant, $e^\pi = (2.71828\cdots)^{(3.14159\cdots)} \approx 23.1407$

6.2.1 Symbolic Differentiation with the Computer

The computer can solve all the symbolic differentiation exercises in this chapter. At first this fact might discourage you, but it should not. We want you to learn to use the rules of differentiation well enough to be confident that you understand them. The computer cannot think or understand the meaning of the result of these computations, and the input syntax needed to make it solve the exercises is troublesome itself.

WARNING: If you do NOT learn to compute derivatives without the computer, you probably will not understand the rules well enough to succeed in calculus, even with a computer. Combining your basic ability with the computer will lead to greater success than used to be possible, because once you understand the rules, the computer can become a mental "lever." It can do complicated symbolic computations for you with great reliability. You are then left with the important and interesting job of formulating the problem, programming it into the computer, and interpreting the result.

You are also welcome to use the computer to check all of your work from this chapter. Use the built-in computer differentiation command (in **DfDx**) once you have learned all the differentiation rules. The **DiffRules** program is only intended to show you what cannot be done without some of the rules. Knowing what cannot be done is part of understanding the strength of the general functional rules.

------- Exercise set 6.2 -------

1. Basic Superposition Drill
  Find $\frac{dy}{dx}$ for each of the following functions $y = y[x]$. (The letters a, b, c, and h denote constants or parameters, and e is the natural base for logs and exponentials.)

a) $y = 7x^4$

b) $y = -5x^2$

c) $y = 7x + x^4$

d) $y = 8x^3 - 5x^2 + 4x + 1$

e) $y = \frac{7}{x^4}$

f) $y = -5\sqrt[3]{x}$

g) $y = 7\sqrt[5]{x} - 3\frac{1}{x^4}$

h) $y = 7\frac{\sqrt{x}}{\sqrt[3]{x}} - 5\frac{1}{\sqrt{x}}$

i) $y = 7e^x - 3\operatorname{Log}[x]$

j) $y = 7\frac{1}{x^e} + 3\operatorname{Log}[1/x]$

k) $y = e\sqrt[5]{x} - \pi\frac{1}{x^4}$

l) $y = 3\operatorname{Sin}[x] - \operatorname{Cos}[7e]$

m) $y = f[x] = a + bx + cx^2 + hx^3$

n) $y = f[x] = 7\operatorname{Cos}[x] - 3\operatorname{Sin}[x]$

The proof of "linearity of differentiation" is easy provided you understand the function notation. Unfortunately, the notation gives many students trouble at first. The next exercise helps you understand the general notation by working a specific example. One payoff to understanding the function formulas is that you will be able to write better computer programs because much of modern computing uses function notation.

2. *Let $f[x] = \sqrt{x}$, let $g[x] = x^2$, let $a = 5$, and let $b = \pi$. Write out each step of the proof of the Superposition Rule that appears above, except write the steps with these specific functions and constants.*

It is important, especially in applications, to be able to apply the differentiation rules to functions defined in terms of letters other than x.

3. Other Variables

a) $v = u^2 + 2u + 2 \quad \Rightarrow \quad \frac{dv}{du} = ?$

b) $u = 4\sqrt{v} - \pi\frac{1}{v^2} \quad \Rightarrow \quad \frac{du}{dv} = ?$

c) $y = 3u\sqrt{u} + 5/u^2 \quad \Rightarrow \quad \frac{dy}{du} = ?$

d) $u = v^{1/4} - \frac{1}{v^{1/3}} \quad \Rightarrow \quad \frac{du}{dv} = ?$

e) $y = \text{Cos}[\theta] - \text{Sin}[\theta] \quad \Rightarrow \quad \frac{dy}{d\theta} = ?$

f) $u = \text{Log}[v] - e^v \quad \Rightarrow \quad \frac{du}{dv} = ?$

The computer program **DiffRules** contains exercises to show you how the rules of differentiation can be used to build a the computer program for differentiation. The computer program **DfDx** shows you how to use the computer's built-in differentiation.

4. Differentiation by Computer

> *Run the **DiffRules** program and work through it line by line so that you can see how the addition of each of the Superposition Rule, the Product Rule, and the Chain Rule makes symbolic differentiation more powerful. We cannot differentiate $\sqrt{x}\,\text{Cos}[x]$ without the Product Rule, and neither can the computer.*

Problem 6.2.

Using only the Superposition Rule and derivatives of sine, cosine, natural log and exponential; find the following or write brief explanations why they cannot be done this way. The letters a, b, c, and m denote parameters (or unknown constants).

a) $y = mx + b$, $y' = ?$

b) $y = uv + w$, $\frac{dy}{dv} = ?$

c) $y = uv + w$, $\frac{dy}{dw} = ?$

d) $f[x] = 1 + \frac{1}{x} + \frac{1}{x^2} + \frac{1}{x^3}$, $f'[x] = ?$

e) $f[x] = a + bx + cx^2 + mx^3$, $f'[x] = ?$

f) $u = 3\,\text{Sin}(\theta) - \text{Cos}(\theta)$, $\frac{du}{d\theta} = ?$

g) $y = \frac{1}{\sqrt[3]{x^5}}$, $\frac{dy}{dx} = ?$

h) $y = \sqrt{x} + \text{Sin}[x]$, $\frac{dy}{dx} = ?$

i) $\sqrt{x}\,\text{Sin}[x]$, $\frac{dy}{dx} = ?$

j) $y = \sqrt{\text{Sin}[x]}$, $\frac{dy}{dx} = ?$

k) $y = \text{Sin}(\sqrt{x})$, $\frac{dy}{dx} = ?$

l) $y = \sqrt{x}$, $f[x] = \frac{dy}{dx}$, $f'[x] = ?$

m) $y = e^x + \text{Log}[x]$, $\frac{dy}{dx} = ?$

n) $y = e^x\,\text{Sin}[x]$, $\frac{dy}{dx} = ?$

What is the slope of each of the above graphs when the independent variable equals -1? 0? (What are the tricks in this question?)

Once you have the Product Rule, you will be able to do some additional parts of the previous problem and when you also have the Chain Rule, you will be able to do all the parts. The computer program **DiffRules** can be used to check your work. Only enter the specific function rules and the Superposition Rule. If the computer cannot find the derivative with these rules, it will return your question as its output.

The next problem shows two practical ways that superposition of derivatives arise. You only need to link those obvious applications with the symbolic expressions to solve the problems.

Problem 6.3.

Express the conditions in the following scenarios in terms of three functions yielding positions as a function of time and the time derivatives of these functions. Write the general function rules that yield the answer to the questions, and verify that the mathematical rules agree with the intuitively obvious answers.

(a) *A man and a woman are riding on a train that is travelling at the rate of 75 mph. Inside the train, the woman is walking forward at the rate of 4 mph and the man is walking backward at the rate of 3 mph. How fast is the man traveling relative to the ground? How fast is the woman traveling relative to the ground?*

Let $T[t]$ equal the distance (in miles) that the train has traveled along the ground. Let $m[t]$ equal the distance the man has traveled forward on the train. Let $w[t]$ equal the distance the woman has traveled forward on the train. How much are $\frac{dT}{dt}$, $\frac{dm}{dt}$, and $\frac{dw}{dt}$, including sign? What does the function $f[t] = T[t] + m[t]$ represent? What is $\frac{df}{dt}$? How does this compare to $\frac{dT}{dt} + \frac{dm}{dt}$? What are the similar constructions for the woman?

(b) *A U.S. tourist is driving in Canada at 90 kilometers per hour, but her odometer and speedometer read in the archaic English units of her home country. We want to see the functional relationship between English and metric measurements of speed and distance. Let the odometer reading be the numerical function $E[t] =$ distance traveled in miles, where $t =$ time in hours. The distance traveled in kilometers is a function $M[t]$. There are approximately 1.609 kilometers in a mile. Express E in terms of a constant and M. Express $\frac{dE}{dt}$ in terms of this same constant and $\frac{dM}{dt}$. What is her speed in miles per hour?*

Problem 6.4. SPHERICAL SHELL

The formula for the volume of a sphere is $V[r] = \pi \frac{4}{3} r^3$ and for the surface area is $S[r] = \pi 4 r^2$. Compute the derivative $\frac{dV}{dr}$ and explain its connection with $S[r]$ by considering small changes in the sphere.

6.3   The Product Rule

The derivative of a product is NOT the product of the derivatives.

Let m, n, b, and c denote unknown constants or parameters. We cannot differentiate the product $y = (mx + b)(nx + c)$ directly using the rules we have so far. However, we could do some algebra,

$$(mx + b)(nx + c) = mnx^2 + mcx + bnx + bc = mnx^2 + (mc + nb)x + bc$$

so,

$$\frac{dy}{dx} = 2mnx + mc + nb$$

To verify that this agrees with the Product Rule given below, take $f[x] = mx + b$ and $g[x] = nx + c$. The derivatives are

$$\frac{df}{dx}[x] = \frac{d(mx + b)}{dx} = m \quad \& \quad \frac{dg}{dx}[x] = \frac{d(nx + c)}{dx} = n$$

The formula from the Product Rule below gives

$$\frac{dy}{dx} = \frac{df[x]}{dx} \cdot g[x] \qquad + f[x] \qquad \cdot \frac{dg[x]}{dx}$$
$$= m \qquad \cdot (nx + c) + (mx + b) \cdot n$$

This is the same answer because

$$m(nx + c) + (mx + b)n = mnx + mc + mnx + nb = 2mnx + mc + nb$$

Algebra can rearrange products of many power functions and linear combinations of power functions into forms we can differentiate, but often it is easier to use the Product Rule. A product like $\sqrt{x}\,\mathrm{Cos}[x]$ really requires a new rule.

Theorem 6.5.   *The Product Rule*
   If $f[x]$ and $g[x]$ are smooth for $\alpha < x < \beta$, then the product $h[x] = f[x] \cdot g[x]$ is also locally linear for $\alpha < x < \beta$ and

$$\frac{d(f[x]\,g[x])}{dx} = \frac{df[x]}{dx} \cdot g[x] + f[x] \cdot \frac{dg[x]}{dx}$$

In words, the Product Rule says, "Differentiate the terms of a product one at a time, multiply by the other undisturbed term, and add these together."

PROOF OF THE PRODUCT RULE

The proof of the product rule is another straightforward computation with tiny increments. All we do is add and subtract a term.

$$h[x + \delta x] - h[x] = f[x + \delta x]g[x + \delta x] - f[x]g[x]$$
$$= f[x + \delta x]g[x + \delta x] - f[x]g[x + \delta x] + f[x]g[x + \delta x] - f[x]g[x]$$
$$= (f[x + \delta x] - f[x])\, g[x + \delta x] + f[x]\, (g[x + \delta x] - g[x])$$
$$= (f'[x]\delta x + \varepsilon_1 \delta x)\, g[x + \delta x] + f[x]\, (g'[x]\delta x + \varepsilon_2 \delta x)$$
$$= (f'[x]\delta x + \varepsilon_1 \delta x)\, (g[x] + \varepsilon_3) + f[x]\, (g'[x]\delta x + \varepsilon_2 \delta x)$$
$$= f'[x]g[x]\delta x + f[x]g'[x]\delta x + \delta x \cdot (\varepsilon_4)$$
$$= (f'[x]g[x] + f[x]g'[x]) \cdot \delta x + \varepsilon_4 \cdot \delta$$

Exercise 6.5 below asks you to show that ε_3 and ε_4 are small. You can do this by simple estimates.

Example 6.9. $\frac{d(\sqrt{x}\,\mathrm{Cos}[x])}{dx}$

Let $f[x] = \sqrt{x}$ and $g[x] = \mathrm{Cos}[x]$, then $\frac{df}{dx} = \frac{1}{2\sqrt{x}}$ and $\frac{dg}{dx} = -\mathrm{Sin}[x]$, so

$$\frac{d(f[x] \cdot g[x])}{dx} = \frac{df[x]}{dx} \cdot g[x] + f[x] \cdot \frac{dg[x]}{dx}$$

$$\frac{d(\sqrt{x}\,\mathrm{Cos}[x])}{dx} = \frac{1}{2\sqrt{x}} \cdot \mathrm{Cos}[x] - \sqrt{x} \cdot \mathrm{Sin}[x]$$

We can also combine the Superposition Rule and Product Rule.

Example 6.10. $\frac{d((\frac{a}{\sqrt{x}} - b\,\mathrm{Log}[x])(e^x - c))}{dx}$

Let $f[x] = \frac{a}{\sqrt{x}} - b\,\mathrm{Log}[x]$ and $g[x] = e^x - c$, for constants a, b, and c. Then superposition says

$$\frac{df}{dx} = a\frac{d(\frac{1}{\sqrt{x}})}{dx} - b\frac{d(\mathrm{Log}[x])}{dx} = a\frac{d(x^{-1/2})}{dx} - b\frac{d\,\mathrm{Log}[x]}{dx}$$
$$= -\frac{a}{2}x^{-3/2} - \frac{b}{x} = -\frac{1}{x}\left(\frac{a}{\sqrt{x}} + b\right)$$

and

$$\frac{dg}{dx} = \frac{d(e^x)}{dx} - \frac{dc}{dx} = e^x + 0 = e^x$$

so

$$\frac{d(f[x] \cdot g[x])}{dx} = \frac{df[x]}{dx} \cdot g[x] + f[x] \cdot \frac{dg[x]}{dx}$$

$$\frac{d(f[x]\,g[x])}{dx} = \left(-\frac{1}{x}\left(\frac{a}{\sqrt{x}} + b\right)\right) \cdot (e^x - c) + \left(\frac{a}{\sqrt{x}} - b\,\mathrm{Log}[x]\right)(e^x)$$

We cannot differentiate $y = \text{Sin}[2\,x]$ with the rules we have developed so far. Even the simple expression $2\,x$ inside the sine makes this problem outside our present rules.

Example 6.11.   *An Impossible Problem Made Possible*

We can use the addition formula for sine to show that $\text{Sin}[2\,x] = 2\,\text{Sin}[x] \times \text{Cos}[x]$,

$$\text{Sin}[\alpha + \beta] = \text{Sin}[\alpha]\,\text{Cos}[\beta] + \text{Sin}[\beta]\,\text{Cos}[\alpha]$$
$$\text{Sin}[x + x] = \text{Sin}[x]\,\text{Cos}[x] + \text{Sin}[x]\,\text{Cos}[x]$$
$$\text{Sin}[2\,x] = 2\,\text{Sin}[x]\,\text{Cos}[x]$$

This form of the expression can be differentiated using the Product Rule as follows:

Example 6.12. $\frac{d(\text{Sin}[x]\,\text{Cos}[x])}{dx}$

Let $f[x] = \text{Sin}[x]$ and $g[x] = \text{Cos}[x]$, so $\frac{df}{dx} = \text{Cos}[x]$ and $\frac{dg}{dx} = -\text{Sin}[x]$, and

$$\frac{d(f[x] \cdot g[x])}{dx} = \frac{df[x]}{dx} \cdot g[x] + f[x] \cdot \frac{dg[x]}{dx}$$

$$\frac{d(\text{Sin}[x]\,\text{Cos}[x])}{dx} = \text{Cos}[x] \cdot \text{Cos}[x] - \text{Sin}[x] \cdot \text{Sin}[x]$$

Together, the two examples mean that

$$\frac{d(\text{Sin}[2x])}{dx} = 2\frac{d(\text{Sin}[x]\,\text{Cos}[x])}{dx}$$
$$= 2\left(\text{Cos}[x]^2 - \text{Sin}[x]^2\right)$$
$$= 2\,\text{Cos}[2x]$$

by the addition formula for cosine, $\text{Cos}[\alpha + \beta] = \text{Cos}[\alpha]\,\text{Cos}[\beta] - \text{Sin}[\alpha]\,\text{Sin}[\beta]$. We can do this directly with the Chain Rule below.

6.3.1   The Microscope Approximation and Rules of Differentiation

The "microscope equation" defining the differentiability (Definition 5.5) of a function $f[x]$,

$$f[x + \delta x] = f[x] + f'[x] \cdot \delta x + \varepsilon \cdot \delta x$$

with $\varepsilon \approx 0$ if $\delta x \approx 0$, is similar to a functional identity in that it involves an unknown function $f[x]$ and its related unknown derivative function $f'[x]$. If we plug in the "input" function $f[x] = x^2$ into this equation, the output is $f'[x] = 2x$. If we plug in the "input" function $f[x] = \text{Log}[x]$, the output is $f'[x] = \frac{1}{x}$.

The microscope equation involves unknown functions, but, strictly speaking, it is not a functional identity because of the error term ε (or the limit that can be used to formalize the error). It is only an approximate identity.

The various "differentiation rules," the Superposition Rule, the Product Rule, and the Chain Rule are functional identities relating functions and their derivatives. For example, the Product Rule states

$$\frac{d(f[x] \cdot g[x])}{dx} = \frac{df[x]}{dx} \cdot g[x] + f[x] \cdot \frac{dg[x]}{dx}$$

We urge you to write out this identity with general functions each time you use the Product Rule in a differentiation computation.

We can think of $f[x]$ and $g[x]$ as "variables" that vary by simply choosing different functions for $f[x]$ and $g[x]$. Then the Product Rule yields an identity by "plugging in" the choices of $f[x]$, $g[x]$, and their derivatives.

Example 6.13. *More Examples of the Product Rule*

Choosing $f[x] = x^2$ and $g[x] = \text{Log}[x]$ and plugging into the Product Rule yields

$$\frac{d(f[x] \cdot g[x])}{dx} = \frac{df[x]}{dx} \cdot g[x] + f[x] \cdot \frac{dg[x]}{dx}$$

$$\frac{d(x^2 \text{Log}[x])}{dx} = 2x \, \text{Log}[x] + x^2 \frac{1}{x}$$

Choosing $f[x] = x^3$ and $g[x] = \text{Exp}[x]$ and plugging into the Product Rule yields

$$\frac{d(f[x] \cdot g[x])}{dx} = \frac{df[x]}{dx} \cdot g[x] + f[x] \cdot \frac{dg[x]}{dx}$$

$$\frac{d(x^3 \text{Exp}[x])}{dx} = 3x^2 \, \text{Exp}[x] + x^3 \, \text{Exp}[x]$$

Example 6.14. *A Half-General Rule*

If we choose $f[x] = x^5$ but do not make a specific choice for $g[x]$, plugging into the Product Rule will yield

$$\frac{d(f[x] \cdot g[x])}{dx} = \frac{df[x]}{dx} \cdot g[x] + f[x] \cdot \frac{dg[x]}{dx}$$

$$\frac{d(x^5 g[x])}{dx} = 5x^4 g[x] + x^5 \frac{dg[x]}{dx}$$

You need to know the Product Rule as a functional identity, not just learn shortcut computation methods that use it. Look at the **DiffRules** program to see a "practical" use of the identity as a computer program command.

Exercise set 6.3

1. Drill on Products

Break each of the following expressions into a product of two terms and apply the Product Rule to find $\frac{dy}{dx}$ for each of the following functions $y = y[x]$. (The letters a, b, c, and h denote constants or parameters, and e is the natural base for logs and exponentials.)

a) $y = (2 x^3 - 4)(3 x + 5)$ b) $y = (x^3 - 2 x^2 + 4)(5 x + 6)$

c) $y = (\frac{2}{x^3} - 4)(3\sqrt{x} + 5)$ d) $y = (x^3 - 2 x^2 + 3 x + 4)\sqrt{x}$

e) $y = (\frac{3}{x^2} - \frac{4}{\sqrt{x}})(5\sqrt{x} + \frac{3}{x^5})$ f) $y = (x + \frac{1}{\sqrt{x}})(\sqrt{x} + \frac{1}{x})$

g) $y = \frac{\text{Log}[x]}{\sqrt[5]{x}}$ h) $y = (e^x - 3)(\frac{1}{x^4})$

i) $y = x^2 e^x$ j) $y = 7 e^x \; \text{Cos}[x]$

k) $y = (a + b x)(c x^2 + h x^3)$ l) $y = 3 \,\text{Sin}[x] \cdot \text{Cos}[7e]$

m) $y = \left(x^{1/2} + x^{1/3} + x^{1/4}\right)\left(x^{-2} + x^{-3} + x^{-4}\right)$

n) $y = \left(x^2 + x^3 + x^4\right)\left(x^{-1/2} + x^{-1/3} + x^{-1/4}\right)$

2. *Which parts of Problem 6.2 can you do now using the Product Rule that you could not do without it? (Again, you can check your work with the **DiffRules** program by entering the rules up to the Product Rule but not entering the Chain Rule.)*

3. *Let $f[x]$ and $g[x]$ be unknown functions that satisfy $f[1] = 2$, $\frac{df}{dx}[1] = 3$, $g[1] = -3$, $\frac{dg}{dx}[1] = 4$. Let $h[x] = f[x]g[x]$. Compute $\frac{dh}{dx}[1]$.*

4. *Show that the derivative of $\text{Sin}[\theta] \times \text{Cos}[\theta]$ is $\text{Cos}[2\theta]$ using the addition formula for cosine. (You can look that formula up in Chapter 28 on the CD.)*
Combine this fact with the previous example to show that

$$\frac{d(\text{Sin}[2 x])}{dx} = \frac{d(2 \,\text{Sin}[x] \,\text{Cos}[x])}{dx} = 2 \,\text{Cos}[2 x]$$

without using the Chain Rule! (We will be able to verify this more simply once we have the Chain Rule. The point is that there are often several ways to apply the rules of algebra and trig in combination with the rules of differentiation.)

It is important to be able to apply the differentiation rules to functions defined in terms of letters other than x.

5. Other Variables

(a) $v = e^u \; \text{Cos}[u] \quad \Rightarrow \quad \frac{dv}{du} = \,?$

(b) $v = (u^3 + 3 u + 3)(\text{Sin}[u] + \text{Cos}[u]) \quad \Rightarrow \quad \frac{dv}{du} = \,?$

(c) $u = (4 \sqrt{v} - \pi \frac{1}{v^2})(v^{1/4} - \frac{1}{v^{1/3}}) \quad \Rightarrow \quad \frac{du}{dv} = ?$

(d) $y = (\text{Cos}[v] - \text{Sin}[v])(\text{Log}[v] - e^v) \quad \Rightarrow \quad \frac{dy}{dv} = \,?$

Problem 6.5.

(a) *Verify that $\varepsilon_3 \approx 0$ and $\varepsilon_4 \approx 0$ in the proof of the product rule above.*

(b) *Let $f[x] = x^2$ and $g[x] = x^3$ and write out the steps of the proof of the product rule given above for these specific functions.*

Problem 6.6.

The "rule" that says the derivative of a product is the product of the derivatives might make things simpler, but it is false. Show this by finding a counterexample from among functions you know how to differentiate directly from increment computations (such as simple powers of x.) For example, let $f[x] = x^2$ and $g[x] = x$, so $\frac{df}{dx} = 2x$ and $\frac{dg}{dx} = 1$. How much is $\frac{df}{dx} \times \frac{dg}{dx}$? How much is $\frac{d(f \cdot g)}{dx} = \frac{d(x^3)}{dx}$? How much is $\frac{df}{dx} \cdot g + f \cdot \frac{dg}{dx}$?

Explain why the derivative of a product is not necessarily the product of the derivatives.

The function notation in the general proof of the Product Rule may seem obscure, but the idea is not. The next problem shows you why.

Problem 6.7. A CONCRETE INSTANCE OF THE PRODUCT RULE

(a) *A contractor's crew is making forms to lay a rectangular concrete floor. One member of the crew measures the length l (feet) and makes an error Δl (feet), and another measures the width w (feet) and makes a separate error Δw (feet). The contracted area is $A = lw$, but the actual area is $A + \Delta A$ where the error in area is $\Delta A = \_$? (Write a formula in terms of l, w, Δl and Δw. See the Figure 6.2 for help.)*

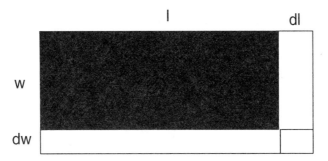

Figure 6.2: *Three error rectangles*

(b) *Check your formula with a numerical example. Suppose that the floor has design dimensions of 20 by 30 feet, the error in length is 2 inches, and the error in width is 1 inch - both too long. How much too large is the floor? Is the error larger or smaller than a desktop?*

(c) *With the same dimensions as in the previous part and same error in length, suppose the error in width is 2 inches too short. Use $\Delta w = -2/12$ in your formula and verify that it gives the correct result. Is the floor too large or too small? What is the error in area?*

(d) *Suppose that the length and width are changing with time (instead of a measurement error), so $\frac{\Delta l}{\Delta t}$ and $\frac{\Delta w}{\Delta t}$ are the rates of change of length and width during the time increment Δt. (Imagine the floor expanding with a change in temperature.) Show that*

$$\frac{\Delta A}{\Delta t} = \frac{\Delta w}{\Delta t} \cdot l + w \cdot \frac{\Delta l}{\Delta t} + \frac{\Delta w}{\Delta t} \cdot \frac{\Delta l}{\Delta t} \cdot \Delta t$$

What is the Product Rule for $A = l[t]\, w[t]$? How do these expressions differ when Δt is very small?

The Scientific Project on the Expanding Economy shows another way that the Product Rule arises in everyday discussions.

6.4   The Chain Rule

The Chain Rule for derivatives shows how to differentiate functions that are "hooked together in a chain." Mathematically this occurs in expressions like $y = \text{Sin}[x^2 + 1]$, with "chain" $u = x^2 + 1$ "linked to" $y = \text{Sin}[u]$.

If we let $u = x^2 + 1$ and let $y = \text{Sin}[u]$, then the original formula, $y = \text{Sin}[x^2 + 1]$ is what we would get if we start with x, compute u, and then use that answer for u to compute y.

$$u = x^2 + 1 \quad \rightarrow \quad y = \text{Sin}[u]$$

The functional notation for this "chaining" if $y = f[u]$ and $u = g[x]$ is

$$y = f[g[x]] = \text{Sin}[x^2 + 1]$$

Of course, we could have broken the final formula down in other ways. For example,

$$v = x^2 \quad \rightarrow \quad y = \text{Sin}[v + 1]$$

or $y = f[g[x]]$, where $f[v] = \text{Sin}[v + 1]$ and $v = g[x] = x^2$. The advantage of the first decomposition is that we can differentiate each of the component pieces with rules already at our disposal,

$$\frac{du}{dx} = 2x \quad \& \quad \frac{dy}{du} = \text{Cos}[u]$$

Notice the importance of being able to differentiate with respect to a letter other than x. This is one reason that we emphasized "other letters" in the early sections of the chapter.

The Chain Rule given next tells us how to use the derivatives of the components to find the derivative of the whole composition. In this case,

$$\frac{dy}{dx} = \frac{dy}{du} \cdot \frac{du}{dx}$$
$$= \text{Cos}[u] \cdot 2x$$
$$= \text{Cos}[x^2 + 1] \cdot 2x$$
$$= 2x\,\text{Cos}[x^2 + 1]$$

We removed the "link" variable u in our final expression for $\frac{dy}{dx}$ because we introduced it only to help solve the problem. (In some applications, the link variables actually have important separate meanings.)

> **Theorem 6.6.** *The Chain Rule*
> *If $y = f[u]$ is smooth on the range of $u = g[x]$ and g is smooth for $\alpha < x < \beta$, then the chained composition $y = h[x] = f[g[x]]$ is smooth for $\alpha < x < \beta$ and $h'[x] = f'[g[x]] \cdot g'[x]$ or*
> $$\frac{dy}{dx} = \frac{dy}{du} \cdot \frac{du}{dx}$$

PROOF OF THE CHAIN RULE

The general proof is only a little more complicated use of the increment approximation.

$$f[g[x + \delta x]] - f[g[x]] = f[g[x] + (g'[x]\delta x + \varepsilon_1 \cdot \delta x)] - f[g[x]]$$
$$= f[g[x] + L_1\delta x] - f[g[x]]$$
$$= f'[g[x]]\, L_1\delta x + \varepsilon_2 \cdot L_1\delta x$$
$$= f'[g[x]]\, (g'[x] + \varepsilon_1)\delta x + \varepsilon_2 \cdot L_1\delta x$$
$$= f'[g[x]]\, g'[x]\, \delta x + (f'[g[x]]\, \varepsilon_1 + \varepsilon_2 \cdot L_1)\delta x$$
$$= f'[g[x]]\, g'[x]\delta x + \varepsilon_3\delta x$$

where $L_1 = (g'[x] + \varepsilon_1)$ is finite so that $L_1\delta x \approx 0$ and we may apply the increment approximation to f at $g[x]$ with change $L_1\delta x$. Also, $\varepsilon_3 = (f'[g[x]]\varepsilon_1) + \varepsilon_2 L_1 \approx 0$ since $f[u]$ is smooth on the range of $g[x]$.

> **Procedure 6.7.** To differentiate an expression like $v = \sqrt{x^2 + 1}$ with the Chain Rule, you need to find a decomposition of the formula satisfying the following
> (a)   Each piece of the decomposition can be differentiated by known rules.
> (b)   When chained back together, the pieces "compose" the original formula.

You can view the 'links' in function notation or by using a new variable.

> **Example 6.15.** $\frac{d(\sqrt{x^2+1})}{dx}$

In this case, we let $u = x^2 + 1$ and $v = \sqrt{u}$ because substituting this value for u into the

u-expression for v makes $v = \sqrt{x^2 + 1}$. The Power Rule and the Superposition Rule tell us

$$v = \sqrt{u} \quad \& \quad u = x^2 + 1$$

$$\frac{dv}{du} = \frac{1}{2\sqrt{u}} \quad \& \quad \frac{du}{dx} = 2x$$

so the Chain Rule above says

$$\boxed{\frac{dv}{dx} = \frac{dv}{du} \cdot \frac{du}{dx}}$$

$$\frac{dv}{dx} = \frac{1}{2\sqrt{u}} \cdot 2x$$

$$= \frac{x}{\sqrt{x^2 + 1}}$$

In function notation, this example is solved as follows: The functions $v = f[u] = u^{1/2}$ and $u = g[x] = x^2 + 1$ have $v = f[g[x]] = \sqrt{x^2 + 1}$. The derivatives are

$$g[x] = x^2 + 1 \quad \& \quad f[u] = \sqrt{u}$$

$$g'[x] = 2x \quad \& \quad f'[u] = \frac{1}{2\sqrt{u}}$$

so

$$(f[g[x]])' = f'[g[x]] \cdot g'[x]$$

$$= \frac{x}{\sqrt{x^2 + 1}}$$

Example 6.16. $\quad \frac{d(\text{Sin}[x^2 + 2x + 1])}{dx}$

In this case, we let $y = \text{Sin}[u]$ and $u = x^2 + 2x + 1$, because substituting this value for u into the u-expression for y makes $y = \text{Sin}[x^2 + 2x + 1]$. The Sine Rule, the Power Rule and the Superposition Rule tell us

$$y = \text{Sin}[u] \quad \& \quad u = x^2 + 2x + 1$$

$$\frac{dy}{du} = \text{Cos}[u] \quad \& \quad \frac{du}{dx} = 2x + 2$$

so the Chain Rule says

$$\boxed{\frac{dv}{dx} = \frac{dv}{du} \cdot \frac{du}{dx}}$$

$$\frac{dy}{dx} = \text{Cos}[u](2x + 2)$$

$$= 2(x + 1)\,\text{Cos}[x^2 + 2x + 1]$$

Example 6.17. *More Links* $y = \text{Log}[\text{Cos}[e^x + x^6]]$

We decompose $y = \text{Log}[\text{Cos}[e^x + x^6]]$ into three pieces below because resubstituting these yields the original expression.

$$y = \text{Log}[u] \quad \& \quad u = \text{Cos}[v] \quad \& \quad v = e^x + x^6$$

$$\frac{dy}{du} = \frac{1}{u} \quad \& \quad \frac{du}{dv} = -\text{Sin}[v] \quad \& \quad \frac{dv}{dx} = e^x + 6x^5$$

The simple two-link Chain Rule says

$$\frac{dy}{dx} = \frac{dy}{du} \cdot \frac{du}{dx}$$

$$\frac{dy}{dx} = \frac{1}{u} \cdot \frac{du}{dx}$$

and

$$\frac{du}{dx} = \frac{du}{dv} \cdot \frac{dv}{dx}$$

$$\frac{du}{dx} = -\text{Sin}[v] \cdot (e^x + 6x^5)$$

so

$$\frac{dy}{dx} = \frac{dy}{du} \cdot \frac{du}{dv} \cdot \frac{dv}{dx}$$

$$\frac{dy}{dx} = \frac{1}{u} \cdot (-\text{Sin}[v]) \cdot (e^x + 6x^5)$$

$$= \frac{1}{\text{Cos}[v]} (-\text{Sin}[e^x + x^6])(e^x + 6x^5)$$

$$= -(e^x + 6x^5) \frac{\text{Sin}[e^x + x^6]}{\text{Cos}[e^x + x^6]}$$

$$= -(e^x + 6x^5) \text{Tan}[e^x + x^6]$$

It is easy to generalize this example to see that if we decompose an expression into three links,

$$\frac{dy}{dx} = \frac{dy}{du} \cdot \frac{du}{dv} \cdot \frac{dv}{dx}$$

Example 6.18. *"Generalized" Differentiation Rules*

Some folks like to remember special cases of the Chain Rule such as

$$\frac{d(e^u)}{dx} = e^u \cdot \frac{du}{dx}$$

which they apply as follows. If we want to differentiate $y = e^{x^3}$, let $u = x^3$, so $\frac{du}{dx} = 3x^2$ and use the "Generalized Derivative" above,

$$\frac{d(e^u)}{dx} = e^u \cdot \frac{du}{dx} = e^u \cdot 3x^2 = 3x^2 \, e^{x^3}$$

Of course, this is just the Chain Rule with $y = e^u$ and $u = x^3$. There is no need to remember a "Generalized" formula, but you may if you wish.

6.4.1   The Everyday Meaning of the Chain Rule

The Scientific Projects contain a chapter on the "Expanding House." The wood in your house expands during the course of a normal day's warming. A 40-foot long house only expands about 0.03 inches on a normal Fall day, but there is a substantial increase in the volume of the house. Do you think it is about a thimble, a bucket, or a bathtub full? The numerical calculation might surprise you, but the ideas of this calculation are similar to the way that the rules of calculus are built. The project starts with simple but surprising arithmetic and progresses to a symbolic formulation of volume expansion. A final section uses the Chain Rule from the next section to give a direct solution. We recommend that you at least skim through the Expanding House project now.

Chapter 7 has several applications of the Chain Rule.

You may want to review chaining or function composition in Exercise 28.7.4 before doing the Chain Rule exercises that follow.

Exercise set 6.4

1. Drill on Chains

Break each of the following expressions into a composition of two functions and apply the Chain Rule to find $\frac{dy}{dx}$. (The letters a, b, c denote constants or parameters, and e is the natural base for logs and exponentials.)

a) $y = (1 + x)^{33}$

b) $y = (a + bx)^{33}$

c) $y = (a + bx)^c$

d) $y = e^{ax}$

e) $y = e^{x^2 + 3}$

f) $y = e^{ax^2 + b}$

g) $y = e^{\mathrm{Cos}[x]}$

h) $y = e^{\mathrm{Sin}[ax] + b}$

i) $y = e^{\mathrm{Log}[x]}$

j) $y = 3(\mathrm{Sin}[x])^3$

k) $y = 3(\mathrm{Sin}[x^3])$

l) $y = 3(\mathrm{Sin}[x^3])^3$

m) $y = \mathrm{Log}[x \sqrt[5]{x}]$

n) $y = \frac{6}{5} \mathrm{Log}[x]$

o) $y = \mathrm{Log}[x^{\frac{6}{5}}]$

p) $y = \mathrm{Log}[x^c]$

q) $y = c \, \mathrm{Log}[x]$

r) $y = \mathrm{Cos}[7 \, e^x]$

2. *The product $\mathrm{Sin}[\theta] \, \mathrm{Cos}[\theta] = \frac{1}{2} \mathrm{Sin}[2\theta]$. Show this using the addition formula for the sine. (HINT: $\mathrm{Sin}[\theta + \theta] = ?$) Use the Chain Rule to show that the derivative of the product* is $\text{Cos}[2\theta]$ by differentiating the right side of the equality. Check your work using the Product Rule on the left side of the equality.

3. The sine function in degrees can be thought of this way

$$sIN[D] = \text{Sin}[\frac{\pi}{180} D]$$

where $\text{Sin}[u]$ denotes the radian measure sine function. Use the Chain Rule to show that the derivative of the sine in degrees is $\frac{\pi}{180}$ times the cosine in degrees,

$$y = sIN[D] \Rightarrow \frac{dy}{dD} = \frac{\pi}{180} cOS[D]$$

4. Which parts of Exercise 6.2 can you now do using the Chain Rule (in addition to the other rules) that you could not do without it?

6.5 General Exponentials

This section uses the Chain Rule to find $\frac{d(a^x)}{dx}$ and the derivatives of more general exponential expressions.

The project on Numerical Differentiation of Exponentials shows you a direct way to approximate the derivative of a general exponential function. The next topic shows you an exact symbolic method. This is important in the theory of calculus because it reduces the calculus of other bases to the natural base.

Suppose we have $y = e^{ct}$ for some constant c. The Chain Rule says

$$\frac{dy}{dt} = \frac{dy}{du} \cdot \frac{du}{dt}$$

$$y = e^u \qquad u = ct$$
$$\frac{dy}{du} = e^u \qquad \frac{du}{dt} = c$$

$$\frac{d(e^{ct})}{dt} = \frac{dy}{du} \times \frac{du}{dt} = c\, e^{ct}$$

If we want to differentiate $y = 2^t$, we can use the preceding Chain Rule computation and two other important facts about exponentials. We know the general exponential law

$$(a^c)^x = a^{cx}$$

so we try to find a constant c that satisfies

$$2 = e^c$$

Once we find this c we have $2^t = e^{ct}$ for all t because of the law of exponents. We know how to differentiate $y = e^{ct}$, and with this value of c, we learn how to differentiate 2^t,

$$\frac{d(2^t)}{dt} = \frac{d(e^{ct})}{dt} = c\,e^{ct} = c\,2^t$$

Now we solve for c using natural log. Natural log and exponential are inverse functions. This simply means

$$\text{Log}[e^t] = t \qquad \text{and} \qquad e^{\text{Log}[s]} = s, \quad s > 0$$

We apply this to our problem by taking logs of both sides of

$$2 = e^c$$

$$\text{Log}[2] = \text{Log}[e^c] = c$$

Thus $c = \text{Log}[2]$ and we see that

$$\frac{d(2^t)}{dt} = \text{Log}[2] \times 2^t$$

In general,

$$\boxed{\frac{d(a^x)}{dx} = \text{Log}[a]\,a^x}$$

but the procedure used above to find the derivative of 2^t is what you should learn. That procedure also applies to formulas like $y = x^x$.

Example 6.19. $\dfrac{d\left((\text{Sin}[x]^{\text{Cos}[x]})\right)}{dx}$

$$\text{Sin}[x] = e^{\text{Log}[\text{Sin}[x]]} \qquad \text{for } \text{Sin}[x] > 0$$

so

$$\text{Sin}[x]^{\text{Cos}[x]} = \left(e^{\text{Log}[\text{Sin}[x]]}\right)^{\text{Cos}[x]} = e^{\text{Cos}[x]\,\text{Log}[\text{Sin}[x]]}$$

and we want to differentiate the chain

$$y = e^u \qquad u = \text{Cos}[x]\,\text{Log}[\text{Sin}[x]]$$

$$\frac{dy}{du} = e^u \qquad \frac{du}{dx} = ??$$

Use of the Product Rule

$$\frac{du}{dx} = \frac{d(\text{Cos}[x]\,\text{Log}[\text{Sin}[x]])}{dx} = \frac{d(\text{Cos}[x])}{dx}\,\text{Log}[\text{Sin}[x]] + \text{Cos}[x]\,\frac{d(\text{Log}[\text{Sin}[x]])}{dx}$$

$$= -\,\text{Sin}[x]\,\text{Log}[\text{Sin}[x]] + \text{Cos}[x]\,\frac{\text{Cos}[x]}{\text{Sin}[x]}$$

since another application of the Chain Rule gives

$$w = \text{Log}[v] \qquad v = \text{Sin}[x]$$

$$\frac{dw}{dv} = \frac{1}{v} \qquad \frac{dv}{dx} = \text{Cos}[x]$$

$$\frac{d(\text{Log}[\text{Sin}[x]])}{dx} = \frac{dw}{dv} \times \frac{dv}{dx} = \frac{\text{Cos}[x]}{\text{Sin}[x]}$$

This shows that

$$\frac{du}{dx} = \frac{\text{Cos}[x]^2}{\text{Sin}[x]} - \text{Sin}[x]\,\text{Log}[\text{Sin}[x]]$$

So finally,

$$\frac{d\left((\text{Sin}[x])^{\text{Cos}[x]}\right)}{dx} = \frac{dy}{du} \times \frac{du}{dx} = \left(\frac{\text{Cos}[x]^2}{\text{Sin}[x]} - \text{Sin}[x]\,\text{Log}[\text{Sin}[x]]\right)\text{Sin}[x]^{\text{Cos}[x]}$$

Exercise set 6.5

1. For $y[t]$ as follows, find $\frac{dy}{dt}[t]$:

a) $y = 3^t$

b) $y = 10^t$

c) $y = a^t$

d) $y = t^t$

e) $y = 2^{\text{Cos}[t]}$

f) $y = te^t$

g) $y = e^{\frac{1}{t}}$

h) $y = 3^{\sqrt{t}}$

i) $y = \sqrt{3^t - 2^t}$

(Check your work with the computer.)

2. Find $\frac{dy}{dx}[x]$

a) $y = x^x$

b) $y = x^{\text{Sin}[x]}$

c) $y = (\text{Cos}[x])^x$

Problem 6.8.

If the number of algae cells is $N[t] = N_0 2^{t/6}$, how long does it take to double the number of cells? How long does it take to triple the number of cells? What is the instantaneous rate of growth of algae at $t = 0$? At $t = 6$? What is the instantaneous rate of growth of algae as a percentage of N at $t = 0$? At $t = 6$?

Problem 6.9.

Let r be a constant. Use rules of calculus to prove that $y = e^{r\,x}$ grows at the instantaneous rate of $r \times 100\%$. (Compare this with the CD section of Chapter 5 on the exponential.)

Problem 6.10. ─── ▼

What is wrong with the following nonsensical differentiation?

$$\frac{dx^x}{dx} = xx^{x-1} = x^1 x^{x-1} = x^{1+x-1} = x^x$$

(HINT: Differentiate with the computer, and try $y = x^x = (e^{Log[x]})^x = e^{xLog[x]}$ yourself.)

── ▲

6.6   Derivative of the Natural Log

$$\frac{d\,Log[u]}{du} = \frac{1}{u}$$

The inverse of $x = Log[y]$ is $y = e^x$ and has derivative $\frac{dy}{dx} = e^x = y$; therefore

$$dy = e^x \, dx$$

or

$$dy = y \, dx$$

and

$$\frac{dx}{dy} = \frac{1}{y}$$

This computation is explored in more detail in the Project on Inverse Functions. The point is that the derivative of the inverse function is the reciprocal of the derivative of the function.

Example 6.20. $\frac{d\,Log[Tan[x]]}{dx}$

Use the chain

$$y = Log[u] \qquad u = Tan[x] = \frac{Sin[x]}{Cos[x]}$$

$$\frac{dy}{du} = \frac{1}{u} \qquad \frac{du}{dx} = \frac{1}{(Cos[x])^2}$$

$$\frac{dy}{dx} = \frac{dy}{du}\frac{du}{dx} = \frac{Cos[x]}{Sin[x]} \times \frac{1}{(Cos[x])^2} = \frac{1}{Cos[x]\,Sin[x]}$$

Compare this with the answer to

$$\frac{d\,Log[Sin[x]]}{dx} - \frac{d\,Log[Cos[x]]}{dx} = \frac{Cos[x]}{Sin[x]} + \frac{Sin[x]}{Cos[x]}$$

$$= \frac{(Cos[x])^2 + Sin[x])^2}{Sin[x]\,Cos[x]}$$

since $Log[Tan[x]] = Log[Sin[x](Cos[x])^{-1}] = Log[Sin[x]] - Log[Cos[x]]$

===== **Exercise set 6.6** =====

1. *For $y[t]$ as follows, find $\frac{dy}{dt}[t]$*

a) $\quad y = (\text{Log}[t])^3$ 　　　　　 *b)* $\quad y = \text{Log}[\text{Cos}[x]]$ 　　　　　 *c)* $\quad y = t\,\text{Log}[t] - t$

d) $\quad y = \text{Log}[\text{Log}[x]]$ 　　　　 *e)* $\quad y = \text{Log}[t^{1/t}]$ 　　　　　 *f)* $\quad y = \text{Log}[t^2 + 2x]$

Assume first that x is independent of t. Second, if x is a function of t but we forgot to give you a formula for $x = x[t]$. Express your answers in terms of x and $\frac{dx}{dt}$. What is $\frac{dx}{dt}$ if x is independent of t?

2. *Differentiate $y = \text{Log}[x^3]$ using $u = x^3$ and $y = \text{Log}[u]$. Also use the log identity, $\text{Log}[x^p] = p\,\text{Log}[x]$ and differentiate $y = \text{Log}[x^3] = 3\,\text{Log}[x]$ without the Chain Rule. Compare the two answers.*

6.7 Combined Symbolic Rules

You can make some additional rules of differentiation for general cases.

Example 6.21. *Custom Rules*

Suppose you often need to differentiate a cube of a product of functions, $y = (f[x] \cdot g[x])^3$, for various smooth functions $f[x]$ and $g[x]$. We use the Chain Rule with unknown functions:

$$y = u^3 \qquad\qquad u = f[x] \cdot g[x]$$

$$\frac{dy}{du} = 3\,u^2 \qquad\qquad \frac{du}{dx} = \frac{df}{dx} \cdot g + f \cdot \frac{dg}{dx}$$

$$\frac{dy}{dx} = \frac{dy}{du} \cdot \frac{du}{dx} = 3\,u^2 \left(\frac{df}{dx} \cdot g + f \cdot \frac{dg}{dx} \right)$$

$$= 3(f[x] \cdot g[x])^2 \left(\frac{df}{dx}[x] \cdot g[x] + f[x] \cdot \frac{dg}{dx}[x] \right)$$

===== **Exercise set 6.7** =====

1. *Use all the rules of differentiation to find the following:*

a) $\quad y = \frac{1}{2x-1},\ \frac{dy}{dx} = ?$ 　　　　　　　　 *b)* $\quad y = \frac{1}{ax+b},\ \frac{dy}{dx} = ?\ \left(\begin{array}{l} u = ax + b \\ y = \frac{1}{u} = u^{-1} \end{array} \right)$

c) $\quad y = \text{Cos}[x^2],\ \frac{dy}{dx} = ?$ 　　　　　　　 *d)* $\quad y = \text{Cos}^2[x],\ \frac{dy}{dx} = ?$ *(unchain two ways)*

e) $y = \sqrt{1 - x^2}$, $\frac{dy}{dx} =$?

f) $y = \frac{1}{\sqrt{1-x^2}}$, $\frac{dy}{dx} =$?

g) $y = (12 - 3x^7)^8$, $\frac{dy}{dx} =$?

h) $y = \text{Sin}[x^2 + x^3]$, $\frac{dy}{dx} =$?

i) $y = (2x^3 + 4)(x^2 - \sqrt{x})$, $\frac{dy}{dx} =$?

j) $y = [(ax + b)^{-1} + c]^{-1}$, $\frac{dy}{dx} =$?

k) $y = \text{Sin}[x]\,\text{Cos}[x]$, $\frac{dy}{dx} =$?

l) $y = \text{Sin}[2x]$, $\frac{dy}{dx} =$?

m) $y = \text{Sin}[3x]$, $\frac{dy}{dx} =$?

n) $y = \frac{1}{\text{Cos}[3x]}$, $\frac{dy}{dx} =$?

o) $y = \text{Sin}[x] \cdot [\text{Cos}[x]]^{-1} = \frac{\text{Sin}[x]}{\text{Cos}[x]} = \text{Tan}[x]$, $\frac{dy}{dx} =$?

p) $y = \frac{2-3x}{1+2x} = (2 - 3x)[(1 + 2x)^{-1}]$, $\frac{dy}{dx} =$?
Check your work with the computer.

2. Differentiate $y = \sqrt{x^2 + 2x + 1}$ using the Chain Rule, the Power Rule and the Superposition Rule. The graph of this function is rather simple, as we saw in Exercise 3.2.4. Where does your symbolic answer not make sense? Can you sketch the graph? Why does the symbolic answer not work at the bad point?

3. Use the Superposition Rule and the Product Rule repeatedly to show in general that if $f[x]$, $g[x]$ and $h[x]$ are smooth on an interval, then so are their sum and product and

$$\frac{d(f[x] + g[x] + h[x])}{dx} = \frac{df}{dx}[x] + \frac{dg}{dx}[x] + \frac{dh}{dx}[x]$$

$$\frac{d[f[x]g[x]h[x]]}{dx} = \frac{df}{dx}[x] \cdot g[x] \cdot h[x] + f[x] \cdot \frac{dg}{dx}[x] \cdot h[x] + f[x] \cdot g[x] \cdot \frac{dh}{dx}[x]$$

(HINT: Let $G[x] = (g[x] \cdot h[x])$ and apply the Product Rule to $f[x] \cdot G[x]$.)

When we compute the derivative of a function $f[x]$ we get a new function $f'[x]$. If we think of this as a function, we could differentiate again: Let $g[x] = f'[x]$ and compute $g'[x]$. The derivative of $g[x]$ is called the second derivative of $f[x]$ and is denoted $g'[x] = f''[x]$ or

General Form	Example
$y = f[x]$	$y = f[x] = x^3$
$\dfrac{dy}{dx} = f'[x]$	$\dfrac{dy}{dx} = f'[x] = 3\,x^2$
$\dfrac{d^2y}{dx^2} = f''[x]$	$\dfrac{d^2y}{dx^2} = f''[x] = 3 \cdot 2\,x^1 = 6x$

4. The Second Derivative Product Rule

(a) Differentiate the general Product Rule identity to get a formula for

$$\frac{d^2(f[x] \cdot g[x])}{dx^2}$$

(b) Let $h[x] = f[x] \cdot g[x]$ and use your rule to compute $\frac{d^2(h)}{dx^2}[1]$ if $f[1] = 2$, $\frac{df}{dx}[1] = 3$, $\frac{d^2(f)}{dx^2}[1] = 5$, $g[1] = -3$, $\frac{dg}{dx}[1] = 4$, and $\frac{d^2(g)}{dx^2}[1] = -2$.

6.7.1   The Quotient Rule

5. The Quotient Rule

Derive the quotient rule: If $q[x] = \frac{f[x]}{g[x]}$, where f and g are smooth and $g[x] \neq 0$ for $\alpha < x < \beta$, then $q[x]$ is also smooth for $\alpha < x < \beta$ and

$$\frac{d(\frac{f[x]}{g[x]})}{dx} = \frac{\frac{df[x]}{dx} g[x] - f[x] \frac{dg[x]}{dx}}{[g[x]]^2}$$

Use the Chain Rule and Product Rule on the formula

$$\frac{f[x]}{g[x]} = f[x] \times (g[x])^{-1} = f[x] \times h[x]$$

You will have to put your answer on a common denominator to get the formula above.

6.7.2   The Relative Growth Rule

We often make relative measurements stating the error (or accuracy) as a fraction of the amount (or stating a percentage). A similar notion is the relative rate of change given by

$$f^*[x] = \frac{1}{f[x]} \frac{df}{dx}[x]$$

6.   (a) *Let $f[x] = e^{rx}$ for a constant r. Compute $f^*[x]$.*
        (b) *Let $f[x] = b^x$ for a constant base b. Compute $f^*[x]$.*

7. *Give a general symbolic rule for the relative rate of change of a product in terms of $f^*[x]$ and $g^*[x]$.*

$$(f[x]\ g[x])^* = ?$$

*(HINT: Substitute a product into the rule for * and rewrite using ordinary rules.)*

6.8   Review - Inside the Microscope

Calculus lets us "see" inside a powerful microscope without actually magnifying the nonlinear graph. We know that the curve looks like its tangent line at high magnification. The "rules" of differentiation are the way we "see." This section combines the rules with "looking."

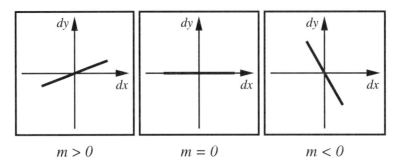

Figure 6.3: Possible microscopic views

6.8.1   Review - Numerical Increments

When a function is smooth, we summarize the local linear approximation by

$$y = f[x] \qquad \Rightarrow \qquad dy = f'[x] \, dx$$

The differential $dy = f'[x] \, dx$ is a linear function of the local variable dx, with dependent variable dy. This is the linear equation in microscope variables. The variable x in $f'[x]$ is considered fixed until we move the point where we focus our microscope. The quantity dy is an approximation to the change $f[x + dx] - f[x]$ in the actual function.

You should memorize the microscope approximation or Definition 5.5 and strive to understand its algebraic and geometric consequences. Functions given by formulas are important in science and mathematics, but they are not the only kind of functions.

The rules of calculus are theorems that guarantee that the local linear approximation is valid. These rules are remarkably easy to use compared with the direct verification of the approximations as in Chapter 5. You simply compute and look at the answers. We used the symbolic approximation in Chapter 5 to estimate Sin[46°].

Contrast what we learn about a function from the approximation with the simplicity of the computation that guarantees that the approximation holds. If

$$y = x^{-3}, \quad \text{then} \quad dy = -\frac{3}{x^4} \, dx$$

according to the rules. Both of these formulas are valid when $x \neq 0$, so the increment approximation defining the derivative holds and the change in f is approximated by dy with $dx = \delta x$. In general,

$$f[x + dx] - f[x] = f'[x] \cdot dx + \varepsilon \cdot dx$$
$$f[x + dx] - f[x] = dy + \varepsilon \cdot dx \approx dy$$

In this case,

$$\frac{1}{(x+dx)^3} - \frac{1}{x^3} = \frac{-3}{x^4} \cdot dx + \varepsilon \cdot dx \approx \frac{-3}{x^4} \cdot dx$$

$$\frac{1}{(3+.01)^3} - \frac{1}{27} \approx -\frac{3}{81} \times 0.01$$

$$\frac{1}{(3.01)^3} - \frac{1}{27} \approx -\frac{.01}{27}$$

so,

$$\frac{1}{(3.01)^3} \approx .99 \times \frac{1}{27} \approx 0.0366667$$

the computer gives $1/(3.01)^3 = 0.0366691$, so the increment approximation is quite close even though the x increment $\delta x = 0.01$ is not infinitesimal.

> To make numerical differential approximations, do the following steps.
> (a) Compute $f'[x]$ by rules. The rules guarantee the approximation when $dx \approx 0$.
> (b) Substitute the fixed x to find the numerical value of the derivative
>
> $$m = f'[x]$$
>
> (c) Compute $df = m\, dx$ when $dx =$ *your perturbation* using the *number m*.
> (d) Compute $f[x+dx]$ using this approximate change and the value of $f[x]$,
>
> $$f[x+\delta x] \approx f[x] + df[x]$$
>
> In short, $f[x+dx] - f[x] \approx df[x]$, with an error small compared to dx when dx is small.

6.8.2   Differentials and the (x,y)-Equation of the Tangent Line

The equation of the tangent line to $y = f[x]$ in local coordinates is simply the differential $dy = f'[x]\, dx$, but there is possible confusion when we try to convert back to regular coordinates because we are treating x as fixed in the local dx-dy-equation. Here is a way to find the equation of the tangent line to $y = x^2$ when $x = -1/3$ as shown in Figure 6.4. We know $dy = 2x\, dx$ and $x = -1/3$, so the slope is $2 \cdot (-1/3) = -2/3$. When $x = -1/3$, $y = x^2 = 1/9$, so the line goes through $(-1/3, 1/9)$ and has slope $-2/3$. Using the change form of a line (or the point-slope formula),

> To find the (x,y) equation of the tangent line, do the following steps.
> (a) Compute $f'[x]$ by rules.
> (b) Substitute the fixed $x = a$ ($x = -1/3$, in this case) to find the numerical value of the slope $m = f'[x]$.
> (c) Compute the specific y value, $y = f[x]$ at the point of tangency. This gives you a specific point $(x,y) = (a,b)$ that lies on both the curve and tangent line.
> (d) Change the local equation of the line $dy = m\, dx$ to the point-slope (x,y) form of a line $\frac{y-b}{x-a} = m$ and simplify to the slope-intercept form $y = mx + i$.

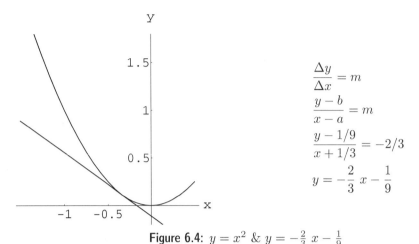

$$\frac{\Delta y}{\Delta x} = m$$

$$\frac{y - b}{x - a} = m$$

$$\frac{y - 1/9}{x + 1/3} = -2/3$$

$$y = -\frac{2}{3}\,x - \frac{1}{9}$$

Figure 6.4: $y = x^2$ & $y = -\frac{2}{3}\,x - \frac{1}{9}$

Exercise set 6.8

1. A Partial View of the "Bell Shaped" Curve
   The derivative of the function $f[x] = e^{-x^2}$ is $f'[x] = -2x \cdot e^{-x^2}$ (as you may verify using rules of differentiation.) This question asks, So what? (or what does this tell us mathematically?) You answer it as follows: Draw microscopic views of the graph $y = e^{-x^2}$ when the microscope is focused on the graph over the x-points, $x = 0, \pm 0.1, \pm 1, \pm 10$. Give the numerical values of the derivatives and sketch the slopes to scale on equal axes.

   The next problem asks you to do all the steps involved in "looking" in an infinitesimal microscope. This is a question that requires you to summarize the steps in writing. This should help you combine the facts you have learned.

2. *Find the (x, y)-equation of the tangent to $y = x^3$ at $x = -2$.*
   Find the (x, y)-equation of the tangent to $y = \text{Sin}[x]$ at $x = \pi/3$.
   Find the (x, y)-equation of the tangent to $y = \text{Log}[x]$ at $x = 1$.
   *Plot these pairs of curves with the computer program **Tangents**.*

3. Differential Approximation
   Approximate $\sqrt[3]{1,000,000,000,000,002} - \sqrt[3]{1,000,000,000,000,000}$ using the differential increment of the function $f[x] = \sqrt[3]{x}$, $x = 1,000,000,000,000,000$ and $\delta x = 2$. (Two is not infinitesimal, but it is small compared to 10^{15}. Computers have a very hard time with this kind of computation because they work in fixed length decimal approximations.) We only need a simple way to estimate $f[x + \delta x] - f[x]$, since we know $f[x] = 100,000$ when $x = 10^{15}$.

   How many decimals of your approximation are accurate in this case? Try it with your calculator or the computer; you'll get the wrong answer unless you work with very very high-precision arithmetic. The differential is very accurate.

Problem 6.11.

You are interested in the accuracy of your speedometer and perform the following experiment on a stretch of flat, straight, deserted Interstate highway. You drive at constant speed with your speedometer reading 60 mph, crossing between two consecutive mile markers in 57 seconds as measured by your quartz watch. For constant speed, we know the formula, "distance equals rate times time," $D = R \times T$, so $R = D/T$ when the units are correct.

    (a) *Express the rate of speed R in miles per hour as a function of the distance D in miles and the time t in **seconds**.*

    (b) *Compute the differential $dR = ? \times dt$ using the appropriate rules, assuming that $D = 1$ is measured exactly.*

    (c) *Approximate the speed of your car in the above experiment using the differential to approximate the increment ΔR. (See Exercise 6.8.3.)*

    (d) *Use the computer to compare the actual rate with the differential approximation using your formula for r and its differential:*

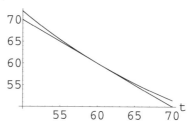

Figure 6.5: *Rate and approximation*

Summarize the idea of this problem in a few sentences.

Problem 6.12.

1) Sketch a pair of (x, y)-axes and plot the point $(1, -1)$. Let x run from 0 to 3, and y run from -2 to 1.

2) The point $(x, y) = (1, -1)$ lies on the explicit curve $y = x^2 \operatorname{Cos}[\pi/x]$. Verify this.

3) Add a pair of (dx, dy)-axes at the (x, y)-point $(1, -1)$. How are these axes related to the (x, y) axes?

4) Use rules of calculus to show that

$$y = x^2 \operatorname{Cos}[\pi/x] \qquad \Rightarrow \qquad dy = (2x \operatorname{Cos}[\pi/x] + \pi \operatorname{Sin}[\pi/x]) \ dx$$

5) Substitute $x = 1$ into your differential to show that

$$dy = -2 \ dx$$

at the (x, y)-point $(1, -1)$ or the (dx, dy)-point $(0, 0)$.

6) Plot the line $dy = -2 \ dx$ on your (dx, dy)-axes.

7) What would you see if you looked at the graph of $y = x^2 \operatorname{Cos}[\pi/x]$ under a very powerful microscope?

*8) Use the computer NoteBook **Micro1D** to plot the function and its differential and to make an animation of a microscope zooming in on the graph at the (x, y)-point $(1, -1)$.*

*9) Explain how the Differential cell of the **Micro1D** NoteBook is actually solving parts (2)
- (7) of this exercise. How does calculus let us "see" a graph in a powerful microscope?*

Problem 6.13.

*What is wrong with the following "general formula" for the tangent to $y = x^2$ at the point
$x = a$? We have $dy = 2x\, dx$ and we know that $dx = x - a$, while $dy = y - a^2$. So (false
conclusion), the equation of the tangent line is $\frac{\Delta y}{\Delta x} = \frac{y-a^2}{x-a} = 2x$ and we can simplify to the
form $y = m\, x + b$.*

The (x, y) equation of the tangent is often not needed (if you plot in (dx, dy) coordinates),
but the idea of working in the correct coordinates is important. Here is another kind of
example of tangency and keeping track of all the variables. A circle of radius 1 centered on
the y axis is moved down the axis until it touches the parabola $y = x^2$, as shown in the next
figure. Since the circle just touches the parabola, both curves are tangent at the point of
contact.

Problem 6.14. TANGENT CURVES

*Write the equation of a circle of radius 1 centered on the y axis in terms of a parameter c
for the unknown y coordinate of the center.*

*Calculate the differential of the equation of your circle, using the unknown parameter.
Solve for $\frac{dy}{dx}$, and write the equation that says*

"the slope of the circle at (x, y) equals the slope of the parabola at (x, y)"

Write the system of three equations that say "(x, y) is the point of tangency:"

$y = x^2$ *"(x, y) lies on the parabola"*
$? =?$ *"(x, y) lies on the circle through $(0, c)$"*
$? =?$ *"the circle and parabola have the same slope at (x, y)"*

Finally, solve the three equations in three unknowns, using the computer if you like.

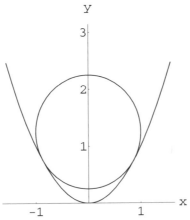

Figure 6.6: A circle tangent to a parabola

Here is a sample of the use of the computer to solve a set of equations:

> **equns = { y == x∧2 , x∧2 + (y - c)∧2 == 1 , x == 2 x (c - y) }**
> **Solve[equns , { x , y , c }]**

Problem 6.15. MORE TANGENT CURVES ◢

A circle with center $(0, 1)$ is expanded until it just touches the parabola $y = x^2$. What is it's radius? Where does it make contact?

A line passes through the origin and is tangent to $y = x^2 + 1$. What is the point of tangency?

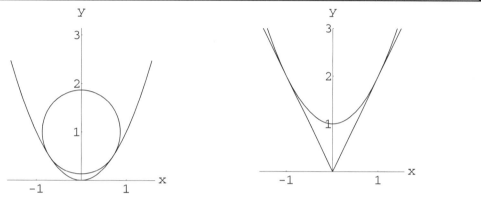

Figure 6.7: Tangent curves

6.9   Projects
6.9.1   Functional Identities

The Mathematical Background chapter on Functional Identities (on the CD) may help you understand the role of unknown functions in mathematics. It is important for you to see the Product Rule and the Chain Rule as identities in unknown functions.

Related Rates and Implicit Derivatives

This chapter gives some basic applications of the Chain Rule but also shows why it is important to learn to work with parameters and variables other than x and y. Most of this chapter is independent of the next few, so it could be skipped now in favor of other topics. If you skip this chapter now, return to implicit differentiation later when it arises in another application.

The main new topic in this chapter is an application of the Chain Rule called "related rates problems" given in the Section 7.4. When functions are chained or composed, the rate of change of the first output variable changes the second output variable: Their rates are related. This idea generalizes to implicitly linked variables.

Implicitly linked variables change with one another, but there are no explicit functions connecting them, only a formula involving both variables. Implicit differentiation is often an easier way to solve related rate, max - min, or other problems later in the course. Essentially, this method is easier because implicit differentiation "treats all variables equally."

7.1 Differentiation with Parameters

You just learned the $\frac{dy}{dx}$ versions of the rules for differentiation. However, it is important to work with parameters (letters you treat as constants) and other variable names. This section has examples to show you why.

In the Chain Rule, we asked you to use a different variable name u and find $\frac{dy}{du}$ with formulas you just learned for $\frac{dy}{dx}$. A few exercises in Chapter 6 were also written in terms of other variables. Usually students do not like this at first which is an understandable reaction. However, there are times mathematically when you have already used the variable x for something but need to vary something else. Here is an oversimplified example to illustrate the point:

Example 7.1. *Approximating a Root x as b Varies*

Suppose we are finding a root of the quadratic equation

$$ax^2 + bx + c = 0$$

where the coefficient b is a measured quantity and not known with perfect accuracy. We want to know how sensitive the largest root of the equation is to errors in measuring b. The largest root of the quadratic equation above can be written as a function of b, including the parameters a and c:

$$x = \frac{-b + \sqrt{b^2 - 4ac}}{2a}$$

The derivative $\frac{dx}{db}$ measures the rate of change of x with respect to b. A small change in b denoted db produces the approximate change in x of

$$dx = x'[b] \, db$$

by the microscope approximation (or meaning of derivative). An exercise below has you explore this approximation. We return to this in the CD section on related rates.

Example 7.2.　*The Role of Parameters*

Much later in the course (in the project on resonance) we will see that the amplitude, A, of a certain kind of oscillation is given by the formula

$$A[\omega] = \sqrt{\frac{1}{(s - m\,\omega^2)^2 + (c\,\omega)^2}}$$

where s, m and c are measured quantities of a particular system. If we think of the oscillator as the front suspension of an old car with weak shocks, m is the mass, s is the strength of the spring, and c measures the strength left in the shocks. These are fixed for any particular car and we want to see how the peak of A depends on these parameters. This will tell us the frequency of the most violent shaking of a car in terms of m, s, and c. The graph typically looks like Figure 7.1

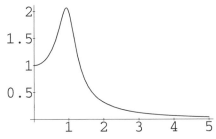

Figure 7.1: Amplitude A as a function of frequency ω

The peak response (called the "resonant frequency") is at the frequency ω_r where $A'[\omega_r] = 0$.

We compute the derivative of A with respect to ω treating the other letters as constants:

$$A'[\omega] = \frac{d(((s - m\,\omega^2)^2 + (c\,\omega)^2)^{(-1/2)})}{d\omega}$$

$$= -\frac{1}{2}((s - m\,\omega^2)^2 + (c\,\omega)^2)^{(-3/2)} \times (2(s - m\,\omega^2) \times 2m\,\omega + 2c^2\,\omega)$$

$$= -\omega \frac{2m^2\,\omega^2 + (c^2 - 2m\,s)}{((s - m\,\omega^2)^2 + (c\,\omega)^2)^{(3/2)}}$$

This is very messy, but we can check our work with the computer.

Notice that this derivative is zero, $A'[\omega] = 0$, when $\omega = 0$ (look at the graph) or ω is a positive root (the peak) of the numerator:

$$2m^2\,\omega^2 + (c^2 - 2m\,s) = 0$$
$$2m^2\,\omega^2 = (2m\,s - c^2)$$
$$\sqrt{2}m\,\omega = \sqrt{2m\,s - c^2}$$
$$\omega = \frac{\sqrt{2m\,s - c^2}}{\sqrt{2}m} = \frac{\sqrt{4m\,s - 2c^2}}{2m}$$

so the resonant frequency is

$$\omega_r = \frac{\sqrt{4m\,s - 2c^2}}{2m}$$

Given m, s, and c, we just "plug in" and find the frequency. The important point is that we find the max BEFORE we know the actual values of these constants.

Exercise set 7.1

The next exercise is solved with implicit differentiation in the CD section of this chapter on implicitly linked variables. Solve it explicitly now so you can compare the explicit and implicit methods.

1. Approximate Roots

   (a) *Compute the derivative $\frac{dx}{db}$ (considering a and c as parameters). When is this defined? In the cases when it is not defined, what is going on in the original root-finding problem? Consider some special cases to help such as $(a, b, c) = (1, -3, 2)$, $(a, b, c) = (1, -2, 1)$, $(a, b, c) = (0, -2, 1)$.*

   (b) *Consider the case $(a, b, c) = (1, -3, 2)$ and denote the error in measuring b by db. Suppose the magnitude of db could be as large as 0.01. Use the differential approximation to estimate the resulting error in the root.*

   The following are for practice differentiating with respect to different letters:

   (c) *Compute the derivative $\frac{dx}{da}$ (considering b and c as parameters). When is this defined?*

   (d) *Compute the derivative $\frac{dx}{dc}$ (considering a and b as parameters). When is this defined?*

   (e) *Check your differentiation with the computer.*

It is customary in physics to let the Greek letter omega, ω, denote frequency and T denote absolute temperature. Certainly, the capital T suggests the word that the variable measures. Whether you like ω or not, it is almost impossible to read the physics literature without working with it. Here is an example to test your skills.

2. *Planck's radiation law can be written*

$$I = \frac{a\omega^3}{e^{b\omega/T} - 1}$$

for constants a and b. This expresses the intensity I of radiation at frequency ω for a body at absolute temperature T. Suppose T is also fixed. Express I as a chain or composition of functions (of the variable ω with parameters) and products of functions to which the rules of this chapter apply. For example, you can use the Exponential Rule, $\frac{d(e^u)}{du} = e^u$. What is the formula for $\frac{dI}{d\omega}$? Check your work with the DfDx program.

There is a project on Planck's law studying the interaction between calculus and graphs and between calculus and maximization. Planck's Law was one of the first big achievements of quantum mechanics.

We postpone the exercises on related rates until we have shown you implicitly linked variables and the method of implicit differentiation. In those problems, you have your choice of solving for explicit nonlinear equations or using implicit differentiation.

7.2   Implicit Differentiation

Implicit equations have many powerful uses. We can differentiate them directly simply by treating all variables equally.

A unit circle in the plane is given by the set of (x, y)-points satisfying the implicit equation

$$x^2 + y^2 = 1$$

This equation is called "implicit" because, if we treat x as given, then y is only implicitly given by the equation. Two values of y satisfy the equation, but the implicit equation does not give a direct "explicit" way to compute them. The explicit equation

$$y = -\sqrt{1 - x^2}$$

gives a direct way to compute y on the lower half-circle, whereas the implicit equation does not but does give the whole circle.

Example 7.3. *Implicit Tangent to the Circle*

The differential of $u^2 + b$, when u is the variable and b is a parameter, is $2\,u\,du$. Treating x as the variable and y^2 as a parameter in $x^2 + y^2$, we have the differential $2\,x\,dx$. Treating y as the variable and x^2 as a parameter in $x^2 + y^2$, we have the differential $2\,y\,dy$. Adding the two, we obtain $2\,x\,dx + 2\,y\,dy$. Since the differential of the constant 1 is zero, the total differential of the implicit equation becomes

$$x^2 + y^2 = 1 \qquad \Rightarrow \qquad 2\,x\,dx + 2\,y\,dy = 0$$

We may view the total differential as an implicit equation for the tangent line to the circle in the local variables (dx, dy) when (x, y) is a fixed point on the circle.

This may be solved for $\frac{dy}{dx}$ as follows:

$$2x\ dx + 2y\ dy = 0$$
$$x\ dx + y\ dy = 0$$
$$y\ dy = -x\ dx$$
$$dy = -\frac{x}{y}dx$$
$$\frac{dy}{dx} = -\frac{x}{y}$$

This is a valid formula for the slope of the tangent to a circle. However, this expression uses both variables so that to use it, we need to know both x and y. For example, the point $(1/2, -\sqrt{3}/2)$ lies on the lower half of the circle as shown in Figure 7.2. At this point the slope is

$$\frac{dy}{dx} = -\frac{x}{y} = \frac{1/2}{\sqrt{3}/2} = \frac{1}{\sqrt{3}}$$

Figure 7.2: $y = -\sqrt{1-x^2}$ and $dy = dx/\sqrt{3} \Leftrightarrow y + \frac{\sqrt{3}}{2} = (x - \frac{1}{2})/\sqrt{3}$

Let us compare this computation of the slope of the circle at $(1/2, -\sqrt{3}/2)$ to the computation with the explicit equation.

$$y = -\sqrt{1-x^2}$$

$$y = -\sqrt{u} = -u^{1/2} \qquad u = 1 - x^2$$

$$\frac{dy}{du} = -u^{-1/2} = -\frac{1}{2\sqrt{u}} \qquad \frac{du}{dx} = -2x$$

so the Chain Rule gives

$$\frac{dy}{dx} = \frac{dy}{du} \cdot \frac{du}{dx} = \frac{x}{2\sqrt{u}} = \frac{x}{2\sqrt{1-x^2}}$$

and when $x = 1/2$

$$\frac{dy}{dx} = \frac{x}{2\sqrt{1-x^2}} = \frac{1/2}{2\sqrt{1-(1/2)^2}} = \frac{1}{\sqrt{3}}$$

The implicit computation

$$\boxed{x^2 + y^2 = 1 \qquad \Rightarrow \qquad 2x\ dx + 2y\ dy = 0}$$

is certainly simpler than solving and performing the four lines of computation above.

The idea of implicit differentiation is to differentiate everything with respect to x and multiply by dx, then to differentiate everything with respect to y and multiply by dy, and finally to add all the differentials together. This description is a little vague, but the drill exercise below should be enough to give you the idea. This works with any variables.

Example 7.4.   *Different Letters*

Find the implicit equation of the tangent to the circle $w^2 + h^2 = 25^2$ at $(20, 15)$.

$$w^2 + h^2 = 25^2$$
$$2\,w\,dw + 2\,h\,dh = 0$$

The equation $w^2 + h^2 = 25^2$ is an implicit equation in w and h. When we consider w and h fixed and located somewhere on the circle, the equation in dw and dh, $w\,dw + h\,dh = 0$, is an implicit equation for a line. The point $(20, 15)$ lies on the circle. The implicit equation of its tangent at this point is shown in Figure 7.3.

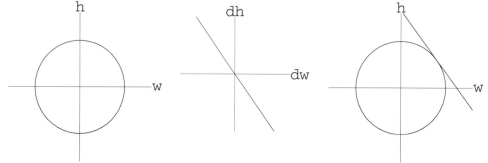

Figure 7.3: $20\,dw + 15\,dh = 0$

Implicit curve	Implicit tangent line
$w^2 + h^2 = 25^2$	$w\,dw + h\,dh = 0$

The formula $w\,dw + h\,dh = 0$ has the advantage that we may think of either variable as the independent variable. When $w = 25$, we have $h = 0$. The circle is smooth, but the tangent is vertical, $25\,dw + 0\,dh = 0$ or $dw = 0$. (This is simply the local variable equation for the dh-axis.) The explicit formula $h = \sqrt{25^2 - w^2}$, with derivative $\frac{dh}{dw} = -w/\sqrt{25^2 - w^2}$, is undefined at $w = 25$.

── **Exercise set 7.2** ──

1. Implicit Drill
   Verify the following implicit total differential calculations:

$$3x^2 + y^2/5 = 1 \qquad \Rightarrow \qquad 6x\,dx + \frac{2}{5}y\,dy = 0$$

$$xy = 1 \qquad \Rightarrow \qquad y\,dx + x\,dy = 0$$

$$x + xy = 2y \qquad \Rightarrow \qquad dx + x\,dy + y\,dx = 2\,dy$$

$$y + \sqrt{y} = x + x^2 \qquad \Rightarrow \qquad dy + \frac{1}{2\sqrt{y}}\,dy = dx + 2x\,dx$$

$$e^x = \text{Log}[y] \qquad \Rightarrow \qquad e^x\,dx = \frac{1}{y}\,dy$$

$$x = \text{Sin}[xy] \qquad \Rightarrow \qquad dx = y\,\text{Cos}[xy]\,dx + x\,\text{Cos}[xy]\,dy$$

$$x = \text{Log}[\text{Cos}[3x + 5y]] \qquad \Rightarrow \qquad dx = -3\,\text{Tan}[3x + 5y]\,dx - 5\,\text{Tan}[3x + 5y]\,dy$$

2. More Implicit Differentiation Drill
   Find the total differential and solve for $\frac{dy}{dx}$

a) $\quad x^2 - y^2 = 3$ 　　　　 b) $\quad y + \sqrt{y} = \frac{1}{x}$ 　　　　 c) $\quad xy = 4$

d) $\quad y = \text{Cos}[xy]$ 　　　　 e) $\quad \text{Sin}[x]\,\text{Cos}[y] = \frac{1}{2}$ 　　　　 f) $\quad y = \text{Sin}[x + y]$

g) $\quad y = e^{xy}$ 　　　　 h) $\quad e^x\,e^y = 1$ 　　　　 i) $\quad e^x = x + y^2$

j) $\quad x = \text{Log}[xy]$ 　　　　 k) $\quad y = \text{Log}[x^2 y]$ 　　　　 l) $\quad x = \text{Log}[x + y]$

7.3   Implicit Tangents and Derivatives

Implicit differentiation can be used directly to find tangents and derivatives.

Example 7.5.   *The General Implicit Slope of a Circle*

The circle of radius r (centered at the origin) is the set of (x, y) points satisfying

$$x^2 + y^2 = r^2$$

Because we are thinking of r as a constant, its differential is zero and

$$2x\,dx + 2y\,dy = 0$$

If we solve the equation $2x\,dx + 2y\,dy = 0$ for the slope of the tangent line,

$$\frac{dy}{dx} = -\frac{x}{y}$$

The slope of a radial line from the center to any point (x, y) on the circle is y/x. The negative reciprocal $-x/y$ is the slope of a perpendicular line. We have just shown that it is the slope of the tangent, so this shows that the tangent to a circle is perpendicular to a radius at a general point as shown on Figure 7.4

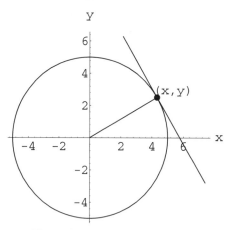

Figure 7.4: Circle and tangent

Example 7.6. *Implicit Tangent to an Ellipse*

The differential of the equation of an ellipse $(\frac{x}{2})^2 + (\frac{y}{3})^2 = 1$ is computed as follows:

$$\frac{x^2}{4} + \frac{y^2}{9} = 1$$
$$\frac{2x\,dx}{4} + \frac{2y\,dy}{9} = 0$$
$$\frac{1}{2}x\,dx + \frac{2}{9}y\,dy = 0$$
$$x\,dx + \frac{4}{9}y\,dy = 0$$

The result of this computation, shown in Figure 7.5, is not geometrically as obvious as the tangent to a circle because a line from the center is no longer perpendicular to the tangent. However, if we want to sketch the tangent at a point, for example where $x = 8/5 \approx 1.6$ and $y = 9/5 \approx 1.8$, so $(\frac{x}{2})^2 + (\frac{y}{3})^2 = \frac{4}{5}^2 + \frac{3}{5}^2 = 1$ is on the ellipse, we can use the local (dx, dy) coordinates (as in Chapter 1) in the implicit form. The specific tangent has equation $\frac{8}{5}\,dx + \frac{4}{9}\frac{9}{5}\,dy = 0$ or $2\,dx + dy = 0$.

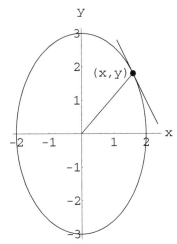

Figure 7.5: $\left(\frac{x}{2}\right)^2 + \left(\frac{y}{3}\right)^2 = 1$ and $2\,dx + dy = 0$ at $(x,y) = \left(\frac{8}{5}, \frac{9}{5}\right)$

7.3.1 The Chain Rule as Substitution in Differentials

Suppose we have a chain

$$y = f[u] \qquad \& \qquad u = g[x]$$

If we calculate the differentials

$$dy = f'[u]\,du \qquad \& \qquad du = g'[x]\,dx$$

and make the substitution for du,

$$dy = f'[u]\,du = f'[u](g'[x]\,dx) = f'[g[x]]g'[x]\,dx$$

we have in effect done a Chain Rule computation. In specific contexts, it looks simpler.

Example 7.7. *The Differential of* $y = e^{x^2}$

We compute the differential of $y = e^{x^2}$ using the decomposition

$$y = e^u \qquad \& \qquad u = x^2$$

The differentials are

$$dy = e^u\,du \qquad \& \qquad du = 2x\,dx$$

Substitution of du gives

$$dy = e^u\,du = e^u\,(2x\,dx) = 2xe^{x^2}\,dx$$

Example 7.8. *The S-I-R Invariant and Differentials*

In Chapter 2, you may recall that we claimed that the quantity

$$s + i - \frac{b}{a}\,\text{Log}[s] = k$$

was constant, where s and i are the susceptible and infectious fraction of a population, wheras a and b are constants. The differential of this equation is

$$ds + di - \frac{b}{a}\frac{1}{s}\, ds = dk$$

$$\left(1 - \frac{b}{a}\frac{1}{s}\right) ds + di = dk$$

In the S-I-R model of an epidemic, the variables s and i are both functions of time,

$$s = s[t] \qquad\qquad i = i[t]$$

$$\&$$

$$ds = s'[t]\, dt \qquad\qquad di = i'[t]\, dt$$

Substituting these in the first differential gives

$$\left(\left(1 - \frac{b}{a}\frac{1}{s}\right) s'[t] + i'[t]\right)\, dt = dk$$

Carrying this one step further, recall that $s'[t] = -a\, s[t]\, i[i]$ and $i'[t] = as[t]i[t] - bi[t]$. Substituting these into the expression gives

$$dk = \left(\left(1 - \frac{b}{a}\frac{1}{s}\right)(-asi) + (asi - bi)\right)\, dt$$

$$dk = ((-asi + bi) + (asi - bi))\, dt$$

$$\frac{dk}{dt} = 0$$

Since the derivative of k with respect to t is zero, k is constant.

7.3.2 Derivatives of Inverse Functions

The method of implicit differentiation applies to inverse functions. This case is treated in detail in the Project on Inverse Functions.

Example 7.9. *Derivative of* $\mathrm{Log}[y]$

Consider an example for the inverse pair of equations

$$y = e^x \qquad \Leftrightarrow \qquad x = \mathrm{Log}[y]$$

The differential of the exponential equation is $dy = e^x\, dx$, but we may use the fact that $e^x = y$ to write

$$dy = y\, dx$$

Solving for the derivative of x with respect to y gives us the derivative of the logarithm,

$$\frac{d(\mathrm{Log}[y])}{dy} = \frac{dx}{dy} = \frac{1}{y}$$

Example 7.10. *Derivative of* ArcTan[y]

Computation of inverse function derivatives this way can present computational difficulties, such as the following:

$$y = \text{Tan}[x] \qquad \Leftrightarrow \qquad x = \text{ArcTan}[y]$$

You computed the derivative of tangent directly in Chapter 5 and using rules in Chapter 6, obtaining the following answer both ways:

$$dy = \frac{1}{(\text{Cos}[x])^2} \, dx$$

Consequently, we have a formula for the derivative of the arctangent

$$\frac{d(\text{ArcTan}[y])}{dy} = \frac{dx}{dy} = (\text{Cos}[x])^2$$

Unfortunately, this form of the equation is in terms of the dependent variable for arctangent, so some trig tricks are needed to put it in the form

$$\frac{d(\text{ArcTan}[y])}{dy} = \frac{1}{1 + y^2}$$

See the project on Inverse Functions. Additional examples are included in that project along with complete justification of this method of computing derivatives of inverse functions. The justification is in the form of a procedure you can use to compute the actual nonlinear inverse function. In other words, it is a "practical" proof.

Exercise set 7.3

1. *Use implicit differentiation to find the equation of the tangent line to*

$$(5x)^2 - (4y)^2 = 3^2$$

at the point $(x, y) = (1, 1)$. *Use the local coordinates* (dx, dy) *centered at* $(x, y) = (1, 1)$.

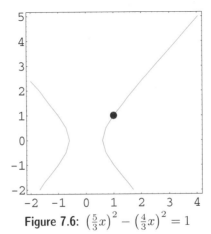

Figure 7.6: $\left(\frac{5}{3}x\right)^2 - \left(\frac{4}{3}x\right)^2 = 1$

7.4   Related Rates - CD

Many applications of calculus involve different quantities that vary with time but are "linked" with one another; the time rate of change of one variable determines the time rate of change of the other. This section illustrates the idea with examples including the falling ladder.

Remainder of this section only on CD

7.5   Implicitly Linked Variables - CD

Implicit solutions of related rates problems often are simpler and more revealing than first solving for a quantity explicitly.

Remainder of this section only on CD

7.6   Projects

7.6.1   Dad's Disaster

Dad is painting the garage when Pooch gets her leash tangled around the bottom of the ladder and starts pulling it away from the wall. The Fast Ladder example suggests that Dad will break the sound barrier before he crashes to the sidewalk. Is this so? The Falling Ladder Project helps you find out.

7.6.2   The Inverse Function Rule

The project on finding the derivative an inverse function such as ArcTan[y] and computing values of the inverse itself. This project relies on basic graphical understanding and the microscope idea.

<table>
<tr><td>

CHAPTER
8

</td><td>

The Natural Log and Exponential

</td></tr>
</table>

This chapter treats the basic theory of logs and exponentials. It can be studied any time after Chapter 6. You might skip it now, but should return to it when needed.

The "natural" base exponential function and its inverse, the natural base logarithm, are two of the most important functions in mathematics. This is reflected by the fact that the computer has built-in algorithms and separate names for them:

$$y = e^x = \operatorname{Exp}[x] \qquad \Leftrightarrow \qquad x = \operatorname{Log}[y]$$

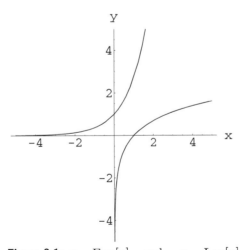

Figure 8.1: $y = \operatorname{Exp}[x]$ and $y = \operatorname{Log}[x]$

We did not prove the formulas for the derivatives of logs or exponentials in Chapter 5. This chapter defines the exponential to be the function whose derivative equals itself. No matter where we begin in terms of a basic definition, this is an essential fact. It is so essential that everything else follows from it. We call this the "official theory."

We already know the differentiation rules for log and exponential, and the basic high school review material about logs and exponentials is contained in Chapter 28. The main facts to memorize are

$$\frac{de^t}{dt} = e^t \qquad\qquad \frac{d\operatorname{Log}[s]}{ds} = \frac{1}{s}$$

$$e^a \times e^b = e^{a+b} \qquad \& \qquad \operatorname{Log}[a \times b] = \operatorname{Log}[a] + \operatorname{Log}[b]$$

$$(e^c)^t = e^{c \cdot t} \qquad\qquad \operatorname{Log}[a^p] = p \times \operatorname{Log}[a]$$

$$\operatorname{Log}[e^t] = t \qquad\qquad e^{\operatorname{Log}[s]} = s, \quad s > 0$$

Some of the graphical properties of these functions are formulated as limits, comparing them to power functions later in the chapter. Section 8.3 explains these "orders of infinity" more technically and shows how to build more limits from them. The basic limits say

- e^t goes to infinity faster than any power as $t \to \infty$.
- $\operatorname{Log}[t]$ tends to infinity slower than any root as $t \to \infty$.
- e^{-t} is positive but tends to zero faster than any reciprocal power as $t \to \infty$.
- $\operatorname{Log}[s] \downarrow -\infty$ as $s \downarrow 0$.

See the the computer programs **ExpGth** and **LogGth** in Chapter 28 for an intuitive explanation of these limits.

8.1   The Official Natural Exponential

One of the most important ways that exponential functions arise in science and mathematics is as the solution to linear growth and decay laws.

The differential equation $\dfrac{dy}{dt} = k\,y$ with a positive constant k represents proportional growth and with a negative constant represents proportional decay. We have already seen a decay law of this form in Newton's Law of Cooling or the Cool Canary Problem 4.1 and a growth law of this form in the first (false) conjecture of Galileo on the Law of Gravity, Problem 4.2. These laws simply say, "The rate of change of a quantity is proportional to the amount present." In Exercise 4.2.1, we solved the differential equation numerically, but now we will be able to solve these problems symbolically (exactly) in Problems 8.1 and 8.5.

The most noteworthy thing about the formulas in this chapter is this: The dependent variable y appears on both sides of the equation.

SOMETHING NEW:

$$\frac{dy}{dt} = k\,y$$

This is an important differential equation, not just another differentiation formula like the ones in Chapter 6. (Those equations all have explicit functions of the independent variable

t on the right hand side, for example, $y = 3\,t^5 \Rightarrow \frac{dy}{dt} = 15\,t^4$.) It might be worthwhile to contrast this situation with the "growth form" of a linear function:

Theorem 8.1. *The Differential Equation of a Linear Function*
For appropriate constants k and Y_0, the following are equivalent:

$$\frac{dy}{dx} = k \quad \Leftrightarrow \quad y = k\,x + Y_0$$

The rate of change of y with respect to x is constant if and only if y varies linearly.

Theorem 8.2. *The Differential Equation of an Exponential Function*
For appropriate constants k and Y_0, the following are equivalent:

$$\frac{1}{y}\frac{dy}{dx} = k \quad \Leftrightarrow \quad y = Y_0\,e^{k\,x}$$

The rate of change of y is a constant percentage if and only if y varies exponentially.
Furthermore, y varies exponentially if and only if $\mathrm{Log}[y]$ varies linearly,

$$y = Y_0\,e^{k\,x} \quad \Leftrightarrow \quad \mathrm{Log}[y] = k\,x + \mathrm{Log}[Y_0]$$

The CD section of Chapter 5 on increments of the exponential and the **PercentGth** program (of Chapter 5) show how to find an exponential given a fixed percentage change for a fixed change in x. The differential equation $\frac{1}{y}\frac{dy}{dx} = k$ simply says y has the fixed instantaneous percentage change $k \times 100\%$.

A differential equation tells us how the quantity changes instantaneously. If we also know an initial value of the quantity, it is intuitively clear that this "start plus change" determines where you go, though it may *not* be entirely clear *how* it determines it.

Mathematically, a continuous dynamical system is the "operation" of going from the "where you start and how you change" to a function of t. We will see that the constant percentage change system has the solution

$$y[0] = Y_0$$
$$dy = k\,y\,dt$$
$$\rightarrow \quad y[t] = Y_0\,e^{k\,t}, \quad t \in [0, \infty)$$

Without knowing this, we can approximate this "operation" by solving a discrete system that moves in small steps of size δt

$$y[0] = Y_0$$
$$y[t + \delta t] \approx y[t] + k\,y[t]\,\delta t$$
$$\rightarrow \quad \{y[0], y[\delta t], y[2\delta t], y[3\delta t], \cdots\}$$

The solution in this case is $y[t] = Y_0(1+\delta t)^{t/\delta t}$ and gives the very fundamental approximation

$$(1 + \delta t)^{t/\delta t} \approx e^t$$

The general idea is called "Euler's Method" of approximating solutions of initial value problems. We have already seen this idea in several places, beginning with the **SecondSIR**

NoteBook in Chapter 2. (Euler's Method for general systems is studied in Chapter 21.) The point of this section is to see that the differential equation gives us a way to work with the function without prior formulas.

You already have an idea of what $y = e^t$ means, so it may seem a little silly to introduce a "definition" for it at this late stage of the game. There may be some gaps in what you know, and we want a definite place to fall back to when the problems get more difficult. For example, later we will want to compute e^{it}, where $i = \sqrt{-1}$. Many important functions in higher mathematics are characterized by their differential equation, so this is the first time you will see something that is quite powerful.

The official theory is only important when we get to a question we cannot answer with facts from high school and simple differentiation formulas. What do we mean by e^π or even $3^{\sqrt{2}}$? Certainly, e^3 means "multiply e times itself 3 times," but you cannot multiply e times itself π times - that makes no sense. You probably do not want to believe that - because you can use your calculator for an approximate answer.

When we use calculators to approximate e^π we raise the approximate base $e \approx 2.71828$ to the approximate power $\pi \approx 3.14159$. This implicitly assumes that the y^x-button on our calculator is continuous in both inputs. In other words, the small errors in both e and π only produce a small error in the approximate output to e^π. Does your calculator produce six significant digits of e^π when you put 6 digits of accuracy in for e and π? This is a tough question because you have to decide what is exact. Similarly, an approach to exponentials based on

$$\lim_{x \to \pi, \ y \to e} y^x = e^\pi$$

as both $y \to e$ and $x \to \pi$ is a very difficult way to build a basic theory. It is "natural" in some ways but technically too hard. (You will use differentials to prove it later.)

Definition 8.3.   *The Official Natural Exponential Function*
   The function
$$y = \text{Exp}[t]$$
   is officially defined to be the unique solution of the initial value problem
$$y[0] = 1$$
$$dy = y \, dt$$

If we use the (unproved) formula for the derivative, we can see that the natural exponential function $y[t] = e^t$ satisfies this differential equation and initial condition because $\frac{dy}{dt} = \frac{de^t}{dt} = e^t = y$ and $e^0 = 1$.

The general Euler's Method is a simple idea once you know the increment approximation from the Definition 5.5 of the derivative. When our function is $f[x]$, we write this approximation

$$f[x + \delta x] = f[x] + f'[x] \, \delta x + \varepsilon \cdot \delta x$$

where the error $\varepsilon \approx 0$ is small when $\delta x \approx 0$ is small. Our function now is $y = y[t]$, so the approximation becomes

$$y[t + \delta t] = y[t] + y'[t] \, \delta t + \varepsilon \cdot \delta t$$

where $\varepsilon \approx 0$ is small when $\delta t \approx 0$ is small. We do not have a formula for $y[t]$, but we do have the value of $y[0] = 1$ and a formula for $y'[t] = y[t]$ given in terms of y. This gives us

APPROXIMATE SOLUTION OF $y[0] = 1$ & $\frac{dy}{dt} = y$:

$$y[t + \delta t] = y[t] + y[t]\ \delta t + \varepsilon \cdot \delta t$$
$$y[t + \delta t] \approx y[t] + y[t]\ \delta t = y[t](1 + \delta t) \qquad \text{when } \delta t \approx 0$$

so

$$y[\delta t] \approx y[0] \cdot (1 + \delta t) = (1 + \delta t)$$
$$y[\delta t + \delta t] \approx y[\delta t] \cdot (1 + \delta t) = (1 + \delta t)^2$$
$$y[2\delta t + \delta t] \approx y[2\delta t] \cdot (1 + \delta t) = (1 + \delta t)^2 \cdot (1 + \delta t) = (1 + \delta t)^3$$

In general, we see that if $y[0] = 1$ and $\frac{dy}{dt} = y$, then

$$y[t] \approx (1 + \delta t)^{(t/\delta t)} \qquad \text{for } t = 0, \delta t, 2\delta t, 3\delta t, \cdots$$

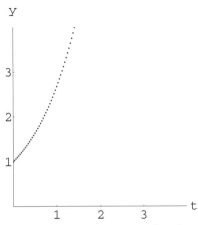

Figure 8.2: Euler's approximation $e^t \approx (1 + \delta t)^{(t/\delta t)}$

A very basic fact of mathematics says

$$\lim_{\delta t \to 0} (1 + \delta t)^{1/\delta t} = e$$

This is a special case of the convergence of the solution of our discrete dynamical system to the solution of the continuous one because $y[1] = e^1 = e$.

We can summarize the section with the formula

$$Y_0(1 + k\,\delta t)^{t/\delta t} \quad \approx \quad Y_0\ e^{k\,t}$$

The formula on the left can be computed by hand for $t = 0, \delta t, 2\delta t, \cdots$, if necessary, and it comes straight from the initial value problem.

Exercise set 8.1

1. *Compare the computer's built in function* Exp[t] *to the Euler approximation of the official definition,* $(1 + \delta t)^{t/\delta t}$*, for* $\delta t = 1/2, 1/4, 1/16, 1/256$*. Graph both and compare*

them numerically. How large is the difference between Exp[1] *and the approximate* $y[1]$
when $\delta t = 1/256$?

Now we want you to use the idea that gave us the approximation $(1 + \delta t)^{t/\delta t} \approx e^t$ to
approximate the solution of a more general exponential law.

2. Approximate Solution of $\dfrac{dy}{dt} = k\,y$ with $y[0] = Y_0$:

    (a) *Show that* $y[t] = Y_0\,(1 + k\,\delta t)^{t/\delta t}$ *(for* $t = 0, \delta t, 2\delta t, 3\delta t, \cdots$) *is an approximate
solution to the initial condition and differential equation*

$$y[0] = Y_0$$
$$\frac{dy}{dt} = k\,y$$

    (b) *Test your approximation numerically and graphically for the special case*

$$y[0] = 3$$
$$\frac{dy}{dt} = -2\,y$$

which has the exact solution $y[t] = 3\,Exp[-2\,t]$.

Here is some help with the exercise. First and foremost, recall the microscope approximation of Definition 5.5 and apply it to the (unknown) function $y[t]$. Discarding the error
term yields an approximation:

$$y[t + \delta t] = y[t] + ?? + \varepsilon \cdot \delta t \approx y[t] + ???$$

Next, use the fact that $y'[t] = k\,y[t]$ and substitute this into the microscope approximation,

$$y[t + \delta t] = y[t] \cdot (??)$$

We know $y[0] = Y_0$, the initial condition. To find $y[\delta t]$, use your approximation

$$y[\delta t] \approx Y_0 \cdot ??$$
$$y[2\delta t] \approx y[\delta t + \delta t] = y[\delta t] \cdot ??$$
$$y[3\delta t] \approx y[2\delta t + \delta t] = y[2\delta t] \cdot ??$$

Simplification yields the desired result.

3. Log Linearity
    Show that if y *grows at a constant percentage rate with respect to* x, *then the quantity*
$z = Log[y]$ *is a linear function of* x, $z = k\,x + Z_0$. *Give the value of* Z_0 *in terms of*
$Y_0 = y[0]$.

8.2   e as a "Natural" Base

The number $a = e$ makes $y = a^x$ satisfy $\frac{dy}{dx} = y$. Similarly, $y = e^{kx}$ satisfies $\frac{dy}{dx} = ky$, and $y = b^x$ satisfies $\frac{dy}{dx} = ky$ provided $b = e^k$.

All exponential bases are not created equal. All exponential functions $y = b^t$ satisfy

$$y[0] = 1$$
$$\frac{dy}{dt} \propto y$$

but the base with constant of proportionality 1 is $b = e$. This makes e the "natural" base from the point of view of calculus.

Exercise set 8.2

---

1. (a) *Let $y = b^t$ for an unknown (but fixed) positive constant b. Use the Chain Rule (see Section 6.5) to show that $y[t]$ satisfies*

    $$y[0] = 1$$
    $$\frac{dy}{dt} = ky$$

    What is the value of the constant k?

    (b) *Show that $y = e^{kt}$ satisfies*

    $$y[0] = 1$$
    $$\frac{dy}{dt} = ky$$

    If the constant k is the same as in the first part, how much is e^k in terms of b?

    (c) *Solve the initial value problem*

    $$y[0] = 5$$
    $$\frac{dy}{dt} = y$$

    (d) *Solve the initial value problem*

    $$y[0] = 5$$
    $$\frac{dy}{dt} = ky$$

    *where $k = \text{Log}[2]$. Show that your solution may also be written as $y = 5 \cdot 2^t$ (See the program **ExpEquns**.)*

The moral of this exercise is this: We *could* write solutions of initial value problems

$$y[0] = Y_0$$
$$\frac{dy}{dt} = k\,y$$

as $y = Y_0 \cdot b^t$, where $b = e^k$ for $\text{Log}[b] = k$; but, for the purposes of calculus, it is "more natural" to write them in the form $y = Y_0 e^{k\,t}$,

$$y[0] = Y_0$$
$$dy = k\,y\,dt$$
$$\longrightarrow \qquad y[t] = Y_0 \cdot e^{k\,t}, \quad t \in [0, \infty)$$

We want you to put this to work in the next problem.

Problem 8.1. THE CANARY'S POSTMORTEM ───────────────────────── ▼

Let $T = T_0 e^{-kt}$ for unknown positive constants T_0 and k. Show that

$$\frac{dT}{dt} = -kT$$

by using the Chain Rule, $T = T_0 e^u$ with $u = -kt$, so $\frac{dT}{du} = T_0 e^u$ and $\frac{du}{dt} = -k$. Express $\frac{dT}{dt}$ in terms of the dependent variable T.

   The value of $e^0 = 1$, so $T = T_0$ when $t = 0$. Show that the function T solves the cooling problem of Exercise 4.2.1 for certain choices of the constants T_0 and k. How much is T_0 in that problem? How could you find the constant k so $T = 60$ when $t = 10$? (HINT: Solve $60 = 75\ e^{-10\,k}$ using log.)

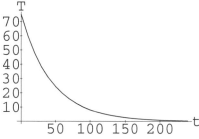

Figure 8.3: *The cooling canary*

   *Graph your specific function $T = T_0\ e^{-k\,t}$ using the computer and verify that the temperature at time 10 is 60. (See the program **ExpEquns**.)*

─── ▲

In the saga of the frozen canary, we let the outside temperature be zero. This simplifies the math (and, of course, freezes the poor dead canary). The next problem has you generalize the law of cooling to an arbitrary ambient temperature. We want you to explain why the initial value problem:

$$T[0] = T_0$$

$$\frac{dT}{dt} = k(T_a - T)$$

is a reasonable model of temperature adjustment toward ambient - either warming or cooling. We also want you to find an analytical solution to the model.

Problem 8.2. NEWTON'S LAW OF COOLING FOR AMBIENT TEMPERATURE T_a ————▼

Suppose a small object is placed in a large room with constant room temperature T_a - the constant ambient temperature. The small inanimate object (that does not heat itself or evaporate, etc.) is placed in the room at a different initial temperature, T_0, and cools or warms toward the room temperature. This problem helps you formulate Newton's law of cooling for non-zero ambient temperature.

(a) *Let T be the (variable) temperature of your object and t be the time. Choose units of your liking. The derivative $\frac{dT}{dt}$ represents the instantaneous rate of increase in the temperature. Why?*

(b) *How do we represent warming and cooling in terms of $\frac{dT}{dt}$?*

(c) *Suppose we put a covered cup of almost boiling water in a normal room, so $T_a = 21°C$ and $T_0 = 100°C$. Initially, the object cools at a fast rate ($\frac{dT}{dt}$ is a large magnitude negative number), while as T approaches $T_a = 21$, the rate of cooling slows.*

    (1) *What is the value of $(T_a - T)$ when $T = 100$? How much "cooling" is happening at this instant if $\frac{dT}{dt} = k(T_a - T)$?*

    (2) *How much "cooling" is happening at the instant when $T = 22$ if $\frac{dT}{dt} = k(T_a - T)$?*

(d) *Suppose we put a covered cup of almost freezing water in a normal room, so $T_a = 21°C$ and $T_0 = 0°C$. Initially, the object warms at a fast rate, while as T approaches $T_a = 21$, the rate of warming slows.*

    (1) *What is the value of $(T_a - T)$ when $T = 0$? How much "warming" is happening at this instant if $\frac{dT}{dt} = k(T_a - T)$?*

    (2) *How much "warming" is happening at the instant when $T = 19$ if $\frac{dT}{dt} = k(T_a - T)$?*

(e) *Use the previous parts of this problem to explain how the initial value problem*

$$T[0] = T_0$$

$$\frac{dT}{dt} = k(T_a - T)$$

describes temperature change. Write something like, "The thing that causes the temperature to change is amount away from ambient temperature of the object ... and the temperature adjusts toward ambient. ... This is because ...."

Now that you understand why the initial value problem describes temperature change, solve the mathematical problem analytically. One approach is to change variables to make the ambient law of cooling mathematically the same as the canary law with zero ambient temperature. The physical meaning of U in the next problem is 'the normalized temperature' or temperature away from ambient.

Problem 8.3. One Solution of $T[0] = T_0$ & $\frac{dT}{dt} = k\,(T_a - T)$ ──────────▼

Let T_a, T_0 and k be constants. Let T and U be dependent variables of the independent variable t.

(a) Let $U = T - T_a$ or $T = U + T_a$ and show that the following are equivalent initial value problems:

$$T[0] = T_0 \qquad\qquad \Leftrightarrow \qquad\qquad U[0] = T_0 - T_a$$

$$\frac{dT}{dt} = k(T_a - T) \qquad\qquad\qquad \frac{dU}{dt} = -k\,U$$

(b) Show that the solution of

$$U[0] = U_0$$

$$\frac{dU}{dt} = -k\,U$$

$$\text{is } U[t] = U_0\,e^{-k\,t}$$

(c) Let $T[t] = U[t] + T_a$, where $U[t]$ is the solution to the previous part of this problem and $T_0 = U_0 + T_a$. Show that $T[t]$ satisfies

$$T[0] = T_0$$

$$\frac{dT}{dt} = k\,(T_a - T)$$

$$\text{and may be written } T[t] = T_a + (T_0 - T_a)\,e^{-k\,t}$$

(d) Prove that $T[0] = T_0$ and $\lim_{t\to\infty} T[t] = T_a$. How could you see the limit directly from the differential equation?

──▲

Another approach to the analytical solution is to just "plug in."

Problem 8.4. The 'Method' of Unknown Constants ─────────────▼

Let a, b, and c be constants.

(a) Substitute the function $y = a + b\,e^{c\,x}$ into the differential equation

$$\frac{dy}{dx} = c\,(y - a)$$

and show that it is a solution for any value of the constant b.

(b) Substitute the function $y = a + b\,e^{c\,x}$ into the condition

$$y[0] = 5$$

and show that you must have $b = 5 - a$.

(c) Substitute the function $y = a + b\,e^{c\,x}$ into the initial value problem

$$y[0] = Y_0$$

$$\frac{dy}{dx} = c\,(y - a)$$

and express b in terms of the constants a, c and Y_0.

──▲

The story of the fallen tourist in Exercise 4.2 led to the differential equation

$$\frac{dD}{dt} = k\,D$$

for some positive constant k, where D is the distance an object has fallen and t is time. This differential equation says, "The farther you fall, the faster you go." in a specific way. In a general sense this is true, but it cannot be specifically a linear function of D. We called this Galileo's first conjecture because it gives rise to the Bugs Bunny Law of Gravity: If you don't look down, you don't fall. We want you to show why.

Problem 8.5. BUGS BUNNY'S LAW OF GRAVITY ⎯⎯⎯⎯⎯⎯⎯⎯⎯⎯⎯⎯⎯⎯⎯▼
*How does Bugs' Law say, "The farther you go, the faster you fall?" Prove that an object released from $D = 0$ at $t = 0$ does not fall under the Bugs Bunny Law of Gravity: $\frac{dD}{dt} = k\,D$, $k > 0$ constant. (See the program **ExpEquns**.)*

⎯⎯⎯⎯⎯⎯⎯⎯⎯⎯⎯⎯⎯⎯⎯⎯⎯⎯⎯⎯⎯⎯⎯⎯⎯⎯⎯⎯⎯⎯⎯⎯⎯▲

In Problem 21.8, we will look at Wiley Coyote's Law of Gravity: $\frac{dD}{dt} = k\sqrt{D}$, $k > 0$ constant. This law gives a correct prediction to the position of an object falling under gravity, but has the strange property that, "you don't fall *until* you look down." (It is not a well-posed physical law because of mathematical non-uniqueness.)

Galileo's simple law $\frac{d^2 D}{dt^2} = g$, g a constant, is studied in Exercise 10.2.1. It is actually simpler than either Bugs' Law or Wiley's. And it's correct, too, in vacuum. The Bungee Diver Project extends this to falling in air by using Newton's extension of Galileo's law.

8.3   Growth of Log, Exp, and Powers

When two functions both tend to infinity as x tends to infinity, one may grow "faster." This section shows how to measure of their "order of infinity" as the limit of their ratio.

Our main aim in this section is to see that logs tend to infinity "much slower" than powers, whereas exponentials tend to infinity "much faster" than powers. First, let us think a little about the "eventual growth" of some familiar high school functions.

Example 8.1. $y = x^2$ *vs.* $y = x^2 + 200x + 3000$

At a modest scale, the function $f[x] = x^2 + 200x + 3000$ is bigger than $g[x] = x^2$. For example, see Figure 8.4.

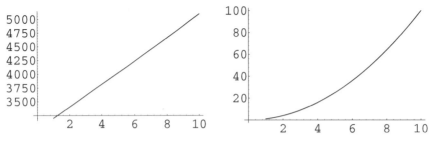

Figure 8.4: $y = x^2 + 200x + 3000$ vs. $y = x^2$ on $1 < x < 10$

At a larger scale, these functions look alike, as in Figure 8.5

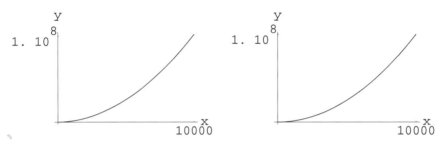

Figure 8.5: $y = x^2 + 200x + 3000$ vs. $y = x^2$ on $1 < x < 10000$

In terms of the "eventual" size of the two functions, we can compare by taking the limit of the ratio

$$
\begin{aligned}
\lim_{x \to \infty} \frac{f[x]}{g[x]} &= \lim_{x \to \infty} \frac{x^2 + 200x + 3000}{x^2} \\
&= \lim_{x \to \infty} \frac{x^2/x^2 + 200x/x^2 + 3000/x^2}{x^2/x^2} \\
&= \lim_{x \to \infty} \frac{1 + 200/x + 3000/x^2}{1} \\
&= \lim_{x \to \infty} 1 + \lim_{x \to \infty} \frac{200}{x} + \lim_{x \to \infty} \frac{3000}{x^2} \\
&= 1 + 0 + 0
\end{aligned}
$$

Figure 8.6 shows the graph of the ratio.

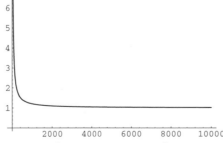

Figure 8.6: $y = (x^2 + 200x + 3000)/x^2$ on $1 < x < 10000$

Definition 8.4. *Order of Infinity*
When two functions tend to infinity, $f[x] \to \infty$ and $g[x] \to \infty$, but the ratio tends to a non-zero amount, $\lim_{x\to\infty} \frac{f[x]}{g[x]} = a$ for $0 < a < \infty$, we say both functions grow at the "same order of infinity."

In general if $f[x] = a\, x^p +$ "lower power terms" the growth at infinity is the same as $g[x] = x^p$. Precisely, the limit of the ratio $f[x]/g[x]$ is just the number a. Different powers, however, grow at different rates.

Example 8.2. $y = 500x$ *vs.* $y = x^2/500$

The linear function $f[x] = 500x$ is much larger than $g[x] = x^2/500$ at small scale,

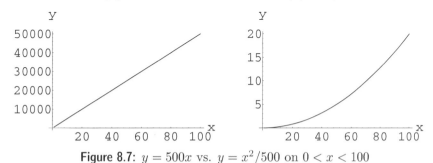

Figure 8.7: $y = 500x$ vs. $y = x^2/500$ on $0 < x < 100$

Figure 8.8 shows that, at a large scale, the quadratic function grows faster.

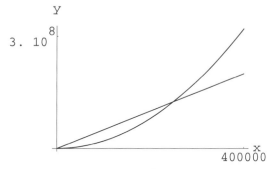

Figure 8.8: $y = 500x$ vs. $y = x^2/500$ on $0 < x < 400000$

The "eventual" size of the two functions can be compared by taking the limit of the ratio

$$
\begin{aligned}
\lim_{x\to\infty} \frac{f[x]}{g[x]} &= \lim_{x\to\infty} \frac{500x}{x^2/500} \\
&= \lim_{x\to\infty} \frac{250000x}{x^2} \\
&= 250000 \lim_{x\to\infty} \frac{1}{x} \\
&= 0
\end{aligned}
$$

In this case, the higher power is "winning" because the denominator is big, thereby forcing the ratio to be small. Figure 8.9 shows the graph of the ratio.

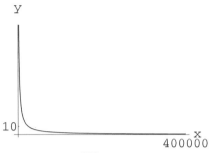

Figure 8.9: $y = \frac{500x}{x^2/500}$ on $0 < x < 400000$

The ratio the other way around gives

$$\lim_{x\to\infty} \frac{g[x]}{f[x]} = \lim_{x\to\infty} \frac{x^2/500}{500x}$$

$$= \lim_{x\to\infty} \frac{x^2}{250000x}$$

$$= \lim_{x\to\infty} \frac{x}{250000}$$

$$= \infty$$

In this example, the function $g[x] = x^2/500$ has a higher order of infinity than $f[x] = 500x$ because "eventually" it is larger in the strong sense that the ratio even tends to infinity. It is eventually twice as big, three times as big, and so forth.

Exercise 8.3.2 shows that higher powers have higher orders of infinity. In particular, roots, $x^{1/n}$ have a lower order of infinity than integer powers x^n, as in Figure 8.10.

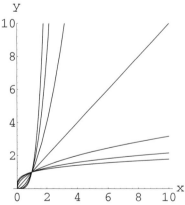

Figure 8.10: $y = x^n$ and $y = x^{1/n}$, $n = 1, 2, 3, 4$

Logs have an order of infinity below *all* roots, while exponentials have an order above every power. Specifically, two important facts about the rate of growth of the natural log and exponential functions are as follows:

Theorem 8.5. *Orders of Infinity*
    Let p be any positive real number. Then

$$\lim_{t\to\infty}\frac{e^t}{t^p}=\infty \qquad and \qquad \lim_{t\to\infty}\frac{t^p}{\text{Log}[t]}=\infty$$

or, equivalently,

$$\lim_{t\to\infty}\frac{t^p}{e^t}=0 \qquad and \qquad \lim_{t\to\infty}\frac{\text{Log}[t]}{t^p}=0$$

The exponential beats any power, and any power beats the logarithm to infinity.

PROOF BASED ON HIGH SCHOOL MATH:

First we will consider the ratio $\frac{t^p}{e^t}$ for integer values of t, $\frac{n^p}{e^n}$, for $n = 1, 2, \cdots$. It is sufficient to show that the pth root of the ratio tends to zero,

$$\frac{n}{e^{n/p}} \to 0$$

by continuity of positive powers at zero. Let $b = e^{1/p}$, and notice that $b > 1$ and so $b^{1/2} > 1$ as well. Write

$$b^{1/2} = 1 + a$$

for $a > 0$.

The binomial theorem says

$$b^{n/2} = (1 + a)^n = 1 + na + \cdots + a^n > na$$

since $a > 0$, so that $e^{n/p} = b^n > n^2 a^2$. Hence, we have

$$\frac{n}{e^{n/p}} < \frac{n}{n^2 a^2} = \frac{1}{na^2} \to 0 \qquad as \quad n \to \infty$$

For continuous values of $t > 0$, there is always an integer satisfying $n - 1 < t \le n$. For these values, we have

$$\frac{t^p}{e^t} < \frac{n^p}{e^{n-1}} = e\frac{n^p}{e^n} \to 0$$

Example 8.3. $2^\infty/\infty^3$

There are many other "orders of infinity" or "orders of infinitesimal" that we can deduce from the basic result given in the preceding theorem. For example, what is

$$\lim_{x\to\infty}\frac{2^x}{x^3}=?$$

First, we can write $2 = e^k$, where $k = \text{Log}[2]$. (Simply take logs of both sides of $2 = e^k$.) Our limit becomes

$$\lim_{x\to\infty}\frac{e^{k\,x}}{x^3}=?$$

for a positive k. ($\text{Log}[2] = k \approx 0.693147806$.)

We consider the kth root of our limit to obtain

$$\lim_{x \to \infty} \sqrt[k]{\frac{e^{k\,x}}{x^3}} = \lim_{x \to \infty} \left(\frac{e^{k\,x}}{x^3}\right)^{1/k}$$

$$= \lim_{x \to \infty} \frac{e^{k\,x/k}}{x^{3/k}}$$

$$= \lim_{x \to \infty} \frac{e^x}{x^p} = \infty$$

with $p = 3/k$, by the theorem. Since positive roots only tend to infinity when the quantity tends to infinity, we have

$$\lim_{x \to \infty} \frac{2^x}{x^3} = \infty$$

Example 8.4. $2^{-\infty}/\infty^3$ *or* $\lim_{x \to -\infty} \frac{2^x}{x^3} = 0$

Consider the limit

$$\lim_{x \to -\infty} \frac{2^x}{x^3} = ?$$

First, change to natural base.

$$\lim_{x \to -\infty} \frac{e^{k\,x}}{x^3} = ?$$

with $k = \text{Log}[2] > 0$. Next, replace x by $u = -x$,

$$\lim_{u \to +\infty} \frac{e^{-k\,u}}{(-u)^3} = -\lim_{u \to +\infty} \frac{e^{-k\,u}}{u^3}$$

Now, notice that the limit is obvious by some arithmetic,

$$-\lim_{u \to +\infty} \frac{e^{-k\,u}}{u^3} = -\lim_{u \to +\infty} \frac{1}{u^3\,e^{k\,u}} = 0$$

because both terms in the denominator tend to infinity. (We can also see this limit by "plugging in" the limiting values.)

Example 8.5. $-\infty^3 \cdot 2^{-\infty}$

The limit

$$\lim_{x \to -\infty} x^3\,2^x = ?$$

has one term growing and the other shrinking ($2^{\text{large negative number}}$). Replace $x = -u$ and $2 = e^k$, for $k = \text{Log}[2] > 0$, so we have

$$\lim_{u \to +\infty} (-u)^3\,2^{-u} = -\lim_{u \to +\infty} u^3\,2^{-u}$$

$$= -\lim_{u \to +\infty} \frac{u^3}{2^u}$$

$$= -\lim_{u \to +\infty} \frac{u^3}{(e^k)^u} = -\lim_{u \to +\infty} \frac{u^3}{e^{k\,u}}$$

$$= 0$$

since exponentials beat powers to infinity.

Similar tricks reduce various limits to the logarithmic comparison of the Orders of Infinity Theorem.

Example 8.6. $\lim_{x \downarrow 0} x^x = 1$

We may write $x = e^{\text{Log}[x]}$, so $x^x = e^{x \, \text{Log}[x]}$ and

$$\lim_{x \downarrow 0} x^x = \lim_{x \downarrow 0} e^{x \, \text{Log}[x]} = e^{\lim_{x \downarrow 0} x \, \text{Log}[x]}$$

if the last limit exists. This is a fundamental limit which we compute next.

Example 8.7. $\lim_{x \downarrow 0} x \, \text{Log}[x] = 0$

Replace $x = 1/u$ in the limit, so $u \to +\infty$ as $x \downarrow 0$, and the limit

$$\lim_{x \downarrow 0} x \, \text{Log}[x]$$

becomes

$$\lim_{u \to \infty} \frac{1}{u} \text{Log}[\frac{1}{u}] = \lim_{u \to \infty} \frac{1}{u} \text{Log}[u^{-1}]$$

$$= -\lim_{u \to \infty} \frac{\text{Log}[u]}{u}$$

$$= 0$$

since powers beat logs to infinity.

Now, apply this result to the x^x limit,

$$\lim_{x \downarrow 0} x^x = e^{\lim_{x \downarrow 0} x \, \text{Log}[x]} = e^0 = 1$$

This limit is a reason to write $0^0 = 1$.

Exercise set 8.3

Use the program **Infinities** to explore orders of infinity and compute limits by computer.

1. Ratios of Powers

   Show that each of the functions $g[x]$ tends to infinity faster than the corresponding $f[x]$:

a) $g[x] = x^5$ and $f[x] = x^3$ b) $g[x] = x^2$ and $f[x] = \sqrt{x}$

c) $g[x] = \sqrt{x}$ and $f[x] = \sqrt[3]{x}$ d) $g[x] = \frac{x^5}{10}$ and $f[x] = 10x^3$

e) $g[x] = \frac{x^2}{2}$ and $f[x] = 2\sqrt{x}$ f) $g[x] = \frac{\sqrt{x}}{2}$ and $f[x] = 3\sqrt[3]{x}$

2. Order of Powers

   Let $f[x] = a_1 x^{p_1} + b_1 x^{q_1} + c_1 x^{r_1}$ and $g[x] == a_2 x^{p_2} + b_2 x^{q_2} + c_2 x^{r_2}$ for positive constants a_i, b_i, p_i, q_i, r_i. Show that $f[x]$ has a higher order of infinity than $g[x]$ when the highest *power of $f[x]$ is greater than the highest power of $g[x]$. What is the limit of the ratio $f[x]/g[x]$? What is the limit if the highest powers agree?*

3. Drill Limits

Compute the limits $(p, k > 0)$

a) $\lim_{x \to 0} x^p \operatorname{Log}[x]$

b) $\lim_{x \to \infty} x^{1/x}$

c) $\lim_{\Delta x \downarrow 0} \Delta x^{1/\operatorname{Log}[\Delta x]}$

d) $\lim_{x \to \infty} x^p \operatorname{Log}[x]$

e) $\lim_{\delta \to 0} \frac{1 + k\,\delta}{\delta}$

f) $\lim_{\delta \downarrow 0} \delta^{\frac{\delta x + k}{Log[\delta]}}$

4. *Suppose $b > a > 0$. Find $\lim_{x \to \infty} \frac{b^x}{a^x}$*

5. Negative? Exponentials

If the base of an exponential function $f[x] = b^x$ satisfies $0 < b < 1$, we think of this as a "negative" exponential in the following sense:

    (a) *Solve for a constant k so that $b^x = e^{k\,x}$ for all x.*

    (b) *Show that k from part (a) is negative when $0 < b < 1$.*

    (c) *Show that $\lim_{x \to \infty} b^x = 0$ when $0 < b < 1$.*

6. *Use logs to prove the other half of the Orders of Infinity theorem above. If we wish to show that $\lim_{x \to \infty} \dfrac{x^p}{\operatorname{Log}[x]} = \infty$, then change variables by letting $u = \operatorname{Log}[x]$, so $x = e^u$. We show show instead that*

$$\lim_{u \to \infty} \frac{(e^u)^p}{u} = \infty$$

by taking p^{th} roots. Why does this establish the result?

7. All Exponentials Beat All Powers

Suppose $b > 1$ and $p > 0$. What are the values of the limits

$$\lim_{x \to \infty} \frac{b^x}{x^p} = ? \qquad and \qquad \lim_{x \to \infty} \frac{x^p}{b^x} = ?$$

8.4   Official Properties - CD

The important functional identity $e^x\, e^y = e^{x+y}$ follows from the differential equation defining the exponential function.

Remainder of this section only on CD

8.5   Projects

8.5.1   Numerical Computation of $\frac{da^t}{dt}$

There are two small Projects that will help you understand the natural base. The first has you use the computer to directly compute the derivative of b^t. You will see that this leads to

$$\frac{db^t}{dt} \propto b^t$$

but it does not give an obvious value to the constant of proportionality. However, you can experiment with your computations and find the value of b that makes this constant one. This is $b = e \approx 2.71828 \cdots$.

8.5.2   The Canary Resurrected - Cooling Data

The second mathematical project in that chapter asks you to compare some actual cooling data with the prediction of Newton's law of cooling. It is an interesting scientific project for you to measure this yourself and we believe that you can do better than the first student data we present if you wish to try. (We are not sure if the students warmed the cup before they started measurements. You will see how this shows up in their data.)

CHAPTER 9

Graphs and the Derivative

The computer draws beautiful graphs and does not force us to think very much, so why use calculus to draw graphs?

The translation between formulas and graphs and the interpretation of graphs themselves are important parts of this course. Graphs often reveal mathematical results simply and clearly, but graphs do this with "trends" or "shapes," not just points. Calculus with the computer will give us information that includes both points and trends.

In real applications, the scale of a plot can be far from obvious. Hot objects radiate; you have heard of "red hot." Planck discovered that the intensity of radiation at frequency ω of a black body at absolute temperature T is

$$I = \frac{\hbar\omega^3}{\pi^2 c^2 (e^{\hbar\omega/(kT)} - 1)}$$

where $\hbar \approx 6.6255 \times 10^{-27}$ (erg sec) is Planck's constant, $c \approx 2.9979 \times 10^{10}$ (cm/sec) is the speed of light, and $k \approx 1.3805 \times 10^{-16}$ (erg/deg) is Boltzman's constant. The frequency ω where this function peaks, is the "color" of the radiation we observe at temperature T. This peak predicts the empirically observed law of radiation discovered earlier by Wein.

If we want to find this peak by graphing, we have a big difficulty. What scale should we use? The constants c, k, and $\hbar$ cover 37 orders of magnitude, whereas a single graph can scarcely show more than two. This is a complicated formula with messy constants; but aside from the technical difficulty, calculus can help, even in just finding the interesting scale of the graph. This is taken up in the project on Planck's Formula.

Calculus helps you set the scales, which usually are not so obvious in real applications. Calculus finds formulas for geometric features of interest. Calculus finds the qualitatively interesting range to plot and once this is found, the computer can make a quantitatively accurate picture. Again, our goal is for you to form a nonlinear combination

Knowledge[calculus + computing] >

Knowledge[calculus] + Knowledge[computing]

The scale of a graph can alter our perception of the behavior of a function. We begin the chapter with a look at graphing without knowledge of shape.

145

9.1   Graphs from Formulas

Two simple approaches to graphing are plotting points and using computer packages. The exercises at the end of this section show some limitations of these methods.

Plotting points alone is usually a bad way to sketch graphs because that information alone requires many points to construct a shape and a leap of faith that we have connected the points correctly. If we have only numerical data, that is all we can do. Later in this chapter, we will learn to use calculus to tell us shape information, such as where the graph is increasing or decreasing, so that only a few points are required to give qualitatively accurate graphs. This first section is about what goes wrong without this information.

Even with the computer, which will plot 1,000 points if you ask it to, we often can use calculus to decide what range of values contains the important information. Poor choice of scales can come up in innocent or simple-minded ways or as a result of large differences in the size of scientific constants as in the Planck's Formula project.

Example 9.1.   *Simple-Minded Scales*

Which of the following is the graph of

$$y = x^5 + 4x^4 + x$$

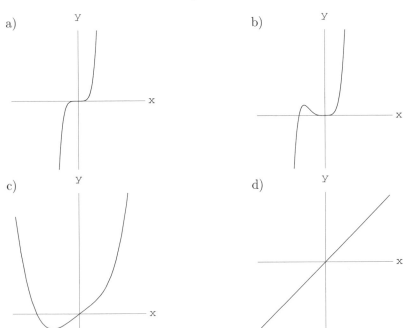

Figure 9.1: $y = x^5 + 4x^4 + x$ on Four different scales

The answer is all of these are graphs of this same polynomial. Graph (a) is for $-100 < x < 100$. Graph (b) is for $-10 < x < 10$. Graph (c) is for $-1 < x < 1$. Graph (d) is for $-0.1 < x < 0.1$. They appear different because the wiggle in the medium-scale graph is an insignificant part of the term x^5 when $x = 100$. The small unit scale misses some of the medium-size 10 wiggle. Why is the tiny scale straight? (What is the view of the graph in a microscope?) These shape differences are obvious if you compute some sizes but could surprise you if you are using the computer and blindly hoping for "good" graphs. Calculus can find the interesting shape information, and the computer can then draw it accurately. The four graphs above are accurate, but they are drawn on different scales. Try these yourself with the computer.

Exercise set 9.1

The next exercise shows you a simple reason why points alone are not enough.

1. Three Points Are Not Enough

   (a) *The cartesian pairs $(-2, -2), (0, 0), (2, 2)$ are recorded in the table below the blank graph in the next figure. Plot these points.*
   (b) *Show that the graph of $y = x$ contains all three points from part (1) and sketch the graph.*
   (c) *Show that the graph of $y = x^3 - 3x$ contains all three points from part (1). Can you sketch it without more points?*
   (d) *Find a point on the graph of $y = x$ that is not on the graph of $y = x^3 - 3x$ and plot it.*
   (e) *Use the graphing program in **aComputerIntro** to plot both functions on the same graph.*

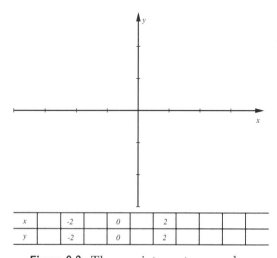

x		-2		0		2			
y		-2		0		2			

Figure 9.2: Three points on two graphs

2. *Find the equations of the lines shown in Figure 9.3.*

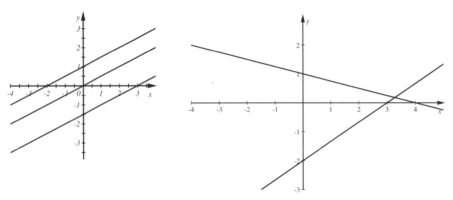

Figure 9.3: Three parallel lines and two intersecting lines

3. Scale of the Plot

Use the computer to make several plots on different scales. First, replot

$$y = x^5 + 4x^4 + x$$

at the four scales described in Example 9.1 above, but leave the default "Ticks" on so that the computer puts the scales on the plots. This will show you more clearly what we described in the previous paragraph.

Second, plot the functions below on the different suggested scales:

$y = x^x$	$0 < x < 2$	*and*	$0 < x < 5$
$y = 3x^4 - 4x^3 - 36x^2$	$-10 < x < 10$	*and*	$-3 < x < 5$

Explain why the pairs of graphs appear to be different even though they are of the same function.

In the next exercise, you are given four choices for the graph and asked which one is best. Identify a shape feature related to the derivative (or what you would see in a microscope), and then check the formula to see if it matches. For example, graph (a) is decreasing for large magnitude negative numbers, and graph (b) is increasing. The derivative, $\frac{dy}{dx} = 15(x^2 - 1)x^2$, is positive if the magnitude of x is large; the squares remove the dependence on sign. What does this say about the choice between (a) and (b)? What else can you eliminate?

4. *Which of the following is the graph of $y = 3x^5 - 5x^3$? Which shape features of the incorrect ones makes each one wrong?*

a)   *b)*

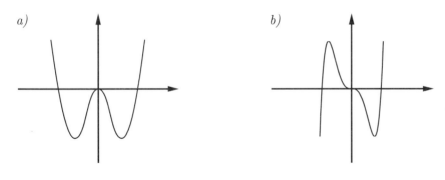

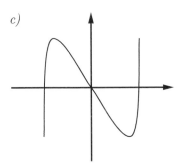

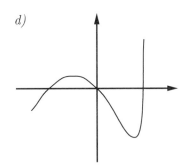

The previous exercise is a little artificial. It really means, "which graph has all the shape information of the algebraic curve." It also implicitly assumes that one of the figures is correct. The point is that calculus tells us the shape or all the "ups and downs" of the curve.

9.2   Graphs Without Formulas

Qualitative information certainly can exist without any formulas.

Graphs are primarily good for qualitative information rather than quantitative accuracy. Graphs readily show where quantities are increasing or decreasing but only give rough approximations to amounts, rates of increase, and so on.

Exercise set 9.2

1. *Make a qualitative rough sketch of a graph of the distance traveled as a function of time on the following hypothetical trip: You travel a total of 100 miles in 2 hours. Most of the trip is on rural interstate highway at the 65 mph speed limit. (What qualitative feature or shape does the graph of distance vs. time have when speed is 65 mph?) You start from your house at rest, gradually increase your speed to 25 mph, slow down, and stop at a stop sign. (What shape is the graph of distance vs. time while you are stopped?) You speed up again to 25 mph, travel a while and enter the interstate. At the end of the trip, you exit, slow to 25 mph, stop at a stop sign, and proceed to your final destination.*

9.3   Ups and Downs of the Derivative

Calculus lets us "look" in a powerful microscope at a graph before we have the whole graph. We must "look" by computing derivatives.

Of course, all we would see in a powerful microscope is the graph of a straight line $dy = m\,dx$ in the microscope (dx, dy)-coordinates (where $m = f'[x]$ with x fixed). This can only be one of the three qualitative shapes shown in Figure 9.4.

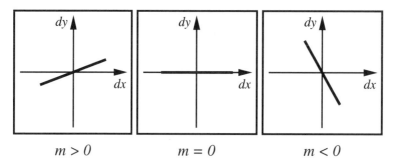

$$m > 0 \qquad\qquad m = 0 \qquad\qquad m < 0$$

Figure 9.4: $dy = m\,dx$ for $m > 0$, $m = 0$, $m < 0$

If $f'[x]$ is undefined, something else may appear in the microscope, and the rules of calculus do not let us "see" in this case. The exact slope can be measured on an auxiliary scale (if necessary), as in Figure 9.5.

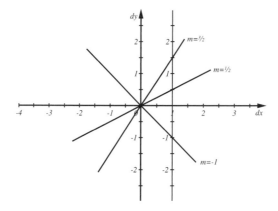

Figure 9.5: A Slope Scale

Our approach to graphing will be to fill out a table that looks like the blank graphing table shown in Figure 9.6.

Procedure 9.1. Graphing $y = f[x]$ with the First Derivative
  (a) Compute $f'[x]$ and find **all** values of x where $f'[x] = 0$ (or $f'[x]$ does not exist). Record these in the microscope row as horizontal line segments (or $*'s$ if the derivative does not exist).
  (b) Check the sign (+) or (-) of $f'[x]$ at values between each of the points from the first part. Record $f'[x] = (+) > 0$ as an upward sloping microscope line and $f'[x] = (-) < 0$ as a downward sloping microscope line.
  (c) Compute a few key (x, y) pairs (using $f[x]$ to find values of y, not $f'[x]$) and record the numbers on the x and y rows. For example, you should at least compute the (x, y) pairs for the x values used in step 1.
  (d) Plot the points and mark small tangent lines.
  (e) Connect the points with a curve matching the tangents as you pass through the points and increasing or decreasing according to the table between the horizontal points.

Do not start by plotting points. Start with the "shape" information of the microscope row so you first find out which points are interesting to plot.

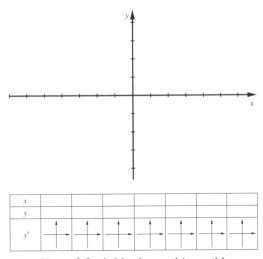

Figure 9.6: A blank graphing table

If you are plotting with the computer, you can just work steps (a) and (b). This will tell you the range of x values to plot.

Example 9.2. *Graphing $y = f[x] = 3x^5 - 5x^3$ from Scratch*

First, compute $f'[x]$,

$$y' = 3 \cdot 5 \cdot x^4 - 5 \cdot 3 \cdot x^2 = 15x^4 - 15x^2$$
$$= 15x^2(x^2 - 1)$$

This derivative is defined for all real values of x.
  Second, find all places, x, where $y' = 0$,

$$0 = 15x^2(x^2 - 1) \qquad \Leftrightarrow \qquad x = 0 \quad \text{or} \quad x = +1 \quad \text{or} \quad x = -1$$

The derivative is always defined and is zero only at these three points - nowhere else.

Third, we check the sign of y' for x-values between the three places where the slope is zero, $x = -2$, $x = -1/2$, $x = 1/2$, and $x = 2$. Notice that we are computing $f'[x]$, not $f[x]$.

$x = -2,$ $\quad y' = 15 \cdot (-2)^2 \cdot [(-2)^2 - 1] = 15 \cdot 4 \cdot [4 - 1] = (+)$ $\quad$ slope up

$x = -\dfrac{1}{2},$ $\quad y' = 15 \cdot (-\dfrac{1}{2})^2 \cdot [(-\dfrac{1}{2})^2 - 1] = 15 \cdot \dfrac{1}{4} \cdot [\dfrac{1}{4} - 1] = (-)$ $\quad$ slope down

$x = \dfrac{1}{2},$ $\quad y' = 15 \cdot (\dfrac{1}{2})^2 \cdot [(\dfrac{1}{2})^2 - 1] = 15 \cdot \dfrac{1}{4} \cdot [\dfrac{1}{4} - 1] = (-)$ $\quad$ slope down

$x = 2,$ $\quad y' = 15 \cdot (2)^2 \cdot [(2)^2 - 1] = 15 \cdot 4 \cdot [4 - 1] = (+)$ $\quad$ slope up

Fourth, we compute the (x, y)-coordinates of several important points. The points with zero slope are important, and we can get a reasonable idea of the shape of the curve with only these three values. Notice that now we are using the formula $y = f[x]$ and not $f'[x]$.

$x = -1,$ $\quad y = 3 \cdot (-1)^5 - 5 \cdot (-1)^3 = -3 + 5 = 2$ $\quad (-1, 2)$

$x = 0,$ $\quad y = 3 \cdot (0)^5 - 5 \cdot (0)^3 = 0$ $\quad (0, 0)$

$x = 1,$ $\quad y = 3 \cdot (1)^5 - 5 \cdot (1)^3 = 3 - 5 = -2$ $\quad (1, -2)$

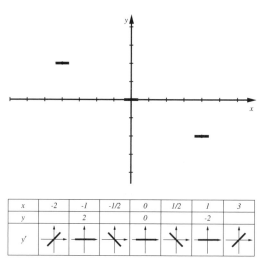

x	-2	-1	-1/2	0	1/2	1	3
y		2		0		-2	
y'							

Figure 9.7: Slope information for $y = 3\,x^5 - 5\,x^3$

Mathematica's version of the graph is shown in Figure 9.8.

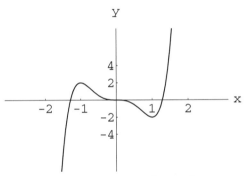

Figure 9.8: $y = 3\,x^5 - 5\,x^3$

Example 9.3. *Graphing $y = x\,e^{-x}$*

First, use the Product Rule and Chain Rule to compute the derivative

$$y = f[x]\,g[x] \qquad\qquad f[x] = x \qquad\qquad \frac{df}{dx} = 1$$

$$g[x] = e^{-x} \qquad g = e^u \qquad\qquad u = -x$$
$$\frac{dg}{du} = e^u \qquad\qquad \frac{du}{dx} = -1$$
$$\frac{dg}{dx} = \frac{dg}{du}\frac{du}{dx}$$
$$= (e^u)(-1) = -e^{-x}$$

So, the final derivative is

$$\frac{dy}{dx} = \frac{df}{dx}\,g + f\,\frac{dg}{dx} = 1\,e^{-x} - x\,e^{-x} = (1-x)\,e^{-x}$$

This derivative is defined for all real x.

We know that $e^u \neq 0$ for any u, so $\frac{dy}{dx} = 0$ only if $1 - x = 0$ or $x = 1$.

Second, we check signs at x values between $-\infty$ and 1 and between 1 and $+\infty$, $x = -1$ satisfies $-\infty < -1 < 1$ and $x = 3$ satisfies $1 < 3 < +\infty$.

$$y'(-1) = (1+1)\,e^{+1} = 2\cdot e \approx 5.4$$
$$y'(+3) = (1-3)\,e^{-3} = -2\,e^{-3} \approx -0.10$$

The values of the y-coordinate at these points are

$$y(-1) = (-1)\,e^{+1} = -e \approx -2.7$$
$$y(+1) = 1\,e^{-1} = 1/e \approx 0.368$$
$$y(+3) = 3\,e^{-3} = 3/e^3 \approx 0.15$$

We also know that

$$\lim_{x\to -\infty} x\,e^{-x} = -\infty \quad \& \quad \lim_{x\to \infty} \frac{x}{e^x} = 0$$

and that $y = x\,e^{-x} > 0$ for $x > 0$. (See Theorem 8.5.)

All this graphing information is recorded in the table of Figure 9.9.

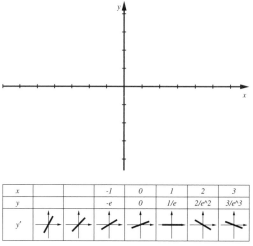

x			-1	0	1	2	3
y			-e	0	1/e	2/e^2	3/e^3
y′							

Figure 9.9: Slope information for $y = x\,e^{-x}$

Exercise set 9.3

The next exercise has a row of microscopic views filled out. Just looking across the y' row we see that the graph goes up - over - down - over -down - over - up. Plot the points given in the x and y row. Add little tangent segments given in the y' row at the (x, y) points. Then fill in a curve that goes through the points and is tangent to the segments.

1. *The table below the axes in Figure* 9.10 *contains* x *and* y *coordinates of the point as well as microscopic views of* $y = f[x]$ *at those dotted points. Sketch the graph.*

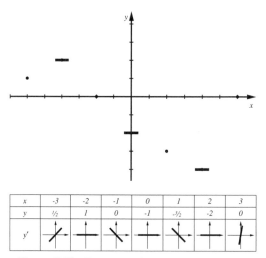

x	-3	-2	-1	0	1	2	3
y	½	1	0	-1	-3/2	-2	0
y′							

Figure 9.10: Points and microscopic views

Reverse the procedure of the last exercise. Look at the next graph, fill out the x and y numbers and make microscopic views of the graph at these points.

2. *Fill out the table in Figure 9.11 to correspond to the given graph.*

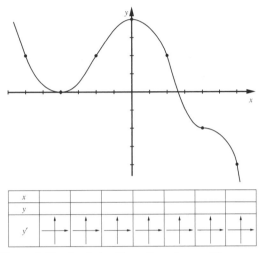

Figure 9.11: *Look in Your Imaginary Microscope*

3. Graphing with Slopes Drill
   Use the above first derivative procedure to sketch the graphs of
   a) $y = f[x] = 2x^2 - x^4$ b) $y = f[x] = x^3 - 3x$
   c) $y = f[x] = 6x^5 - 10x^3$ d) $y = (x + 2)^{\frac{2}{3}}$
   e) $y = \mathrm{Sin}[x] + \mathrm{Cos}[x]$ f) $y = \mathrm{Sin}[x] \times \mathrm{Cos}[x]$
   g) $y = x\, e^x$ h) $y = e^{-x^2}$
   i) $y = x\, \mathrm{Log}[x]$ j) $y = x - \mathrm{Log}[x]$

You may check your graphs with the computer after you sketch by hand. Later, you will need the skills you develop in this exercise together with the computer in order to understand complicated graphs like Planck's Formula in the projects.

9.3.1   The Theorems of Bolzano and Darboux

How do we know that it is sufficient to just check one point between the zeros of $f'[x]$ in the graphing procedure of the last section? This is because derivatives have the property that they cannot change sign without being zero, provided that they are defined on an interval. If $f'[x]$ is not zero in an interval $a < x < b$, then $f'[x]$ cannot change sign. This is taken up in the Mathematical Background Chapter on Bolzano's Theorem, Darboux' Theorem, and the Mean Value Theorem.

9.4   Bending & the Second Derivative

If the slope gets steeper and steeper, the curve bends up. The derivative $f'[x]$ is just a function and its derivative must be positive if it is increasing, $f''[x] = (+) > 0$.

A negative second derivative make the first derivative decrease. If the slope decreases, the curve bends down.

These two facts can be summed up in the diagram of Figure 9.12.

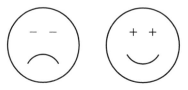

Figure 9.12: *$f''[x]$ Negative on frown - positive on smile*

This section shows you how to use this diagram.

Which is a better graph of graph of $y = 3x^5 - 5x^3$?

a)   b)

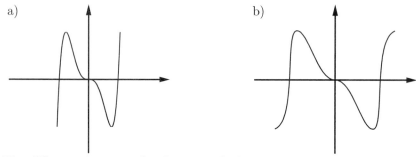

The difference between the choices is the bending, not the slope. This information can be obtained by asking whether the first derivative is increasing or decreasing.

Example 9.4.   *The Bends of $y = 3\,x^5 - 5\,x^3$*

We computed the slope information of $y = 3\,x^5 - 5\,x^3$ in Example 9.2 above. The derivatives are

$$y = 3\,x^5 - 5\,x^3$$
$$y' = 3 \cdot 5\,x^4 - 5 \cdot 3\,x^2 = 15\,(x^4 - x^2)$$
$$y'' = 15\,(4\,x^3 - 2\,x) = 30\,x\,(2\,x^2 - 1)$$

We already found the slope table and (x, y)-points at the places where the slope is zero. Now we want to find out where the curve bends up (or looks like part of a smile) and where it bends down.

$$y'' = 0 \quad \Leftrightarrow \quad 0 = 30\,x\,(2\,x^2 - 1) \quad \Leftrightarrow \quad x = 0 \quad \text{or} \quad 2\,x^2 - 1 = 0$$

$$\Leftrightarrow \quad x = 0 \quad \text{or} \quad x = \frac{1}{\sqrt{2}} \approx 0.707 \quad \text{or} \quad x = -\frac{1}{\sqrt{2}} \approx -0.707$$

The second derivative is always defined and only is zero at these three points.

Now we check the signs of the second derivative at values between the zeros.

$$x = -1, \qquad y'' = 30 \cdot (-1) \cdot [2(-1)^2 - 1] = -30 \cdot [4 - 1] = (-) \qquad \text{frown}$$

$$x = -\frac{1}{2}, \qquad y'' = 30 \cdot (-\tfrac{1}{2}) \cdot [2(-\tfrac{1}{2})^2 - 1] = -15 \cdot [\tfrac{1}{2} - 1] = (+) \qquad \text{smile}$$

$$x = \frac{1}{2}, \qquad y'' = 30 \cdot (\tfrac{1}{2}) \cdot [2(\tfrac{1}{2})^2 - 1] = 15 \cdot [\tfrac{1}{2} - 1] = (-) \qquad \text{frown}$$

$$x = 1, \qquad y'' = 30 \cdot (1) \cdot [2(1)^2 - 1] = 30 \cdot [4 - 1] = (+) \qquad \text{smile}$$

So we see that graph (a) above is a better representation of the curve $y = 3\,x^5 - 5\,x^3$, because the extra bends on graph (b) have the bending sequence smile - frown - smile - frown - smile - frown. We could eliminate graph (b) because the second derivative at a large positive number would need to be negative in order to have the right-most downward (frown) bend shown on graph (b). Similarly, the left-most upward bend would require that the second derivative is positive. Graph (a) above has the (-) - (+) - (-) - (+) or frown - smile - frown - smile sequence of signs to its second derivative.

Sketching curves with both the slope and bend information amounts to filling out the table of Figure 9.13 according to the following procedure.

Procedure 9.2.   Plotting with the First and Second Derivatives
  (a) Compute $f'[x]$ and find **all** values of x where $f'[x] = 0$ (or $f'[x]$ does not exist). Record these in the microscope row as horizontal line segments (or $*$s if the derivative does not exist).
  (b) Check the sign (+) or (-) of $f'[x]$ at values between each of the points from the first part. Record $f'[x] = (+) > 0$ as an upward sloping microscope line and $f'[x] = (-) < 0$ as a downward sloping microscope line.
  (c) Compute $f''[x]$ and find **all** values of x where $f''[x] = 0$ (or $f''[x]$ does not exist). Record these in the y'' row of the table as 0s (or $*$s if the derivative does not exist).
  (d) Check the sign (+) or (-) of $f''[x]$ at values between each of the points from the first part. Record $f''[x] = (+) > 0$ as a smile and $f''[x] = (-) < 0$ as a frown.
  (e) Compute a few key (x, y) pairs (using $f[x]$ to find values of y, not $f'[x]$ or $f''[x]$) and record the numbers on the x and y rows. For example, you should at least compute the (x, y) pairs for the x values used in steps (a) and (c).
  (f) Plot the points and mark small tangent lines.
  (g) Connect the points with a curve matching the tangents as you pass through the points and increasing or decreasing and bending according to the table.

Do not start by plotting points. Start with the "shape" information of the slope row and then do the bend row so you first find out which points are interesting to plot.

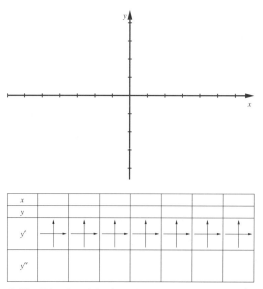

Figure 9.13: Blank table for plotting with x, y, y' and y''

Example 9.5. $y = x^3 + x$

The graph of $y = x^3 + x$ illustrates the additional information of the second derivative. We have $\frac{dy}{dx} = 3x^2 + 1$ which is always positive. The first derivative slope information just says the graph is increasing. However, $\frac{d^2y}{dx^2} = 6x$ which is zero at $x = 0$ and only there. When $x < 0$, $\frac{d^2y}{dx^2} < 0$, so the graph bends downward, but slopes upward. The left half of a frown slopes up, but bends down. When $x > 0$, $\frac{d^2y}{dx^2} > 0$, so the graph bends upward and slopes upward. The right half of the smile slopes up, and bends up. Fill out the slope and bend shape tables for this graph which is given next.

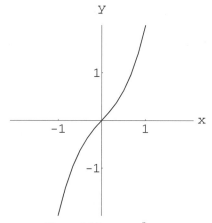

Figure 9.14: $y = x^3 + x$

Example 9.6. *Graphing $y = x\,e^{-x}$ - (con't).*

We now add the second derivative information to the sketch of the graph of Example 9.3. We know from that example that $\frac{dy}{dx} = (1 - x)\, e^{-x}$, so the second derivative is

$$\frac{d^2 y}{dx^2} = (x - 2)\, e^{-x}$$

which is defined everywhere.

The second derivative is zero only if $x = 2$, since $e^u > 0$ for all u.

We check the values $x = 0$ with $-\infty < 0 < 2$ and $x = 3$ with $2 < 3 < +\infty$,

$$y''[0] = -2\, e^0 = -2 \qquad \text{frown}$$

$$y''(3) = (3 - 2)\, e^{-3} = 1/e^3 > 0 \qquad \text{smile}$$

The slope and bend information we have computed is recorded in Figure 9.15.

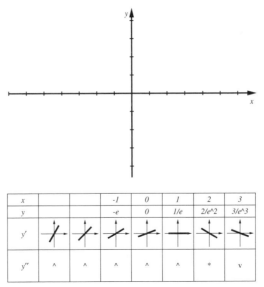

x			-1	0	1	2	3
y			-e	0	1/e	2/e^2	3/e^3
y'							
y''	^	^	^	^	^	*	v

Figure 9.15: Slope and bend information for $y = x\, e^{-x}$

The computer graph is in Figure 9.16.

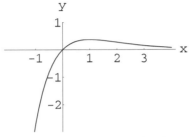

Figure 9.16: The computer graph of $y = x\, e^{-x}$

Example 9.7. *Graph* $y = e^{-x^4}$

Begin with the Chain Rule,

$$y = e^u \qquad\qquad\qquad u = -x^4$$

$$\frac{dy}{du} = e^u \qquad\qquad\qquad \frac{du}{dx} = -4x^3$$

so $\qquad\qquad \dfrac{dy}{dx} = \dfrac{dy}{du} \cdot \dfrac{du}{dx} = -4x^3\, e^{-x^4}$

We have $\frac{dy}{dx} = 0$ only if $x = 0$ and the derivative is defined everywhere. Check $\frac{dy}{dx}$ at $x = \pm1$ to see that the slope table is up - over - down.

Now use the Product Rule on $\frac{dy}{dx} = f[x]\,g[x]$,

$$f = -4x^3 \qquad\qquad\qquad g = e^{-x^4}$$

$$\frac{df}{dx} = -12x^2 \qquad\qquad\qquad \frac{dg}{dx} = -4x^3 e^{-x^4}$$

so

$$\frac{d^2y}{dx^2} = \frac{df}{dx}\,g + f\,\frac{dg}{dx} = -12x^2\, e^{-x^4} + 16x^6\, e^{-x^4}$$

$$\frac{d^2y}{dx^2} = \left(16x^6 - 12x^2\right) e^{-x^4}$$

The second derivative is defined for all x and only equals zero when $16x^6 - 12x^2 = 0$. This happens at $x = 0$ and $16x^4 = 12$ or $x = \pm\sqrt[4]{\frac{3}{4}} \approx \pm0.9306$. Checking values, we see that the bending table is smile - frown - frown - smile.

The limiting values as $x \to \pm\infty$ are simple since $-x^4 < 0$ tends to $-\infty$, $y \to 0$.

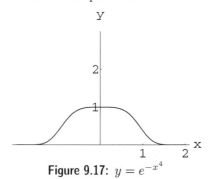

Figure 9.17: $y = e^{-x^4}$

Example 9.8.   *Graph* $y = x \operatorname{Log}[x]$

SOLUTION: The Product Rule makes the derivative

$$\frac{dy}{dx} = \operatorname{Log}[x] + x\frac{1}{x} = 1 + \operatorname{Log}[x]$$

Setting this equal to zero, we find that

$$\frac{dy}{dx} = 0 \quad \Leftrightarrow \quad x = e^{-1} = \frac{1}{e}$$

and the slope table on $(0, \infty)$ is down-over-up. This makes the minimum occur at $x = 1/e$. We also know that $x \operatorname{Log}[x]$ is negative for $x < 1$.

The rest of the graphing information is easy to get,

$$\frac{d^2 y}{dx^2} = \frac{1}{x} > 0$$

for all $x > 0$, so the curve always bends upward.

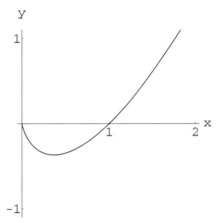

Figure 9.18: $y = x \operatorname{Log}[x]$

When $x = H$ an infinitely large number, $H \operatorname{Log}[H]$ is also infinitely large, so

$$\lim_{x \to \infty} x \operatorname{Log}[x] = \infty$$

The question is, "What value does $x \operatorname{Log}[x]$ increase to as x decreases to zero?"

$$\lim_{x \downarrow 0} x \operatorname{Log}[x] =?$$

To compute the limit at zero, we make a change of variables. Let $z = 1/x$ and rewrite our problem

$$\lim_{x \downarrow 0} x \operatorname{Log}[x] = \lim_{z \to \infty} \frac{\operatorname{Log}[1/z]}{z} = \lim_{z \to \infty} \frac{-\operatorname{Log}[z]}{z} = 0$$

(See Example 8.6.)

Graphical analysis is very useful and often only a rough idea is enough. Here is an example of using one graph to help find another.

Exercise set 9.4

Identify bends in the figures of the next exercise and then compute y'' from the given formula to see which graph is right.

1. *Which graph is nearest* $y = 4x^2 - \frac{x^5}{5}$?

a)
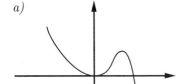

b)
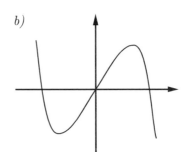

c)
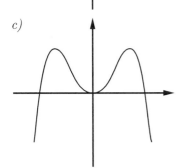

d)
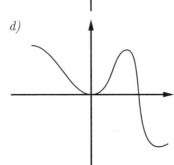

2. *Use first and second derivative procedure to sketch graphs of*

a) $y = f[x] = 2x^2 - x^4$

b) $y = f[x] = x^3 - 3x$

c) $y = f[x] = 6x^5 - 10x^3$

d) $y = (x+2)^{\frac{2}{3}}$

e) $y = \mathrm{Sin}[x] + \mathrm{Cos}[x]$

f) $y = \mathrm{Sin}[x] \times \mathrm{Cos}[x]$

g) $y = x\, e^x$

h) $y = e^{-x^2}$

i) $y = \dfrac{\mathrm{Log}[x]}{x}$

j) $y = x - \mathrm{Log}[x]$

k) $y = \dfrac{1}{1+x^2}$

l) $y = \dfrac{1}{1-x^2}$

Check your graphs with the computer, but remember that these are simple practice problems for you to work by hand. Later, messy real-world problems (like Planck's Law in the Projects) will require this calculus effort before you start the computer.

Problem 9.1.

Graph $y = (x^3 - x^2 + 1)^2$. How does the graph of the simpler equation $z = x^3 - x^2 + 1$ help in finding all the places where $y'(x) = 0$?

9.5   Graphing Differential Equations

A differential equation, such as

$$\frac{dy}{dt} = y\,(3 - y)$$

can be thought of as a description of the slope of the curve $y = y[t]$, given that you know y. This is enough to sketch a graph.

Example 9.9.   *The Slopes of $y[t]$ When $\frac{dy}{dt} = y\,(3-y)$*

In this case, if we start at $y = 1$ when $t = 0$, the initial slope is $\frac{dy}{dt} = y\,(3-y) = 1(3-1) = 2$. We can begin to sketch the curve by putting our pencil at $(0,1)$ and moving up along a line of slope 2. After we have sketched a small distance, both t and y will be larger and the slope will change accordingly.

We can move a specific small amount to y_1 and recompute the slope from the differential equation, $y_1\,(3 - y_1)$, as we have the computer do in the **SecondSIR** program and the computations on the dead canary, Exercise 4.2.1. The specific amounts are not essential to sketch the curve. Whenever y is between 1 and 3, the slope, $\frac{dy}{dt} = y\,(3 - y)$ is positive. This means that as t increases, y increases. However, as we approach $y = 3$ from below, the term $3 - y$ tends to zero and the slope $\frac{dy}{dt} = y\,(3 - y)$ also tends to zero. This means that the rate of increase slows down as y approaches 3. If we ever get to $y = 3$, the slope becomes zero and the curve ceases to increase. This rough reasoning gives a sketch of the solution of the differential equation when the solution starts at $y[0] = 1$, as shown as the lowest graph in Figure 9.19. It does not require a formula for the solution.

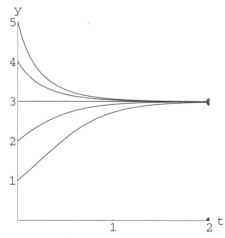

Figure 9.19: Solutions to $\frac{dy}{dt} = y\,(3 - y)$, starting at $y[0] = 1, 2, 3, 4, 5$

This is a graphical version of Euler's Method or the idea of the **SecondSIR** program. Although it is not very accurate, it does give the behavior of solutions. We will return to this kind of curve sketching when we study differential equations later in the course.

Example 9.10. *The Bends of $y[t]$ When $\frac{dy}{dt} = y(3-y)$*

We can also ask where the concavity of the curve changes. We use the Product Rule and general Chain Rule to compute

$$\frac{d}{dt}\frac{dy}{dt} = \frac{d}{dt}(y(3-y))$$

$$\frac{d^2y}{dt^2} = \frac{dy}{dt}(3-y) + y\,\frac{d(3-y)}{dt}$$

$$= \frac{dy}{dt}(3-y) - y\,\frac{dy}{dt} = (3-2y)\,\frac{dy}{dt}$$

Now, substitute the formula for $\frac{dy}{dt} = y(3-y)$, to obtain

$$\frac{d^2y}{dt^2} = (3-2y)\,\frac{dy}{dt} = y(3-y)(3-2y)$$

The second derivative equals $y(3-y)(3-2y)$, so $y''[t] > 0$ for $0 < y < 3/2$ and $y > 3$, whereas $y''[t] < 0$ for $y < 0$ or $\frac{3}{2} < y < 3$. Notice the change of concavity in Figure 9.20.

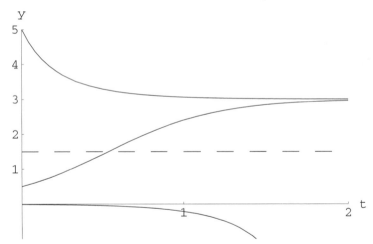

Figure 9.20: Concavity of logistic solutions

Exercise set 9.5

A fundamental "decay law" says that the rate of decrease of a quantity is proportional to the amount that is left. Radioactive substances have this property. You can sketch the amount of such a substance as a function of time without any formula.

1. Exponential Decay Without Formulas
   Given that a quantity q satisfies $q[0] = 4$ and $\frac{dq}{dt} = -\frac{1}{2}\,q$, sketch the graph of q vs t. What happens as the quantity as it gets close to $q = 0$?
   You can find an analytical solution to the conditions $q[0] = 4$ and $\frac{dq}{dt} = -\frac{1}{2}\,q$? (See Chapter 8. This is not needed for the sketch.)

2. *Sketch the solutions of $\frac{dy}{dt} = y(4-y)$ that begin with*
a) $y[0] = 1$ b) $y[0] = 2$ c) $y[0] = 6$ d) $y[0] = 4$ e) $y[0] = 0$
    Sketch the solutions of $\frac{dy}{dt} = 10y(4-y)$ that begin with
a) $y[0] = 1$ b) $y[0] = 2$ c) $y[0] = 6$ d) $y[0] = 4$ e) $y[0] = 0$

Euler's approximation uses the tangent line for a small step and then recomputes the slope. The increment equation says, $y[t+\delta t] = y[t] + y'[t] \cdot \delta t + \varepsilon \cdot \delta t$, so $y[t+\delta t] \approx y[t] + y'[t] \cdot \delta t$. If $y[t]$ satisfies $y[0] = y_0$ and $y'[t] = f[y[t]]$, Euler's approximation is given recursively by

$$y_{approx}[0] = y_0$$
$$y_{approx}[t + \delta t] = y_{approx}[t] + f[y_{approx}[t]] \cdot \delta t$$

for $t = 0, \delta t, 2\delta t, 3\delta t, \dots$. Notice that the equation of the tangent at a fixed point $y_1 = y[t]$ with slope $m = y'[t]$ is $dy = m \cdot dt$ in local coordinates, where $dy = y - y_1$. So $y = y_1 + m \cdot \delta t$ is the point on the tangent line of slope m at time δt past t.

$$y = y_1 + m \cdot \delta t$$
$$y[t + \delta t] = y[t] + y'[t] \cdot \delta t$$

If the second derivative $y''[t]$ is positive and $y'[t]$ is increasing in y, Euler's approximation is always low. We can see this just from the relation between the graph and the derivatives.

3. *Suppose $y[0] = 1/4$ and $\frac{dy}{dt} = \sqrt{y}$. Why is the tangent line to the exact solution $y = y[t]$ below the curve at $t = 0$? If Euler's approximate solution $y_{approx}[\delta t]$ is low at the first step, why is the slope $f[y_{approx}[\delta t]]$ below the slope of $y[t]$ at $t = \delta t$? (HINT: Compute $\frac{d^2y}{dt^2} = \frac{1}{2}$ using the Chain Rule. Where does the tangent lie in relation to a curve with a positive second derivative? Draw a tangent on a smiley face.) Explain why the Euler approximation is always below the true solution $y[t]$.*

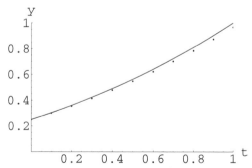

Figure 9.21: Euler's approximation to $y[0] = 1/4$ and $y'[t] = \sqrt{y[t]}$

4. Solutions to the S-I-S Equations
An S-I-S disease is one that does not confer immunity; you are either susceptible or infectious. If s is the fraction of the population that is susceptible, the spread of the disease in time t is given by

$$\frac{ds}{dt} = b(1-s)(1-cs)$$

where b is the reciprocal of the infectious period and c is the contact number. Assume that $c > 1$ (so that each infectious person contacts more than one other person)

and sketch the graphs of the solutions for various initial conditions between 0 and 1.
(HINTS: Show that s has a positive derivative for $0 < s < 1/c$ and a negative deriva-
tive for $1/c < s < 1$. The second derivative $\frac{d^2 s}{dt^2} = b^2(s-1)(1-cs)((1+c)-2cs)$, so
concavity changes at the average of 1 and $1/c$.)

9.6 Projects
9.6.1 Planck's Formula & Wein's Law

Planck won a Nobel prize for a formula that tells the intensity of radiation as a function
of temperature. His formula predicts the empirically observed law of radiation discovered
earlier by Wein. The peak in Planck's formula gives Wein's law. This was an important
early discovery in quantum mechanics because classical thermodynamics does not predict
Wein's Law.

This project shows that graphing alone is not enough to find the peak but that calculus
and the computer together make the story clear.

9.6.2 Algebraic Formulations of Increasing and Bending

The project on "Taylor's Formula," can be used to give algebraic proofs of the meaning
of the slope and bending icons used in graphing.

9.6.3 Bolzano, Darboux, and the Mean Value Theorem

The Mathematical Background chapter on these three theorems provides complete justification for our simple method of finding the slope and bend tables.

9.6.4 Horizontal and Vertical Asymptotes

The limits at infinity that we computed in some of our graphs show that the graph is
tending toward a line called an "asymptote." This project explores this topic in greater
detail.

CHAPTER

10

Velocity, Acceleration, and Calculus

The first derivative of position is velocity, and the second deriva-
tive is acceleration. These derivatives can be viewed in four ways:
physically, numerically, symbolically, and graphically.

The ideas of velocity and acceleration are familiar in everyday experience, but now we want you to connect them with calculus. We have discussed several cases of this idea already. For example, recall the following (restated) Exercise 9.2.1 from Chapter 9.

Example 10.1. *Over the River and Through the Woods*

We want you to sketch a graph of the distance traveled as a function of elapsed time on your next trip to visit Grandmother.

Make a qualitative rough sketch of a graph of the distance traveled, s, as a function of time, t, on the following hypothetical trip. You travel a total of 100 miles in 2 hours. Most of the trip is on rural interstate highway at the 65 mph speed limit. You start from your house at rest and gradually increase your speed to 25 mph, slow down and stop at a stop sign. You speed up again to 25 mph, travel for a while and enter the interstate. At the end of the trip you exit and slow to 25 mph, stop at a stop sign, and proceed to your final destination.

The correct "qualitative" shape of the graph means things like *not* crashing into Grandmother's garage at 50 mph. If the end of your graph looks like the one on the left in Figure 10.1, you have serious damage. Notice that Leftie's graph is a straight line, the rate of change is constant. He travels 100 miles in 2 hours, so that rate is 50 mph. Imagine Grandmother's surprise as he arrives!

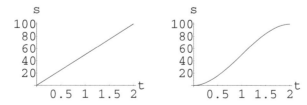

Figure 10.1: Leftie and Rightie go to Grandmother's

The graph on the right slows to a stop at Grandmother's, but Rightie went though all the stop signs. How could the police convict her using just the graph?

She passed the stop sign 3 minutes before the end of her trip, 2 hours less 3 minutes = 2 - 3/60 = 1.95 hrs. Graphs of her distance for short time intervals around $t = 1.95$ look like

Figure 10.2

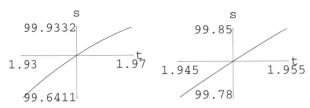

Figure 10.2: Two views of Rightie's moving violation

Wow, the smallest scale graph looks linear? Why is that? Oh, yeah, microscopes. What speed will the cop say Rightie was going when she passed through the intersection?

$$\frac{99.85 - 99.78}{1.955 - 1.945} = 7 \quad \text{mph}$$

He could even keep up with her on foot to give her the ticket at Grandmothers. At least Rightie will not have to go to jail like Leftie did.

You should understand the function version of this calculation:

$$\frac{99.85 - 99.78}{1.955 - 1.945} = \frac{s[1.955] - s[1.945]}{1.955 - 1.945} = \frac{s[1.945 + 0.01] - s[1.945]}{(1.945 + 0.01) - 1.945}$$

$$= \frac{s[t + \Delta t] - s[t]}{\Delta t} = \frac{0.07}{0.01}$$

Exercise set 10.0

1. *Look up your solution to Exercise 9.2.1 or resolve it. Be sure to include the features of stopping at stop signs and at Grandmother's house in your graph. How do the speeds of 65 mph and 25 mph appear on your solution? Be especially careful with the slope and shape of your graph. We want to connect slope and speed and bend and acceleration later in the chapter and will ask you to refer to your solution.*

2. *A very small-scale plot of distance traveled vs. time will appear straight because this is a magnified graph of a smooth function. What feature of this straight line represents the speed? In particular, how fast is the person going at $t = 0.5$ for the graph in Figure 10.3? What feature of the large-scale graph does this represent?*

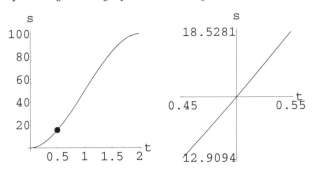

Figure 10.3: *A microscopic view of distance*

Velocity and the First Derivative

Physicists make an important distinction between speed and velocity. A speeding train whose speed is 75 mph is one thing, and a speeding train whose velocity is 75 mph on a vector aimed directly at you is the other. Velocity is speed plus direction, while speed is only the instantaneous time rate of change of distance traveled. When an object moves along a line, there are only two directions, so velocity can simply be represented by speed with a sign, $+$ or $-$.

3. *An object moves along a straight line such as a straight level railroad track. Suppose the time is denoted t, with $t = 0$ when the train leaves the station. Let s represent the distance the train has traveled. The variable s is a function of t, $s = s[t]$. We need to set units and a direction. Why? Explain in your own words why the derivative $\frac{ds}{dt}$ represents the instantaneous velocity of the object. What does a negative value of $\frac{ds}{dt}$ mean? Could this happen? How does the train get back?*

4. *Krazy Kousin Keith drove to Grandmother's, and the reading on his odometer is graphed in Figure 10.4. What was he doing at time $t = 0.7$? (HINT: The only way to make my odometer read less is to back up. He must have forgotten something.)*

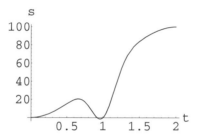

Figure 10.4: *Keith's regression*

5. *Portions of a trip to Grandmother's look like the next two graphs.*

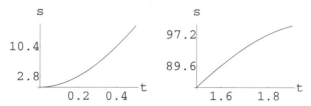

Figure 10.5: *Positive and negative acceleration*

Which one is "gas," and which one is "brakes"? Sketch two tangent lines on each of these graphs and estimate the speeds at these points of tangency. That is, which one shows slowing down and which speeding up?

10.1   Acceleration

Acceleration is the physical term for "speeding up your speed... " Your car accelerates when you increase your speed.

Since speed is the first derivative of position and the derivative of speed tells how it "speeds up." In other words, the second derivative of position measures how speed speeds up... We want to understand this more clearly. The first exercise at the end of this section asks you to compare the symbolic first and second derivatives with your graphical trip to Grandmother's. A numerical approach to acceleration is explained in the following examples. You should understand velocity and acceleration numerically, graphically, and symbolically.

Example 10.2.   *The Fallen Tourist Revisited*

Recall the tourist of Exercise 4.2. He threw his camera and glasses off the Leaning Tower of Pisa in order to confirm Galileo's Law of Gravity. The Italian police videotaped his crime and recorded the following information:

$t =$ time (seconds)	$s =$ distance fallen (meters)
0	0
1	4.90
2	19.6
3	44.1

We want to compute the average speed of the falling object during each second, from 0 to 1, from 1 to 2, and from 2 to 3? For example, at $t = 1$, the distance fallen is $s = 4.8$ and at $t = 2$, the distance is $s = 18.5$, so the change in distance is $18.5 - 4.8 = 13.7$ while the change in time is 1. Therefore, the average speed from 1 to 2 is 13.7 m/sec,

$$\text{Average speed} = \frac{\text{change in distance}}{\text{change in time}}$$

Time interval	Average speed $= \dfrac{\Delta s}{\Delta t}$
$[0,1]$	$v_1 = \dfrac{4.90 - 0}{1 - 0} = 4.90$
$[1,2]$	$v_2 = \dfrac{19.6 - 4.90}{2 - 1} = 14.7$
$[2,3]$	$v_3 = \dfrac{44.1 - 19.6}{3 - 2} = 24.5$

Example 10.3.   *The Speed Speeds Up*

These average speeds increase with increasing time. How much does the speed speed up during these intervals? (This is not very clear language, is it? How should we say, "the speed speeds up"?)

Interval to interval	Rate of change in speed
$[0,1]$ to $[1,2]$	$a_1 = \dfrac{14.7 - 4.90}{?} = \dfrac{9.8}{?}$
$[1,2]$ to $[2,3]$	$a_2 = \dfrac{24.5 - 14.7}{?} = \dfrac{9.8}{?}$

The second speed speeds up 9.8 m/sec during the time difference between the measurement of the first and second average speeds, but how should we measure that time difference since the speeds are averages and not at a specific time?

The tourist's camera falls "continuously." The data above only represent a few specific points on a graph of distance vs. time. Figure 10.6 shows continuous graphs of time vs. height and time vs. $s = $ distance fallen.

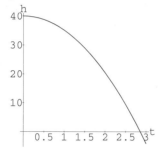 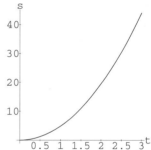

Figure 10.6: Continuous fall of the camera

The computation of the Example 10.2 finds $s[1] - s[0]$, $s[2] - s[1]$, and $s[3] - s[2]$. Which continuous velocities do these best approximate? The answer is $v[\frac{1}{2}]$, $v[\frac{3}{2}]$, and $v[\frac{5}{2}]$ - the times at the midpoint of the time intervals. Sketch the tangent at time 1.5 on the graph of s vs. t and compare that to the segment connecting the points on the curve at time 1 and time 2. In general, the symmetric difference

$$\frac{f[x + \frac{\delta x}{2}] - f[x - \frac{\delta x}{2}]}{\delta x} \approx f'(x)$$

gives the best numerical approximation to the derivative of

$$y = f[x]$$

when we only have data for f. The difference quotient is best as an approximation at the midpoint. The project on Taylor's Formula shows algebraically and graphically what is happening. Graphically, if the curve bends up, a secant to the right is too steep and a secant to the left is not steep enough. The average of one slope below and one above is a better approximation of the slope of the tangent. The average slope given by the symmetric secant, even though that secant does not pass through $(x, f[x])$. The general figure looks like Figure 10.7.

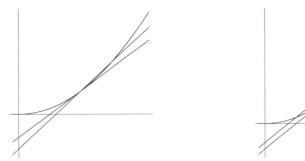

Figure 10.7: $[f(x + \delta x) - f(x)]/\delta x$ vs. $[f(x + \delta x) - f(x - \delta x)]/\delta x$

The best times to associate to our average speeds in comparison to the continuous real fall are the midpoint times:

Time	Speed $= \dfrac{\Delta s}{\Delta t}$
0.50	$v_1 = v[0.50] = 4.90$
1.50	$v_2 = v[1.50] = 14.7$
2.50	$v_3 = v[2.50] = 24.5$

This interpretation gives us a clear time difference to use in computing the rates of increase in the acceleration:

Time	Rate of change in speed
Ave[0.50&1.50] = 1	$a[1] = \dfrac{14.7 - 4.90}{1.50 - 0.50} = 9.8$
Ave[1.50&2.50] = 2	$a[2] = \dfrac{24.5 - 14.7}{2.50 - 1.50} = 9.8$

We summarize the whole calculation by writing the difference quotients in a table opposite the various midpoint times as follows:

First and Second Differences of Position Data			
Time	Position	Velocity	Acceleration
0.00	0.00		
0.50		4.90	
1.00	4.90		9.8
1.50		14.7	
2.00	19.6		9.8
2.50		24.5	
3.00	44.1		

Table 10.3: One-second position, velocity, and acceleration data

Exercise set 10.1

The first exercise seeks your everyday interpretation of the positive and negative signs of $\frac{ds}{dt}$ and $\frac{d^2s}{dt^2}$ on the hypothetical trip from Example 10.1. We need to understand the mechanical interpretation of these derivatives as well as their graphical interpretation.

1. *Look up your old solution to Exercise 9.2.1 or Example 10.1 and add a graphing table like the ones from the Chapter 9 with slope and bending. Fill in the parts of the table corresponding to $\frac{ds}{dt}$ and $\frac{d^2s}{dt^2}$ using the microscopic slope and smile and frown icons including $+$ and $-$ signs.*

Remember that $\frac{d^2s}{dt^2}$ is the derivative of the function $\frac{ds}{dt}$; so, for example, when it is positive, the function $v[t] = \frac{ds}{dt}$ increases, and when it is negative, the velocity decreases.

We also need to connect the sign of $\frac{d^2s}{dt^2}$ with physics and the graph of $s[t]$.

Use your solution graph of time, t, vs. distance, s, to analyze the following questions.

    (a) *Where is your speed increasing? Decreasing? Zero? If speed is increasing, what geometric shape must that portion of the graph of $s[t]$ have? (The graph of $v[t]$ has upward slope and positive derivative, $\frac{dv}{dt} = \frac{d^2s}{dt^2} > 0$, but we are asking how increase in $v[t] = \frac{ds}{dt}$ affects the graph of $s[t]$.)*

    (b) *Is $\frac{ds}{dt}$ ever negative in your example? Could it be negative on someone's solution? Why does this mean that you are backing up?*

    (c) *Summarize both the mechanics and geometrical meaning of the sign of the second derivative $\frac{d^2s}{dt^2}$ in a few words. When $\frac{d^2s}{dt^2}$ is positive .... When $\frac{d^2s}{dt^2}$ is negative ....*

    (d) *Why must $\frac{d^2s}{dt^2}$ be negative somewhere on everyone's solution?*

There are more accurate data for the fall of the camera in half-second time steps:

Accurate Position Data			
Time	Position	Velocity	Acceleration
0.000	0.000		
0.500	1.233		
1.000	4.901		
1.500	11.03		
2.000	19.60		
2.500	30.63		
3.000	44.10		

Table 10.4: Half-second position data

2. Numerical Acceleration

*Compute the average speeds corresponding to the positions in Table 10.4 above and write them next to the correct midpoint times so that they correspond to continuous velocities at those times. Then use your velocities to compute accelerations at the proper times. Simply fill in the places where the question marks appear in the velocity and acceleration Tables 10.5 and 10.6 following this exercise. The data are also contained in the **Gravity** program so you can complete this arithmetic with the computer in Exercise 10.2.2.*

HINTS: We begin the computation of the accelerations as follows. First, add midpoint times to the table and form the difference quotients of position change over time change:

Differences of the Half-Second Position Data			
Time	Position	Velocity	Acceleration
0.000	0.000		
0.250		$\frac{1.223-0}{0.5} = 2.446$	
0.500	1.233		
0.750		$\frac{4.901-1.223}{1.0-0.5} = 7.356$	
1.000	4.901		
1.250		$\frac{?-?}{?} = ?$	
1.500	11.03		
1.750		?	
2.000	19.60		
2.250		?	
2.500	30.63		
2.750		?	
3.000	44.10		

Table 10.5: *Half-second velocity differences*

Next, form the difference quotients of velocity change over time change:

Second Differences of the Half-Second Position Data			
Time	Position	Velocity	Acceleration
0.000	0.000		
0.250		2.446	
0.500	1.233		$\frac{7.356-2.446}{0.75-0.25} = 9.820$
0.750		7.356	
1.000	4.901		$\frac{?-?}{?} = ?$
1.250			
1.500	11.03		?
1.750			
2.000	19.60		?
2.250			
2.500	30.63		?
2.750			
3.000	44.10		

Table 10.6: *Half-second acceleration differences*

10.2   Galileo's Law of Gravity

The acceleration due to gravity is a universal constant, $\frac{d^2 s}{dt^2} = g$.

Data for a lead cannon ball dropped off a tall cliff are contained the the computer program

Gravity. The program contains time-distance pairs for $t = 0$, $t = 0.5$, $t = 1.0$, $\cdots$, $t = 9.5$, $t = 10$. A graph of the data is included in Figure 10.8.

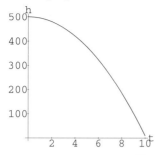 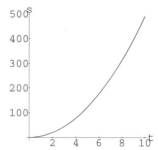

Figure 10.8: Free fall without airfriction

Galileo's famous observation turns out to be even simpler than the first conjecture of a linear speed law (which you will reject in an exercise below). He found that as long as air friction can be neglected, the rate of increase of speed is constant. Most striking, the constant is universal - it does not depend on the weight of the object.

> **Galileo's Law of Gravity**
> The acceleration due to gravity is a constant, g, independent of the object.

The value of g depends on the units of time and distance, $9.8 m/sec^2$ or $32 ft/sec^2$.

"Fluency" in calculus means that you can express Galileo's Law with the differential equation

$$\frac{d^2 s}{dt^2} = g$$

The first exercise for this section asks you to clearly express the law with calculus.

Exercise set 10.2

1. Galileo's Law and $\frac{d^2 s}{dt^2}$
   Write Galileo's Law, "The rate of increase in the speed of a falling body is constant." in terms of the derivatives of the distance function s[t]. What derivative gives the speed? What derivative gives the rate at which the speed increases?

We want you to verify Galileo's observation for the lead ball data in the **Gravity** program.

2. Numerical Gravitation

    (a) *Use the computer to make lists and graphs of the speeds from 0 to $\frac{1}{2}$ second, from $\frac{1}{2}$ to 1, and so forth, using the data of the **Gravity** program. Are the speeds constant? Should they be?*

    (b) *Also use the computer to compute the rates of change in speed. Are these constant? What does Galileo's Law say about them?*

3. Galileo's Law and the Graph of $\frac{ds}{dt}$
   *Galileo's Law is easiest to confirm with the data of the **Gravity** program by looking at*

the graph of $\frac{ds}{dt}$ (because error measurements are magnified each time we take differences of our data). What feature of the graph of velocity is equivalent to Galileo's Law?

In the Exercise 4.2 you formulated a model for the distance an object has fallen. You observed that the farther an object falls, the faster it goes. The simplest such relationship says, "The speed is proportional to the distance fallen." This is a reasonable first guess, but it is not correct. We want you to see why. (Compare the next problem with Problem 8.5.)

Problem 10.1.

Try to find a constant to make the conjecture of Exercise 4.2 match the data in the program **Gravity**, *that is, make the differential equation*

$$\frac{ds}{dt} = k\,s$$

predict the position of the falling object. This will not work, but trying will show why.

There are several ways to approach this problem. You could work first from the data. Compute the speeds between 0 and $\frac{1}{2}$ seconds, between $\frac{1}{2}$ and 1 second, and so forth, then divide these numbers by s and see if the list is approximately the same constant. The differential equation $\frac{ds}{dt} = k\,s$ says it should be, because

$$v = \frac{ds}{dt} = k\,s \qquad \Leftrightarrow \qquad \frac{v}{s} = k \quad \text{is constant.}$$

There is some help in the **Gravity** *program getting the computer to compute differences of the list. Remember that the time differences are $\frac{1}{2}$ seconds each. You need to add a computation to divide the speeds by s,*

$$\frac{s[t + \frac{1}{2}] - s[t]}{\frac{1}{2}} \approx \frac{ds}{dt} \quad at\ t + \frac{1}{4}$$

each divided by s[t]. Notice that there is some computation error caused by our approximation to $\frac{ds}{dt}$ actually being best at $t + \frac{1}{4}$ but only having data for s[t]. Be careful manipulating the lists with Mathematica because the velocity list has one more term than the acceleration list.

Another approach to rejecting Galileo's first conjecture is to start with the differential equation. We can solve $\frac{ds}{dt} = k\,s$ with the initial s[0] = 0 by methods of Chapter 8, obtaining $s[t] = S_0\,e^{k\,t}$. What is the constant S_0 if s[0] = 0. How do you compute s[0.01] from this? See Bugs Bunny's Law of Gravity, Problem 8.5 and Exercise 8.1.

The zero point causes a difficulty as the preceding part of this problem shows. Let's ignore that for the moment. If the data actually are a solution to the differential equation, $s = S_0\,e^{k\,t}$, then

$$\text{Log}[s] = \text{Log}[S_0\,e^{k\,t}] = \text{Log}[S_0] + \text{Log}[e^{k\,t}] = \sigma_0 + k\,t$$

so the logarithms of the positions (after zero) should be linear. Compute the logs of the list of (non-zero) positions with the computer and plot them. Are they linear?

10.3   Projects

Several Scientific Projects go beyond this basic chapter by using Newton's far-reaching extension of Galileo's Law. Newton's Law says $F = m\,a$, the total applied force equals mass times acceleration. This allows us to find the motion of objects that are subjected to several forces.

10.3.1   The Falling Ladder

An example in the CD section on related rates introduces a simple mathematical model for a ladder sliding down a wall. The rate at which the tip resting against the wall slides tends to infinity as the tip approaches the floor. Could a real ladder's tip break the sound barrier? The speed of light? Of course not. That model neglects the physical mechanism that makes the ladder fall - Galileo's Law of Gravity. The project on the ladder asks you to correct the physics of the falling ladder model.

10.3.2   Linear Air Resistance

A feather does not fall off a tall cliff as fast as a bowling ball does. The acceleration due to gravity is the same, but air resistance plays a significant role in counteracting gravity for a large, light object. A basic project on Air Resistance explores the path of a wooden ball thrown off the same cliff as the lead ball we just studied in this chapter.

10.3.3   Bungee Diving and Nonlinear Air Resistance

Human bodies falling long distances are subject to air resistance, in fact, sky jumpers do not keep accelerating but reach a "terminal velocity." Bungee jumpers leap off tall places with a big elastic band hooked to their legs. Gravity and air resistance act on the jumper in his initial "flight," but once he reaches the length of the cord, it pulls up by an amount depending on how far it is stretched. The Bungee Jumping Project has you combine all these forces to find out if a jumper hits the bottom of a canyon or not.

10.3.4   The Mean Value Math Police

The police find out that you drove from your house to Grandmother's, a distance of 100 miles in 1.5 hours. How do they know you exceeded the maximum speed limit of 65 mph? The Mean Value Theorem Project answers this question.

10.3.5   Symmetric Differences

The Taylor's Formula (from the project of that name) shows you why the best time for the velocity approximated by $(s[t_2] - s[t_1])/(t_2 - t_1)$ is at the midpoint, $v[(t_2 + t_1)/2]$. This is a general numerical result that you should use any time that you need to estimate a derivative from data. The project shows you why.

CHAPTER 11
Maxima and Minima in One Variable

Finding a maximum or a minimum clearly is important in everyday experience. A manufacturer wants to maximize her profits, a contractor wants to minimize his costs subject to doing a good job, and a physicist wants to find the wavelength that produces the maximum intensity of radiation.

Even a manufacturer with a monopoly cannot maximize her profits by charging a very high price because at some point consumers will stop buying. She seeks an optimal balance between supply and demand. In everyday experience, these optima are sought in intuitive ways, probably incorrectly in many cases. The problems are not usually simple, and often they are not even clearly formulated. Calculus can help. It can solve closed-form problems and offer guidance when the mathematical models are incomplete. Much of the success of science and engineering is based on finding symbolic optima for accurate models, but no one pretends to know closed-form models for the national health profile and similar interesting, but complicated facts of everyday life.

This chapter starts with some basic mathematical theory and then looks at some 'simple' applications. They are simple in terms of life's big questions but still offer some challenge mathematically. They may not be major "real world" revelations but are a start and have some practical significance. We are not retreating from our basic philosophy in this course, which is to show you what good calculus is. We think you know that optimization is important, so we will begin with the easy part: the mathematical theory. (Ugh, you say, but it really is the easy part.) Read the theory quickly, try the applications, and come back to the theory as needed.

11.1   A Graphical Minimum

We begin with a graphical maximization.

We want to make a box with a square base and no top. We need the box to hold 100 cubic inches and want to make it out of the least possible amount of cardboard. What dimensions should we use?

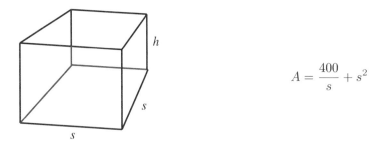

$$A = \frac{400}{s} + s^2$$

The graph of area as a function of the length of the side is shown in Figure 11.1.

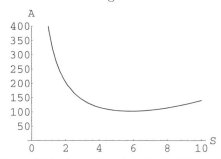

Figure 11.1: Area as a function of the side

It is clear on the graph of Figure 11.1 that the function decreases as s increases from 0 to about 5.8 and then the function increases as s increases beyond this. Hence, the minimum occurs when $s \approx 5.8$ and $h \approx 2.9$. The exercise below asks you to work through this reasoning.

This example is fine as far as it goes, but we will see that calculus can tell us more. The minimizing s and A are readily seen on the graph but values can only be read approximately. As you work through the exercise, think about how calculus could help you determine the shape and especially the point at which the graph changes from going down to going up.

Exercise set 11.1

1. Graphical Minimum Box
   (a) *Show that the area of a box with an open top and a square base of volume 100 cubic inches is given by $A = \frac{400}{s} + s^2$, where s is the length of the edge of the base. (HINTS: The volume of 100 cubic inches can be expressed in terms of length, width and height, $V = lwh$, but $l = w = s$ and $V = 100$, so $100 = hs^2$. This makes $h = \frac{100}{s^2}$. The area of each side piece is $hs = \frac{100}{s}$, substituting $h = \frac{100}{s^2}$.)*
   (b) *Sketch the graph of $A = A[s]$. One approach is to sketch $y = \frac{400}{s}$ and $z = s^2$, then sketch the graph of the sum $y + z$ on the same graph. Use the same scales $0 < s < 10$. You may use the computer to check your graphs if you wish. (With or without the computer, how do we know which scales to use?)*
   (c) *Locate the side s_m that minimizes area on the graph of A vs. s.*
   (d) *What is the slope of the graph and the value of $\frac{dA}{ds}[s_m]$ at s_m?*

11.2   Critical Points

You should think of the theory of max-min as a lazy approach to graphing. If you had a "complete" graph, you could look and see where the maximum and minimum occurred (assuming all features occur on the same scale).

Even with a properly scaled graph, you might want to compute the symbolic formula for the maximum or minimum. Many scientific results amount to the symbolic way a max or a min depends on some parameter. We will see examples at the end of this chapter and several projects use this idea. (See Planck's derivation of Wien's Law or Resonant Frequency or the Notch Filter in the accompanying book of projects.)

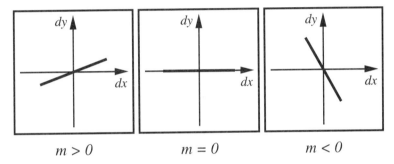

Figure 11.2: Up, over, down

Usually you do not need the whole graph (which is frequently hard to compute) to find a max or a min. For example, if you knew that a function $f[x]$ had a microscopic slope table starting at $x = 1$ with an upward slope that continued until $x = 2$ and then had a downward slope until $x = 3$, you would know that the maximum of $f[x]$ for $1 \leq x \leq 3$ occurs at $x = 2$. Use the microscope table to sketch such a curve.

Where does the minimum of this function occur? You need more information but only a little. It could be at either $x = 1$ or $x = 3$, but it could not be at any other x. Why? This is obvious. Think about it and sketch two graphs with the microscope table up-over-down, one with a min at $x = 1$ and the other with a min at $x = 3$. Endpoints or the lack of endpoints play an important role in max - min theory. You will learn to look for them in applications, because they simplify such problems.

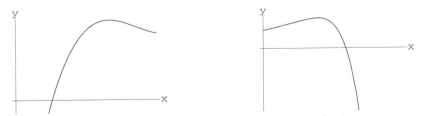

Figure 11.3: Minima at different endpoints

Mathematicians are fussy about the exact meaning of technical terms. When we say the function $f[x]$ attains its maximum for all real x at $x = 0$, we mean that $f[0] \geq f[x]$ for all x.

That is reasonable enough. The graph in Figure 11.4 of $f[x] = \frac{1}{x^2+1}$ has its max at $x = 0$. Everyday talk might say that the minimum of this function is $y = 0$, but this is **not** the custom in mathematics. It is true that $f[x]$ tends to zero as x tends to either $+\infty$ or $-\infty$, and it is true that $f[x] > 0$ for all x, but since $f[x]$ never actually takes the value zero, we say $f[x]$ does not have a minimum. Perhaps it would be better to say it does not attain its minimum.

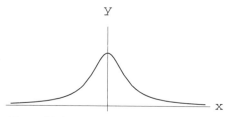

Figure 11.4: Max, but no min attained

Limiting values are not entirely the difficulty. The graph of $g[x] = \frac{x}{x^2+1}$ in Figure 11.5 has both a max and a min and also has limiting values. Precisely speaking, there are points x_{min} and x_{MAX} such that for all other x, $y_{min} = g[x_{min}] < g[x] < g[x_{MAX}] = y_{MAX}$. Find these points (x_{min}, y_{min}) and (x_{MAX}, y_{MAX}) on the graph.

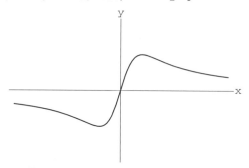

Figure 11.5: Max and min attained

Having pre-assigned endpoints simplifies max-min theory because endpoints eliminate limiting behavior. We discuss this in the next section. For now, we want to explore the "critical point condition." The question for this section is, "What slope could the graph have at a max or min?" The answer is clear graphically, except for a proviso.

The next result rules out many possible places that might be maxima or minima. That's right - rules out points. It does not tell you where they are but where they are not. It is usually stated in the following logically correct but confusing way, but it means that we need to examine only points x_0 such that $f'[x_0] = 0$ for the possibility that they are a max or min. These points are "critical" in our investigation of extrema, but they may or may not be extrema.

Theorem 11.1. *Interior Critical Points*
Suppose $f[x]$ is a smooth function on some interval. If $f[x]$ has a maximum or a minimum at a point x_0 inside the interval, then $f'[x_0] = 0$.

PROOF:
We give a geometric proof. Suppose $f'[x_0] \neq 0$. We know that we will see a straight

line of slope $f'[x_0] \neq 0$ if we look at the graph under a powerful microscope. Now, **if** we can move even a small amount to either side of x_0, **then** we can make $f[x]$ both larger and smaller than $f[x_0]$ by moving both ways. If x_0 is inside an interval over which we maximize or minimize, then x_0 cannot be the max or the min.

(This proof is correct, but we can elaborate on it some more. If $f'[x_0] \neq 0$, then $f[x]$ is either increasing or decreasing in an interval containing x_0. This is intuitively clear if you think of looking in a microscope and seeing a linear graph of slope $f'[x_0]$, but you can prove it algebraically in the project on Taylor's Formula.)

This completes the proof by use of the contrapositive. Yuck! The contrapositive? Why not find $f'[x]$, set it equal to zero, and solve. Then the solutions will be the max and min. Why not? Because that is wrong. Notice that the proof of the theorem is little more than a negative statement of where the extrema cannot be. Implications can be confusing and we want to make it clear that the theorem does **not** say that "if $f'[x_0] = 0$, then x_0 is a max or a min." Logically, A implies B does **not** mean that B implies A.

Correct reasoning is important in everyday life as well as in mathematics. The logic of implication can be confusing and that is what is behind the erroneous everyday conclusion that A implies B means that B also implies A. Let's see why this is so. "If it is raining, then there are clouds." is a correct statement.

$$A \Rightarrow B$$

where A stands for "it is raining" and B stands for "there are clouds." Certainly

$$B \Rightarrow A \quad \text{is false}$$

because $B \Rightarrow A$ says "if there are clouds, then it is raining."

A way to explain implication is

$$A \Rightarrow B \quad \equiv \quad (\text{not} A) \text{ or } B$$

This is a funny way to think of it, but it is equivalent. When A is true, the implication means that B must also be true. When A is false, B can be either true or false. The other statement has the same properties, so it is equivalent to the implication. If A is true, $\text{not} A$ is false so the other part of the "or" must be true, B. If A is false, $\text{not} A$ is true; so B may be either true or false.

Now, we can do some logical calculus to see the contrapositive

$$A \Rightarrow B \quad \equiv \quad (\text{not} B) \Rightarrow (\text{not} A)$$

because $(\text{not} A)$ or $B \equiv (\text{not not} B)$ or $(\text{not} A)$.

Exercise set 11.2

1. *Find the maximum and minimum values of the function $f[x] = 3 - 2x$ for $-1 \leq x \leq 2$. What is the slope of the graph $y = f[x]$ at these points? Sketch the graph and mark the max and min.*

2. *Find the max and min of $f[x] = x^3$ for $-1 \leq x \leq 2$. Graph the function. Is $f'[x] = 0$ at the max or at the min? Where is $f'[x] = 0$?*

3. *In the Interior Critical Point Theorem 11.1, A stands for the statement "$f[x]$ has an extremum at x_0 inside the interval" and B stands for "$f'[x_0] = 0$." The logical statement is of the form $A \Rightarrow B$. Give an example function to show that $B \Rightarrow A$ is false in this case.*

4. *Let B be the statement "$f'[x_0] = 0$." Let A be the statement "$f[x_0]$ is an interior extremum for the interval I." The interior critical point theorem above is the statement $A \Rightarrow B$. State the contrapositive theorem $(notB) \Rightarrow (notA)$ in English.*

5. Symbolic Minimum Box
    In Exercise 11.1.1, you showed that the area A of a square-base box with no top is

$$A = \frac{400}{s} + s^2$$

   (a) *Show symbolically that the one and only (real) value of s where the slope of the A vs. s graph is zero is $s_0 = \sqrt[3]{200} \approx 5.84804$.*
   (b) *Make a slope table for the graph (as in Section 9.3).*
   (c) *Explain why your computation and the shape table PROVES that the minimum occurs when $s = s_0 = \sqrt[3]{200}$ and $A = 60\sqrt[3]{5} \approx 102.599$*

11.3   Max - min with Endpoints

A smooth function $f[x]$ on a compact interval $[a, b]$ must have both a max and a min. This fact makes the job easier when we have endpoints.

Various intervals with and without endpoints arise in max-min problems. It is convenient to have some notation for the various cases. The basic notation is that round brackets, (or), cut off "just before" the endpoint, while square brackets, [or], include the endpoint. Here we will use the love knot symbol ∞ to mean intervals "keep going." (The infinity symbol ∞ cannot stand for an extremely large hyperreal number because it **violates** rules of algebra, contrary to the Algebra Axiom for hyperreals given in the background material. For example, $\infty + \infty = \infty$ and $\infty \times \infty = \infty$ are false for both real and hyperreal numbers.) It is not a number, but ∞ is a convenient symbol.

Definition 11.2. *Notation: If a and b are numbers,*
   a) $[a, b] = \{x : a \leq x \leq b\}$ *b)* $(a, b) = \{x : a < x < b\}$
   c) $[a, b) = \{x : a \leq x < b\}$ *d)* $(a, b] = \{x : a < x \leq b\}$
   e) $(-\infty, b] = \{x : x \leq b\}$ *f)* $(-\infty, b) = \{x : x < b\}$
   g) $[a, \infty) = \{x : a \leq x\}$ *h)* $(a, \infty) = \{x : a < x\}$

This is simple. Try it out in Exercise 11.3.1.

The most important intervals are the ones of finite length that include their endpoints, $[a, b]$, for numbers a and b. These intervals are sometimes described as "closed and bounded," because they have the endpoints and have bounded length. A shorter name is "compact" intervals.

Theorem 11.3.   *The Extreme Value Theorem*
If $f[x]$ is a continuous real function on the real compact interval $[a, b]$, then f attains its maximum and minimum; that is, there are real numbers x_m and x_M such that $a \leq x_m \leq b$, $a \leq x_M \leq b$, and for all x with $a \leq x \leq b$

$$f[x_m] \leq f[x] \leq f[x_m]$$

PROOF:

We will show how to locate the maximum. You can find the minimum. We begin with an approximate maximum. Partition the interval into steps of size Δx,

$$a < a + \Delta x < a + 2\Delta x < \cdots < b$$

and define a function

$$M[\Delta x] = \text{the x of the form } x_1 = a + k\Delta x$$

so that $f[x]$ is maximal *for the partition* at $x = M[\Delta x]$,

$$f[M[\Delta x]] = f[x_1] = max[f[x] : x = a + h\Delta x,\ h = 0, 1, \cdots, n]$$

This function is the discrete maximum from among a finite number of possibilities, so that $M[\Delta x]$ has two properties: (1) $M[\Delta x]$ is one of the partition points. (2) All other partition points $x = a + h\Delta x$ satisfy $f[x] \leq f[M[\Delta x]]$.

Next, we partition the interval into tiny steps,

$$a < a + \delta x < a + 2\delta x < \cdots < b$$

with $\delta x \approx 0$, and consider the discrete maximizing function $M[\delta x]$ along the fine partition. We know that (1) $x_1 = M[\delta x]$ is one of the points in the small partition. (2) $f[x] \leq f[x_1]$ for all other partition points x. (3) Every other number x_2 in $[a, b]$ is within δx of some partition point, $x_2 \approx x$.

Continuity of f means that $f[x] \approx f[x_2]$, so we have

$$f[x_2] \approx f[x] \leq f[x_1]$$

which says f almost attains its maximum at x_1. This completes the intuitive proof. In the Mathematical Background on CD, we show how to complete it. (Essentially, we only need to fix the location of x_1.)

You can make the computer mimic the proof for partitions of step Δx. You might hope to take Δx smaller and smaller and see what $M[\Delta x]$ converges toward. There really are serious problems with using this as a general purpose algorithm. It is very inefficient. It can oscillate as you make Δx smaller if there are several places near the maximum so the process might not converge. Still, you might want to try the proof directly on some simple functions.

11.3.1 Summary of the Theory on Compact Intervals

I am not a candidate for governor. That means that you do not have to consider me in making your vote or in your betting pool. We have two theorems about differentiable functions $f[x]$. One says that interior points of $[a, b]$ cannot be extrema unless $f'[x] = 0$, and the other says that there must be a max and a min. So what is left?

CANDIDATES:

If $f[x]$ is a differentiable function on the real compact interval $[a, b]$, then there is a max and a min amongt the points:

1) Endpoints, $x = a$ and $x = b$

2) Critical points $x = c$ such that $f'[c] = 0$

To find a max and a min, isolate the candidates, examine their record and vote. That is,

Procedure 11.4. Isolate the Candidates:

If you have a differentiable function $f[x]$ to extremize over a compact interval $[a, b]$:

(a) Compute $f'[x]$. (Be sure it is defined on all of $[a, b]$.)

(b) Find the critical points, that is, all solutions c of $f'[c] = 0$ with $a < c < b$.

(c) Make a table of the values $f[x]$ at $x =$ endpoints and critical points.

(d) Select the largest and smallest values of the function at the candidate points.

It is often helpful to compute the microscopic slope table after you have found all the critical points in step (2). This is not essential when you have endpoints, but shows clearly where the graph is increasing and decreasing. The Extreme Value Theorem guarantees that there will be both a max and a min, whereas the Interior Critical Point Theorem says that we only need to check critical points and endpoints.

Example 11.1. *Find the maximum and minimum of*

$$f[x] = x^3 - 6x^2 + 9x + 1$$

on the interval $[0, 5]$.

SOLUTION:

First, we isolate the possible candidates. The endpoints are

$$x = 0 \quad \text{and} \quad x = 5$$

The interior critical points are found by first computing $f'[x]$ and then finding all solutions of the equation $f'[x] = 0$.

$$\frac{df}{dx} = f'[x] = 3x^2 - 12x + 9$$
$$= 3(x - 1)(x - 3)$$

The derivative is always defined, so $f[x]$ is continuous and differentiable on $[0, 5]$.

The solutions of $f'[x] = 0$ are

$$3(x - 1)(x - 3) = 0 \quad \Leftrightarrow \quad x = 1 \quad \text{or} \quad x = 3$$

This isolates the candidates, so we compute their values:

Candidate $x =$	Value $f[x] =$
0	1
1	5
3	1
5	21

We see from the table of values that the maximum of f is 21 and that occurs at $x = 5$, $f(5) = 21$. The minimum is 1 and it occurs at two places, $f[0] = f(3) = 1$.

It is instructive to go on and compute the microscopic slope table. The formula $f'[x] = 3x^2 - 12x + 9 = 3(x-1)(x-3)$ makes $f'[x] > 0$ on $(-\infty, 1)$, $f'[x] < 0$ on $(1,3)$ and $f'[x] > 0$ on $(3,5)$. This is shown on Figure 11.6.

x	0	1		3		5
y	1	5		1		21
y'	╱	─	╲	─	╱	╱

Figure 11.6: Slope Table for $f = x^3 - 6x^2 + 9x + 1$

The slope table shows that the left endpoint is a possible min, the critical point $x = 1$ is a possible max, the critical point $x = 3$ is another possible min, and the endpoint $x = 5$ is a possible max. The table of values is needed to decide which possibilities occur. A the computer graph is shown in Figure 11.7.

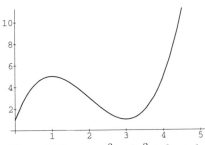

Figure 11.7: $y = x^3 - 6x^2 + 9x + 1$

It might be helpful to see how endpoint conditions arise in a simple application. We take that up in the next example, but first review a procedure for word problems:

Procedure 11.5. To solve applied max-min exercises:

(a) Read the question and decide on a list of variables - main variables and auxiliary ones that help in translation. List the variables with units.

(b) Translate each phrase of the problem into a statement about your variables.

(c) State the question in terms of your variables. Try to give a compact interval over which you seek the max or min. Look hard for endpoints because they make the problem much easier. Unfortunately, it is not always possible to find a compact interval. In that case, you have less theory to help you solve your problem and it may not even have a solution.

(d) Apply the max-min theory that you can, computing limiting values and graphs as needed. When you have a compact interval, you only need to isolate the candidates as above. In the noncompact case, you will have to examine slopes and perhaps even limits.

(e) Interpret your solution.

To solve real-world max-min problems, you often need to formulate a clear statement of the question before you can begin to translate the information and state the problem mathematically. Often a drawing helps in the formulation.

Example 11.2. *We have a 12 inch square piece of thin material and want to make an open box by cutting small squares from the corners of our material and folding the sides up. The question is, "Which cut produces the box of maximum volume?"*

Figure 11.8: Max of volume

First, we draw the diagram Figure 11.8 and assign variables

$$x = \text{ the length of the cut on each side of the little squares}$$
$$V = \text{ the volume of the folded box}$$

The length of the base after two cuts along each edge of size x is $12 - 2x$. The depth of the box after folding is x, so the volume is

$$V = x(12 - 2x)(12 - 2x) = 4x^3 - 48x^2 + 144x$$

Endpoints arise mathematically from the geometry. A cut must have positive length, $0 \le x$, and even $x = 0$ means we have nothing to fold up. Two equal length cuts from 12 inches can each have a maximum length of 6 inches, and again $x = 6$ means we have no base left on our box.

MAXIMIZE:

$$V = 4x^3 - 48x^2 + 144x \qquad x \in [0, 6]$$

The two endpoints $x = 0$ and $x = 6$ make $V[x] = 0$. Mathematically, these are minimum points, and so, are of no interest in the application. All we are left with for maximum candidates are interior critical points.

$$\frac{dV}{dx} = 12x^2 - 96x + 144 = 12(x - 2)(x - 6)$$

and

$$\frac{dV}{dx} = 0 \qquad \Leftrightarrow \qquad x = 2 \quad \text{or} \quad x = 6$$

There is only one interior critical point, $x = 2$, where $V = 128$. This must be the maximum, because we have already checked all other candidates.

Exercise set 11.3

Finding the interior critical points requires that we find all solutions of

$$f'[x] = 0$$

Our textbook exercises are contrived to make this problem fairly easy. Factor or use the quadratic formula, something basic usually works. In "real-world" problems, this step can be quite difficult, but remember that the computer has FindRoot[.] and Solve[.] to help you.

1. *Describe of the following intervals and sketch them as segments on the real line:*

a) $(0, \infty)$ b) $(0, 1)$ c) $(-1, 3]$ d) $[-2, 5]$

2. *Find the maximum and minimum of each of the following functions over the specified interval.*

(a) $f[x] = x^3 - 3x^2 + 3x + 2, \quad x \in [-2, 2]$

(b) $f[x] = x^3 - 6x + 4, \quad x \in [-3, 3]$

(c) $f[x] = x^3 + 7x + 4, \quad x \in [-10, 10]$

(d) $f[x] = x^3 + 7x + 4, \quad x \in [0, 10]$

(e) $f[x] = \frac{x}{x^2+1}, \quad x \in [-5, 5]$

3. *A homeowner has a long strip of 10 inch wide metal and wants to make a rain gutter by folding the sides up to form a rectangular cross-section with an open top shown in Figure 11.9. Where should she fold in order to get the maximum cross sectional area?*

Figure 11.9: Folded gutter

4. *Find the dimensions of the right circular cylinder with the largest volume that can be enclosed in a sphere of radius 24 inches. (What are the endpoints of the intervals for your variables? Is this question a continuous max-min on a compact interval?)*

5. Geometric Endpoints?

In Exercise 11.2.5 *you PROVED that the function* $A = \frac{400}{s} + s^2$ *with derivative* $\frac{dA}{ds} = 2s - \frac{400}{s^2}$ *has a minimum at* $s = \sqrt[3]{200}$. *There is no maximum for the area of this box and we want you to understand this both intuitively and mathematically.*

   (a) *Why are there no endpoints to the interval of possible s-values in this problem?*

   (b) *What happens to A as s approaches "limiting" values? (That is, what is the mathematical limit of A as s approaches each of the meaningful limits of the problem?)*

   (c) *Why does A have no maximum? For example, how could you build a box with a volume of 100 cubic inches and an area of 1 square mile (4014489600 square inches)?*

6. Fictitious Farming

   (a) *A farmer has 100 feet of fence and wants to make a rectangular holding pen. What dimensions should he make it in order to maximize the area?*

   (b) *Suppose the farmer makes the rectangular holding pen using an existing fence for one side. What dimensions should he use then to maximize the area within his 100 feet of new fence?*

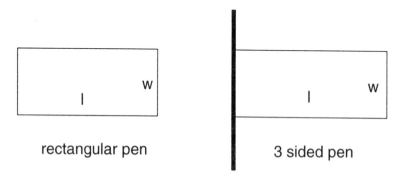

rectangular pen   3 sided pen

Figure 11.10: Holding Pens

7. Mathematical Farming

A math professor moves to the country and wants to fence some property so that he will enclose a rectangle of exactly 625 square feet of land. He likes fencing, so find the maximum amount of fence he can use in doing this. In particular, what are the dimensions of the math professor's 625 square foot tract if he uses 1 mile of fence?

(What's wrong with this question intuitively and mathematically? Real farmers would have more sense, of course. What are the hypotheses of our max-min procedure?)

8. *Find the max and min of*

$$f[x] = \frac{\sqrt{x^2 - \frac{x^4}{2}}}{x} \qquad for \qquad x \in [-1, 1]$$

Be careful and use the computer to plot the graph if you have trouble analyzing the function. What are the hypotheses of our max-min procedure?

9. *Write a the computer program to compute the maximizing function in the proof of the Extreme Value Theorem, $M[\Delta x]$. There is a start for you in the program **ExtremeValue**.*

11.4   Max - min without Endpoints

The moral of this section is that you need to investigate more graphical properties when you have less theory.

The math professor who tries farming in Exercise 11.3.7 is an example of a max-min problem over a non-compact interval. The problem is silly in order to help get your attention and to help you reason intuitively about what is wrong with the optimization question. If the length of the field is l (feet) and the width is w, then the area of 625 makes $l = 625/w$ and the length of the fence $p = 2l + 2w = 2(w + 625/w)$. In other words, we are asking to MAXIMIZE $f[w] = w + 625/w$ for all positive values of w. (Of course, the function $f[w] = w + 625/w$ is not continuous at $w = 0$ and in fact, $f[w]$ tends to infinity as w tends to zero. We solve this in Example 11.3.)

There is no maximum. The silly math professor can use as much fence as he wants and still only enclose 625 square feet. (One student suggested that the solution was to have the math professor move back to town and rent an apartment.) The point is that lack of an endpoint, whether it is at infinity or zero, causes extra mathematical problems. The Extreme Value Theorem does not apply, and we need to use additional information from microscopic shape tables and the resulting graphs.

Without the hypotheses of the candidates, Procedure 11.4 - that we are working with a differentiable function $f[x]$ on a compact interval $[a, b]$ - there may not be either a max or a min. If the function is not differentiable, it might not be continuous and then there might not be a max or min. If the problem does not have the variable restricted to a compact interval, there may be problems as you approach an endpoint that is not there.

Example 11.3.   *Find the max and min of*

$$f[x] = x + \frac{625}{x}$$

for all positive values, $x \in (0, \infty)$.

SOLUTION:

We know that we cannot have an interior max or min unless $f'[x] = 0$ or $f'[x]$ does not exist. In this problem,

$$f'[x] = 1 - \frac{625}{x^2}$$

which is defined on the open interval $(0, \infty)$. The positive critical value is

$$f'[x] = 0 \qquad \Leftrightarrow \qquad x = 25$$

When $0 < x < 25$, $f'[x] < 0$ and when $x > 25$, $f'[x] > 0$, so the microscopic slope table is down-over-up as shown in Figure 11.11.

x	0		25			+ infinity
y						
y'	*	\	—	/	/	*

Figure 11.11: Slope table for $y = x + \frac{625}{x}$

The slope table proves that $f[25] = 50$ is a minimum. What about a maximum? There is not any and now we have to reason directly. We need to know

$$\lim_{x \downarrow 0} f[x] \qquad \text{and} \qquad \lim_{x \to \infty} f[x]$$

because the function increases as we move from 25 in either of these directions. How can we compute these limits? The idea of $\lim_{x \to 0} f[x]$ is to see what happens as x gets smaller and smaller. We cannot just plug in $x = 0$, because the function is undefined. We can take a positive tiny $0 < \delta \approx 0$ and plug that in:

$$f[\delta] = \delta + \frac{625}{\delta} \approx \frac{625}{\delta} = 625\frac{1}{\delta}$$

The reciprocal of a positive tiny number is an extremely large positive number, so $f[\delta]$ is positive and extremely large. In other words, $f[x]$ can be made as large as we please by taking x closer and closer to 0. We write this result

$$\lim_{x \downarrow 0} f[x] = \infty$$

What happens as x gets larger and larger? We write this $x \to \infty$, but recall that ∞ is not an ordinary (real or hyperreal) number. It is only a symbol to indicate "larger and larger." To compute $\lim_{x \to \infty} f[x]$, take an extremely large positive number such as $H = \frac{1}{\delta}$ and compute

$$f[H] = H + \frac{625}{H} \approx H$$

which is extremely large again, so

$$\lim_{x \to \infty} f[x] = \infty$$

This function has no maximum - you can make it as large as you please by either taking x near enough to zero or taking x very large. A computer graph of $y = f[x]$ is shown on Figure 11.12.

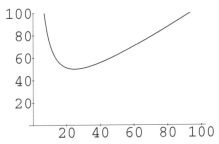

Figure 11.12: Min only attained

Our next example comes up as part of an economic model later in this chapter.

Example 11.4.   *Find the max and min of*

$$T[p] = (p - 0.5)\left(\frac{1000}{1 + p^2}\right)$$

for $0.5 \leq p$ or $p \in [0.5, \infty)$

SOLUTION:

The first step is to find the derivative of $T[p] = 500(2p - 1)(1 + p^2)^{-1}$ using the Product Rule and Chain Rule (or quotient rule, if you prefer):

$$T'[p] = 500 \frac{d(2p - 1)}{dp}(1 + p^2)^{-1} + 500(2p - 1)\frac{d\left([1 + p^2]^{-1}\right)}{dp}$$

$$= 1000(1 + p^2)^{-1} + 500(2p - 1)(-1)(1 + p^2)^{-2}(2p)$$

$$= 1000 \frac{1 + p^2}{(1 + p^2)^2} - 500 \frac{(2p - 1)(2p)}{(1 + p^2)^2}$$

$$= (-p^2 + p + 1)\frac{1000}{(1 + p^2)^2}$$

The critical points are where $-p^2 + p + 1 = 0$, so by the quadratic formula,

$$p = \frac{-1 \pm \sqrt{1 + 4}}{-2} = \frac{1}{2} \pm \frac{\sqrt{5}}{2}$$

or $p \approx 1.618$ and $p \approx -0.618$. We are only interested in values $p \in [0.5, \infty)$, so our only critical point is $p = \frac{1}{2} + \frac{\sqrt{5}}{2} \approx 1.618$.

The sign of $T'[p]$ for $0.5 \leq p < 1.618$ is $T'[p] > 0$. The sign for $p > 1.62$ is $T'[p] < 0$, so the microscopic slope table is up-over-down.

p	0.5		1.62			+ infinity
T						
T'	/	/	—	\	\	*

Figure 11.13: Slope table for $T = (p - 0.5)(\frac{1000}{1+p^2})$

This proves that the maximum occurs at the critical value, but what about the minimum? At the endpoint $p = 0.5$, $T[0.5] = 0$. At all other values of $p \in [0.5, \infty)$, $T[p] > 0$, so zero is the minimum. What happens as p tends to infinity?

$$\lim_{p \to \infty} T[p] = ?$$

Take $p = P$ extremely large and estimate

$$T[p] = (2P - 1)\frac{500}{1 + P^2}$$

$$= 500\frac{\left(\frac{2}{P} - \frac{1}{P^2}\right)}{\frac{1}{P^2} + 1}$$

The quantities $\frac{2}{P}$ and $\frac{1}{P^2}$ are tiny when P is extremely large, so

$$T[p] = 500\frac{\varepsilon}{\delta + 1}$$

for two tiny numbers, ε and δ. The product $500\varepsilon \approx 0$ and $\frac{1}{\delta+1} \approx 1$, so

$$T[P] \approx 0 \qquad \text{and} \qquad \lim_{p \to \infty} T[p] = 0$$

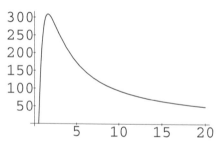

Figure 11.14: Min and max

Exercise set 11.4

1. *Find the maximum and the minimum (if they exist) of the following functions for all real x where they are defined. Are the extrema attained or only limiting values?*

a) $f[x] = 3x^4 - 4x^3$

b) $f[x] = x^3 + 3x^2 + 4x + 5$

c) $f[x] = \dfrac{x}{x^2 + 4}$

d) $f[x] = \dfrac{x^2}{x^2 + 4}$

e) $f[x] = \dfrac{x^2}{x^3 + 4}$

f) $f[x] = \dfrac{x^3}{x^4 + 27}$

2. *Find the maximum and the minimum of the following functions for all real x where they are defined. Are the extrema attained or only limiting values? Use the computer if you have technical difficulties. You should be able to find the shapes, but perhaps not the exact limiting values.*

a) $f[x] = \dfrac{\text{Sin}[x]}{x}$

b) $f[x] = \dfrac{\text{Cos}[x]}{x}$

c) $f[x] = \dfrac{\text{Log}[x]}{x}$

d) $f[x] = x\,\text{Log}[x]$

e) $f[x] = xe^x$

f) $f[x] = xe^{-x^2}$

3. *Complete the analysis of the max and min of*

$$f[x] = \frac{\sqrt{x^2 - x^4/2}}{x}$$

Where is it defined? What are its max and min, if any? Are they attained or only approached?

11.5   Supply and Demand - CD

In this section, we study simplified monopoly economies and show how a manufacturer maximizes profit.

Remainder of this section only on CD

11.6   Constrained Max-Min - CD

This section solves max-min problems with a constraint. For example, we might want to know the minimum distance between two curves. We are constrained to choose points on the curves.

Remainder of this section only on CD

11.7   Max-min with Parameters - CD

This section looks at max-min problems with a parameter. Often scientific max-min problems ask how the maximum depends on one or more parameters. This is how Wein's Temperature Law is derived from Planck's Radiation Law in the projects. Sharing half the cost of a price increase for linear demand was proved with parameters in the economics section of this chapter.

Remainder of this section only on CD

11.8   Projects
11.8.1   Optics and Least Time

Many interesting basic results in optics can be proved using minimization. Fermat's Principle says that light travels along the path that requires the least time. From this and max-min theory we can show that light reflects off mirrors at equal angles and is refracted according to Snell's Law which can be expressed:

$$\frac{\mathrm{Sin}[\alpha]}{\mathrm{Sin}[\beta]} = \frac{u}{v}$$

Using the computer, we can add many beautiful pictures showing light focusing at a point when reflecting off a parabolic mirror or creating a spherical aberration, or light caustic, when reflecting off a spherical mirror.

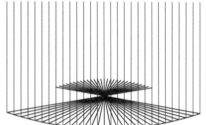 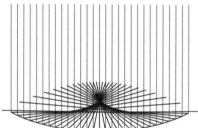

Figure 11.15: Parabolic and Spherical Mirrors

11.8.2   Monopoly Pricing

Optimal pricing is studied in a model of a monopoly which can charge different kinds of customers different prices. This is based on cable TV prices for separate homes and apartments.

11.8.3   Max-min in S-I-R Epidemics

We cannot find explicit formulas for the functions $s[t]$ and $i[t]$ of the epidemic model of Chapter 2, and this project shows how you can find max-min information anyway.

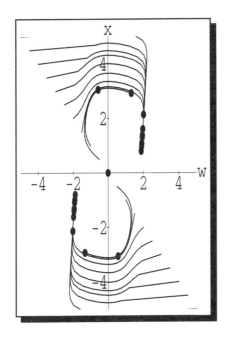

Part 2

Integration in One Variable

CHAPTER
12

Basic Integration

This chapter contains the fundamental theory of integration. We begin with some problems to motivate the main idea: approximation by a sum of slices. The chapter confronts this squarely, and Chapter 13 concentrates on the basic rules of calculus that you use after you have found the integrand.

Definite integrals have important uses in geometry and physics. Both the geometric and the physical integral formulas are derived in the following way: First, find a formula for the quantity (volume, area, length, distance, etc.) in the case of either a constant or a linear function. Next, approximate the nonlinear quantity by a sum of "slices" using the constant or linear formula for each slice. The primary difficulty is usually in expressing the variable sizes of the approximating pieces. In the geometric problems, this step is analytical geometry - finding the formula that goes with the picture you want. The Fundamental Theorem of Integral Calculus 12.14 gives us a simple way to compute the limit of the sum approximations exactly once we have the integral formula.

In order for sum approximations to tend to an integral we need to write them in the form

$$f[a]\,\Delta x + f[a + \Delta x]\,\Delta x + f[a + 2\Delta x]\,\Delta x + \cdots + f[b - \Delta x]\,\Delta x$$

where Δx is the "thickness of the slice" and $f[x]$ is the variable "base of the slice," so that $f[x]\,\Delta x$ is the variable amount of the slice at x. This symbolic expression is an important part of the way the formulas are expressed in integration; without the symbolic expression the more or less obvious approximations could not be computed exactly in a common way. The computer has a sum command that computes this expression

$$\sum_{\substack{x=a \\ \text{step } \Delta x}}^{b-\Delta x} [f[x]\Delta x]$$

Computer summation is studied in detail in Section 12.4 below. The integral is the limit of this sum as Δx tends to zero,

$$\int_a^b f[x]\,dx = \lim_{\Delta x \to 0} \sum_{\substack{x=a \\ \text{step } \Delta x}}^{b-\Delta x} [f[x]\Delta x]$$

The slicing approximation is the forest that you must strive to see through a tangle of technical trees. The first problem is in finding the symbolic sum.

12.1   Geometric Slice Approximations

The volume of a right circular cone with height h and base of radius r is

$$V = \frac{\pi}{3} r^2 h$$

*We want **you** to derive this formula by approximating a cone with a sum of cylindrical "disks." Once you understand the step from a constant radius to a varying radius, you will be able to find integral formulas for general solids of revolution. This important generalization illustrates the power of integration theory.*

The volume of a right cylinder is

$$V = A h$$

the area of the base times the height. If the base is a circle of radius r, as shown in Figure 12.1, the formula becomes

$$V = \pi r^2 h$$

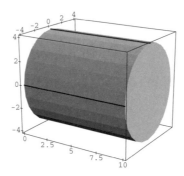

Figure 12.1: $V = \pi r^2 h$

Our next task is to let the radius vary with x, $r = R[x]$, and not be the same all along a cylinder.

12.1.1   The Volume of a Cone

Now we think of a cone as a figure with circular cross-sections that vary linearly, starting with zero radius and increasing to radius r when the distance from the tip is h. Let the variable x run down the axis of the cone with $x = 0$ at the tip as shown in Figure 12.2. The expression for the radius of the cross-section can be expressed as a function of x.

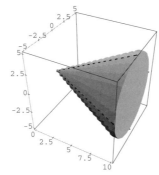

Figure 12.2: A cone along the x-axis

Example 12.1. *The Varying Radius of the Cross-Section of a Cone.*

We want a formula $R[x] =?$ for the radius of the cross-section at x for a cone with tip at $x = 0$ and base radius r at height $x = h$. Since the side of the cone is a straight line, we want a function such that

$$R[x] = m\,x + b, \quad \text{a linear function}$$
$$R[0] = 0, \quad \text{zero cross-section at } x = 0, \text{ so } b = 0$$
$$R[h] = r, \quad \text{or } m\,x = r, \text{ when } x = h, \quad \text{so} \quad m = \frac{r}{h}$$

So, the radius of the cross-section at x of a cone of height h and base radius r is $R[x] = \frac{r}{h}x$.

Example 12.2. *The Volume of a Particular Cone*

Our approximation of the volume can be obtained by slicing the cone in steps of Δx, using the formula for the volume of a cylinder for each slice, $V_{slice} = \pi R^2[x]\Delta x$. For example, suppose we have a cone of height 10 and radius 5 at the base (in the same length units). In this case the formula for the radius x units below the apex is

$$R[x] = \frac{1}{2}\,x$$

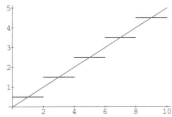

Figure 12.3: Generating lines for the cone approximation

If we let $\Delta x = 2$, or we slice the cone 5 times, we obtain 5 disks with radii at $x = 1$, $x = 3, \cdots, x = 9$. The disks are generated by the horizontal segments shown in Figure 12.3. These segments generate the rough approximating pile of disks shown in Figure 12.4, and have volumes given as follows:

(a) x from 0 to 2, $R(1) = \frac{1}{2}$, thickness of cylindrical disk $= \Delta x = 2$,

  $V_{slice1} = \pi \left[\frac{1}{2}\right]^2 2$

(b) x from 2 to 2, $R(3) = \frac{3}{2}$, thickness of cylindrical disk $= \Delta x = 2$,

  $V_{slice2} = \pi \left[\frac{3}{2}\right]^2 2$

(c) x from 4 to 6, $R(5) = \frac{5}{2}$, thickness of cylindrical disk $= \Delta x = 2$,

  $V_{slice3} = \pi \left[\frac{5}{2}\right]^2 2$

(d) x from 6 to 8, $R(7) = \frac{7}{2}$, thickness of cylindrical disk $= \Delta x = 2$,

  $V_{slice4} = \pi \left[\frac{7}{2}\right]^2 2$

(e) x from 8 to 10, $R(9) = \frac{9}{2}$, thickness of cylindrical disk $= \Delta x = 2$,

  $V_{slice5} = \pi \left[\frac{9}{2}\right]^2 2$

The total volume approximation is

$$\frac{\pi}{2}(1^2 + 3^2 + \cdots + 9^2) = \frac{\pi\,165}{2} \approx 259.$$

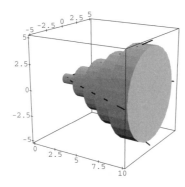

Figure 12.4: A cone approximated by 5 "disks"

We need to find the formula that expresses the sum of the volumes of the slices in terms of a variable for the thickness of the slice, Δx. The previous examples suggest an interesting special-purpose algebraic approach, but that is likely to work only for this specific cone. We can express the approximation in terms of the formula for the radius $R[x]$ of the cross-section at x and a thickness Δx as

$$\pi[(R[\Delta x/2])^2\Delta x + (R[3\Delta x/2])^2\Delta x + (R[5\Delta x/2])^2\Delta x + \cdots + (R[10 - \Delta x/2])^2\Delta x]$$

$$= \pi \sum_{\substack{x=\Delta x/2 \\ \text{step } \Delta x}}^{10-\Delta x/2} [(R[x])^2\Delta x]$$

12.1.2 A Parabolic Nose Cone

The simple linear formula for $R[x]$ in the case of the cone is easily generalized to more complicated shapes. A parabolic rocket "nose cone" may be described by the volume swept out as we rotate the curve $y = \sqrt{x}$ for $0 \le x \le 10$ about the x-axis shown in Figure 12.5.

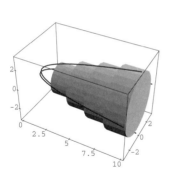

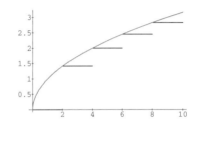

Figure 12.5: A parabolic nose cone

The formula for the radius of a cross-section is $R[x] = \sqrt{x}$, and the volume of a slice becomes $\pi R^2[x]h = \pi(\sqrt{x})^2 \Delta x = \pi \, x \, \Delta x$, making the approximating sum with the radius at the left side of the slices

$$\pi\left(0\,\Delta x + \Delta x\,\Delta x + 2\Delta x\,\Delta x + \cdots + (10 - \Delta x)\,\Delta x\right)$$

$$= \pi \sum_{\substack{x=0 \\ \text{step } \Delta x}}^{10-\Delta x} [x\,\Delta x]$$

12.1.3   A Spear Tip

A slender spear tip may be described by the volume swept out by the region between $y = \frac{x^2}{50}$ and the x-axis as the region is revolved about the x-axis. In this case, the radius of a cross-section at x is $R[x] = \frac{x^2}{50}$, the volume of one slice is $\pi R^2[x]h = \pi\left[\frac{x^2}{50}\right]^2 \Delta x$, and the left radius sum by approximating disks is

$$\pi\left(0\,\Delta x + \frac{\Delta x^4}{2500}\,\Delta x + \frac{(2\Delta x)^4}{2500}\,\Delta x + \cdots + \frac{(10-\Delta x)^4}{2500}\,\Delta x\right)$$

$$= \frac{\pi}{2500} \sum_{\substack{x=0 \\ \text{step } \Delta x}}^{10-\Delta x} [x^4\,\Delta x]$$

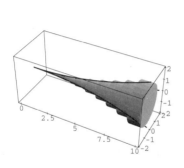

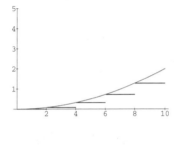

Figure 12.6: A cusped spear point and slice generators

12.1.4   The Area Between Two Curves

We need two formulas to build an approximation to the complicated areas between curves. First, the distance between points with real coordinates y_1 and y_2, is $|y_2 - y_1|$. The signed quantity $y_2 - y_1$ gives the directed distance from y_1 to y_2 where a minus sign means moving opposite the direction from zero to one. (The absolute value will cause us technical difficulties in calculus.) Second, the area of a rectangle is the height times the width.

Example 12.3.   *Area Between Sinusoids*

Slice the region between the sine and cosine curves from $-\pi$ to π into strips of width Δx as shown in Figure 12.7.

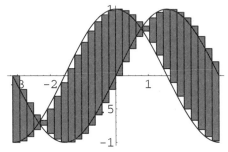

Figure 12.7: The area between the sine and cosine

The height of the slice with left point at x is given by $|\operatorname{Sin}[x] - \operatorname{Cos}[x]| = |\operatorname{Cos}[x] - \operatorname{Sin}[x]|$. The area of the slice is $|\operatorname{Cos}[x] - \operatorname{Sin}[x]|\,\Delta x$. The sum of all the approximating slices is

$$
(|\operatorname{Cos}[-\pi] - \operatorname{Sin}[-\pi]|\,\Delta x +
$$
$$
|\operatorname{Cos}[-\pi + \Delta x] - \operatorname{Sin}[-\pi + \Delta x]|\,\Delta x +
$$
$$
|\operatorname{Cos}[-\pi + 2\Delta x] - \operatorname{Sin}[-\pi + 2\Delta x]|\,\Delta x +
$$
$$
\cdots + |\operatorname{Cos}[\pi - \Delta x] - \operatorname{Sin}[\pi - \Delta x]|\,\Delta x) =
$$
$$
\sum_{\substack{x=-\pi \\ \text{step } \Delta x}}^{\pi - \Delta x} [|\operatorname{Cos}[x] - \operatorname{Sin}[x]|\,\Delta x]
$$

12.1.5   More Slices

In Chapter 14, we study geometric slicing approximations in greater detail. The first approximation in that chapter, the length of the sine curve, shows why extra care must sometimes be taken to find a correct approximation.

(Exercise set 12.1)

1. *Show that the volume approximation for the cone of height 10 and base radius 5 when cut into 10 slices with thickness 1 and radii at the midpoints, $\frac{1}{2}, \frac{3}{2}, \cdots, \frac{19}{2}$ is*

$$
\frac{\pi}{4^2}(1^2 + 3^2 + 5^2 + \cdots + 19^2) = \frac{\pi\,1330}{4^2} \approx 261.
$$

*Begin by writing the volume approximation as a sum of the volumes of approximating disks as in the above list for 5 slices. See the program **Sums**.*

2. *Verify that your 10-slice approximation sum is a special case of this formula*

$$\pi \sum_{\substack{x=a \\ step\ \Delta x}}^{b} \left[(R[x])^2 \Delta x, \{x, \Delta x/2, 10 - \Delta x/2, \Delta x\} \right]$$

in the case where $\Delta x = 1$ by writing out the steps of the sum command.

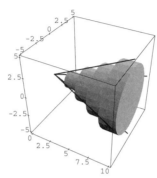

Figure 12.8: A cone approximated by 5 "disks"-radii at left

3.   (a) *Use the program **Cone Vol** to draw a cone sliced into 20 slices.*

(b) *Modify the program **Cone Vol** to draw a sphere sliced into 20 slices.*

If we measured our radii at the left end of the slice points instead of the midpoint, the approximation would be less accurate, because the slices fit completely inside the cone but the formula would be simpler,

$$\pi[(R[0])^2 \Delta x + (R[1\Delta x])^2 \Delta x + (R[2\Delta x])^2 \Delta x + \cdots + (R[10 - \Delta x])^2 \Delta x]$$

$$= \pi \sum_{\substack{x=a \\ step\ \Delta x}}^{b} \left[(R[x])^2 \Delta x, \{x, 0, 10 - \Delta x, \Delta x\} \right]$$

This has the form $f[a]\,\Delta x + f[a + \Delta x]\,\Delta x + f[a + 2\Delta x]\,\Delta x + \cdots + f[b - \Delta x]\,\Delta x$ where $f[x] = \pi R^2[x]$, $a = 0$ and $b = 10$.

4. *The approximation for the area between the sine and cosine curves shown on Figure 12.7 is*

$$\sum_{\substack{x=-\pi \\ step\ \Delta x}}^{\pi - \Delta x} \left[|\operatorname{Cos}[x] - \operatorname{Sin}[x]|\ \Delta x \right]$$

What function $f[x]$ and numbers a and b makes this sum have the form

$$\sum_{\substack{x=a \\ step\ \Delta x}}^{b - \Delta x} \left[f[x]\Delta x \right]$$

12.2   Distance When Speed Varies

If we drive 50 mph for an hour and a half, we compute the distance traveled by $D = R \cdot T$, "distance equals the rate times the time," $D = 50 \cdot \frac{3}{2} = 75$ miles. If we vary our speed, we cannot use this formula. We can compute the distance traveled as an integral. This section shows how.

Suppose we get on the highway and accelerate. We speed up cautiously and drive 25 mph for 1 minute, 26 mph for 1 minute, 27 mph for 1 minute, $\cdots$, 49 mph for 1 minute and 50 mph for the remainder of the hour and a half as shown on Figure 12.9. How far do we go?

The distance traveled at 25 mph must be computed in the correct units,

$$D_1 = 25 \cdot \frac{1}{60} \approx 0.4167 \qquad (\text{mph} \times \text{hours} = \text{miles})$$

The distance traveled at 26 mph is

$$D_2 = 26 \cdot \frac{1}{60} \approx 0.4333 \qquad (\text{miles})$$

The distance traveled at 27 mph is

$$D_3 = 27 \cdot \frac{1}{60} \approx 0.45 \qquad (\text{miles})$$

Each minute's distance can be computed for 25 minutes, giving a sum of

$$\frac{25}{60} + \frac{26}{60} + \cdots + \frac{49}{60} \approx 15.42$$

The last part of the trip is $\frac{50 \cdot (90-25)}{60} \approx 54.17$ for a total of 69.59 miles.

There are several questions:

(a) What have we done symbolically?
(b) How can we interpret the computation geometrically?
(c) Why should we interpret the distance computation geometrically?
(d) How can we extend this to continuously varying speed?

Symbolically, we have speed as a function of time in minutes, $S[1] = 25$, $S[2] = 26$, $S[3] = 27, \cdots$ and

$$\text{distance traveled during acceleration} = \sum_{\substack{x=1 \\ \text{step } 1}}^{25} [S[m] \cdot \frac{1}{60}]$$

whereas we have

$$\text{distance traveled at 50 mph} = 50 \cdot \frac{65}{60}$$

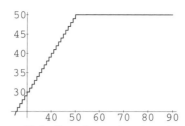

Figure 12.9: Speed vs. time in minutes

We may also think of this as a sum with a rate function defined at each minute, $S[m] = 50$, for all the minutes from 26 to 90. However, it is better to also use the correct units, $t =$ elapsed time in hours, and give the speed by the function

$$R[t] = 25 \qquad \text{for } 0 \le t < \frac{1}{60}$$

$$R[t] = 26 \qquad \text{for } \frac{1}{60} \le t < \frac{2}{60}$$

$$R[t] = 27 \qquad \text{for } \frac{2}{60} \le t < \frac{3}{60}$$

$$R[t] = 28 \qquad \text{for } \frac{3}{60} \le t < \frac{4}{60}$$

$$\vdots$$

$$R[t] = 50 \qquad \text{for } \frac{3}{60} \le t \le \frac{90}{60}$$

In this case, we can combine the two pieces into one sum

$$\text{distance traveled} = \sum_{\substack{t=0 \\ \text{step } \frac{1}{60}}}^{\frac{3}{2} - \frac{1}{60}} [R[t] \cdot \frac{1}{60}]$$

$$= \sum_{\substack{x=a \\ \text{step } \Delta t}}^{b - \Delta t} [R[t] \cdot \Delta t]$$

with $\Delta t = \frac{1}{60}$, $a = 0$ and $b = \frac{3}{2}$.

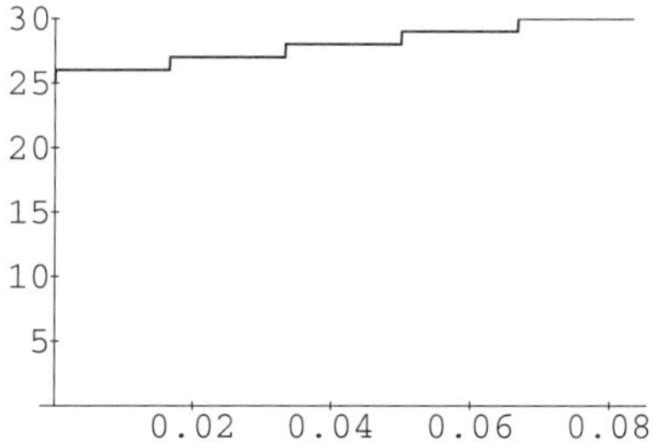

Figure 12.10: Speed vs. time in hours for five minutes

The products $25 \cdot \frac{1}{60}$, $26 \cdot \frac{1}{60}$, $27 \cdot \frac{1}{60}$, etc., can be associated with rectangles of height 25, 26, 27, etc., and width $\frac{1}{60}$ hours or 1 minute. Distance is not an area; but, in this case, the distance traveled at 25 mph is a product and that product is also an area of the rectangle under the segment on the graph of speed vs. time. This means that we can represent the distance traveled as the total area between the speed curve and the x-axis - provided our units are hours on the x-axis and miles per hour on the y-axis.

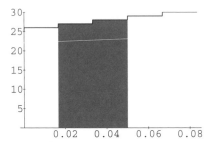

Figure 12.11: Distance represented as area

The area representation helps us see what happens if we accelerate continuously, instead of going exactly 25 mph for exactly 1 minute, 26 mph for another minute, and so forth Suppose we accelerate linearly from 25 mph to 50 mph in 25 minutes, so that our speed is given by

$$R[t] = 25 + 60\,t \qquad \text{for } 0 \le t < \frac{25}{60}$$

and

$$R[t] = 50 \qquad\qquad \text{for } \frac{3}{60} \le t < \frac{90}{60}$$

We could calculate the distance traveled in each second using the speed at the beginning of the second,

$$25 \cdot \frac{1}{3600} = \frac{1}{144} \approx 0.006944$$
$$(25 + 60\,\frac{2}{3600}) \cdot \frac{1}{3600} = \frac{1501}{216000} \approx 0.006949$$
$$(25 + 60\,\frac{3}{3600}) \cdot \frac{1}{3600} = \frac{751}{108000} \approx 0.006954$$
$$(25 + 60\,\frac{4}{3600}) \cdot \frac{1}{3600} = \frac{167}{24000} \approx 0.006958$$

This computation is no fun to do by hand, but the computer gives

$$\sum_{\substack{x=0 \\ \text{step } 1/3600}}^{1.5} [R[t] \cdot \Delta t] = \frac{20099}{288} \approx 69.788$$

when $\Delta t = 1/3600$.

The the computer computation is a waste in this case, however, because we can associate the distance traveled with the area under the speed curve in Figure 12.12 and use the formula for a trapezoid for $0 \le t < 25/60$ and for a rectangle for $25/60 \le t \le 90/60$,

$$\begin{aligned} \text{area} &= \frac{1}{2}(25 + 50) \cdot \frac{25}{60} + 50 \cdot \frac{65}{60} \\ &= \frac{1675}{24} \\ &\approx 69.7917 \end{aligned}$$

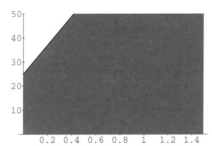

Figure 12.12: Linear acceleration and distance

If speed varies by a more complicated rule than linearly in pieces, we can still associate the area under the speed curve with the distance traveled, but now we have no formula to find the area. We will have to find that area and distance by definite integration. For a short interval of time, where speed does not change very much, the distance is approximately the product $R[t] \cdot \Delta t$.

$$\text{distance traveled from time } t \text{ to time } t + \Delta t = R[t] \cdot \Delta t + \varepsilon \cdot \Delta t$$

The approximation is actually close, even compared to Δt, which makes total distance an integral. We will see exactly what the approximation is and verify that it holds as long as $R[t]$ is a continuous function. This should be plausible, because we could also approximate by

$$\text{speed traveled from time } t \text{ to time } t + \Delta t \approx R[t']$$

for any t' between t and $t + \Delta t$. Continuity of $R[t]$ means that $R[t] \approx R[t']$, so $R[t] \, \Delta t = R[t'] \, \Delta t + \varepsilon \cdot \Delta t$, with $\varepsilon \approx 0$.

Exercise set 12.2

1. The Shortcut to Grandmother's
You're late and rush to get to Thanksgiving dinner at Grandmother's house in 2 hours. Your speed is recorded on Figure 12.13. How far away does Grandmother live? (Note: 10 minutes = 0.16667 hours.) How much was the ticket?

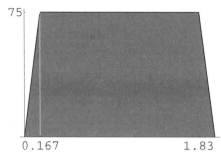

Figure 12.13: The speed on a trip to grandmother's

2. The Return from Grandmother's
A graph of your trip odometer is shown on Figure 12.14 on the return from Grand-mother's to school. Graph your speed and show the distance as an area.

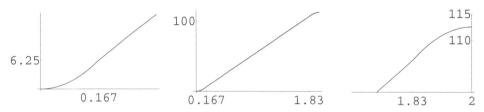

Figure 12.14: The speed on the trip back from grandmother's

12.3   The Definite Integral

This section records the definition of the integral that was motivated by ideas like the problem of computing distance from variable speeds and by the geometric area, length, and volume "slicing" problems.

Definition 12.1.   *The Definite Integral in One Variable*
Let $f[x]$ be a continuous function defined on the interval $[a, b]$. The definite integral of $f[x]$ over $[a, b]$ is given by the following limit:

$$\int_a^b f[x]\,dx = \lim_{\Delta x \to 0} \sum_{\substack{x=a \\ step\ \Delta x}}^{b-\Delta x} [f[x]\Delta x]$$

$$\approx \sum_{\substack{x=a \\ step\ \delta x}}^{b-\delta x} [f[x]\,\delta x], \quad when \ \ \delta x, \approx 0$$

We need to use the algebraic properties of the summation function to deduce results about the integral. In either the limiting case $\Delta x \to 0$ or when we extend the $\sum_{\substack{x=a \\ \text{step } \Delta x}}^{b}$ function to tiny increments, $\delta x \approx 0$, algebra of general sums extends to integrals. The most basic question is, "What is the role of continuity of $f[x]$ in the definition of $\int_a^b f[x]\,dx$?" The answer is, "Continuity guarantees us that the limit exists or, equivalently, that we get nearly the same value for each tiny δx." Extension of the integral to certain discontinuous functions is possible, but involves extra mathematical complications. Four steps in an example of this limit are shown in Figure 12.15.

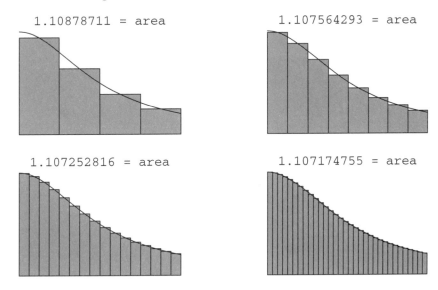

Figure 12.15: Approximations to the Area ArcTan[2] ≈ 1.10715

One of the most important algebraic properties of summation, the Telescoping Sum Theorem 12.2, plus the differential approximation (or microscope equation) for a smooth function will give us half of the Fundamental Theorem of Integral Calculus 12.16. This theorem tells us how to find exact (symbolic) integrals without summing or taking a limit. We build lots of "techniques of integration" around that theorem because it gives us these exact answers. Summation is still important, however, because the "sums of little slices" idea is where nearly all applications of integration come from.

Exercise set 12.3

1. *Run the computer program **IntegrAprx**. Compare the convergence of the slice approximations with the exact area of a circle described there.*

12.4 Computer Summation

Suppose that $F[x]$ is a function of x. To form the sum

$$F[a] + F[a + \Delta x] + F[a + 2\Delta x] + \cdots + F[b]$$

we may use a single computer command

$$\sum_{\substack{x=a \\ step \; \Delta x}}^{b} F[x]$$

This section studies this operator.

For example, if $F[x] = 1$, $a = 0$, $b = 1$ and $\Delta x = 1$

$$\sum_{\substack{x=a \\ step \; \Delta x}}^{b} 1 = F[0] + F[1] = 1 + 1 = 2$$

$$\sum_{\substack{x=a \\ step \; \Delta x}}^{b} 1$$

If we change to $\Delta x = \frac{1}{2}$, then

$$\sum_{\substack{x=a \\ step \; \Delta x}}^{b} 1 = F[0] + F\left[\frac{1}{2}\right] + F[1] = 3$$

The way $\sum_{\substack{x=a \\ step \; \Delta x}}^{b} F[x]$ works on the computer is as follows: We start with $x = a$ and start with $\sum_{\substack{x=a \\ step \; \Delta x}}^{b} = F[x] = F[a]$. Next, we add Δx to x and ask whether $x > b$. If not, we add $F[x] = F[a + \Delta x]$ to $\sum_{\substack{x=a \\ step \; \Delta x}}^{b}$, increment x again, check $x > b$, add $F[x]$ to $\sum_{\substack{x=a \\ step \; \Delta x}}^{b}$, and continue until $x > b$.

Suppose we have $F[x] = x$, $a = 0$, $b = 1$ and $\Delta x = \frac{1}{2}$, then

$$\sum_{\substack{x=a \\ step \; \Delta x}}^{b} F[x] = F[0] + F\left[\frac{1}{2}\right] + F[1] = 0 + \frac{1}{2} + 1 = \frac{3}{2}$$

Keeping the other values, but changing to $\Delta x = \frac{1}{4}$ gives

$$\sum_{\substack{x=a \\ step \; \Delta x}}^{b} x = F[0] + F\left[\frac{1}{4}\right] + F\left[\frac{2}{4}\right] + F\left[\frac{3}{4}\right] + F[1]$$

$$= 0 + \frac{1}{4} + \frac{2}{4} + \frac{3}{4} + 1 = \frac{10}{4}$$

Values of Δx that do not exactly divide $b - a$ are also allowed, so when $\Delta x = .15$,

$$\sum_{\substack{x=a \\ step\ \Delta x}}^{b} x = 0 + .15 + .30 + .45 + .60 + .75 + .90 = 3.15$$

In integration, we will take Δx smaller and smaller and still reach a limiting value in our sums by having a factor Δx in the summand. The next exercise shows you how this works.

Exercise set 12.4

1. *Write out all the terms of the sums* $\sum_{\substack{x=a \\ step\ \Delta x}}^{b} F[x]$ *defined by the following functions and increments, always using* $a = 0$ *and* $b = 1$
   (a) $F[x] = x$, $\Delta x = 1/5$
   (b) $F[x] = x\Delta x$, $\Delta x = 1/2$
   (c) $F[x] = x\Delta x$, $\Delta x = 1/3$
   (d) $F[x] = x\Delta x$, $\Delta x = 1/4$
   (e) $F[x] = x\Delta x$, $\Delta x = 1/5$

2. *Use the computer program* **Sums** *to compare the summed values of* $\sum_{\substack{x=a \\ step\ \Delta x}}^{b} x$ *and* $\sum_{\substack{x=a \\ step\ \Delta x}}^{b} x\Delta x$ *for* $\Delta x = \frac{1}{2}$, $\Delta x = \frac{1}{3}$, $\Delta x = \frac{1}{4}$, $\Delta x = \frac{1}{5}$.

12.5   The Algebra of Summation

This section writes simple properties of summation in terms of the summation operator $\sum_{\substack{x=a \\ step\ \Delta x}}^{b}$. *These properties allow us to develop an algebra of integration in the following section.*

Our first result is an algebraic method of computing exact sums.

Theorem 12.2.   *The Telescoping Sum Theorem*
If $F[x]$ is defined for $a \leq x \leq b$, then the sum of differences below equals the last value minus the first,

$$\sum_{\substack{x=a \\ step\ \Delta x}}^{b} F[x + \Delta x] - F[x] = F[b'] - F[a]$$

where b' is the last value of x of the form $a + n\Delta x$, which is less than or equal to b. If Δx divides $b - a$ exactly, $b' = b$.

PROOF:

$$\sum_{\substack{x=a \\ \text{step } \Delta x}}^{b} F[x + \Delta x] - F[x] =$$

$$(F[a + \Delta x] - F[a]) + (F[a + 2\Delta x] - F[a + \Delta x])$$
$$+ (F[a + 3\Delta x] - F[a + 2\Delta x]) + (F[a + 4\Delta x] - F[a + 3\Delta x])$$
$$+ \cdots + (F[b'] - F[b' - \Delta x])$$

The part $F[a + \Delta x]$ from the first term cancels the part $-F[a + \Delta x]$ from the second term. The part $F[a + 2\Delta x]$ from the second term cancels the part $-F[a + 2\Delta x]$ from the third term, and so on. Each term has a positive and part that cancels the corresponding negative part from the next term. The negative part from the first term and the positive part from the last term are never canceled, so the sum "telescopes" to $F[b'] - F[a]$.

Notice that we sum to $b - \Delta x$ (in steps of Δx), so that the last term is $F[b]$ when Δx divides $b - a$ (rather than $(F[b + \Delta x] - F[b])$)

Example 12.4. *Finding the Difference*

The difficulty in using the Telescoping Sum Theorem is in finding an expression of the form $F[x + \Delta x] - F[x]$ for the summand. Consider the sum

$$\sum_{\substack{x=a \\ \text{step } \Delta x}}^{b} \left[\frac{\Delta x}{x(x + \Delta x)} \right]$$

Assuming that neither $x = 0$ nor $x + \Delta x = 0$, we can write

$$\frac{\Delta x}{x(x + \Delta x)} = \frac{1}{x} - \frac{1}{x + \Delta x}$$

because putting the right-hand side on a common denominator gives,

$$\frac{1}{x} - \frac{1}{x + \Delta x} = \frac{x + \Delta x}{x(x + \Delta x)} - \frac{x}{x(x + \Delta x)}$$
$$= \frac{x + \Delta x - x}{x(x + \Delta x)} = \frac{\Delta x}{x(x + \Delta x)}$$

This makes computation of the sum easy,

$$\sum_{\substack{x=a \\ \text{step } \Delta x}}^{b-\Delta x} \left[\frac{\Delta x}{x(x + \Delta x)} \right] = \sum_{\substack{x=a \\ \text{step } \Delta x}}^{b-\Delta x} \left[\frac{1}{x} - \frac{1}{x + \Delta x} \right]$$

$$= \sum_{\substack{x=a \\ \text{step } \Delta x}}^{b-\Delta x} \left[-\left(\frac{1}{x + \Delta x} - \frac{1}{x} \right) \right]$$

$$= F[b'] - F[a] = -\left(\frac{1}{b'} - \frac{1}{a} \right)$$

$$= \frac{1}{a} - \frac{1}{b'}$$

Theorem 12.3. *Superposition of Summation*
*Let α and β be constants, and let $F[x]$ and $G[x]$ be functions defined on $[a, b]$.
Then*

$$\sum_{\substack{x=a \\ step\ \Delta x}}^{b-\Delta x} [\alpha F[x] + \beta G[x]] = \alpha \sum_{\substack{x=a \\ step\ \Delta x}}^{b-\Delta x} [F[x]] + \beta \sum_{\substack{x=a \\ step\ \Delta x}}^{b-\Delta x} [G[x]]$$

PROOF:

$$[\alpha F[a] + \beta G[a]] + [\alpha F[a + \Delta x] + \beta G[a + \Delta x]]$$
$$+ [\alpha F[a + 2\Delta x] + \beta G[a + 2\Delta x]] + \cdots + [\alpha F[b'] + \beta G[b']]$$
$$=$$
$$\alpha[F[a] + F[a + \Delta x] + F[a + 2\Delta x] + \cdots + F[b']]$$
$$+ \beta[G[a] + G[a + \Delta x] + G[a + 2\Delta x] + \cdots + G[b']]$$

Example 12.5. $\sum_{\substack{x=a \\ step\ \Delta x}}^{b} [x\Delta x] = \frac{b^2 - a^2}{2} - \Delta x \frac{b - a}{2}$

Given that

$$\sum_{\substack{x=a \\ step\ \Delta x}}^{b-\Delta x} [\Delta x] = \sum_{\substack{x=a \\ step\ \Delta x}}^{b-\Delta x} [([x + \Delta x] - x)] = b - a$$

$$\sum_{\substack{x=a \\ step\ \Delta x}}^{b-\Delta x} [[x + \Delta x]^2 - x^2] = b^2 - a^2$$

$$\text{and } [x + \Delta x]^2 - x^2 = 2x\Delta x + \Delta x^2$$

we can use superposition to find

$$\sum_{\substack{x=a \\ step\ \Delta x}}^{b-\Delta x} [x\Delta x] = ?$$

We have

$$b^2 - a^2 = \sum_{\substack{x=a \\ \text{step } \Delta x}}^{b-\Delta x} [[x+\Delta x]^2 - x^2]$$

$$= \sum_{\substack{x=a \\ \text{step } \Delta x}}^{b-\Delta x} [2x\,\Delta x + \Delta x^2]$$

$$= 2 \sum_{\substack{x=a \\ \text{step } \Delta x}}^{b-\Delta x} [x\,\Delta x] + \Delta x \sum_{\substack{x=a \\ \text{step } \Delta x}}^{b-\Delta x} [\Delta x]$$

$$= 2 \sum_{\substack{x=a \\ \text{step } \Delta x}}^{b-\Delta x} [x\,\Delta x] + \Delta x\,(b-a)$$

Solving for the unknown sum,

$$2 \sum_{\substack{x=a \\ \text{step } \Delta x}}^{b} [x\,\Delta x] = (b^2 - a^2) - \Delta x\,(b-a)$$

Dividing by 2, we have our formula:

$$\sum_{\substack{x=a \\ \text{step } \Delta x}}^{b-\Delta x} [x\,\Delta x] = \frac{1}{2}\left((b^2-a^2) - \Delta x\,(b-a)\right)$$

Notice that if Δx is small, this formula says,

$$\sum_{\substack{x=a \\ \text{step } \Delta x}}^{b-\Delta x} [x\,\Delta x] \approx \frac{1}{2}(b^2 - a^2)$$

Theorem 12.4. *The Triangle Inequality*
For $F[x]$ defined on $[a,b]$,

$$\left| \sum_{\substack{x=a \\ \text{step } \Delta x}}^{b-\Delta x} [F[x]] \right| \leq \sum_{\substack{x=a \\ \text{step } \Delta x}}^{b-\Delta x} [|F[x]|]$$

PROOF:
  Terms of opposite sign inside the sum $|F[a]+F[a+\Delta x]+\cdots+F[b]|$ could cancel and make that sum smaller than the sum with all positive terms, $|F[a]| + |F[a+\Delta x]| + \cdots + |F[b]|$.

Theorem 12.5. *Additivity of Summation*
    Suppose $F[x]$ is defined on $[a, c]$, Δx divides $b - a$ and $a < b < c$. Then

$$\sum_{\substack{x=a \\ step \ \Delta x}}^{b-\Delta x} [F[x]] + \sum_{\substack{x=b \\ step \ \Delta x}}^{c-\Delta x} [F[x]] = \sum_{\substack{x=a \\ step \ \Delta x}}^{c-\Delta x} [F[x]]$$

PROOF:

$$[F[a] + F[a + \Delta x] + \cdots + F[b - \Delta x]] + [F[b] + F[b + \Delta x] + \cdots + F[c']]$$
$$= [F[a] + F[a + \Delta x] + \cdots + F[b - \Delta x] + F[b] + F[b + \Delta x] + \cdots + F[c']]$$

Theorem 12.6. *Monotony of Summation*
    If $F[x] \le G[x]$, then

$$\sum_{\substack{x=a \\ step \ \Delta x}}^{b-\Delta x} [F[x]] \le \sum_{\substack{x=a \\ step \ \Delta x}}^{b-\Delta x} [G[x]]$$

PROOF:
    This is obvious because each term in the left sum is smaller than the corresponding term in the right sum.

Theorem 12.7. *Orientation of Summation*
    Suppose $f[x]$ is defined on $[a, b]$ with $a < b$ and Δx divides $b - a$, then

$$\sum_{\substack{x=b \\ step \ -\Delta x}}^{a} [f[x](-\Delta x)] = - \sum_{\substack{x=a \\ step \ \Delta x}}^{b} [f[x]\Delta x]$$

PROOF: To go from b to a, we must take negative steps.

Exercise set 12.5

1. What's the Difference?
    Use the Telescoping Sum Theorem to give exact answers to the following sums in the case where Δx divides $b - a$.
    (a) $\sum_{\substack{x=a \\ step \ \Delta x}}^{b-\Delta x} [\Delta x]$
    (b) $\sum_{\substack{x=a \\ step \ \Delta x}}^{b-\Delta x} [2x\Delta x + \Delta x^2]$
    (c) $\sum_{\substack{x=a \\ step \ \Delta x}}^{b-\Delta x} [3x^2\Delta x + 3x\Delta x^2 + \Delta x^3]$
    The question really is, "What $F[x + \Delta x] - F[x]$ equals the expressions that you are asked to sum?"

2.   (a) *Use telescoping sums, superposition, and known sums to find*

$$\sum_{\substack{x=a \\ step\ \Delta x}}^{b-\Delta x} [x^2 \Delta x] = ?$$

  Notice that

$$\sum_{\substack{x=a \\ step\ \Delta x}}^{b-\Delta x} [\Delta x] = (b - a)$$

$$\sum_{\substack{x=a \\ step\ \Delta x}}^{b-\Delta x} [x\Delta x] = \frac{1}{2}\left((b^2 - a^2) - \Delta x\ (b - a)\right)$$

$$and\ (x + \Delta x)^3 - x^3 = 3\,x^2\,\Delta x + 3\,x\,\Delta x^2 + \Delta x^3$$

   (b) *Use telescoping sums, superposition and known sums to find*

$$\sum_{\substack{x=a \\ step\ \Delta x}}^{b-\Delta x} [x^3 \Delta x] = ?$$

3. *Give an example of a sum with the cancellation described in the proof of the triangle inequality.*

4. *Write out the three sums in the proof of Additivity for $a = 0$, $b = 1$, $c = 2$, $\Delta x = \frac{1}{2}$, and $F[x] = x\Delta x$.*

5. *Write out the two sums in the proof of Monotony for $a = 0$, $b = 1$, $\Delta x = \frac{1}{2}$, $F[x] = x^2\Delta x$, and $G[x] = x\Delta x$.*

6. *Write out the two sums in the proof of Orientation for $a = 0$, $b = 1$, $\Delta x = \frac{1}{2}$, and $F[x] = x\Delta x$.*

12.6   The Algebra of Integration

For a fixed choice of the continuous real function $f[x]$, we know that

$$\int_a^b f[x]\,dx \approx \sum_{\substack{x=a \\ step\ \delta x}}^{b-\delta x} [f[x]\,\delta x]$$

when $\delta x \approx 0$. The sum properties of the previous section hold for all Δx, so, in particular, they hold when $\Delta x = \delta x \approx 0$ is small. This shows that integrals have the following properties.

Proofs of the properties of integrals are given in the Mathematical Background.

Theorem 12.8. *Superposition*

$$\int_a^b [\alpha f[x] + \beta g[x]]\ dx = \alpha \int_a^b f[x]\ dx + \beta \int_a^b g[x]\ dx$$

Theorem 12.9. *Triangle Inequality*

$$|\int_a^b f[x]\ dx| \le \int_a^b |f[x]|\ dx, (a < b)$$

Theorem 12.10. *Monotony*

$$\int_a^b f[x]\ dx \le \int_a^b g[x]\ dx, \quad if f[x] \le g[x]$$

Theorem 12.11. *Additivity*

$$\int_a^b f[x]\ dx + \int_b^c f[x]\ dx = \int_a^c f[x]\ dx$$

Theorem 12.12. *Orientation*

$$\int_b^a f[x]\ dx = -\int_a^b f[x]\ dx$$

Note: The "dx" in the reverse-oriented integral can be thought of as negative to move us from b back to a.

Example 12.6. *Integral Computation the Hard Way*

We have seen that

$$\sum_{\substack{x=a \\ \text{step } \Delta x}}^{b-\Delta x} [x\Delta x] = \frac{1}{2}\left((b^2 - a^2) - \Delta x(b - a)\right)$$

so, when $\delta x \approx 0$,

$$\int_a^b x\ dx \approx \sum_{\substack{x=a \\ \text{step } \delta x}}^{b-\delta x} [x\delta x]$$

$$= \frac{1}{2}\left((b^2 - a^2) - \Delta x(b - a)\right)$$

$$\approx \frac{1}{2}\left(b^2 - a^2\right)$$

and the two fixed quantities must be equal,

$$\int_a^b x\,dx = \frac{1}{2}\left(b^2 - a^2\right)$$

Theorem 12.13. *Estimation of Sums*

$$\sum_{\substack{x=a \\ step\ \delta x}}^{b-\delta x} [|F[x,\delta x]|\,\delta x] \leq |F[x_{Max},\delta x]| \times [b-a]$$

In particular,

$$\sum_{\substack{x=a \\ step\ \delta x}}^{b-\delta x} [\varepsilon[x,\delta x]\,\delta x] \approx 0$$

if $\varepsilon \approx 0$ *for all* x *in* $[a,b]$ *when* $\delta x \approx 0$.

PROOF:

This is the last technical result we need from algebra. We use the finite maximum function. Define a function like the computer's maximum

$$\text{Max}[F[x,\Delta x],\{x=a,b-\Delta x,\Delta x\}] =$$
$$\text{maximum}[F[a,\Delta x],F[a+\Delta x,\Delta x],F[a+2\Delta x,\Delta x],\ldots,F[b',\Delta x]]$$

We know that

$$\text{Max}[F[x,\Delta x],\{a,b-\Delta x,\Delta x\}] = F[x_M,\Delta x]$$

for some real x_M of the form $x_M = a + n\Delta x$, $a \leq x_M \leq b - \Delta x$. Apply the Max function to a small $\delta x \approx 0$,

$$\text{Max}[F[x,\delta x],\{a,b-\delta x,\delta x\}] = F[x_M,\delta x]$$

for some x_M of the form $x_M = a + n\delta x$, $a \leq x_M \leq b$.

The formula we need is the formal statement that we can estimate by making all the terms of a sum larger,

$$\sum_{\substack{x=a \\ step\ \delta x}}^{b-\delta x} [|F[x,\delta x]|\,\delta x]$$

$$\leq \text{Max}[|F[x,\delta x]|,\{a,b-\delta x,\delta x\}] \times \sum_{\substack{x=a \\ step\ \delta x}}^{b-\delta x} [\delta x]$$

$$= |F[x_M,\delta x]| \times \sum_{\substack{x=a \\ step\ \delta x}}^{b-\delta x} [\delta x] = |F[x_M,\delta x]| \times (b-a)$$

Exercise set 12.6

1. (a) *Compute $\int_a^b x^2\,dx$ by extending the exact formula you computed earlier for*

$$\sum_{\substack{x=a \\ step\ \Delta x}}^{b-\Delta x} [x^2 \Delta x] \qquad to\ \delta x \approx 0$$

(b) *Compute $\int_a^b x^3\,dx$ by extending the exact formula for*

$$\sum_{\substack{x=a \\ step\ \Delta x}}^{b-\Delta x} [x^3 \Delta x] \qquad to\ \delta x \approx 0$$

(c) *Compute $\int_a^b x^n\,dx$ the hard way for integer n by showing*

$$(x+\delta x)^n - x^n = n\,x^{n-1}\cdot \delta x + \varepsilon\cdot \delta x$$

with $\varepsilon = \varepsilon[x,\delta x] \approx 0$. Note that

$$\sum_{\substack{x=a \\ step\ \delta x}}^{b-\delta x} [(x+\delta x)^n - x^n] \approx b^n - a^n$$

2. Estimation for $\int_a^b \frac{1}{x^2}\,dx$
  We showed above that

$$\sum_{\substack{x=a \\ step\ \Delta x}}^{b-\Delta x} [\frac{\Delta x}{x(x+\Delta x)}] = \frac{1}{a} - \frac{1}{b'}$$

provided that neither $x=0$ nor $x+\Delta x = 0$ for any term of the sum.
Compute the integral

$$\int_a^b \frac{1}{x^2}\,dx = \frac{1}{a} - \frac{1}{b}$$

by estimating the difference between the unknown sum

$$\sum_{\substack{x=a \\ step\ \delta x}}^{b-\delta x} [\frac{\delta x}{x^2}]$$

and the sum above,

$$\sum_{\substack{x=a \\ step\ \delta x}}^{b-\delta x} [\frac{\delta x}{x(x+\delta x)}] = \frac{1}{a} - \frac{1}{b'}$$

HINT: Calculate the difference $\left[\frac{1}{x^2} - \frac{1}{x(x+\delta x)}\right]\delta x$ to estimate

$$\sum_{\substack{x=a \\ step\ \delta x}}^{b-\delta x} [\frac{\delta x}{x^2}] - \sum_{\substack{x=a \\ step\ \delta x}}^{b-\delta x} [\frac{\delta x}{x(x+\delta x)}]$$

Is your computation valid if $a < 0 < b$?

3. Estimation for $\int_a^b \frac{1}{x^3}\,dx$
  Show that

$$\int_a^b \frac{1}{x^3}\,dx = \frac{1}{2a^2} - \frac{1}{2b^2}$$

as follows:
First, by putting the difference on a common denominator, show that

$$\frac{1}{x^3} = \frac{x}{(x^2 - \delta x^2)^2} + \varepsilon$$

with $\varepsilon \approx 0$ if $\delta x \approx 0$.

$$\varepsilon = \frac{1}{x^3} - \frac{x}{(x^2 - \delta x^2)^2}$$

Second, show that

$$\frac{4\,x\,\delta x}{(x^2 - \delta x^2)^2} = \frac{1}{(x - \delta x)^2} - \frac{1}{(x + \delta x)^2}$$

by putting the right-hand side on a common denominator.
Third, show that

$$\frac{4\,x\,\delta x}{(x^2 - \delta x^2)^2} = \left(\frac{1}{(x - \delta x)^2} + \frac{1}{x^2} \right) - \left(\frac{1}{(x + \delta x)^2} + \frac{1}{x^2} \right)$$

and that

$$\sum_{\substack{x=a \\ \text{step } \delta x}}^{b-\delta x} \left[\frac{4\,x\,\delta x}{(x^2 - \delta x^2)^2} \right] = \left(\frac{1}{a^2} + \frac{1}{(a - \delta x)^2} \right) - \left(\frac{1}{b^2} + \frac{1}{(b - \delta x)^2} \right)$$

4. Estimation for $\int_a^b \frac{1}{x^7}\,dx$
  You can sum $\frac{1}{(x-\delta x)^3} - \frac{1}{(x+\delta x)^3}$. What integral does that tell you?

5. Estimation for $\int_a^b \frac{1}{x^{n+1}}\,dx$
  Show that

$$\int_a^b \frac{1}{x^{n+1}}\,dx = \frac{1}{n} \left(\frac{1}{a^n} - \frac{1}{b^n} \right)$$

as follows. First, show that

$$\frac{1}{x^n} - \frac{1}{(x + \delta x)^n} = \frac{n\,x^{n-1}\delta x}{x^n(x + \delta x)^n} + \varepsilon_1 \cdot \delta x$$

with $\varepsilon_1 \approx 0$ when $\delta x \approx 0$. Second,

$$\frac{x^{n-1}}{x^n(x + \delta x)^n} = \frac{1}{x^{n+1}} + \varepsilon_2$$

with $\varepsilon_2 \approx 0$ when $\delta x \approx 0$. Can we have $a < 0 < b$?

6. Estimation for $\int_a^b \frac{1}{2\sqrt{x}}\, dx$

You can sum $\sqrt{x + \delta x} - \sqrt{x}$ in general for $a \le x \le b$. Multiply this expression by $\sqrt{x + \delta x} + \sqrt{x}$

$$\sqrt{x + \delta x} - \sqrt{x} = \frac{(\sqrt{x + \delta x} - \sqrt{x})(\sqrt{x + \delta x} + \sqrt{x})}{\sqrt{x + \delta x} + \sqrt{x}}$$

to find

$$\int_a^b \frac{1}{2\sqrt{x}}\, dx$$

7. Estimation for $\int_a^b \sqrt{x}\, dx$

Sum $\sqrt{(x + \delta x)^3} - \sqrt{x^3}$ to find

$$\int_a^b \sqrt{x}\, dx$$

12.7   Fundamental Theorem, Part 1

This section gives the theory to find integrals by antiderivatives without taking limits or approximating with tiny increments. (Chapter 13 gives techniques for using the theory.) The theory results from an estimate and a telescoping sum. The general estimate is even simpler than the specific telescoping sum exercises of the previous section.

Recall the differential approximation for a smooth function 5.5 at this time, because we are about to use it to prove

Theorem 12.14. *First Half of the Fundamental Theorem*
Given an integrand $f[x]\, dx$, suppose that we can find a function $F[x]$ such that $dF[x] = f[x]\, dx$ (or $\frac{dF}{dx} = f(x)$) for all $a \le x \le b$, then

$$\int_a^b f[x]\, dx = F[b] - F[a]$$

We also sometimes write

$$\int_a^b f[x]\, dx = \int_a^b dF[x] = F[x]\big|_a^b = F[b] - F[a]$$

The notation $F[x]\big|_a^b$ simply means $F[b] - F[a]$.

PROOF:
The differential approximation 5.5 (microscope equation) for $F[x]$ is

$$F[x + \delta x] - F[x] = f[x]\delta x + \varepsilon \cdot \delta x$$

for all x satisfying $a \le x \le b$, where $\varepsilon \approx 0$ when $\delta x \approx 0$. (This forces $f[x]$ to be a continuous function for the ordinary derivative defined in Definition 5.5. See the Mathematical Background Chapter on "Epsilons and Deltas" for further details.)

The telescoping sum and superposition properties say

$$F[b] - F[a] = \sum_{\substack{x=a \\ \text{step } \delta x}}^{b-\delta x} [F[x + \delta x] - F[x]] = \sum_{\substack{x=a \\ \text{step } \delta x}}^{b-\delta x} [f[x]\delta x + \varepsilon \cdot \delta x]$$

$$= \sum_{\substack{x=a \\ \text{step } \delta x}}^{b-\delta x} [f[x]\delta x] + \sum_{\substack{x=a \\ \text{step } \delta x}}^{b-\delta x} [\varepsilon \cdot \delta x]$$

$$\approx \int_a^b f[x]\, dx + \sum_{\substack{x=a \\ \text{step } \delta x}}^{b-\delta x} [\varepsilon \cdot \delta x]$$

We conclude the proof by using Theorem 12.13, which shows

$$\sum_{\substack{x=a \\ \text{step } \delta x}}^{b-\delta x} [\varepsilon \cdot \delta x] \approx 0$$

We have shown that the fixed quantities satisfy

$$F[b] - F[a] \approx \int_a^b f[x]\, dx$$

which forces them to be equal and proves the theorem.

Strange though it sounds, we have shown two things with the proof of the first half of the Fundamental Theorem: 1) the integral exists! and 2) its value is $F[b] - F[a]$. (Instructor Note: The usual continuity hypothesis is hidden in our ability to find a function $F[x]$ satisfying the uniform microscope equation 5.5 with $f[x] = F'[x]$ for all x in $[a, b]$.)

Example 12.7. *Computation of $\int_a^b x^2\, dx$, the Easy Way*

Compare the following use of the Fundamental Theorem with your direct computation from the last section. We have $f[x] = x^2$; so, if we take $F[x] = x^3/3$, then $dF[x] = x^2\, dx$ and

$$\int_a^b x^2\, dx = \frac{1}{3} x^3 \big|_a^b = \frac{1}{3} \left(b^3 - a^3 \right)$$

Example 12.8. *Integral of Cosine*

$$\int_a^b \mathrm{Cos}\,[\theta]\, d\theta = \mathrm{Sin}\,[\theta] \big|_a^b = \mathrm{Sin}\,[b] - \mathrm{Sin}\,[a]$$

It would be quite difficult to compute this integral without the help of the Fundamental Theorem.

────────────────── **Exercise set 12.7** ──────────────────

1. Fundamental Drill

Show that $dF[x] = f[x]\,dx$ for the given pairs of functions $F[x]$ and $f[x]$. Use this to compute the integrals.

(a) $F[x] = \frac{1}{4}x^4$, $f[x] = x^3$, $\int_0^2 x^3\,dx =$

(b) $F[x] = \sqrt{x}$, $f[x] = \frac{1}{2\sqrt{x}}$, $\int_4^9 \frac{1}{\sqrt{x}}\,dx =$

(c) $F[x] = \frac{2}{3}\sqrt{x^3}$, $f[x] = \sqrt{x}$, $\int_4^9 \sqrt{x}\,dx =$

(d) $F[x] = -\frac{1}{x^n}$, $f[x] = \frac{n}{x^{n+1}}$, $\int_2^3 \frac{1}{x^3}\,dx =$

2. *Can you use the formula $F[x] = -\frac{1}{x^n}$ implies $dF[x] = \frac{1}{x^{n+1}}\,dx$ to find $\int_2^3 \frac{1}{x}\,dx$? What function $F[x]$ has $dF[x] = \frac{1}{x}\,dx$?*

If $f[x]$ is not continuous everywhere on $[a,b]$, the Fundamental Theorem does not apply and we shall see that using $F[b] - F[a]$ anyway will lead to errors, even if $\frac{dF}{dx} = f[x]$ at all but one point of $[a,b]$.

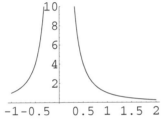
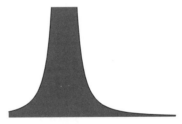

Figure 12.16: Area under the curve $y = 1/x^2$

3. *Is the following computation correct?*

$$\int_{-1}^2 \frac{1}{x^2}\,dx = \int_{-1}^2 x^{-2}\,dx$$

$$= -x^{-1}\big|_{-1}^2 = \frac{1}{x}\big|_2^{-1}$$

$$= -1 - \frac{1}{2} = -\frac{3}{2}$$

Note that if $F[x] = -x^{-1}$, then $dF[x] = x^{-2}\,dx$, but how could the "area" under a positive curve be a negative number?

Draw a picture and compute the areas $\int_{.0001}^1 \frac{1}{x^2}\,dx$, $\int_{-1}^{-.0001} \frac{1}{x^2}\,dx$ and $\int_{+1}^2 \frac{1}{x^2}\,dx$. What do you think the area $\int_{-.0001}^{.0001} \frac{1}{x^2}\,dx$ should be?

Incorrect programs can produce incorrect results.

4. Garbage In, Garbage Out

What is wrong with the following "computer computation" of $\int_0^2 \frac{1}{\mathrm{Cos}^2[x]}\,dx$?

In[1]

    $f := 1/(\mathrm{Cos}[x]) \wedge 2;$

    $Integrate[f, \{x, 0, 2\}]$ $< Enter >$

Out[1]

$$\frac{\text{Sin}[2]}{\text{Cos}[2]} \approx -2.18504$$

Can the integral of a positive function be negative? Is the function $F[x] = \text{Sin}[x]/\text{Cos}[x]$ an antiderivative for the function $f[x] = 1/\left[\text{Cos}[x]\right]^2$ over the whole interval $0 \le x \le 2$? That is, does $dF[x] = f[x]\,dx$ for all these values of x? Use the computer to plot the integrand function over the interval $0 \le x \le 2$.

12.8   Fundamental Theorem, Part 2

The second part of the Fundamental Theorem of Integral Calculus says that the derivative of an integral of a continuous function is the integrand,

$$\frac{d}{dX} \int_a^X f[x]\,dx = f[X]$$

The function $A[X] = \int_a^X f[x]\,dx$ can be thought of as the "accumulated area" under the curve $y = f[x]$ from a to X shown in Figure 12.17. The "accumulation function" can also be thought of as the reading of your odometer at time X for a given speed function $f[x]$.

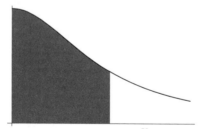

Figure 12.17: $A[X] = \int_a^X f[x]\,dx$

When we cannot find an antiderivative $F[x]$ for a given $f[x]$, we sometimes still want to work directly with the definition of the integral. A number of important functions are given by integrals of simple functions. The logarithm and arctangent are elementary examples. Some other functions like the probability function of the "bell-shaped curve" of probability and statistics do not have elementary formulas but do have integral formulas. The second half of the Fundamental Theorem justifies the integral formulas, which are useful as approximations. The **NumIntAprx** computer program shows us efficient ways to estimate the limit of sums directly.

Continuity of $f[x]$ is needed to show that the limit actually "converges." With discontinuous integrands, it is possible to make the sums oscillate as $\Delta x \to 0$. In these cases, numerical integration by computer will likely give the wrong answer.

We need naked existence of the limit for the second half of the Fundamental Theorem. The first half of the Fundamental Theorem is used often, whereas the second half is seldom used. This is why we have postponed a general proof of existence (without antiderivatives). (Our proof of the First Half of the Fundamental Theorem gave existence and a formula at the same time.)

Theorem 12.15.   *Existence of the Definite Integral*

   Let $f[x]$ be a continuous function on the interval $[a,b]$. Then there is a real number I such that

$$\sum_{\substack{x=a \\ step\ \Delta x}}^{b-\Delta x} [f[x]\Delta x] \to I \ as\ \Delta x \to 0$$

or, equivalently,

$$\sum_{\substack{x=a \\ step\ \delta x}}^{b-\delta x} [f[x]\,\delta x] \approx I \ for\ any\ \delta x \approx 0$$

PROOF:

   First, by the Extreme Value Theorem for Continuous Functions, $f[x]$ has a min, m, and a Max, M, on the interval $[a,b]$. Monotony of summation tells us

$$m \times [b-a] \le \sum_{\substack{x=a \\ step\ \delta x}}^{b-\delta x} [f[x]\,\delta x] \le M \times [b-a]$$

so that $\sum_{\substack{x=a \\ step\ \delta x}}^{b-\delta x} [f[x]\,\delta x]$ is a finite number and thus near some fixed value $I\,[\delta x] \approx \sum_{\substack{x=a \\ step\ \delta x}}^{b-\delta x} [f[x]\,\delta x]$. What we need to show is that if we choose a different increment, for example $\delta u \approx 0$, then $I\,[\delta x] = I\,[\delta u]$ or

$$\sum_{\substack{x=a \\ step\ \delta x}}^{b-\delta x} [f[x]\,\delta x] \approx \sum_{\substack{x=a \\ step\ \delta x}}^{b-\delta u} [f\,[u]\,\delta u]$$

Draw your own picture of two different rectangular approximations to clarify this idea. If we superimpose the different partitions and consider two overlapping rectangles, the areas only differ by an small amount on a scale of the increments because $f[x] \approx f[u]$ by continuity for $x \approx u$. The proof that we do get the same number is taken up in detail in the Mathematical Background.

Theorem 12.16.   *Second Half of the Fundamental Theorem*

   Suppose that $f[x]$ is a continuous function on an interval containing a and we define a new function by "accumulation,"

$$A\,[X] = \int_a^X f[x]\,dx$$

Then $A\,[X]$ is smooth and $\frac{dA}{dX}\,[X] = f\,[X]$; in other words,

$$\frac{d}{dX} \int_a^X f[x]\,dx = f\,[X]$$

PROOF:

We show that $A[X]$ satisfies the differential approximation $A[X + \delta X] - A[X] = f[X]\delta X + \varepsilon \cdot \delta X$ with $\varepsilon \approx 0$ when $\delta X \approx 0$. This proves that $f[X]$ is the derivative of $A[X]$.

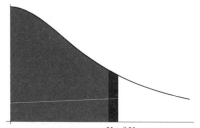

Figure 12.18: $\int_a^{X+\delta X} f[x]\, dx$

By definition of A and the additivity property of integrals, we have

$$A[X + \delta X] = \int_a^{X+\delta X} f[x]\, dx = \int_a^X f[x]\, dx + \int_X^{X+\delta X} f[x]\, dx$$

$$= A[X] + \int_X^{X+\delta X} f[x]\, dx$$

So, we need to show that

$$\int_X^{X+\delta X} f[x]\, dx = f[X]\delta X + \varepsilon \cdot \delta X$$

with $\varepsilon \approx 0$, when $\delta X \approx 0$.

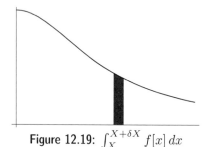

Figure 12.19: $\int_X^{X+\delta X} f[x]\, dx$

This is when we use the continuity hypothesis about $f[x]$. The Extreme Value Theorem for Continuous Functions 11.3 says that $f[x]$ has a max and a min on the interval $[X, X+\delta X]$, $m = f[X_m] \leq f[x] \leq M = f[X_M]$ for all $X \leq x \leq X + \delta X$. Monotony of the integral gives us the estimates

$$m\delta X = \int_X^{X+\delta X} m\, dx \leq \int_X^{X+\delta X} f[x]\, dx \leq \int_X^{X+\delta X} M\, dx = M \cdot \delta X$$

Since both X_m and X_M lie in the interval $[X, X + \delta X]$, we know that $X_m \approx X_M$ when $\delta X \approx 0$. Continuity of $f[x]$ means that $f[X_m] \approx f[X_M] \approx f[X]$ in this case, so

$$\int_X^{X+\delta X} f[x]\, dx = f[X]\delta X + \varepsilon \cdot \delta X$$

using upper and lower estimates of the integral based on the max and the min of the function over the small subinterval.

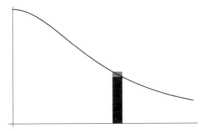

Figure 12.20: Upper and lower estimates

This proves the theorem because we have verified the microscope equation from Definition 5.5

$$A[X + \delta X] = A[X] + A'[X]\delta X + \varepsilon \cdot \delta X$$

with $A'[X] = f[X]$.

Exercise set 12.8

1. *Use the Second Half of the Fundamental Theorem to explain how YOU could compute* $\text{Log}[x]$ *and* $\text{ArcTan}[x]$ *directly using a numerical integration program. We know that*

$$\frac{d\,\text{Log}[x]}{dx} = \frac{1}{x} \quad and \quad \frac{d}{dX}\int_1^X \frac{1}{x}\,dx = \frac{1}{X}$$

and

$$\frac{d\,\text{ArcTan}[x]}{dx} = \frac{1}{1+x^2} \quad and \quad \frac{d}{dX}\int_0^X \frac{1}{1+x^2}\,dx = \frac{1}{1+X^2}$$

*Use the Numerical Integration program **NumIntAprx** to compute the various numerical integral approximations for these functions, and compare your results with the computer's built-in algorithms for these functions. For example, how do the following compare:*

$$\text{Log}[7.38905] \approx NIntegrate[1/x, \{x, 1, 7.38905\}]$$

and

$$4 \times \text{ArcTan}[1] \approx 4\,NIntegrate[1/(1 + x \wedge 2), \{x, 0, 1\}]$$

*Explain in terms of the exact symbolic computations why we have chosen these particular numbers (7.38905 in log and the 4 in arctangent). See the basic left- right- and midpoint-approximations in the program **IntegrAprx**.*

Sometimes it is not possible to find an antiderivative in terms of elementary functions, but even in these cases, the integral defines a function and the numerical approximations give a direct way to compute these functions.

2. Non-elementary Integrals

Use the computer's numerical integral numerical integration command to approximate the following integrals:

a) $\int_0^1 e^{-x^2}\,dx =$ b) $\int_0^1 \frac{\text{Sin}[x]}{x}\,dx =$ c) $\int_0^1 \text{Sin}[x^2]\,dx =$

*Also use the computer's symbolic integration to find expressions for these integrals. These expressions are not combinations of elementary functions. See the symbolic integration program **SymbolicIntegr** for details.*

13 Symbolic Integration

This chapter contains the basic tricks of the "symbolic integration" trade. The goal of this chapter is not to make you a slow innacurate integration software, but rather to help you understand the basics well enough to use modern integration software effectively.

The basic methods of integration are important, but there are many more tricks that are useful in special situations. The computer knows most of the special tricks but sometimes a basic preliminary hand computation allows the computer to calculate a very complicated integral that it cannot otherwise do. A change of variables often clarifies the meaning of an integral. Integration by parts is theoretically important in both math and physics.

We will encounter a few of the special tricks later when they arise in important contexts. The cable of a suspension bridge can be described by a differential equation, which can be antidifferentiated with the "hyperbolic cosine." If you decide to work on that project, you will want to learn that trick. "Partial fractions" is another integration trick that arises in the logistic growth model, the S-I-S disease model, and the linear air resistance model in a basic form. We will take that method up when we need it.

The Fundamental Theorem of Integral Calculus 12.14 gives us an indirect way to exactly compute the limit of approximations by sums of the form

$$f[a]\,\Delta x + f[a + \Delta x]\,\Delta x + f[a + 2\Delta x]\,\Delta x + \cdots + f[b - \Delta x]\,\Delta x = \sum_{\substack{x=a \\ \text{step } \Delta x}}^{b-\Delta x} [f[x]\Delta x]$$

We have

$$\int_a^b f[x]\,dx = \lim_{\Delta x \to 0} \sum_{\substack{x=a \\ \text{step } \Delta x}}^{b-\Delta x} [f[x]\Delta x] \approx \sum_{\substack{x=a \\ \text{step } \delta x}}^{b-\delta x} [f[x]\,\delta x]$$

but the limit can be computed without forming the sum. The Fundamental Theorem says:

If we can find $F[x]$ so that $dF[x] = f[x]\,dx$, for all $a \leq x \leq b$, then

$$\int_a^b f[x]\,dx = \lim_{\Delta x \to 0} \sum_{\substack{x=a \\ \text{step } \Delta x}}^{b-\Delta x} [f[x]\Delta x] = F[b] - F[a]$$

Finding an "antiderivative" lets us skip from the left side of the above equations to the right side without going through the limit in the middle.

This indirect computation of the integral works any time we can find a trick to figure out an antiderivative. There are many such tricks, or "techniques," and the computer knows them all. Your main task in this chapter is to understand the fundamental techniques and their limitations, rather than to develop skill at very elaborate integral computations.

The rules of integration are more difficult than the rules of differentiation because they amount to trying to use the rules of differentiation in reverse. You must learn all the basic techniques to understand this, but we will not wallow very deep into the swamp of esoteric techniques. That is left for the computer.

13.1 Indefinite Integrals

The first half of the Fundamental Theorem means that we can often find an integral in two steps: 1) Find an antiderivative, 2) compute the difference in values of the antiderivative. It makes no difference which antiderivative we use. We formalize notation that breaks up these two steps.

One antiderivative of $3x^2$ is x^3, since $\frac{dx^3}{dx} = 3x^2$, but another antiderivative of $3x^2$ is $x^3 + 273$, since the derivative of the constant 273 (or any other constant) is zero. The integral $\int_a^b 3x^2 dx$ may be computed with either antiderivative.

$$\int_a^b 3x^2 dx = x^3 \big|_a^b = b^3 - a^3$$
$$= [x^3 + 273] \big|_a^b = [b^3 + 273] - [a^3 + 273] = b^3 - a^3$$

Our next result is the converse of the result that says the derivative of a constant is zero. It says that if the derivative is zero, the function must be constant. This is geometrically obvious - draw the graph of a function with a zero derivative!

Theorem 13.1. *The Zero-Derivative Theorem*
Suppose $F[x]$ and $G[x]$ are both antiderivatives for the same function $f[x]$ on the interval $[a, b]$; that is, $\frac{dF}{dx}[x] = \frac{dG}{dx}[x] = f[x]$ for all x in $[a, b]$. Then $F[x]$ and $G[x]$ differ by a constant for all x in $[a, b]$.

PROOF:
   The function $H[x] = F[x] - G[x]$ has zero derivative on $[a, b]$. We have

$$H[X] - H[a] = \int_a^X 0 \, dx = 0$$

by the first half of the Fundamental Theorem and direct computation of the integral of zero.
   If X is any real value in $[a, b]$, $H[X] - H[a] = 0$. This means $H[X] = H[a]$, a constant. In turn, this tells us that $F[X] - G[X] = F[a] - G[a]$, a constant for all X in $[a, b]$.

Definition 13.2. *Notation for the Indefinite Integral*
The indefinite integral of a function f [x] denoted

$$\int f[x]\,dx$$

is equal to the collection of all functions F [x] with differential

$$dF[x] = f[x]\,dx$$

or derivative $\frac{dF}{dx}[x] = f[x]$.

We write "+c" after an answer to indicate all possible antiderivatives. For example,

$$\int \mathrm{Cos}[\theta]\,d\theta = \mathrm{Sin}[\theta] + c$$

Exercise set 13.1

1. *Verify that both x^2 and $x^2 + \pi$ equal $\int 2x\ dx$?*

13.2 Specific Integral Formulas

$$\int x^p\,dx = \frac{1}{p+1}x^{p+1} + c, \quad p \neq -1$$

$$\int \frac{1}{x}\,dx = \mathrm{Log}[x] + c, \quad x > 0$$

$$\int e^x\,dx = e^x + c$$

$$\int \mathrm{Sin}[x]\,dx = -\,\mathrm{Cos}[x] + c$$

$$\int \mathrm{Cos}[x]\,dx = \mathrm{Sin}[x] + c$$

Example 13.1. *Guess and Correct x^6*

Suppose we want to find

$$\int 3\,x^5\,dx = F[x]?$$

We know that if we differentiate a power function, we reduce the exponent by 1, so we guess and check our answer,

$$F_1[x] = x^6 \qquad\qquad F_1'[x] = 6\,x^5$$

The constant is wrong but would be correct if we chose

$$F[x] = \frac{1}{2}\,x^6 \qquad\qquad F'[x] = \frac{1}{2}\cdot 6\,x^5 = 3\,x^5$$

Here is another example of guessing and correcting the guess by adjusting a constant.

Example 13.2.   *Guess and Correct* Sin[3 x]

Find

$$\int 7\,\mathrm{Cos}[3\,x]\,dx = G[x]$$

Begin with the first guess and check

$$G_1[x] = \mathrm{Sin}[3\,x] \qquad\qquad G_1'[x] = 3\,\mathrm{Cos}[3x]$$

Adjusting our guess gives

$$G[x] = \frac{7}{3}\,\mathrm{Sin}[3\,x] \qquad\qquad G'[x] = \frac{7}{3}\,3\,\mathrm{Cos}[3x] = 7\,\mathrm{Cos}[3\,x]$$

It is best to check your work in any case so you only need to remember the five specific basic formulas above and use them to adjust your guesses. We will learn general rules based on each of the rules for differentiation but used in reverse. Here is some basic drill work.

────────────────── (**Exercise set 13.2**) ──────────────────

1. Basic Drill on Guessing and Correcting

a) $\int 7\sqrt{x}\,dx = ?$

b) $\int 5\,x^3\,dx = ?$

c) $\int \frac{3}{x^2}\,dx = ?$

d) $\int x^{\frac{3}{2}}\,dx = ?$

e) $\int (5-x)^2\,dx = ?$

f) $\int \mathrm{Sin}[3\,x]\,dx = ?$

g) $\int e^{2\,x}\,dx = ?$

h) $\int \frac{-7}{x}\,dx = ?$

You cannot do anything you like with indefinite integrals and expect to get the intended function. In particular, the "dx" in the integral tells you the variable of differentiation for the intended answer.

2. *Explain what is wrong with the following nonsense:*

$$\int x^2 \, dx = \int x \cdot x \, dx = x \int x \, dx$$

$$= x \left[\frac{1}{2} x^2 + c \right] = \frac{1}{2} x^3 + c \, x$$

and

$$\int_0^1 x^2 \, dx = \int_0^1 x \cdot x \, dx = x \int_0^1 x \, dx$$

$$= x \left[\frac{1}{2} x^2 |_0^1 \right] = \frac{1}{2} x$$

13.3   Superposition of Antiderivatives

$$\int a \, f[x] + b \, g[x] \, dx = a \int f[x] \, dx + b \int g[x] \, dx$$

Example 13.3.   *Superposition of Derivatives in Reverse*

We prove the superposition rule

$$\int a \, f[x] + b \, g[x] \, dx = a \int f[x] \, dx + b \int g[x] \, dx$$

by letting $F[x] = \int f[x] \, dx$, $G[x] = \int g[x] \, dx$ and writing out what the claim for indefinite integrals means in terms of these functions:

$$F'[x] = f[x]$$
$$G'[x] = g[x]$$
$$(a \, F[x] + b \, G[x])' = a \, F'[x] + b \, G'[x] = a \, f[x] + b \, g[x]$$
$$\left(a \int f[x] \, dx + b \int g[x] \, dx \right)' = a \, f[x] + b \, g[x]$$

and

$$\left(\int a \, f[x] + b \, g[x] \, dx \right)' = a \, f[x] + b \, g[x]$$

Do the arbitrary constants of integration matter? No, as long as we interpret the sum of two arbitrary constants as just another arbitrary constant.

Example 13.4. *Superposition for Integrals*

$$\int \left(2\operatorname{Sin}[x] - 3\frac{1}{x}\right) dx = 2\int \operatorname{Sin}[x] \, dx - 3\int \frac{1}{x} \, dx = -2\operatorname{Cos}[x] - 3\operatorname{Log}[x] + c$$

Now, use your rule to break linear combinations of integrands into simpler pieces.

----------------------- **Exercise set 13.3** -----------------------

1. Superposition of Antiderivatives Drill

a) $\int 5x^3 - 2 \, dx = 5 \int x^3 \, dx - 2 \int 1 \, dx = $?

b) $\int \frac{5}{x^3} - 2\sqrt{x} \, dx = 5 \int \frac{1}{x^3} \, dx - 2 \int \sqrt{x} \, dx = $?

c) $\int \frac{3}{x^2} - \sqrt[3]{x} + \frac{1}{\sqrt{x}} \, dx = $?   d) $\int 5\operatorname{Sin}[x] - e^{2x} \, dx = $?

e) $\int \operatorname{Sin}[5x] - 5\operatorname{Sin}[x] \, dx = $?   f) $\int \operatorname{Cos}[5x] - \frac{5}{x} \, dx = $?

Remember that the computer can be used to check your work on basic skills.

2. *Run the computer program **SymbolicIntegr**, and use the computer to check your work from the previous exercise.*

13.4 "Substitution" for Integrals

One way to find an indefinite integral is to change the problem into a simpler one. Of course, you want to change it into an equivalent problem.

Change of variables can be done legitimately as follows. First, let $u = $ part of the integrand. Next, calculate $du = \cdots$. If the remaining part of the integrand is du, make the substitution and; if it is not du, try a different substitution. The point is that we must look for both an expression and its differential. Here is a very simple example.

Example 13.5. *A Change of Variable and Differential*

Find

$$\int 2x \sqrt{1+x^2} \, dx = \int \sqrt{[1+x^2]}\{2x \, dx\}$$

Begin with

$$u = [1+x^2] \qquad\qquad du = \{2x \, dx\}$$

We replace the expression for u and du, thereby obtaining the simpler problem: Find

$$\int \sqrt{[u]} \, \{du\} = \int u^{\frac{1}{2}} \, du$$

$$= \frac{1}{1 + \frac{1}{2}} \, u^{1 + \frac{1}{2}} + c$$

$$= \frac{2}{3} \, u^{\frac{3}{2}} + c$$

The expression $\frac{2}{3} u^{\frac{3}{2}} + c$ is not an acceptable answer to the question, "What functions of x have derivative $2x\sqrt{1 + x^2}$?" However, if we remember that $u = 1 + x^2$, we can express the answer as

$$\frac{2}{3} \, u^{\frac{3}{2}} + c = \frac{2}{3} \, [1 + x^2]^{\frac{3}{2}} + c$$

Checking the answer will show why this method works. We use the Chain Rule:

$$y = \frac{2}{3} \, u^{\frac{3}{2}} \qquad\qquad\qquad u = 1 + x^2$$

so

$$\frac{dy}{du} = \frac{2}{3} \frac{3}{2} \, u^{\frac{3}{2} - 1} \qquad\qquad\qquad \frac{du}{dx} = 2x$$

and

$$\frac{dy}{dx} = \frac{dy}{du} \frac{du}{dx} = u^{\frac{1}{2}} \, 2x = 2x\sqrt{1 + x^2}$$

or

$$dy = 2x\sqrt{1 + x^2} \, dx$$

Example 13.6.   *Another Change*

You *must* substitute for the differential du associated with your change of variables $u = \cdots$. Sometimes this is a little complicated. For example, suppose we try to compute

$$\int 2x\sqrt{1 + x^2} \, dx$$

with the change of variable and differential

$$v = \sqrt{1 + x^2} \qquad\qquad\qquad dv = \frac{x}{\sqrt{1 + x^2}} \, dx$$

Our integral becomes

$$\int 2v [\text{ rest of above}]$$

with the rest of the above equal to $x\,dx$. We can find $dv = \frac{x}{\sqrt{1+x^2}}dx$ by multiplying numerator and denominator by $\sqrt{1+x^2}$, so our integral becomes

$$\int 2\sqrt{1+x^2}\,x\,\frac{\sqrt{1+x^2}}{\sqrt{1+x^2}}\,dx = \int 2\left(\sqrt{1+x^2}\right)^2 [\frac{x}{\sqrt{1+x^2}}\,dx]$$

$$= 2\int v^2 dv = \frac{2}{3}v^3 + c$$

$$= \frac{2}{3}\left[1+x^2\right]^{\frac{3}{2}}$$

This is the same answer as before, but the substitution was more difficult on the dv piece.

Example 13.7. *A Failed Attempt*

Sometimes an attempt to simplify an integrand by change of variables will lead you to either a more complicated integral or a situation in which you cannot make the substitution for the differential. In these cases, scratch off your work and try another change.

Suppose we try a grand simplification of

$$\int 2x\,\sqrt{1+x^2}\,dx$$

taking

$$w = x\,\sqrt{1+x^2} \qquad\qquad dw = \left(\sqrt{1+x^2} + \frac{x^2}{\sqrt{1+x^2}}\right)dx$$

We might substitute w for the whole integrand, but there is nothing left to substitute for dw and we cannot complete the substitution. We simply have to try a different method.

Example 13.8. *A Less Obvious Substitution with* $u = \sqrt{x}$

We may be slipping into the symbol swamp, but a little wallowing can be fun. Here is a change of variable and differential with a twist:

$$\int \frac{\sqrt{x}}{1+x}\,dx$$

$$u = \sqrt{x} \quad\Leftrightarrow\quad u^2 = x \qquad (x > 0)$$

$$du = \frac{1}{2\sqrt{x}}\,dx \quad\Leftrightarrow\quad 2u\,du = dx$$

$$\int \frac{u}{1+u^2}\,2u\,du = 2\int \frac{u^2}{1+u^2}\,du = 2\left[\int \left(1 - \frac{1}{1+u^2}\right)du\right]$$

$$= 2\int du - 2\int \frac{1}{1+u^2}\,du = 2u - 2\,\mathrm{ArcTan}[u] + c$$

$$\int \frac{\sqrt{x}}{1+x}\,dx = 2\sqrt{x} - 2\,\mathrm{ArcTan}[\sqrt{x}] + c$$

Now, you try it.

────────────────── (**Exercise set 13.4**) ──────────────────

1. Change of Variable and Differential Drill

a) $\int \frac{1}{(3x-2)^2} \, dx = \,?$

b) $\int \frac{2t}{\sqrt{1-t^2}} \, dt = \,?$

c) $\int x(3+7x^2)^3 \, dx = \,?$

d) $\int \frac{3y}{(2+2y^2)^2} \, dy = \,?$

e) $\int (\text{Cos}[x])^3 \, \text{Sin}[x] \, dx = \,?$

f) $\int \frac{\text{Cos}[\text{Log}[x]]}{x} \, dx = \,?$

g) $\int e^{\text{Cos}[\theta]} \, \text{Sin}[\theta] \, d\theta = \,?$

h) $\int \text{Sin}[a\,x + b] \, dx = \,?$

──

If you have checked several indefinite integration problems that you computed with a change of variable and differential, you probably realize that the Chain Rule lies behind the method. Perhaps you can formalize your idea.

Problem 13.1. THE CHAIN RULE IN REVERSE ──────────────────────────▼
Use the Chain Rule for differentiation to prove the indefinite integral form of "Integration by Substitution." Once you have this indefinite rule, use the Fundamental Theorem to prove the definite rule.

──▲

13.5   Change of Limits of Integration

When we want an antiderivative such as $\int 2x\sqrt{1+x^2} \, dx$, we have no choice but to re-substitute the expressions for new variables back into our answer. In the first example of the previous section,

$$\int 2x\sqrt{1+x^2} \, dx = \frac{2}{3} u^{\frac{3}{2}} + c = \frac{2}{3}[1+x^2]^{\frac{3}{2}}$$

However, when we want to compute a definite integral such as

$$\int_5^7 2x\sqrt{1+x^2} \, dx$$

we can change the limits of integration along with the change of variables. For example,

$$u = 1 + x^2 \qquad\qquad du = 2x\,dx$$
$$u[5] = 26 \qquad\qquad u[7] = 50$$

so the new problem is to find

$$\int_{26}^{50} u^{\frac{1}{2}} \, du$$

which equals

$$\frac{2}{3} u^{\frac{3}{2}} \big|_{26}^{50} = \frac{2}{3} [50^{\frac{3}{2}} - 26^{\frac{3}{2}}] \approx 235.702 - 88.383 = 147.319$$

We recommend that you do your definite integral changes of variable this way:

Procedure 13.3. To compute

$$\int_a^b f[x] \, dx$$

(a) Set a portion of your integrand $f[x]$ equal to a new variable $u = u[x]$.
(b) Calculate $du = ?? \, dx$.
(c) Calculate $u[a] = \alpha$ and $u[b] = \beta$.
(d) Substitute both the function and the differential in $f[x] \, dx$ for $g[u] \, du$.
(e) Compute

$$\int_\alpha^\beta g[u] \, du$$

PAY ME NOW OR PAY ME LATER:

You *can* compute the antiderivative in terms of x and then use the original limits of integration, but there is a danger that you will lose track of the variable with which you started. If you are careful, both methods give the same answer; for example,

$$\int_5^7 2x\sqrt{1+x^2} \, dx = \frac{2}{3}[1+x^2]^{\frac{3}{2}}\big|_5^7 = \frac{2}{3}\left([1+7^2]^{\frac{3}{2}} - [1+5^2]^{\frac{3}{2}}\right) \approx 147.319$$

Exercise set 13.5

1. Change of Variables with Limits Drill

a) $\int_0^1 \frac{x}{1+x^2} \, dx = ?$

b) $\int_0^{\pi/6} \text{Sin}[3\theta] \, d\theta = ?$

c) $\int_2^3 \frac{2x}{(x^2-3)^2} \, dx = ?$

d) $\int_0^1 a x + b \, dx = ?$

e) $\int_0^1 \frac{3x^2-1}{1+\sqrt{x-x^3}} \, dx = 0$ *because of the new limits.*

f) $\int_0^1 x e^{-x^2} \, dx = ?$

g) $\int_0^{\pi/2} \text{Cos}[\theta] \, \text{Sin}[\theta] \, d\theta = ?$

2. *Use the change of variable $u = \sqrt{x}$ and associated change of differential to convert $\int \sqrt{x} \, \text{Cos}[\sqrt{x}] \, dx$ into a multiple of $\int u^2 \, \text{Cos}[u] \, du$.*

13.5.1 Integration with Parameters

Change of variables can make integrals with parameters, which are important in many scientific and mathematical problems, into specific integrals. For example, suppose that ω

and a are constant. Then

$$\int \text{Sin}[\omega\, t]\; dt$$

$$u = \omega\, t$$

$$du = \omega\; dt \quad \Leftrightarrow \quad \frac{1}{\omega}\, du = dt$$

$$\int \text{Sin}[\omega\, t]\; dt = \int \text{Sin}[u]\, \frac{1}{\omega}\, du = \frac{1}{\omega} \int \text{Sin}[u]\; du$$

and

$$\int \frac{1}{a^2 + x^2}\; dt = \int \frac{1}{a^2(1 + (x/a)^2)}\; dx = \frac{1}{a^2} \int \frac{1}{1 + (x/a)^2}\; dx$$

$$u = \frac{x}{a}$$

$$du = \frac{1}{a}\, dx \quad \Leftrightarrow \quad a\, du = dx$$

$$\int \frac{1}{a^2 + x^2}\; dt = \frac{1}{a^2} \int \frac{1}{1 + u^2}\, a\; du = \frac{1}{a} \int \frac{1}{1 + u^2}\; du$$

This type of change of variable and differential may be the most important kind for you to think about because the computer can give you specific integrals, but you may want to see how an integral depends on a parameter.

Problem 13.2. PULL THE PARAMETER OUT ⎯⎯⎯⎯⎯⎯⎯⎯⎯⎯⎯⎯⎯⎯⎯▼
Show that

$$\int_0^3 \sqrt{9 - x^2}\; dx = 3 \int_0^3 \sqrt{1 - (x/3)^2}\; dx = 9 \int_0^1 \sqrt{1 - u^2}\; du$$

Show that the area of a circle of radius r is r^2 times the area of the unit circle. First, change variables to obtain the integrals below and then read on to interpret your computation.

$$\int_0^r \sqrt{r^2 - x^2}\; dx = r^2 \int_0^1 \sqrt{1 - u^2}\; du$$

▲

13.6   Trig Substitutions - CD

Study this section if you have a personal need to compute integrals with one of the following expressions (and do not have your computer):

$$\sqrt{a^2 - x^2}, \qquad a^2 + x^2 \quad or \quad \sqrt{a^2 + x^2}$$

You should skim read this section even if you do not wish to develop this skill because the positive sign needed to go between

$$(\text{Cos}[\theta])^2 = 1 - (\text{Sin}[\theta])^2 \quad \text{and} \quad \text{Cos}[\theta] = \sqrt{1 - (\text{Sin}[\theta])^2}$$

can cause errors in the use of a symbolic integration package. In other words, the computer may use the symbolic square root when you intend for it to use the negative.

Remainder of this section only on CD

13.7   Integration by Parts

Integration by Parts is important theoretically, but it appears to be just another trick. It is more than that, but first you should learn the trick.

The formulas for the "technique" are

$$\int_a^b u[x]\ dv[x] = u[x]\,v[x]\,|_a^b - \int_a^b v[x]\ du[x]$$

or, suppressing the x dependence,

$$\int_{x=a}^b u\ dv = uv|_{x=a}^b - \int_{x=a}^b v\ du$$

The idea is to break up an integrand into a function $u[x]$ and a differential $dv[x]$ where you can find the differential $du[x]$ (usually easy), the antiderivative $v[x]$ (sometimes harder), and, finally, where $\int v[x]\ du[x]$ is an easier problem. Unfortunately, often the only way to find out if the new problem is easier is to go through all the substitution steps for the terms.

Helpful Notation

We encourage you to block off the four terms in this formula.

First break up your integrand into u and dv,

$$u = \qquad\qquad\qquad\qquad dv =$$

and then compute the differential of u and the antiderivative of dv

$$du = \qquad\qquad\qquad\qquad v =$$

Here is an example of the use of Integration by Parts.

Example 13.9.   *Integration by Parts for $\int_a^b x \, \text{Log}\,[x]\ dx = ?$*

Use the "parts"

$$u = \text{Log}\,[x] \qquad\qquad\qquad\qquad dv = x\,dx$$

$$\text{and}$$

$$du = \frac{1}{x}\,dx \qquad\qquad\qquad\qquad v = \frac{1}{2}x^2$$

making the integrals

$$\int_a^b u\,dv = uv\big|_a^b - \int_a^b v\,du$$

$$\int_a^b x\,\text{Log}\,[x]\,dx = \frac{1}{2}x^2\ \text{Log}\,[x]\,\big|_a^b - \int_a^b \frac{1}{2}x^2\,\frac{dx}{x}$$

$$\int_a^b x\,\text{Log}\,[x]\,dx = \frac{1}{2}x^2\ \text{Log}\,[x]\,\big|_a^b - \int_a^b \frac{1}{2}x\,dx$$

$$\int_a^b x\,\text{Log}\,[x]\,dx = \frac{1}{2}x^2\ \text{Log}\,[x]\,\big|_a^b - \frac{x^2}{4}\big|_a^b$$

$$\int_a^b x\,\text{Log}\,[x]\,dx = \frac{1}{2}[b^2\,\text{Log}\,(b) - a^2\,\text{Log}\,[a]] - \frac{1}{4}[b^2 - a^2]$$

provided that both a and b are positive. (Otherwise, Log is undefined).

CHECK:

Notice that the indefinite integral of the calculation above is

$$\int x\,\text{Log}\,[x]\,dx = \frac{1}{2}\,x^2\ \text{Log}[x] - \frac{1}{4}\,x^2 + c$$

We check the correctness of this antiderivative by differentiating the right side of the equation. First, we use the Product Rule

$$\frac{d(f[x]\cdot g[x])}{dx} = \frac{df}{dx}\cdot g + f\cdot\frac{dg}{dx}$$

$$f[x] = x^2 \qquad\qquad g[x] = \text{Log}[x]$$
$$\frac{df}{dx} = 2\,x \qquad\qquad \frac{dg}{dx} = \frac{1}{x}$$

$$\frac{d(x^2\ \text{Log}[x])}{dx} = 2\,x\cdot\text{Log}[x] + x^2\cdot\frac{1}{x}$$
$$= 2\,x\cdot\text{Log}[x] + x$$

Next, we differentiate the whole expression

$$\frac{d(\frac{1}{2}\,x^2\ \text{Log}[x] - \frac{1}{4}\,x^2 + c)}{dx} = x\,\text{Log}[x] + \frac{1}{2}\,x - \frac{2}{4}\,x + 0 = x\,\text{Log}[x]$$

This verifies that the indefinite integral is correct.

Example 13.10. $\int x\,e^x\,dx$

To compute this integral, use integration by parts with

$$u = x \qquad\qquad dv = e^x \ dx$$

so

$$du = dx \qquad\qquad v = e^x$$

This gives $\boxed{\displaystyle\int u \ dv = u\,v - \int v \ du =}$

$$\int x\,e^x \ dx = x\,e^x - \int e^x \ dx$$
$$= x\,e^x - e^x = (x-1)\,e^x + c$$
$$\int x\,e^x \ dx = (x-1)\,e^x + c$$

CHECK:
Differentiate using the Product Rule:

$$\frac{d(f[x] \cdot g[x])}{dx} = \frac{df}{dx} \cdot g + f \cdot \frac{dg}{dx}$$

$$f[x] = (x-1) \qquad\qquad g[x] = e^x$$
$$\frac{df}{dx} = 1 \qquad\qquad \frac{dg}{dx} = e^x$$

$$\frac{d((x-1)\,e^x)}{dx} = e^x + (x-1)\,e^x$$
$$= x\,e^x$$

Example 13.11. *A Reduction of One Integral to a Previous One*

The integral $\int x^2\,e^x \ dx$ can be done "by parts" in two steps. First, take the parts

$$u = x^2 \qquad\qquad dv = e^x \ dx$$

so

$$du = 2x \ dx \qquad\qquad v = e^x$$

This gives $\boxed{\displaystyle\int u \ dv = u\,v - \int v \ du =}$

$$\int x^2\,e^x \ dx = x^2\,e^x - 2\int x\,e^x \ dx$$
$$= x^2\,e^x - 2(x-1)\,e^x + c$$

Note that the second integral was computed in the previous example, so

$$\int x^2\, e^x\ dx = (x^2 - 2\, x + 2)\, e^x + c$$

CHECK:

Differentiate using the Product Rule:

$$\frac{d(f[x] \cdot g[x])}{dx} = \frac{df}{dx} \cdot g + f \cdot \frac{dg}{dx}$$

$$f[x] = (x^2 - 2\, x + 2) \qquad\qquad g[x] = e^x$$
$$\frac{df}{dx} = 2\, x - 2 \qquad\qquad \frac{dg}{dx} = e^x$$

$$\frac{d((x^2 - 2\, x + 2)\, e^x)}{dx} = (2\, x - 2)\, e^x + (x^2 - 2\, x + 2)\, e^x$$
$$= x^2\, e^x$$

Example 13.12. *Circular Parts Still Gives an Answer*

We compute the integral $\int e^{2\, x}\, \mathrm{Sin}[3\, x]\ dx$ by the parts

$$u = e^{2\, x} \qquad\qquad dv = \mathrm{Sin}[3\, x]\ dx$$

so

$$du = 2\, e^{2\, x}\ dx \qquad\qquad v = -\frac{1}{3}\, \mathrm{Cos}[3\, x]$$

This gives $\displaystyle \int u\ dv = u\, v - \int v\ du =$

$$\int e^{2\, x}\, \mathrm{Sin}[3\, x]\ dx = -\frac{1}{3}\, e^{2\, x}\, \mathrm{Cos}[3\, x] + \frac{2}{3} \int e^{2\, x}\, \mathrm{Cos}[3\, x]\ dx$$

Now, use the parts on the second integral,

$$w = e^{2\, x} \qquad\qquad dz = \mathrm{Cos}[3\, x]\ dx$$

so

$$dw = 2\, e^{2\, x}\ dx \qquad\qquad z = \frac{1}{3}\, \mathrm{Sin}[3\, x]$$

This gives $\displaystyle \int w\ dz = w\, z - \int z\ dw =$

$$\int e^{2\, x}\, \mathrm{Cos}[3\, x]\ dx = \frac{1}{3}\, e^{2\, x}\, \mathrm{Sin}[3\, x] - \frac{2}{3} \int e^{2\, x}\, \mathrm{Sin}[3\, x]\ dx$$

Substituting this into the second integral above, we obtain

$$\int e^{2x}\,\text{Sin}[3\,x]\,dx = -\frac{1}{3}\,e^{2x}\,\text{Cos}[3\,x] + \frac{2}{3}\left[\frac{1}{3}\,e^{2x}\,\text{Sin}[3\,x] - \frac{2}{3}\int e^{2x}\,\text{Sin}[3\,x]\,dx\right]$$

$$= -\frac{1}{3}\,e^{2x}\,\text{Cos}[3\,x] + \frac{2}{9}\,e^{2x}\,\text{Sin}[3\,x] - \frac{4}{9}\int e^{2x}\,\text{Sin}[3\,x]\,dx$$

Bringing the like integral to the left side, we obtain

$$\int e^{2x}\,\text{Sin}[3\,x]\,dx + \frac{9}{4}\int e^{2x}\,\text{Sin}[3\,x]\,dx = \frac{2}{9}\,e^{2x}\,\text{Sin}[3\,x] - \frac{1}{3}\,e^{2x}\,\text{Cos}[3\,x] + c$$

$$\frac{4+9}{4}\int e^{2x}\,\text{Sin}[3\,x]\,dx = \frac{2}{9}\,e^{2x}\,\text{Sin}[3\,x] - \frac{1}{3}\,e^{2x}\,\text{Cos}[3\,x] + c$$

$$\int e^{2x}\,\text{Sin}[3\,x]\,dx = \frac{2}{13}\,e^{2x}\,\text{Sin}[3\,x] - \frac{3}{13}\,e^{2x}\,\text{Cos}[3\,x] + c$$

Example 13.13. *A Two-Step Computation of $\int (Cos[x])^2\,dx$*

This integral can also be computed without the trig identities from the CD section. Use the parts

$$u = \text{Cos}[x] \qquad\qquad dv = \text{Cos}[x]\,dx$$

so

$$du = -\text{Sin}[x]\,dx \qquad\qquad v = \text{Sin}[x]$$

which yields the integration formula

$$\int (Cos[x])^2\,dx = -\text{Sin}[x]\,\text{Cos}[x] + \int (\text{Sin}[x])^2\,dx$$

$$= -\text{Sin}[x]\,\text{Cos}[x] + \int [1 - (\text{Cos}[x])^2]\,dx$$

$$= -\text{Sin}[x]\,\text{Cos}[x] + \int 1\,dx - \int (Cos[x])^2\,dx$$

so $2\int (Cos[x])^2\,dx = x - \text{Sin}[x]\,\text{Cos}[x] + c$ and

$$\int (Cos[x])^2\,dx = \frac{x}{2} - \frac{1}{2}\,\text{Sin}[x]\,\text{Cos}[x] + c$$

Here are some practice problems. (Remember that you can check your work with the computer).

--- **Exercise set 13.7** ---

1. Drill on Integration by Parts

a) $\int \theta\,\text{Cos}[\theta]\,d\theta = ?$

b) $\int \theta^2\,\text{Sin}[\theta]\,d\theta = ?$

c) $\int_0^{\pi/2} \theta \, \text{Sin}[\theta] \, d\theta = ?$

d) $\int_0^{\pi/2} \theta \, \text{Cos}[\theta] \, d\theta = ?$

e) $\int xe^{2x} \, dx = ?$

f) $\int x^2 e^{2x} \, dx = ?$

g) $\int e^{5x} \, \text{Sin}[3\,x] \, dx = ?$

h) $\int e^{3x} \, \text{Cos}[5\,x] \, dx = ?$

2. *Compute $\int \text{Log}[x] \, dx$ using integration by parts with $u = \text{Log}[x]$ and $dv = dx$. Check your answer by differentiation.*

3. *Check the previous indefinite integral from Example 13.12 by using the Product Rule to differentiate $\frac{2}{13} e^{2x} \, \text{Sin}[3\,x] - \frac{3}{13} e^{2x} \, \text{Cos}[3\,x]$.*
Compute the integral $\int e^{2x} \, \text{Sin}[3\,x] \, dx$ by the parts

$$u = \text{Sin}[3\,x] \qquad\qquad dv = e^{2x} \, dx$$

so

$$du = 3 \, \text{Cos}[3\,x] \, dx \qquad\qquad v = \frac{1}{2} e^{2x}$$

4. *Calculate $\int (\text{Sin}[x])^2 \, dx$.*

There is something indefinite about these integrals.

5. *What is wrong with the equation*

$$\int \frac{dx}{x} = \int \frac{x}{x^2} \, dx = -1 + \int \frac{dx}{x}$$

when you use integration by parts with $u = x$, $dv = \frac{dx}{x^2}$, $du = dx$, and $v = \frac{-1}{x}$?
Subtracting $\int \frac{dx}{x}$ from both sides of the equations above yields

$$0 = -1$$

The proof of the Integration by Parts formula is actually easy.

6. The Product Rule in Reverse
Use the Product Rule for differentiation to prove the indefinite integral form of Integration by Parts. Notice that if $H[x]$ is any function, $\int dH[x] = H[x] + c$, by definition. Let $H = u\,v$ and show that $dH = u\,dv + v\,du$. Indefinitely integrate both sides of the dH equation,

$$u \cdot v = H[x] = \int dH = \int u\,dv + \int v\,du$$

Once you have this indefinite rule, use the Fundamental Theorem to prove the definite rule.

Here are some tougher problems in which you need to use more than one method at a time.

7. (a) $\int x \, \text{Cos}[x] \, \text{Sin}[x] \, dx = ?$
Notice that $\int (\text{Cos}[\theta])^2 \, d\theta$ is done above two ways.

(b) $\int \theta \, (\text{Cos}[\theta])^2 \, d\theta = ?$
Use integrals from the previous drill problems.

(c) $\int \frac{x^3}{\sqrt{x^2-1}}\, dx = ?$

   HINT: Use parts $u = x^2$ and $dv = \frac{x\, dx}{\sqrt{x^2-1}}$ and compute the dv integral.

(d) $\int \frac{1}{x^3} \sqrt{\frac{1}{x} - 1}\, dx = ?$

   Use parts $u = \frac{1}{x}$ and $dv = \frac{1}{x^2}\sqrt{\frac{1}{x} - 1}$. Compute the dv integral.

(e) $\int_0^1 \text{ArcTan}[x]\, dx = ?$
   Use parts $u = \text{ArcTan}[x]$ and $dv = dx$.

(f) $\int_0^1 x\, \text{ArcTan}[x]\, dx = ?$
   Use parts and the previous exercise.

(g) $\int_0^1 \text{ArcTan}[\sqrt{x}]\, dx = ?$
   Use parts $u = \text{ArcTan}[\sqrt{x}]$ and change variables in the resulting integral.

(h) $\int_4^9 \text{Sin}[\sqrt{x}]\, dx = ?$
   Use parts $u = \text{Sin}[\sqrt{x}]$ and $dv = dx$. Then change variables with $w = \sqrt{x}$.

(i) $\int (\text{Log}[x])^2\, dx = ?$
   Use the parts $u = \text{Log}[x]$ and $dv = \text{Log}[x]\, dx$. Calculate the dv integral.

13.8   Impossible Integrals

There are important limitations to symbolic integration that go beyond the practical difficulties of learning all the tricks. This section explains why.

Integration by parts and the change of variable and differential are important ideas for the theoretical transformation of integrals. In this chapter, we tried to include just enough drill work for you to learn the basic methods. Before the practical implementation of general antidifferentiation algorithms on computers, development of human integration skills was an important part of the training of scientists and engineers. Now, the computer can makes this skill easier to master. The skill has always had limitations.

Early in the days of calculus, it was quite impressive that integration could be used to learn many many new formulas such as the classical formulas for the area of a circle or volume of a sphere. We saw how easy it was to generalize the integration approach to the volume of a cone. However, some simple-looking integrals have no antiderivative what so ever. This is not the result of peculiar mathematical examples.

The arclength of an ellipse just means the length measured as you travel along an ellipse. Problem 14.12 asks you to find integral formulas for this arclength. Early developers of calculus must have tried very hard to compute those integrals with symbolic antiderivatives, but after more than a century of trying, Liouville proved that there is no analytical expression for that antiderivative in terms of the classical functions.

The fact that the antiderivative has no expression in terms of old functions does not mean that the integral does not exist. If you find the following integral with the computer, you will see a peculiar result:

$$\int \text{Cos}[x^2]\ dx$$

The innocent-looking integral $\int \text{Cos}[x^2]\ dx$ is not innocent at all. The function $\text{Cos}[x^2]$ is perfectly smooth and well behaved, but it does not have an antiderivative that can be expressed in terms of known functions. The bottom line is this: Integrals are used to *define* and numerically compute important new functions in science and mathematics, even when they do not have expressions in terms of elementary functions. Functions given by integral formulas can still be computed just as you did in Exercise 12.8.1.

―――――――――――――――――――(**Exercise set 13.8**)―――――――――――――――――――

C

1. *Run the **SymbolicIntegr** program.*

CHAPTER 14

Applications of Integration

This chapter explores deeper applications of integration, especially integral computation of geometric quantities.

The most important parts of integration are setting the integrals up and understanding the basic techniques of Chapter 13. Proficiency at basic techniques will allow you to use the computer to correctly perform complicated symbolic integration, but the computer cannot tell if the integral formula is a correct approximation. Chapter 12 began with this "slicing" approximation and this chapter returns to it in more detail. The problems at the end of the chapter ask *you* to derive some integral formulas yourself. The problems are hard because of the analytical geometry, but solving them will give you important insight.

14.1   The Length of a Curve

We begin by showing that some care must be taken in the more delicate geometric approximation problems in order to be sure that a sum of many small errors does not build up. The secret in all the integral approximations is to measure the error on a scale of the *increment* of the independent variable, Δx. (See Theorem 14.1.) It is *not* sufficient to have the error of each slice tend to zero.

The way you "slice" sometimes matters. The approximation to the length of a curve by sloping line segments that connect $(x, f[x])$ and $(x+\Delta x, f[x+\Delta x])$ is given in Example 14.1. This is a "good" approximation that converges to the actual length. "Slices" that just run horizontally out from the slice points do not approximate the length even though the smaller and smaller segments do get closer and closer to the curve.

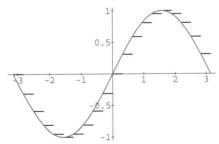

Figure 14.1: An incorrect "approximation" to the length

Horizontal slices are a good approximation to the area, but not the length. (See Figure 12.7.)

Example 14.1.   *Length* $\approx \sum_{\substack{x=a \\ step\ \Delta x}}^{b-\Delta x} [\sqrt{(\Delta x)^2 + (f[x+\Delta x] - f[x])^2}]$

The Pythagorean Theorem shows that the length of a line segment from the point $(x, \mathrm{Sin}[x])$ to the point $(x+\Delta x, \mathrm{Sin}[x+\Delta x])$ is

$$\sqrt{(x+\Delta x - x)^2 + (\mathrm{Sin}[x+\Delta x] - \mathrm{Sin}[x])^2} = \sqrt{\Delta x^2 + (\mathrm{Sin}[x+\Delta x] - \mathrm{Sin}[x])^2}$$

The sum of the lengths of the segments in Figure 14.2 is

$$\sum_{\substack{x=-\pi \\ step\ \Delta x}}^{\pi-\Delta x} [\sqrt{\Delta x^2 + (\mathrm{Sin}[x+\Delta x] - \mathrm{Sin}[x])^2}]$$

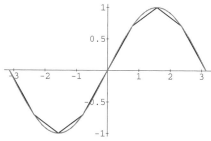

Figure 14.2: Approximate Length of the Sine Curve

More generally, if the curve is $y = f[x]$, the sum of the lengths of connecting segments is

$$\sum_{\substack{x=a \\ step\ \Delta x}}^{b-\Delta x} [\sqrt{\Delta x^2 + (f[x+\Delta x] - f[x])^2}]$$

This is a correct approximation but it is not in the form of an integral approximation

$$\sum_{\substack{x=a \\ step\ \Delta x}}^{b-\Delta x} [F[x]\Delta x]$$

However, when we replace Δx by a tiny increment δx, we can use the differential approximation of Definition 5.5 to write this sum in the form of a definite integral. First,

$$f[x+\delta x] - f[x] = f'[x] \cdot \delta x + \varepsilon \cdot \delta x$$

with $\varepsilon \approx 0$ for each x in $[a, b]$. This means

$$\sum_{\substack{x=a \\ step\ \delta x}}^{b-\delta x} [\sqrt{(\delta x)^2 + (f[x+\delta x] - f[x])^2}] = \sum_{\substack{x=a \\ step\ \delta x}}^{b-\delta x} [\sqrt{(\delta x)^2 + (f'[x] + \varepsilon)^2 (\delta x)^2}]$$

$$= \sum_{\substack{x=a \\ step\ \delta x}}^{b-\delta x} [\sqrt{1 + (f'[x] + \varepsilon)^2}\ \delta x]$$

where $\varepsilon \approx 0$. In Problem 14.1, you can use properties of summation to show that

$$\sum_{\substack{x=a \\ \text{step } \delta x}}^{b-\delta x} [\sqrt{1+(f'[x]+\varepsilon)^2} \ \delta x] \approx \sum_{\substack{x=a \\ \text{step } \delta x}}^{b-\delta x} [\sqrt{1+(f'[x])^2} \ \delta x]$$

We return to the arclength formula below in Example 14.4.

Exercise set 14.1

1. Pythagoras meets Descartes

Given two points (a, b) and (u, v) in the (x, y)-plane, show that the Pythagorean Theorem gives

$$\text{the length of the line segment connecting the points} = \sqrt{(u-a)^2 + (v-b)^2}$$

HINT: Make the segment the hypotenuse of a right triangle with horizontal and vertical legs. What are the lengths of these legs?

Now apply your formula to a function graph.

2. Pythagoras and the Increment

This exercise asks you to find a certain formula on a general function graph $y = f[x]$. If you wish, you can begin with a specific nonlinear function like $f[x] = x^2$, but the goal is an expression in terms of a general $f[x]$.

    (a) *Sketch a figure showing a graph $y = f[x]$ in the (x, y)-plane.*

    (b) *Put a dot on the graph of your function at one point $(x, y) = (x, f[x])$.*

    (c) *Put a second dot on your graph of $y = f[x]$ at a nearby point $(x + \Delta x, ??)$.*

    (d) *Express the y-value of the point on your graph above $x + \Delta x$ in terms of $f[\cdot]$.*

    (e) *Draw the straight line segment connecting the two dots on your graph. We want a formula for the length of this segment.*

    (f) *Apply the formula from Exercise 14.1.1 to show that the length of the segment connecting your two points is*

$$\sqrt{(\Delta x)^2 + (f[x + \Delta x] - f[x])^2}$$

The next exercise has you show what can go wrong when an "approximation" is not accurate.

3. Horizontal Slices Do Not Approximate Length

This exercise has you find a sum expression for the attempt at approximating the length of a curve by horizontal slices. Then it has you explain why it is a bad approximation.

    (a) *Sketch your general curve $y = f[x]$ for $a \le x \le b$.*

    (b) *Draw a dot on your curve at a particular point $(x, f[x])$.*

    (c) *Draw a small horizontal segment of length Δx beginning at $(x, f[x])$. This should look similar to one of the ones shown on Figure 14.3 below. Your segment begins at the point $(x, f[x])$ and ends at a point a distance Δx to the right and at the same y-height, so its coordinates are $(x + \Delta x, f[x])$.*

    (d) *Sketch a sequence of these horizontal segments beginning at $(a, f[a])$, each having length Δx and continuing until one segment goes beyond $x = b$. (The specific curve $y = \text{Sin}[x]$ for $-\pi \le x \le \pi$ is shown in Figure 14.3.)*

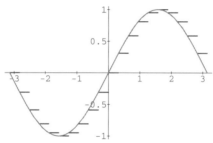

Figure 14.3: *An incorrect "approximation" to the length*

(e) *The length of one segment is Δx because that was the instruction. The sum of the lengths of the segments is Δx times the number of segments, or, to be fancy,*

$$\sum_{\substack{x=a \\ step\ \Delta x}}^{b-\delta x} \Delta x = \sum_{\substack{x=a \\ step\ \Delta x}}^{b-\delta x} [(x+\Delta x) - (x)]$$

$$= \sum_{\substack{x=a \\ step\ \Delta x}}^{b-\delta x} [F[x+\Delta x] - F[x]]$$

with $F[x] = x$. Show that the sum of the lengths of the horizontal segments is approximately $b - a$.

$$\sum_{\substack{x=a \\ step\ \Delta x}}^{b-\delta x} \Delta x \approx b - a$$

(The answer $b - a$ is exact if Δx divides the interval $[a,b]$ exaclty. Think about how many terms there are or use Theorem 12.2.)

(f) *According to your correct result from the last part of this exercise, the sum of the lengths of the horizontal segments does not depend on the particular function $f[x]$. In particular, show that the sum is 2π, for Figure 14.3, the function $f[x] = 0$ for $-\pi \le x \le \pi$, and the function $f[x] = x$ for $-\pi \le x \le \pi$.*

Why can't the sum of the horizontal slices be used to find the length of a graph?

Problem 14.1. ──▼

Let $f[x]$ be a smooth function on an interval $[a, b]$. Show that

$$\sum_{\substack{x=a \\ step\ \delta x}}^{b-\delta x} [\sqrt{(\delta x)^2 + (f[x+\delta x] - f[x])^2}] \approx \sum_{\substack{x=a \\ step\ \delta x}}^{b-\delta x} [\sqrt{1 + (f'[x])^2}] \delta x$$

so that length is given by the integral

$$\int_a^b \sqrt{1 + (f'[x])^2}\ dx$$

Use the following hints.

First, use Definition 5.5 to show that

$$\sqrt{(\delta x)^2 + (f[x + \delta x] - f[x])^2} = \sqrt{1 + (f'[x] + \varepsilon)^2}\,\delta x$$

with $\varepsilon \approx 0$ when $\delta x \approx 0$.

Next, show that

$$\sqrt{1 + (f'[x] + \varepsilon)^2} = \sqrt{1 + (f'[x])^2} + \iota$$

with $\iota \approx 0$. Because

$$\sqrt{1 + (f'[x] + \varepsilon)^2} - \sqrt{1 + (f'[x])^2} =$$

$$= \frac{\left(\sqrt{1 + (f'[x] + \varepsilon)^2} - \sqrt{1 + (f'[x])^2}\right)\left(\sqrt{1 + (f'[x] + \varepsilon)^2} + \sqrt{1 + (f'[x])^2}\right)}{\sqrt{1 + (f'[x] + \varepsilon)^2} + \sqrt{1 + (f'[x])^2}}$$

$$= \frac{\varepsilon(2f'[x] + \varepsilon)}{\sqrt{1 + (f'[x] + \varepsilon)^2} + \sqrt{1 + (f'[x])^2}}$$

Finally, show that a sum with $\iota \approx 0$, for all x, satisfies

$$\sum_{\substack{x=a \\ step\ \delta x}}^{b-\delta x} [\iota \cdot \delta x] \approx 0$$

(See Theorem 12.13.)

14.2 Duhamel's Principle

Duhamel's Principle gives us an accuracy test for integral formulas.

In order to give a general result, we need to formulate the problem in terms of an "additive" quantity. Additivity expresses a simple property that many geometric functions such as accumulated area, length, volume, and surface area, have. This means that we can apply Duhamel's Principle to finding integral formulas of many geometric quantities. For example, the accumulated area used in the second half of the Fundamental Theorem of Integral Calculus is additive. Figure 14.4 shows the area accumulated from a to x:

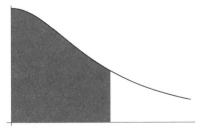

Figure 14.4: $A[a, x]$, the area from a to x

Additivity of $A[a, b]$ means that the area from a to b plus the area from b to c equals the whole area from a to c (where $a < b < c$) as shown in Figure 14.5:

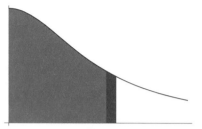

Figure 14.5: $A[a,c] = A[a,b] + A[b,c]$

Additivity is an important property of the integral. If we define a function of two variables, $I[u,v] = \int_u^v f[x]\; dx$ (for $f[x]$ continuous on $[a,b]$), then we have the additivity property

$$\int_u^v f[x]\; dx + \int_v^w f[x]\; dx = \int_u^w f[x]\; dx$$
$$I[u,v] \qquad +I[v,w] \qquad =I[u,w]$$

for any $u < v < w$ in $[a,b]$.

Theorem 14.1. *Keisler's Infinite Sum Theorem or Duhamel's Principle*
Let $Q[u,v]$ be an additive quantity of a real variable, that is, satisfy

$$Q[u,v] + Q[v,w] = Q[u,w]$$

for $u < v < w$ in $[a,b]$. Suppose $f[x]$ is a continuous real function on $[a,b]$, such that for any tiny subinterval $[x, x+\delta x] \subseteq [a,b]$, with $\delta x \approx 0$,

$$Q[x, x+\delta x] = f[x]\; \delta x + \varepsilon \cdot \delta x$$

for a small error $\varepsilon \approx 0$. In other words, suppose $f[x]\; \delta x$ approximates a small "slice" of Q on a scale of δx. Then,

$$Q[a,b] = \int_a^b f[x]\, dx$$

First, notice that $Q[a,x] + Q[x, x+\delta x] = Q[a, x+\delta x]$ by additivity, so

$$Q[a, x+\delta x] - Q[a, x] = Q[x, x+\delta x]$$

Duhamel's approximation formula in the theorem above becomes

$$Q[a, x+\delta x] - Q[a, x] = f[x] \cdot \delta x + \varepsilon \cdot \delta x$$

So, if we let $F[x] = Q[a,x]$, Duhamel's approximation is the differential approximation 5.5 for $F[x]$. This shows that when we find an integral formula, we are "writing a differential equation for $Q[a,x]$" in terms of the slicing variable x.
PROOF:

By repeated use of additivity

$$Q[a, b] = Q[a, a + \delta x] + Q[a + \delta x, a + 2\delta x] + \ldots + Q[b - \delta x, b]$$

$$= \sum_{\substack{x=a \\ \text{step } \delta x}}^{b - \delta x} [Q[x, x + \delta x]]$$

$$= \sum_{\substack{x=a \\ \text{step } \delta x}}^{b - \delta x} [f[x]\,\delta x] + \sum_{\substack{x=a \\ \text{step } \delta x}}^{b - \delta x} [\varepsilon \cdot \delta x]$$

$$\approx \int_a^b f[x]\,dx + \sum_{\substack{x=a \\ \text{step } \delta x}}^{b - \delta x} [\varepsilon \cdot \delta x]$$

and

$$\sum_{\substack{x=a \\ \text{step } \delta x}}^{b - \delta x} [\varepsilon \cdot \delta x] \approx 0$$

by the triangle inequality estimate in Theorem 12.13. We have shown that the two fixed quantities satisfy $Q[a, b] \approx \int_a^b f[x]\,dx$, forcing them to be equal and proving the theorem.

Our first two examples of the use of Duhamel's Principle are simple and could be done directly by sandwiching the approximating sums between upper and lower estimates instead of using the general principle. These estimates do give an error of the form $\varepsilon \cdot \delta x$, too, however.

Example 14.2. *Distance Is an Integral*

We return to the distance example of Section 12.2 above. Suppose that $R[t]$ is any continuous rate (speed) function. By the Extreme Value Theorem 11.3, $R[t]$ has a max and a min over the interval $[t, t + \delta t]$. We denote these

$$R_m \le R[s] \le R_M \qquad \text{for} \quad t \le s \le t + \delta t$$

The distance traveled during this time interval must lie between the extremes

$$R_m \cdot \delta t \le \text{actual distance traveled} \le R_M \cdot \delta t$$

However, since $R[t]$ is continuous,

$$R_m \approx R[t] \approx R_M$$

so that

$$\text{actual distance traveled from time } t \text{ to time } t + \delta t = R[t] \cdot \delta t + \varepsilon \cdot \delta t$$

with $\varepsilon \approx 0$.

Finally, let $D[t_1, t_2]$ denote the distance traveled between the times t_1 and t_2. This is additive because if $t_1 < t_2 < t_3$, $D[t_1, t_3] = D[t_1, t_2] + D[t_2, t_3]$. Duhamel's Principle shows that

$$\text{total distance traveled} = \int_a^b R[t]\,dt$$

Example 14.3. *The General Disk Method*

When the graph of a positive continuous function, $y = f[x]$ for $a \leq x \leq b$, is revolved about the x-axis, the volume of the resulting solid is

$$V = \pi \int_a^b (f[x])^2 \, dx$$

The general formula is explained in the Mathematical Background by an upper and lower estimate similar to the previous example.

Definite integrals *must* be written in the form $\sum_{\substack{x=a \\ \text{step } \delta x}}^{b-\delta x} [F[x]\delta x]$ for some function $F[x]$; so although

$$\sum_{\substack{x=a \\ \text{step } \delta x}}^{b-\delta x} [\sqrt{(\delta x)^2 + (f[x+\delta x] - f[x])^2}]$$

is a valid approximation for length, this approximation cannot be computed as an integral. The derivative formula or "microscope equation" allows us to express approximate length in the integral form using the "integrand" $F[x] = \sqrt{1 + (f'[x])^2}$. This means that the arclength can be expressed by

$$\sum_{\substack{x=a \\ \text{step } \delta x}}^{b-\delta x} [\sqrt{1 + (f'[x])^2} \, \delta x] \approx \int_a^b \sqrt{1 + (f'[x])^2} \, dx = \text{length of the curve } y = f[x]$$

Example 14.4. *Length Formula* $L = \int_a^b \sqrt{1 + (f'[x])^2} \, dx$

Example 14.1, Exercise 14.1.2 and Problem 14.1 symbolically derive the arclength formula for a general explicit curve $y = f[x]$. A simpler way to summarize that work is to imagine measuring the length of a segment of the curve viewed inside a powerful microscope. We will show now that what we see in the microscope gives the same answer as the symbolic approach.

By Theorem 14.1, the formula for the length of the curve is

$$L = \int_a^b \sqrt{1 + (f'[x])^2} \, dx$$

provided that we can show that the length of a tiny segment between x and $x + \delta x$ satisfies

$$\text{arclength between } x \text{ and } x + \delta x = \sqrt{1 + (f'[x])^2} \, \delta x + \iota \, \delta x$$

for $\iota \approx 0$ when $\delta x \approx 0$.

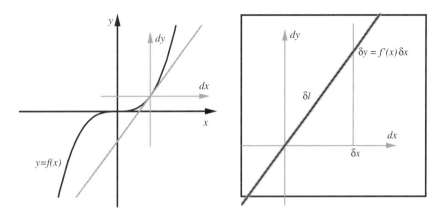

Figure 14.6: Microscopic view of length

We know that the line we see in a microscopic view of a smooth function has equation $dy = f'[x] \, dx$, so the height of the triangle with vertices $(x, f[x])$, $(x + \delta x, f[x + \delta x])$, and $(x+\delta x, f[x])$, is $f'[x] \, \delta x$, for a base of length δx. The Pythagorean Theorem says the square of the length is the sum of the squares of the sides,

$$\sqrt{(\delta x)^2 + (f'[x])^2 \delta x^2} = \sqrt{1 + (f'[x])^2} \; \delta x$$

This is the length of a segment of the curve except for errors that are tiny when viewed in the microscope of power $1/\delta x$. Errors that are small after magnification by $1/\delta x$ have actual magnitude $\varepsilon \cdot \delta x$ with $\varepsilon \approx 0$. Microscopic errors are exactly what Duhamel's Principle 14.1 allows, so the Theorem says that we can add the linear pieces we see in the microscopic views to obtain the full nonlinear quantity.

Example 14.5.   *Parametric Length $L = \int_a^b \sqrt{[dx[t]]^2 + [dy[t]]^2}$*

The simple shortcut for finding the length integral with the view in a microscope also works for parametric curves. Sometimes parametric integral formulas are better behaved than explicit formulas, and sometimes curves do not even have single explicit formulas so that we must use parametric formulas.

What are parametric formulas for curves? Parametric curves are given by functions $x = f[t]$ and $y = g[t]$ where we plot (x, y) but not t. This is taken up in more detail in Chapter 16, but you should be familiar with the following example. A circle is given by the equations

$$x[\theta] = \mathrm{Cos}[\theta]$$
$$y[\theta] = \mathrm{Sin}[\theta]$$

These are parametric equations for a circle. We measure a distance θ along the unit circle starting at $(1, 0)$, and the point has coordinates $(x, y) = (\mathrm{Cos}[\theta], \mathrm{Sin}[\theta])$. See the section on trig functions in Chapter 28. The value of θ is not plotted (althought it appears as the length measured along the circle.) Parametric equations describing an ellipse are given in Problem 14.12.

We used this idea in Chapter 5 when we computed the increments of sine and cosine and proved the differentiation formulas for sine and cosine. A tiny increment of the circle looks as follows under a powerful microscope:

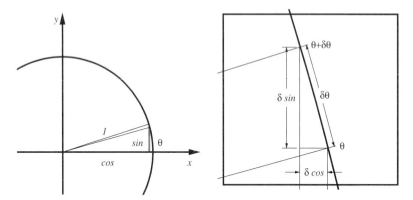

Figure 14.7: Increments of Sine and Cosine

The small triangle in the magnified view has

$$\text{Horizontal base} = -\delta \operatorname{Cos}[\theta] = -(\operatorname{Cos}[\theta + \delta\theta] - \operatorname{Cos}[\theta])$$
$$\text{Vertical side} = \delta \operatorname{Sin}[\theta] = \operatorname{Sin}[\theta + \delta\theta] - \operatorname{Sin}[\theta]$$
$$\text{Hypotenuse} = \delta\theta$$

The length of the "hypotenuse" is $\delta\theta$ because radian measure is defined to be the length measured along the unit circle. However, the side that looks like a hypotenuse is actually a magnified small piece of a circle. Intuitively, the error between the length of the circular arc and the approximating straight line is small compared to $\delta\theta$, so, by Theorem 14.1, the length of the circle is given by the integral

$$\int_0^{2\pi} d\theta$$

It is still worthwhile to see how this is related to the computation with increments.

Definition 5.5 and the formulas for the derivatives of sine and cosine give us

$$\delta x = \delta \operatorname{Cos}[\theta] = \operatorname{Cos}[\theta + \delta\theta] - \operatorname{Cos}[\theta] = -\operatorname{Sin}[\theta] \cdot \delta\theta + \iota_1 \cdot \delta\theta$$
$$\delta y = \delta \operatorname{Sin}[\theta] = \operatorname{Sin}[\theta + \delta\theta] - \operatorname{Sin}[\theta] = \operatorname{Cos}[\theta] \cdot \delta\theta + \iota_2 \cdot \delta\theta$$

with $\iota_j \approx 0$ whenever $\delta\theta \approx 0$, $j = 1, 2$.

The Pythagorean Theorem says that the length of the true straight hypotenuse is the square root of the sum of the squares of the lengths of the legs,

$$\text{Small hypotenuse} = \sqrt{[\delta x]^2 + [\delta y]^2}$$

$$= \sqrt{(\text{Cos}[\theta + \delta\theta] - \text{Cos}[\theta])^2 + (\text{Sin}[\theta + \delta\theta] - \text{Sin}[\theta])^2}$$

$$= \sqrt{(-\text{Sin}[\theta]\delta\theta + \iota_1\delta\theta)^2 + (\text{Cos}[\theta]\delta\theta + \iota_2\delta\theta)^2}$$

$$= \sqrt{(-\text{Sin}[\theta]\delta\theta + \iota_1)^2 + (\text{Cos}[\theta]\delta\theta + \iota_2)^2} \ \delta\theta$$

$$= \sqrt{(-\text{Sin}[\theta])^2 + (\text{Cos}[\theta])^2} \ \delta\theta + \iota_3\delta\theta$$

$$= \delta\theta + \iota_3\delta\theta$$

since $(\text{Cos}[\theta])^2 + (\text{Sin}[\theta])^2 = 1$. By algebraic approximations, $\iota_3 \approx 0$ whenever $\delta\theta \approx 0$. This means that we may compute the length by the integral of $d\theta$ by Duhamel's Principle 14.1.

The general approximation idea for smooth functions $x[t]$, $y[t]$ is

$$\text{Small hypotenuse} = \sqrt{[\delta x]^2 + [\delta y]^2}$$

$$= \sqrt{(x'[\theta]\delta\theta + \iota_1\delta\theta)^2 + (y'[\theta]\delta\theta + \iota_2\delta\theta)^2}$$

$$= \sqrt{(x'[\theta] + \iota_1)^2 + (y'[\theta] + \iota_2)^2} \ \delta\theta$$

$$= \sqrt{(x'[\theta])^2 + (y'[\theta])^2} \ \delta\theta + \iota_3\delta\theta$$

In the integral for the length of an ellipse in Problem 14.12, the expression $\sqrt{(x')^2 + (y')^2}$ is more complicated, and we cannot do the final step of replacing $\sqrt{(-\text{Sin}[\theta])^2 + (\text{Cos}[\theta])^2}$ by 1 as above.

Example 14.6. *A Simple Parametric Length*

Let us consider another example, the length of the parametric curve

$$x = t^2$$
$$y = t^3$$

for $0 \le t \le 1$. We have the increments

$$\delta x = x'[t]\delta t + \iota_1\delta t \qquad\qquad \delta y = y'[t]\delta t + \iota_2\delta t$$
$$= 2t\delta t + \iota_1\delta t \qquad\qquad\qquad = 3t^2 \ \delta t + \iota_2\delta t$$

so that the length of the hypotenuse of the triangle we would see in an tiny microscope is

$$\delta l = \sqrt{(\delta x)^2 + (\delta y)^2} = \sqrt{(2t)^2 + (3t^2)^2} \ \delta t + \iota_3\delta t$$

and Duhamel's Principle says,

$$L = \int_0^1 \sqrt{(2t)^2 + (3t^2)^2} \ dt = \frac{(13)^{3/2} - 8}{27} \approx 1.4971$$

Example 14.7.   *Volume of a Half Bagel*

The upper half disk of radius 1 centered at $(2, 0)$ is revolved about the y-axis. The equation of the semicircular boundary of the half disk is

$$y = \sqrt{1 - (x - 2)^2}$$

The resulting figure looks like the top half of a sliced bagel as in Figure 14.9.

We approximate the accumulated volume between radius x and radius $x + \delta x$ by the cylindrical "shell" with inner radius x, outer radius $x + \delta x$, and height $ht\,[x] = \sqrt{1 - (x - 2)^2}$.

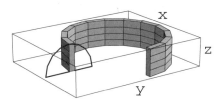

Figure 14.8: Half bagel shells

$$\text{Outer cylinder} \; - \; \text{inner cylinder} \; =$$
$$\pi R^2 ht - \pi r^2 ht = \pi((x + \delta x)^2 - x^2)ht$$
$$= \pi(2x\delta x + \delta x^2)ht$$
$$= \pi(2x\delta x + \delta x^2)\sqrt{1 - (x - 2)^2}$$

The δx^2 term can be neglected in our integral formula by Duhamel's Principle because it produces a term of the form $\varepsilon \cdot \delta x$ with $\varepsilon \approx 0$.

$$\pi(2x\delta x + \delta x^2)\sqrt{1 - (x - 2)^2} =$$
$$= \left(2\pi\, x\sqrt{1 - (x - 2)^2} \right) \delta x + \left(\delta x\pi\sqrt{1 - (x - 2)^2} \right) \delta x$$
$$= \left(2\pi\, x\sqrt{1 - (x - 2)^2} \right) \delta x + \varepsilon \cdot \delta x$$

The flat top on our cylinder produces another error between the accumulated volume from x to $x + \delta x$. If we are on the rising side of the circle, the left height, $ht\,[x] = \sqrt{1 - (x - 2)^2}$ produces a shell that lies completely inside the bagel while the right height, $ht\,(x + \delta x) =$

$\sqrt{1 - ([x + \delta x] - 2)^2}$, produces a shell that includes the cylindrical slice of the bagel (with its curved top.) This means

$$\pi 2x\delta x \sqrt{1 - (x - 2)^2} + \varepsilon_1 \delta x \leq V[x, x + \delta x] \leq \pi 2x\delta x \sqrt{1 - ([x + \delta x] - 2)^2} + \varepsilon_2 \delta x$$

On the falling part of the circle, the outside terms in this inequality are interchanged - the left height is above and the right height below. In either case,

$$V[x, x + \delta x] = \pi 2x\delta x \sqrt{1 - ([x + \delta x] - 2)^2} + \varepsilon_3 \delta x$$

with $\varepsilon_3 \approx 0$, so we have

$$V[a, b] = 2\pi \int_1^3 x\sqrt{1 - (x - 2)^2}\, dx$$

Example 14.8. *Computation of* $2\pi \int_1^3 x\sqrt{1 - (x - 2)^2}\, dx$

This integral can be computed with the computer or by hand with a trig substitution,

$$x - 2 = \operatorname{Sin}[\theta] \qquad dx = \operatorname{Cos}[\theta]\, d\theta$$
$$x = 1 \Leftrightarrow \theta = -\pi/2 \qquad x = 3 \Leftrightarrow \theta = \pi/2$$

so the integral becomes

$$V[a, b] = 2\pi \int_{-\pi/2}^{\pi/2} (\operatorname{Sin}(\theta) + 2)\sqrt{1 - \operatorname{Sin}^2[\theta]} \operatorname{Cos}[\theta]\, d\theta$$

$$= 2\pi \int_{-\pi/2}^{\pi/2} (\operatorname{Sin}(\theta) + 2)\operatorname{Cos}^2(\theta)\, d\theta$$

$$= 2\pi \int_{-\pi/2}^{\pi/2} \operatorname{Sin}(\theta)\operatorname{Cos}^2(\theta)\, d\theta + 2\pi \int_{-\pi/2}^{\pi/2} 2\operatorname{Cos}^2[\theta]\, d\theta$$

$$= 2\pi \left(\frac{-1}{3}\operatorname{Cos}^3(\theta)\right)|_{-\pi/2}^{\pi/2} + 2\pi \int_{-\pi/2}^{\pi/2} (1 + \operatorname{Cos}(2\theta))\, d\theta$$

$$= 0 + 2\pi[\theta + \frac{1}{2}\operatorname{Sin}(2\theta)]|_{-\pi/2}^{\pi/2}$$

$$= 2\pi^2$$

Example 14.9. *Trig Substitutions and Parametric Forms*

The change of variables in the Example 14.8 can be viewed as parametric equations for the surface of the bagel. A unit semicircle is given by the parametric equations

$$x = \operatorname{Sin}[\theta]$$
$$y = \operatorname{Cos}[\theta]$$

where x goes from -1 to $+1$ as θ goes from $-\pi/2$ to $+\pi/2$. You should verify that y takes only positive values in this range and that this pair traces out the top half circle of the unit radius centered at zero.

Adding 2 to the value of x moves the center of the semicircle to $(2, 0)$, so it forms the outline semicircle of our half bagel,

$$x = 2 + \text{Sin}[\theta]$$
$$y = \text{Cos}[\theta]$$

The integral change of variables formula for dx can be thought of as expressing the thickness of the shells in terms of θ. It is also perfectly OK to think of the change of variables as a formal manipulation that makes the integral easier to compute.

Example 14.10. *The General Shell Method - Explicit Form*

If the region below the graph of a positive continuous function, $y = f[x]$ for $0 \leq a \leq x \leq b$, is revolved about the y-axis, the volume of the solid obtained is

$$V = 2\pi \int_a^b x\, f[x]\, dx$$

This general result is given in the Mathematical Background.

Example 14.11. *The Icing on the Donut*

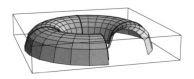

Figure 14.9: Area of donut icing

The surface of the top half of the torus is obtained by rotating the semicircle

$$y = \sqrt{1 - (x - 2)^2}$$

about the y axis. We will approximate the area by considering what happens to a tiny segment of the semicircle as we rotate that segment about the y axis. The slope of the segment matters in calculating the area of one of these 'barrel hoops.'

Figure 14.10: Hoops of different slant, but same radius and dx-thickness

The area is approximately the total length of the generating segment, δl, times the distance through which it travels. It is rotated about the y-axis, so it goes around a circle of radius x. The circumference of a circle is $C = 2\pi r = 2\pi x$, in this case. This makes the area of the hoop at x

$$\delta A = 2\pi x \ \delta l$$

We need an expression for δl in terms of the x variable so that we can form an integral.

The length of a tiny segment of the curve is

$$\delta l = \sqrt{\delta x^2 + \delta y^2} + \varepsilon_1 \cdot \delta x$$

with an error that is small compared to δx. Since we have $y = y[x]$ as an explicit function of x, we can write

$$\delta y = y'[x] \ \delta x + \varepsilon_2 \cdot \delta x$$

so that our length becomes $\delta l = \sqrt{\delta x^2 + (y'[x])^2 \delta x^2} + \varepsilon_3 \cdot \delta x = \sqrt{1 + (y'[x])^2} \ \delta x + \varepsilon_3 \cdot \delta x$ and our approximating area is

$$\delta A = 2\pi x \ \sqrt{1 + (y'[x])^2} \ \delta x + \varepsilon \cdot \delta x$$

Finally, by Duhamel's Principle,

$$A = 2\pi \int_1^3 x \ \sqrt{1 + (y'[x])^2} \ dx$$

$$= 2\pi \int_1^3 x \ \sqrt{1 + \frac{(x-2)^2}{1 - (x-2)^2}} \ dx$$

$$= 2\pi \int_1^3 \frac{x}{\sqrt{1 - (x-2)^2}} \ dx$$

This is a nasty integral, when $x = 1$ or $x = 3$. Of course, the computer might do the integration for us, but there is a more geometric way to see how to proceed.

Example 14.12. *A Second Approach to the Area of Revolution*

This time represent the generating semicircle of radius 1 centered at 2 on the x axis by parametric equations:

$$x[t] = 2 + \mathrm{Sin}[t]$$
$$y[t] = \mathrm{Cos}[t] \qquad \text{for } -\pi/2 \le t \le \pi/2$$

We go back to the length portion of the area of the approximating hoops. The length of a tiny segment of the curve is approximately

$$\delta l = \sqrt{\delta x^2 + \delta y^2} + \varepsilon_1 \cdot \delta t$$

but, in this case, the increments satisfy

$$\delta x[t] = x[t + \delta t] - x[t] = x'[t] \; \delta t + \varepsilon_2 \; \delta t$$
$$= \text{Cos}[t] \; \delta t + \varepsilon_2 \cdot \delta t$$
$$\delta y[t] = y[t + \delta t] - y[t] = y'[t] \; \delta t + \varepsilon_3 \; \delta t$$
$$= -\text{Sin}[t] \; \delta t + \varepsilon_3 \cdot \delta t$$

so the length is $\delta l = \sqrt{\delta x^2 + \delta y^2} + \varepsilon \cdot \delta t = \sqrt{\text{Cos}^2[t] + \text{Sin}^2[t]} \; \delta t + \varepsilon \cdot \delta t = \delta t + \varepsilon \cdot \delta t$. We must express the increment of area in terms of t,

$$\delta A = 2\pi \, x \; \delta l = 2\pi \, x \; \delta t + \varepsilon \cdot \delta t$$
$$= 2\pi(2 + \text{Sin}[t]) \; \delta t + \varepsilon \cdot \delta t$$
$$= 2\pi(2 + \text{Sin}[t]) \; \delta t + \varepsilon \cdot \delta t$$

and the area is given by

$$A = 2\pi \int_{-\pi/2}^{\pi/2} (2 + \text{Sin}[t]) \; dt$$
$$= 2\pi \int_{-\pi/2}^{\pi/2} 2 \; dt + 2\pi \int_{-\pi/2}^{\pi} \text{Sin}[t] \; dt$$
$$= 4\pi^2 + 0$$

since the integral of sine over a half period is zero.

Not only is this parametric integral for the area easy to compute, it also does not have the mathematical singularities of the cartesian form above when $x = 1$ and $x = 3$.

Example 14.13. *General Area of Revolution Formulas*

If the graph of a positive smooth function $y = f[x]$ for $a \leq x \leq b$ is revolved about the x-axis, the area of the surface obtained is

$$A = 2\pi \int_a^b f[x]\sqrt{1 + (f'[x])^2} \; dx$$

If a smooth parametric graph, $x[t] = f[t]$, $y[t] = g[t]$ for $a \leq x \leq b$, with $g[t] > 0$, is revolved about the x-axis the area of the surface so obtained is

$$A = 2\pi \int_a^b x[t]\sqrt{(x'[t])^2 + (y'[t])^2} \; dt$$

These results are explained in the Mathematical Background.

Exercise set 14.2

1. *Define the accumulated energy function*

$$E\left[t_1, t_2\right]$$

to be the amount of energy a household consumes from time t_1 to time t_2. Explain why $E\left(s, t\right)$ is additive. (This is just a matter of saying what the terms in the formula mean.)

2. Explicit Length
Verify the parametric computation of Example 14.6 by using the explicit equation for arclength, $L = \int_a^b \sqrt{1 + [f'[x]]^2}\ dx$, on the curve $y = x^{3/2}$. This is the same as the parametric curve above, since $t = \sqrt{x} = x^{1/2}$, so $y = t^3 = (x^{1/2})^3 = x^{3/2}$.

14.3   A Project on Geometric Integrals

It is important for you to try "slicing" on your own. These problems are hard, but you need to do several to understand integration.

Each problem has the following steps

Procedure 14.2.
1. Slice the figure and find an approximate formula for one slice of the form 'a function of the slice at x' times the 'thickness,' $f[x]\ \delta x$.
2. Find the limits of integration.
3. Compute using rules or numerical integration with or without the computer.

You have some freedom in your approximation as long as the error for one slice is small compared to the 'thickness' of the slice,

$$\text{amount of one slice } = f[x]\ \delta x + \varepsilon \cdot \delta x$$

GENERAL INSTRUCTIONS FOR THE PROBLEMS:

Find integral formulas for the following quantities. Use Duhamel's Principle, Theorem 14.1, explicitly to verify the correctness of your formulas. If possible, compute your integral symbolically; otherwise use numerical integration. Use the computer if you wish.

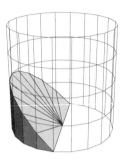

Figure 14.11: A sliced pyramid and a round wedge

Problem 14.2. DISCS ⎯⎯⎯⎯⎯⎯⎯⎯⎯⎯⎯⎯⎯⎯⎯⎯⎯⎯⎯⎯⎯⎯⎯⎯⎯⎯⎯⎯⎯⎯⎯⎯▼

Find the volume of a right circular cone of base radius r and height h.

⎯⎯⎯▲

Problem 14.3. SQUARE SLICES ⎯⎯⎯⎯⎯⎯⎯⎯⎯⎯⎯⎯⎯⎯⎯⎯⎯⎯⎯⎯⎯⎯⎯⎯▼

Find the volume of a square pyramid with base area B and height h. Does it matter whether or not it is a right pyramid or slant pyramid? Does the base have to be square?

⎯⎯⎯▲

Problem 14.4. TRIANGLES ⎯⎯⎯⎯⎯⎯⎯⎯⎯⎯⎯⎯⎯⎯⎯⎯⎯⎯⎯⎯⎯⎯⎯⎯⎯⎯⎯▼

A wedge is cut from a (cylindrical) tree trunk of radius r by cutting the tree with two planes meeting on a diameter. One plane is perpendicular to the axis and the other makes an angle θ with the first. Find the volume of the wedge.

⎯⎯⎯▲

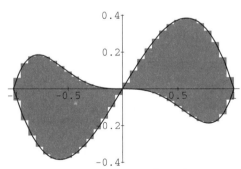

Figure 14.12: Area between $y = x^5 - x^3$ and $y = x - x^3$

Problem 14.5. AREA BETWEEN CURVES ⎯⎯⎯⎯⎯⎯⎯⎯⎯⎯⎯⎯⎯⎯⎯⎯⎯⎯⎯▼

*Find the area of the bounded regions between the curves $y = x^5 - x^3$ and $z = x - x^3$. Notice that the curves define two regions, one with z on top and the other with y on top. (See the program **AreaBetween**.)*

⎯⎯⎯▲

Problem 14.6.

Find the area between the curves of Problem 6.14.

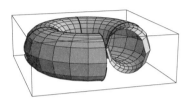

Figure 14.13: A torus

Problem 14.7. SLICE BY SHELLS

Find the volume of the solid torus (donut)

$$(x - R)^2 + y^2 \leq r^2 \qquad (0 < r < R)$$

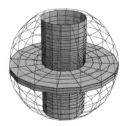 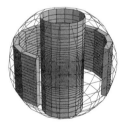

Figure 14.14: Slice by washers or slice by shells

Problem 14.8. SLICE BY WASHERS OR SHELLS

A (cylindrical) hole of radius r is bored through the center of a sphere of radius R. Find the volume of the remaining part of the sphere.

 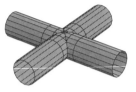

Figure 14.15: A Spherical cap and crossing cylinders

Problem 14.9. PART OF A SPHERE ——————————————▼
Find the volume of the portion of a sphere that lies above a plane a distance c above the center of the sphere for $0 < c < r$, where r is the radius of the sphere.
▲

Problem 14.10. INTERSECTING CYLINDERS ——————————▼
Two circular cylinders of equal radii r intersect through their centers at right angles. Find the volume of the common part. (HINT: The intersection can actually be sliced into square cross sections.)
▲

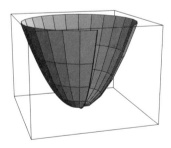

Figure 14.16: A parabolic antenna

Problem 14.11. SURFACE AREA ——————————————▼
*Find the surface area of the surface obtained by revolving $y = \frac{1}{2}x^2$ about the y-axis. (See the program **AntennaArea**.)*
▲

Problem 14.12. ARCLENGTH OF AN ELLIPSE —————————————————————————————— ▼

An ellipse is given parametrically by the pair of equations

$$x = 3\operatorname{Cos}[\theta]$$
$$y = 2\operatorname{Sin}[\theta]$$

with $-\pi < \theta < \pi$.

Find an integral formula for the length of the curve by "looking in a powerful microscope," as we did in the computation of the length of a parametric curve above. In a microscope we will see a right triangle with the change in x on the horizontal leg, the change in y on the vertical leg and the length along the hypotenuse. The Pythagorean Theorem says that the corresponding increment of length is given by

$$\delta l = \sqrt{\delta x^2 + \delta y^2}$$

This time, the length will NOT be equal to the change in the angle, $\delta\theta$, because an ellipse is a circle that has been stretched different amounts in the x and y directions. The changes in the coordinates are function changes:

$$\delta x[\theta] = x(\theta + \delta\theta) - x[\theta]$$
$$\delta y[\theta] = y(\theta + \delta\theta) - y[\theta]$$

Use the increment approximation for these changes to express them approximately in terms of $\delta\theta$. Substitute the approximations into the Pythagorean expression above.

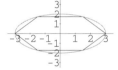 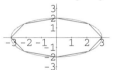 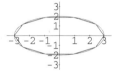

Figure 14.17: *Equal θ partition of the ellipse*

Test your formula on the circle (where you know the answer) as well as the "parametric" equations

$$x = \theta$$

$$y = 2\sqrt{1 - \left(\frac{\theta}{3}\right)^2}$$

(which can be compared to the explicit arclength formula $L = \int_a^b \sqrt{1 + [f'[x]]^2}\, dx$.)

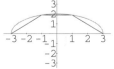 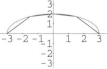 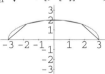

Figure 14.18: *Equal x partition of the ellipse*

The (correct) arclength formula for the ellipse cannot be computed by antidifferentiation so you must use numerical integration such as the computer NIntegrate[.]. Why is the parametric integral for the ellipse better behaved than the explicit formula?

——— ▲

14.4   Improper Integrals

"Improper" integrals, such as $\int_0^1 \frac{1}{\sqrt{x}}\, dx$, whose integrand tends to infinity and is discontinuous at $x = 0$ or $\int_1^\infty \frac{1}{x^2}\, dx$, which is integrated over an infinite interval, are studied in the book of projects and used in Chapter 18 on infinite series.

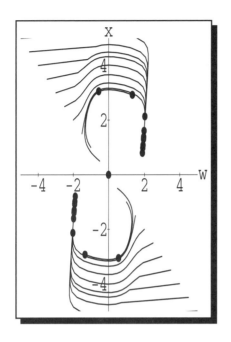

Part 3

3-D Geometry

Basic Vector Geometry

*We live in three space dimensions with changing time, so we expect
some applications of calculus take place in more than one variable.
This chapter gives the basics of 3-dimensional analytical geometry in its powerful vector formulation. These tools help us extend
calculus to two and three dimensions.*

Geometrically, vectors are "arrows," but these arrows can be measured in various ways.
The simplest way to measure vectors is with cartesian coordinates, but then the challenge
becomes, "How do I calculate a geometric quantity I need from the components?" or
the inverse problem, "How do I formulate the geometric condition analytically in terms of
coordinates?" The real topic of this chapter is basic translation back and forth between
pictures and formulas.

Translation is hard. At first, you will understand the pictures and the formulas separately.
In order to have a working knowledge of 3-dimensional mathematics, it is not sufficient just
to memorize formulas. You must learn to translate between the languages of algebra and
geometry. It will pay off in many ways.

Vectors arise in physics as (nongeometrical) forces or electrical fields and in many other
areas. We will use them mathematically in total derivatives of functions of several variables
and in differential equations. Our translation ability allows us to apply both geometry and
analysis to all of these problems.

15.1 Cartesian Coordinates

*This section reviews cartesian coordinates in one, two, and three dimen-
sions. Later, we view the coordinates as (one representation of) a vector.*

Cartesian coordinates are a way to measure the location of a point on a line, in a plane,
or in three space.

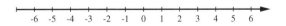

Figure 15.1: A number line

To locate a point on a line, we select one point to be the origin, select a unit scale and select a positive direction (usually to the right on a horizontal line). Any point can be located by measuring the real number distance associated with that point. Positive numbers are to the right, and negative numbers are measured to the left of the origin. This real number is called the coordinate of the geometrical point.

Example 15.1. *Number Line*

Draw a number line and graph the coordinate 5.2.
SOLUTION:
    A line is shown below with origin 0, scale 1 and a rightward positive direction. Simply measure 5.2 units to the right.

Figure 15.2: A coordinate point

Example 15.2. *Number Line*

Draw a number line and graph the coordinate -3.5.
SOLUTION:
    Measure 3.5 units in the negative direction to the left.

Figure 15.3: A negative coordinate

Example 15.3. *Distance between numbers*

How far apart are -3.5 and 5.2?
SOLUTION:
    Begin at the point 3.5 units to the left of the origin and measure all the way to the point 5.2 units to the right of the origin. This is a total of $3.5 + 5.2 = 8.7$ units.

Example 15.4. *Directed Distance Between Numbers*

Once we have decided to use positive and negative to denote right and left on our line, we can also ask: How far is it from -3.5 to 5.2 as a directed distance? And how far is it from 5.2 to -3.5 as a directed distance?
SOLUTION:
    Beginning at the point 3.5 units to the left of the origin, we move to the right to get to the point 5.2 units to the right of the origin. Moving to the right makes the direction of the distance positive, so the answer is $+8.7$.
    To move from 5.2 to -3.5 we go to the left, so the distance is negative and the answer from 5.2 to -3.5 is -8.7.

15.1.1   Cartesian Coordinates in Two Dimensions

To locate a point in a plane, we place two perpendicular number lines on the plane, as in

Figure 15.4. We call the point of intersection the origin of a cartesian coordinate system. The number lines are the axes of the system. When viewed from the usual perspective, the x_1-axis extends positively to the right and the x_2-axis extends positively upward, with both axes having the same scale. This is a counterclockwise orientation from the first to the second axis. We associate a point with the pair of numbers - in order - that it takes to measure their location. If the point is X and the numbers are x_1 and x_2, we denote the point $X(x_1, x_2)$

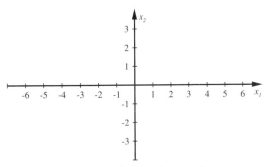

Figure 15.4: A coordinate plane

If we want to locate a point $X(x_1, x_2)$ in the plane, we start at the origin and move a directed distance x_1 along the first axis. Next, we move from this point along a line parallel to the second axis a directed distance x_2, as shown in Figure 15.5.

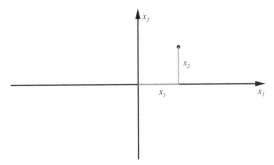

Figure 15.5: A coordinate pair

Example 15.5. *Plane Coordinates*

Use a 2-dimensional Cartesian coordinate system to locate the point $P(4, 5.3)$.

SOLUTION:

First we move 4 units to the right of the origin along the x_1-axis. Then we move from this point 5.3 units up, parallel to the x_2-axis.

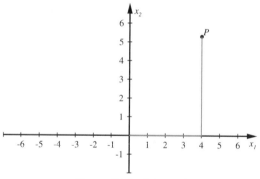

Figure 15.6: $P(4, 5.3)$

Given the location of a point, we may find its coordinates by drawing a line through the point parallel to the x_2-axis, as shown in Figure 15.6. The directed distance on this line from its intersection with the x_1-axis to the point is its x_2 coordinate. The directed distance from the origin to the intersection of this line with the x_1-axis is the x_1 coordinate.

Example 15.6. *Coordinates from Points*

Use the 2-dimensional graph in Figure 15.7 to find the coordinates of point A.

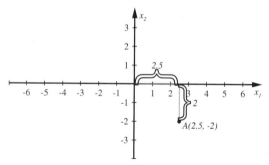

Figure 15.7: $A(2.5, -2)$

SOLUTION:
    Drop a perpendicular to the x-axis. Notice that the distance from the origin to the base of the perpendicular is 2.5 units to the right. The distance along the perpendicular from the x-axis to the point is 2 units down. Combining these measurements, we have $A(2.5, -2)$

15.1.2   Cartesian Coordinates in Three Dimensions

To locate a point in space, we place three mutually perpendicular number lines in space. Each number line is perpendicular to the other two, and the lines intersect each other at the point we call the origin. The number lines are the axes of the system. When viewed from the usual perspective, the x_1-axis extends positively out toward our observing position, the x_2-axis extends positively to the right, and the x_3-axis extends positively upward; all axes have the same scale, sa shown in Figure 15.8. This system is right-hand oriented; that is, if you align your right thumb with the positive x_1-axis, your first finger with the positive x_2-axis, and your second finger with the x_3-axis, then your second finger points in the positive x_3 direction. Your left hand would have your middle finger pointing in the negative x_3-direction.

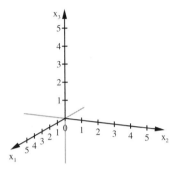

Figure 15.8: Space coordinates

To locate a particular point from its three coordinates (x_1, x_2, x_3), first move a directed distance x_1 from the origin along the x_1-axis, then move a directed distance x_2 from this point along a line parallel to the x_2-axis, and, finally, move from this point along a line parallel to the x_3-axis. The final location is the point $X(x_1, x_2, x_3)$.

Example 15.7. *Points in Three-Space*

Locate the point in three-space with coordinates X(3,-2,4).

SOLUTION:

Move three units out along the first axis, which is shown pointing toward us. Then move 2 units negatively along a line parallel to the second axis. This is left 2 units. Finally, move 4 units up along a line parallel to the third axis. The point **X** is at the top of the final segment shown in Figure 15.9.

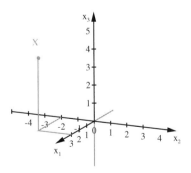

Figure 15.9: $X(3, -2, 4)$

Exercise set 15.1

1. Distance and Directed Distance in 1-D

(a) *Give a general formula for the distance between a point on the line with number coordinate a and a point with coordinate b. Explain how your formula covers the various cases of positive and negative signs for a and b.*

(b) *Give a general formula for the directed distance from a point with coordinate a to a point with coordinate b. Explain how your formula covers the various cases of positive and negative signs for a and b.*

(HINT: How much is: $a - b$?, $b - a$?, $|b - a|$?)

2. Coordinates
2-D: Locate the points $X(-3, 4)$, $Y(3, -4)$ and $Z(3, 4)$.
3-D: Locate the points $X(-3, 2, 4)$, $Y(3, 2, -4)$ and $Z(3, 2, 4)$.

15.2 Position Vectors

Coordinates may be thought of as position vectors in 1, 2 and 3 dimensions.

A position vector is the "arrow" that points from the origin in a coordinate system to the point with those coordinates. This arrow gives a direction and a magnitude starting at the origin of a coordinate system. We will compute these geometric quantities from cartesian coordinates. The necessary formulas appear below.

Example 15.8. *Vectors in 1-D*

Graph the 1-dimensional position vector 3.
SOLUTION:

Figure 15.10: A 1-D vector

Example 15.9. *Vectors in 2-D*

Graph the 2-dimensional position vector $\begin{bmatrix} 4 \\ -2 \end{bmatrix}$.
SOLUTION:

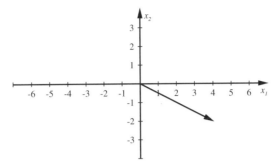

Figure 15.11: A 2-D vector

Example 15.10.   *Vectors in 3-D*

Graph the 3-dimensional position vector $\begin{bmatrix} 3 \\ -2 \\ 4 \end{bmatrix}$.

SOLUTION:

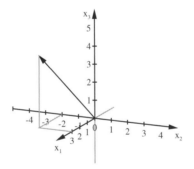

Figure 15.12: A 3-D vector

A position vector will be labeled with a boldface letter, **X**, a horizontal ordered triple (or pair), $X(2,1,3)$, or a vertical ordered triple (or pair), $\begin{bmatrix} 2 \\ 1 \\ 3 \end{bmatrix}$. Notice that a point and a position vector are denoted in the same way because we want to treat them the same way whether or not we draw the arrow connecting the origin and the point.

The length of a position vector is a unique non-negative real number that gives the distance from the origin to the tip of the position vector. This length is denoted $|\mathbf{X}|$ and is sometimes also called the norm of the vector. A one-dimensional vector b has length $|b|$. In other words, a negative or a positive number b is a distance $|b|$ from the origin.

The length in 2 dimensions is given by $|\mathbf{X}| = \sqrt{x_1^2 + x_2^2}$. To prove this formula, we apply the Pythagorean Theorem to the right triangle with sides parallel to the coordinate axes shown in Figure 15.13.

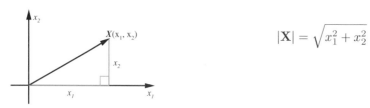

$$|\mathbf{X}| = \sqrt{x_1^2 + x_2^2}$$

Figure 15.13: The length formula for 2-D vectors

Example 15.11. *Length of 2-D Vectors*

The length of the position vector $\begin{bmatrix} 2 \\ -3 \end{bmatrix}$ is

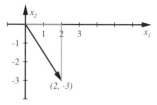

$$\sqrt{2^2 + (-3)^2}$$
$$\sqrt{4 + 9}$$
$$\sqrt{13}$$

Example 15.12. *Length in Three Dimensions*

To find the length of a position vector in three dimensions, we have to apply the Pythagorean Theorem twice. Consider position vector $\mathbf{X} = \begin{bmatrix} x_1 \\ x_2 \\ x_3 \end{bmatrix}$ as shown in Figure 15.14.

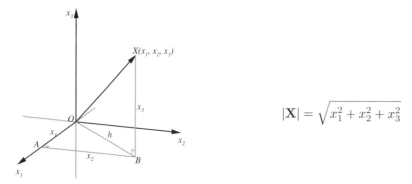

$$|\mathbf{X}| = \sqrt{x_1^2 + x_2^2 + x_3^2}$$

Figure 15.14: The length formula for 3-D vectors

First, we apply the Pythagorean Theorem to find h, the length of the hypotenuse of right triangle AOB, which lies in the x_1x_2-plane and has legs x_1 and x_2.

$$h^2 = x_1^2 + x_2^2$$
$$h = \sqrt{x_1^2 + x_2^2}$$

Now, we can apply the Pythagorean Theorem in triangle XOB with an hypotenuse along **X**, so having the same length as **X**, and legs h and x_3.

$$|\mathbf{X}| = \sqrt{h^2 + x_3^2}$$

Substituting the value of h^2 from above, we obtain the length of a 3-dimensional position vector $\mathbf{X} = \begin{bmatrix} x_1 \\ x_2 \\ x_3 \end{bmatrix}$:

$$|\mathbf{X}| = \sqrt{x_1^2 + x_2^2 + x_3^2}$$

Example 15.13.   *Length of 3-D Vectors*

Find the length of the position vector $\begin{bmatrix} 3 \\ -2 \\ 4 \end{bmatrix}$

SOLUTION:
$$\sqrt{3^2 + (-2)^2 + 4^2}$$
$$\sqrt{9 + 4 + 16}$$
$$\sqrt{29}$$

Procedure 15.1.   A general d-dimensional position vector $\mathbf{X} = \begin{bmatrix} x_1 \\ \vdots \\ x_d \end{bmatrix}$ has length

$$|\mathbf{X}| = \sqrt{x_1^2 + \ldots + x_d^2}$$

where d is any dimension (usually two or three).

Position vectors always start at the origin. This is very important. As far as the computations going from algebra to geometry are concerned, arrows that do not start at the origin are "displacements," not position vectors, and the formulas do not apply to displacement arrows. We will still draw them but will figure out ways to find computations on position vectors associated with shifted displacement arrows.

Exercise set 15.2

C

1. *Here is some practice with position vectors.*

1-D: Graph the following position vectors and compute their lengths.

a) $A(-3)$ b) $B(6)$ c) $C(5.333)$

2-D: Graph the following position vectors and compute their lengths.

a) $A(3, 4)$ b) $B(-3, -2)$ c) $C(-2, 6)$

d) $D(5, -3)$ e) $E(4.1, 2.7)$ f) $F(-3.5, 1)$

*Check your drawings by plotting with the program **Posit Vect**.*
3-D: Graph the following position vectors and compute their lengths.

a) $A(2,1,5)$ b) $B(-3,4,2)$ c) $C(-2,-4,-1)$
d) $D(3,-2,-5)$ e) $E(3.25,2.1,-6.5)$ f) $F(8.3,0,7)$

*Check your drawings by plotting with the program **Posit Vect**.*

2. *2-D: Draw the following vectors and find their lengths.*

a) $\begin{bmatrix} 4 \\ 2 \end{bmatrix}$ b) $\begin{bmatrix} -5 \\ -2 \end{bmatrix}$ c) $\begin{bmatrix} -3 \\ 7 \end{bmatrix}$ d) $\begin{bmatrix} 6 \\ -2 \end{bmatrix}$

3-D: Draw the following vectors and find their lengths.

a) $\begin{bmatrix} 4 \\ 6 \\ 5 \end{bmatrix}$ b) $\begin{bmatrix} -4 \\ 7 \\ -2 \end{bmatrix}$ c) $\begin{bmatrix} 6 \\ 8 \\ -10 \end{bmatrix}$ d) $\begin{bmatrix} -6 \\ -9 \\ -1 \end{bmatrix}$

*Check your drawings and length computations with the program **Posit Vect**.*

3. *Draw a 3-D rectangular box with sides parallel to the x, y, z axes and the vector $X(1,3,2)$ pointing across its diagonal.*

4. *Find the ordered pairs or triples designating the following position vectors.*

a)

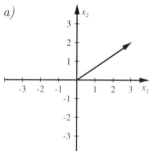

b)

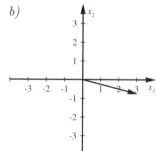

c)

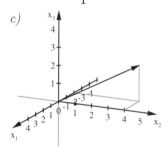

d)

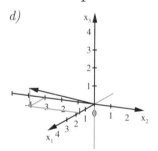

5.   (a) *Use the floor of your classroom as a 2-dimensional cartesian coordinate system with a corner as the origin. Form a system with equal scales using the edge of the floor as the positive axes. Locate and state the coordinates of a point at the base of*
      (1) *the bottom of your desk*
      (2) *the bottom of the doorway*
      (3) *the bottom of the instructor's desk*
   (b) *In a rectangular room, choose a corner to be the origin. Label the three edges at the corner between the walls and between the walls and floor so that they form the positive axes of a right-handed coordinate system. Use equal scales along the axes and measure the coordinates of the position vector pointing to the following:*
      (1) *the top of your desk*

(2) *the light switch*
(3) *the top of the instructor's desk*
(4) *the corner at the ceiling diagonally opposite your origin corner*
Locate a point for each of the following and estimate its coordinates
(1) *the front door of the building*
(2) *your parent's home*

15.3   Geometry of Vector Addition

From computer graphics to high-tech theory, analytical geometry is a key ingredient of mathematics in two and more dimensions. These applications are built on algebraic formulas with geometric counterparts. This section gives the symbolics and graphics of vector addition. Adding like components algebraically corresponds to geometrically sliding tips to tails of vectors. The goal of the section is to have you understand the two interpretations of this single operation.

Vector addition is algebraically easy, but to add vectors geometrically, we slide one parallel to itself until its tail is at the tip of the other. The translated vector is a displacement arrow (not a position vector starting at the origin), but its tip points to the sum position vector that starts at the origin. This geometric interpretation of the vector sum is a key to handling displacement arrows and understanding more advanced analytical geometry.

Suppose we wish to add two 2-dimensional position vectors, $\mathbf{R} = R(r_1, r_2)$ and $\mathbf{S} = S(s_1, s_2)$.

Procedure 15.2.   Vector Sum in 2-D:
Algebraically,
$$\mathbf{R} + \mathbf{S} = \begin{bmatrix} r_1 \\ r_2 \end{bmatrix} + \begin{bmatrix} s_1 \\ s_2 \end{bmatrix} = \begin{bmatrix} r_1 + s_1 \\ r_2 + s_2 \end{bmatrix}$$

Now, we will work through a geometric version of this operation.

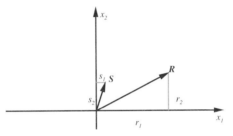

Figure 15.15: Vectors and components

The idea of position vector addition is to translate $\mathbf{S}$ to the tip of the position vector $\mathbf{R}$. The arrow parallel to $\mathbf{S}$ starting at the tip of $\mathbf{R}$ is displacement, but it can be measured by putting parallel coordinate axes at the tip of $\mathbf{R}$ and using the coordinates of $\mathbf{S}$ in the translated coordinate system.

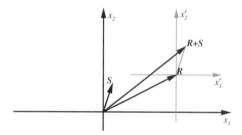

Figure 15.16: Tips to tails vector sum

The point at the tip of the oriented displacement parallel to $\mathbf{S}$ is the position vector for the sum $\mathbf{R} + \mathbf{S}$.

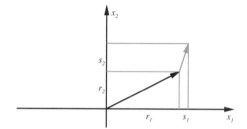

Figure 15.17: Sum components

Let us see why this is called "addition." From the diagram, the x_1-coordinate of the position vector $\mathbf{R} + \mathbf{S}$ is $r_1 + s_1$, the sum of the x_1-coordinate of $\mathbf{R}$ and the x_1 -coordinate of $\mathbf{S}$. Similarly the x_2 coordinate of $\mathbf{R} + \mathbf{S}$ is $r_2 + s_2$, the sum of the x_2-coordinate of $\mathbf{R}$ and the x_2-coordinate of $\mathbf{S}$. Algebraically the position vector sum in two dimensions is:

$$\mathbf{R} + \mathbf{S} = \begin{bmatrix} r_1 \\ r_2 \end{bmatrix} + \begin{bmatrix} s_1 \\ s_2 \end{bmatrix} = \begin{bmatrix} r_1 + s_1 \\ r_2 + s_2 \end{bmatrix}$$

Example 15.14. *Geometric Summing in the Other Order*

We could also move from the origin to the tip of position vector $\mathbf{S}$. Then, move to the tip of an oriented displacement parallel to $\mathbf{R}$ starting at the tip of $\mathbf{S}$.

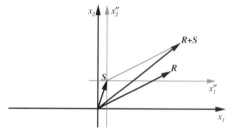

Figure 15.18: Tips to tails the other way

The point at the tip of this oriented displacement is the position vector for the sum $\mathbf{S} + \mathbf{R}$.

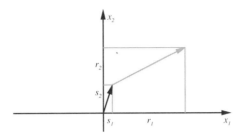

Figure 15.19: Sum components

From the diagram, the x_1 coordinate of the position vector $\mathbf{S} + \mathbf{R}$ is $s_1 + r_1$ and the x_2 coordinate is $s_2 + r_2$. Algebraically,

$$\mathbf{S} + \mathbf{R} = \begin{bmatrix} s_1 \\ s_2 \end{bmatrix} + \begin{bmatrix} r_1 \\ r_2 \end{bmatrix} = \begin{bmatrix} s_1 + r_1 \\ s_2 + r_2 \end{bmatrix} = \begin{bmatrix} r_1 + s_1 \\ r_2 + s_2 \end{bmatrix}$$

So $\mathbf{R} + \mathbf{S} = \mathbf{S} + \mathbf{R}$, and we see that it does not matter from which position vector we start.

Example 15.15.   *Tips to Tails as a Parallelogram*

Figure 15.20 combines both oriented tips to tails displacements in one diagram, showing the sum as a diagonal of a parallelogram with $\mathbf{R}$ and $\mathbf{S}$ as sides.

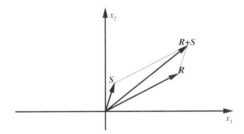

Figure 15.20: Parallelogram vector addition

The position vectors $\mathbf{R}$ and $\mathbf{S}$ form two adjacent sides of a parallelogram with one vertex at the origin. The sum $\mathbf{R} + \mathbf{S}$ is the directed diagonal from the origin of this parallelogram. This is the geometrical way to see that $\mathbf{R} + \mathbf{S} = \mathbf{S} + \mathbf{R}$.

Notice the sum of two position vectors is another position vector with its tail fixed at the origin. The displaced arrows parallel to $\mathbf{R}$ and $\mathbf{S}$ are used to geometrically construct the sum, but they are not the sum vector.

Example 15.16.   *2-D Vector Sum*

Given

$$\mathbf{X} = \begin{bmatrix} 3 \\ 1 \end{bmatrix} \qquad \text{and} \qquad \mathbf{Y} = \begin{bmatrix} -5 \\ 6 \end{bmatrix}$$

geometrically and algebraically find $\mathbf{X} + \mathbf{Y}$.

SOLUTION, GEOMETRIC PART:

Draw the position vectors $\begin{bmatrix} 3 \\ 1 \end{bmatrix}$ and $\begin{bmatrix} -5 \\ 6 \end{bmatrix}$. Next, draw an oriented displacement from the tip of $\begin{bmatrix} 3 \\ 1 \end{bmatrix}$ parallel to $\begin{bmatrix} -5 \\ 6 \end{bmatrix}$. You are now at the point (-2,7). Draw the position vector to that point from the origin. This is the position vector $\mathbf{X} + \mathbf{Y}$.

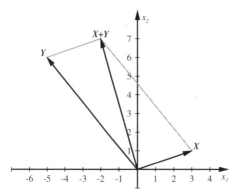

Figure 15.21: Another vector sum

SOLUTION, ALGEBRAIC PART:

$$\mathbf{X} + \mathbf{Y} = \begin{bmatrix} 3 \\ 1 \end{bmatrix} + \begin{bmatrix} -5 \\ 6 \end{bmatrix} = \begin{bmatrix} 3 + (-5) \\ 1 + 6 \end{bmatrix} = \begin{bmatrix} -2 \\ 7 \end{bmatrix}$$

Example 15.17. *2-D Vector Sum*

Given

$$\mathbf{X} = \begin{bmatrix} 2 \\ -1 \end{bmatrix} \qquad \text{and} \qquad \mathbf{Y} = \begin{bmatrix} -3 \\ -4 \end{bmatrix}$$

geometrically and algebraically find $\mathbf{Y} + \mathbf{X}$.

GEOMETRIC SOLUTION:

Draw the position vectors $\begin{bmatrix} 2 \\ -1 \end{bmatrix}$ and $\begin{bmatrix} -3 \\ -4 \end{bmatrix}$. Draw an oriented displacement from the tip of $\begin{bmatrix} -3 \\ -4 \end{bmatrix}$ parallel to $\begin{bmatrix} 2 \\ -1 \end{bmatrix}$. Draw a position vector from the origin to this point $(-1, -5)$. This vector is $\mathbf{X} + \mathbf{Y}$.

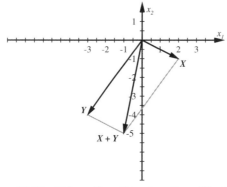

Figure 15.22: $X(2, -1) + Y(-3, -4) = Z(-1, -5)$

ALGEBRAIC SOLUTION:

$$\mathbf{Y} + \mathbf{X} = \begin{bmatrix} -3 \\ -4 \end{bmatrix} + \begin{bmatrix} 2 \\ -1 \end{bmatrix} = \begin{bmatrix} -3+2 \\ -4+(-1) \end{bmatrix} = \begin{bmatrix} -1 \\ -5 \end{bmatrix}$$

Procedure 15.3.   Vector Sum in 3-D:

Algebraically

$$\mathbf{R} + \mathbf{S} = \begin{bmatrix} r_1 \\ r_2 \\ r_3 \end{bmatrix} + \begin{bmatrix} s_1 \\ s_2 \\ s_3 \end{bmatrix} = \begin{bmatrix} r_1+s_1 \\ r_2+s_2 \\ r_3+s_3 \end{bmatrix} = \begin{bmatrix} s_1+r_1 \\ s_2+r_2 \\ s_3+r_3 \end{bmatrix}$$

We will work through a geometric example of adding the vectors

$$\mathbf{R} = R(r_1, r_2, r_3) = R(-6, 8, 2) \qquad \text{and} \qquad \mathbf{S} = S(s_1, s_2, s_3) = S(2, -3, 3)$$

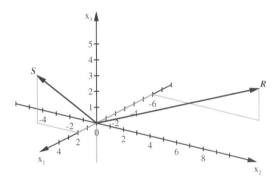

Figure 15.23: Vectors **R** and **S**

Let us form the geometric sum $\mathbf{R} + \mathbf{S}$. We move from the origin to the tip of position vector $\mathbf{R}$ first. Then, move to the tip of an oriented displacement parallel to $\mathbf{S}$ starting at the tip of $\mathbf{R}$.

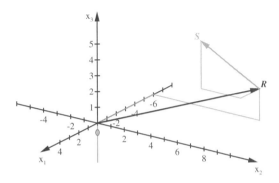

Figure 15.24: **S** with tail displaced to tip of **R**

The point at the tip of this oriented displacement is the position vector for the sum $\mathbf{R} + \mathbf{S}$. From the diagram, the x_1 coordinate of the position vector $\mathbf{R} + \mathbf{S}$ is $r_1 + s_1$, the x_2

coordinate is $r_2 + s_2$, and the x_3 coordinate is $r_3 + s_3$. Algebraically,

$$\mathbf{R} + \mathbf{S} = \begin{bmatrix} r_1 \\ r_2 \\ r_3 \end{bmatrix} + \begin{bmatrix} s_1 \\ s_2 \\ s_3 \end{bmatrix} = \begin{bmatrix} r_1 + s_1 \\ r_2 + s_2 \\ r_3 + s_3 \end{bmatrix} = \begin{bmatrix} s_1 + r_1 \\ s_2 + r_2 \\ s_3 + r_3 \end{bmatrix}$$

Example 15.18. *Geometric Summation in the Other Order*

Now, suppose we wish to add the two 3-dimensional position vectors in the other order. Geometrically, to form $\mathbf{S} + \mathbf{R}$, we translate $\mathbf{R}$ so that its tail begins at the tip of $\mathbf{S}$. This oriented displacement or displacement arrow parallel to $\mathbf{R}$ can be measured from the components of $\mathbf{R}$ by using a translated coordinate system with its origin at the tip of $\mathbf{S}$.

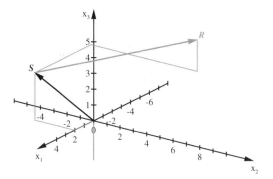

Figure 15.25: $\mathbf{R}$ with tail displaced to tip of $\mathbf{S}$

The point at the tip of the oriented displacement is the position vector for the sum $\mathbf{S} + \mathbf{R}$.

The net movement on the x_1 axis for the position vector $\mathbf{S} + \mathbf{R}$ is $s_1 + r_1$, the sum of the x_1 coordinate of $\mathbf{S}$ and the x_1 coordinate of $\mathbf{R}$. Similarly, the x_2 coordinate of $\mathbf{S} + \mathbf{R}$ is $s_2 + r_2$ and the x_3 coordinate of $\mathbf{S} + \mathbf{R}$ is $s_3 + r_3$. Algebraically, the position vector sum in three dimensions is

$$\mathbf{S} + \mathbf{R} = \begin{bmatrix} s_1 \\ s_2 \\ s_3 \end{bmatrix} + \begin{bmatrix} r_1 \\ r_2 \\ r_3 \end{bmatrix} = \begin{bmatrix} s_1 + r_1 \\ s_2 + r_2 \\ s_3 + r_3 \end{bmatrix} = \begin{bmatrix} -4 \\ 5 \\ 5 \end{bmatrix}$$

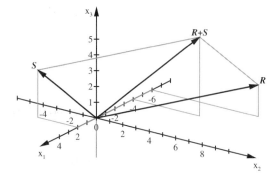

Figure 15.26: The vector sum $\mathbf{R} + \mathbf{S}$

Again, $\mathbf{S} + \mathbf{R} = \mathbf{R} + \mathbf{S}$. We see that it does not matter from which position vector we begin.

Example 15.19. *Both Tips to Tails as a Parallelogram*

Geometrically, we can see this from a plot of both additions. The vector sum is the diagonal through the origin of a parallelogram with $\mathbf{R}$ on one edge and $\mathbf{S}$ on another.

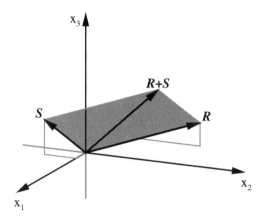

Figure 15.27: 3-D vector parallelogram sum

Notice the sum of two position vectors is another position vector with its tail fixed at the origin. The displaced arrows parallel to $\mathbf{R}$ and $\mathbf{S}$ are used to geometrically construct the sum, but they are not the sum vector.

Example 15.20. *3-D Vector Sum*

Given

$$\mathbf{X} = \begin{bmatrix} 5 \\ 2 \\ 3 \end{bmatrix} \qquad \text{and} \qquad \mathbf{Y} = \begin{bmatrix} -3 \\ 4 \\ 2 \end{bmatrix}$$

find $\mathbf{X} + \mathbf{Y}$ with geometry and with algebra.

GEOMETRIC SOLUTION:

Draw the position vectors $\begin{bmatrix} 5 \\ 2 \\ 3 \end{bmatrix}$ and $\begin{bmatrix} -3 \\ 4 \\ 2 \end{bmatrix}$. Draw an oriented displacement from the tip

of $\begin{bmatrix} -3 \\ 4 \\ 2 \end{bmatrix}$ parallel to $\begin{bmatrix} 5 \\ 2 \\ 3 \end{bmatrix}$. Draw a position vector from the origin to this point $P(2, 6, 5)$.

This position vector is $\mathbf{X} + \mathbf{Y}$.

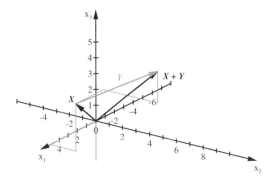

Figure 15.28: The vector sum $\mathbf{X} + \mathbf{Y}$

ALGEBRAIC SOLUTION:

$$\mathbf{X} + \mathbf{Y} = \begin{bmatrix} 5 \\ 2 \\ 3 \end{bmatrix} + \begin{bmatrix} -3 \\ 4 \\ 2 \end{bmatrix} = \begin{bmatrix} 5 + (-3) \\ 2 + 4 \\ 3 + 2 \end{bmatrix} = \begin{bmatrix} 2 \\ 6 \\ 5 \end{bmatrix}$$

Exercise set 15.3

1. *For each problem, use geometry and algebra to find the sum* $\mathbf{Z} = \mathbf{X} + \mathbf{Y}$ *of the two position vectors.*

a) $\mathbf{X} = \begin{bmatrix} 3 \\ 5 \end{bmatrix}$ $\mathbf{Y} = \begin{bmatrix} 2 \\ 1 \end{bmatrix}$ b) $\mathbf{X} = \begin{bmatrix} -3 \\ 4 \end{bmatrix}$ $\mathbf{Y} = \begin{bmatrix} -2 \\ -5 \end{bmatrix}$

c) $\mathbf{X} = \begin{bmatrix} -3 \\ -4 \end{bmatrix}$ $\mathbf{Y} = \begin{bmatrix} 4 \\ -2 \end{bmatrix}$ d) $\mathbf{X} = \begin{bmatrix} -3 \\ -4 \end{bmatrix}$ $\mathbf{Y} = \begin{bmatrix} 5 \\ 4 \end{bmatrix}$

e) $\mathbf{X} = \begin{bmatrix} 5 \\ 3 \\ 4 \end{bmatrix}$ $\mathbf{Y} = \begin{bmatrix} 1 \\ 6 \\ 2 \end{bmatrix}$ f) $\mathbf{X} = \begin{bmatrix} 4 \\ 6 \\ 2 \end{bmatrix}$ $\mathbf{Y} = \begin{bmatrix} -5 \\ 2 \\ 3 \end{bmatrix}$

g) $\mathbf{X} = \begin{bmatrix} -2 \\ -4 \\ 2 \end{bmatrix}$ $\mathbf{Y} = \begin{bmatrix} 2 \\ 6 \\ 3 \end{bmatrix}$ h) $\mathbf{X} = \begin{bmatrix} 2 \\ 2 \\ 1 \end{bmatrix}$ $\mathbf{Y} = \begin{bmatrix} -2 \\ 1 \\ 2 \end{bmatrix}$

Check your work with the program **VectSum.**

2. *Use a diagram and explain that the sum* $\mathbf{X} + \mathbf{Y}$ *of two 3-dimensional position vectors* $\mathbf{X}$ *and* $\mathbf{Y}$ *is the directed diagonal of a parallelogram with* $\mathbf{X}$ *and* $\mathbf{Y}$ *as two adjacent sides.*

3. *What geometric property must* $\mathbf{X}$ *and* $\mathbf{Y}$ *share for the formula* $|\mathbf{X} + \mathbf{Y}| = |\mathbf{X}| + |\mathbf{Y}|$ *to be true?*

15.4   Scalar Multiplication

This section gives the algebraic and geometric interpretations of scalar multiplication. Multiplying components by a factor corresponds to stretching, compressing, or reversing vectors. Conversely, two position vectors are parallel only if they are scalar multiples.

Procedure 15.4. Scalar Multiplication in 2-D

The algebraic operation of multiplication of a number (or scalar) c times a 2-D vector $\mathbf{X} = \begin{bmatrix} x_1 \\ x_2 \end{bmatrix}$ is to multiply each component by the scalar,

$$c\,\mathbf{X} = c \begin{bmatrix} x_1 \\ x_2 \end{bmatrix} = \begin{bmatrix} c \cdot x_1 \\ c \cdot x_2 \end{bmatrix}$$

Geometrically, a scalar multiple of one vector points along the same line through the origin but is stretched, compressed, or reversed (and possibly also stretched or compressed). There are three basic geometric cases of scalar multiplication for a scalar c and a position vector $\mathbf{X}$, where $c\,\mathbf{X} = c \begin{bmatrix} x_1 \\ x_2 \end{bmatrix} = \begin{bmatrix} c \cdot x_1 \\ c \cdot x_2 \end{bmatrix}$:

$$
\begin{array}{lll}
c > 1 & \Rightarrow & c\mathbf{X} \text{ is } \mathbf{X} \text{ stretched in length by a factor of } c \\
0 < c < 1 & \Rightarrow & c\mathbf{X} \text{ is } \mathbf{X} \text{ compressed in length by a factor of } c \\
c = -1 & \Rightarrow & c\mathbf{X} \text{ is } \mathbf{X} \text{ reversed in direction}
\end{array}
$$

We also have the degenerate cases:

$$
\begin{array}{lll}
c = 1 & \Rightarrow & c\mathbf{X} \text{ is } \mathbf{X} \\
c = 0 & \Rightarrow & c\mathbf{X} \text{ is } \mathbf{X} \text{ annihilated}
\end{array}
$$

We want to examine scalar multiplication geometrically.

Example 15.21. *Geometric Scalar Multiplication* $\mathbf{Y} = 2\mathbf{X}$

Let us begin by looking at the position vectors $\mathbf{X} = \begin{bmatrix} 5 \\ 2 \end{bmatrix}$ and $\mathbf{Y} = \begin{bmatrix} 10 \\ 4 \end{bmatrix}$ shown in Figure 15.29. Notice that position vector $\mathbf{Y}$ has twice the length of position vector $\mathbf{X}$.

$$|\mathbf{Y}| = \sqrt{10^2 + 4^2} = 2\sqrt{5^2 + 2^2} \qquad\qquad |\mathbf{X}| = \sqrt{5^2 + 2^2}$$

The position vector $\mathbf{Y} = \begin{bmatrix} 10 \\ 4 \end{bmatrix}$ could be rewritten as $\begin{bmatrix} 2 \cdot 5 \\ 2 \cdot 2 \end{bmatrix}$ or $2 \begin{bmatrix} 5 \\ 2 \end{bmatrix}$, hence $\mathbf{Y} = 2\mathbf{X}$. This is consistent with our observation about the length of $\mathbf{Y}$ compared to the length of $\mathbf{X}$.

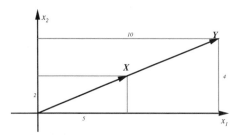

Figure 15.29: X and Y

Example 15.22. *Geometric Scalar Multiplication* $\mathbf{U} = 3\mathbf{T}$

Let us test our observations with another example. Examine the position vectors $\mathbf{T} = \begin{bmatrix} 1 \\ 2 \end{bmatrix}$ and $\mathbf{U} = \begin{bmatrix} 3 \\ 6 \end{bmatrix}$ shown in Figure 15.30.

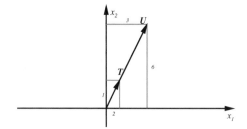

Figure 15.30: T and U

This time position vector $\mathbf{U}$ has three times the length of position vector $\mathbf{T}$. Position vector $\mathbf{U} = \begin{bmatrix} 3 \\ 6 \end{bmatrix}$ can be rewritten as $\begin{bmatrix} 3 \cdot 1 \\ 3 \cdot 2 \end{bmatrix}$ or $3 \begin{bmatrix} 1 \\ 2 \end{bmatrix}$, and it is true that $\mathbf{U} = 3\mathbf{T}$.

Example 15.23. *Geometric Scalar Multiplication, Compression* $\mathbf{R} = \frac{1}{2}\mathbf{S}$

What is the geometric situation when $0 < c < 1$? The above algebraic computation still works, but what does it show? Let us examine the position vectors $\mathbf{S} = \begin{bmatrix} 8 \\ 2 \end{bmatrix}$ and $\mathbf{R} = \begin{bmatrix} 4 \\ 1 \end{bmatrix}$ shown in Figure 15.31.

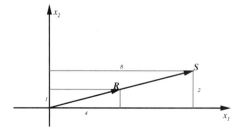

Figure 15.31: R and S

Here we observe that the length of $\mathbf{R}$ is one half the length of $\mathbf{S}$. So, $\mathbf{R} = \begin{bmatrix} 4 \\ 1 \end{bmatrix}$ may be rewritten $\begin{bmatrix} \frac{1}{2} \cdot 8 \\ \frac{1}{2} \cdot 2 \end{bmatrix}$ or $\frac{1}{2} \begin{bmatrix} 8 \\ 2 \end{bmatrix}$, which is $\frac{1}{2}\mathbf{S}$ and $\mathbf{R} = \frac{1}{2}\mathbf{S}$. In this case, the position vector $\mathbf{S}$ shrinks to a new position vector $\mathbf{R}$ when multiplied by the scalar $\frac{1}{2}$.

Example 15.24. *Algebraic Length of Scalar Multiples*

When a position vector $\mathbf{X}$ is multiplied by a number (a scalar) $c > 0$, the result is another position vector obtained by multiplying each element of the original position vector by the scalar. This multiplication stretches or compresses the position vector into a new, vector pointing in the same direction. The length formula says the length of $\mathbf{Y}$ is c times the length of $\mathbf{X}$ when $\mathbf{Y} = c\,\mathbf{X}$,

$$|\mathbf{Y}| = \left| \begin{bmatrix} c\,x_1 \\ c\,x_2 \end{bmatrix} \right| = \sqrt{(c\,x_1)^2 + (c\,x_2)^2}$$

$$= \sqrt{c^2 x_1^2 + c^2 x_2^2} = \sqrt{c^2(x_1^2 + x_2^2)}$$

$$= c\sqrt{x_1^2 + x_2^2}$$

$$|c\,\mathbf{X}| = c\sqrt{x_1^2 + x_2^2} = c\,|\mathbf{X}|$$

Example 15.25. *Algebraic Negative Multiples*

Consider the algebraic proof that scalar multiplication changes length by the factor c,

$$|\mathbf{Y}| = \left| \begin{bmatrix} c\,x_1 \\ c\,x_2 \end{bmatrix} \right| = \sqrt{(c\,x_1)^2 + (c\,x_2)^2} = \sqrt{c^2(x_1^2 + x_2^2)}$$

The conclusion $|c\,\mathbf{X}| = c\,|\mathbf{X}|$ is not correct if $c < 0$ because $\sqrt{c^2} = |c| \neq c$ in that case. However, it does show that

$$|c\,\mathbf{X}| = \sqrt{c^2(x_1^2 + x_2^2)} = |c|\sqrt{x_1^2 + x_2^2}$$

$$|c\mathbf{X}| = |c|\,|\mathbf{X}|$$

The length of a negative multiple is multiplied by the absolute value of the scalar.

Example 15.26. *Geometric Negative Multiples*

Now compare the position vectors $\mathbf{R} = \begin{bmatrix} 4 \\ 3 \end{bmatrix}$ and $\mathbf{S} = \begin{bmatrix} -8 \\ -6 \end{bmatrix}$ shown in Figure 15.32.

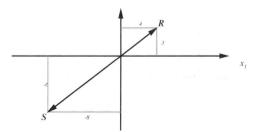

Figure 15.32: $\mathbf{R}$ and $\mathbf{S}$

Here, position vector $\mathbf{S}$ has twice the length of position vector $\mathbf{R}$ but is oriented in exactly the opposite direction. So, $\mathbf{S} = \begin{bmatrix} -8 \\ -6 \end{bmatrix}$ or $\begin{bmatrix} -2 \cdot 4 \\ -2 \cdot 3 \end{bmatrix}$ or $-2 \begin{bmatrix} 4 \\ 3 \end{bmatrix}$ which is $-2\mathbf{R}$. So, $\mathbf{S} = -2\mathbf{R}$.

Procedure 15.5. Scalar Multiplication in 3-D
The algebraic operation of multiplication of a number (or scalar) c times a vector
$\mathbf{X} = \begin{bmatrix} x_1 \\ x_2 \\ x_3 \end{bmatrix}$ is to multiply each component by the scalar,

$$c\,\mathbf{X} = c \begin{bmatrix} x_1 \\ x_2 \\ x_3 \end{bmatrix} = \begin{bmatrix} c \cdot x_1 \\ c \cdot x_2 \\ c \cdot x_3 \end{bmatrix}$$

There are three basic geometric cases of scalar multiplication for a scalar c and a position vector $\mathbf{X}$ as follows. These are the same in two and three dimensions.

$$
\begin{array}{lcl}
c > 1 & \Rightarrow & c\mathbf{X} \text{ is } \mathbf{X} \text{ stretched in length by a factor of } c \\
0 < c < 1 & \Rightarrow & c\mathbf{X} \text{ is } \mathbf{X} \text{ compressed in length by a factor of } c \\
c = -1 & \Rightarrow & c\mathbf{X} \text{ is } \mathbf{X} \text{ reversed in direction}
\end{array}
$$

Following are general illustrations of the cases in 3-D.

STRETCHING:

If $c > 1$ then $c\,\mathbf{X}$ is a position vector in the *same* direction as $\mathbf{X}$ but of longer length; in fact, the length is c times longer.

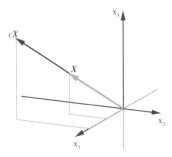

Figure 15.33: Stretching by $c > 1$

COMPRESSING:

If $0 < c < 1$, then $c\,\mathbf{X}$ is a position vector in the same direction as $\mathbf{X}$, but of shorter length.

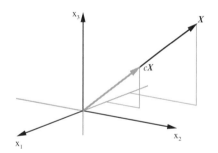

Figure 15.34: Compressing by $0 < c < 1$

REVERSING:

If $c < 0$, then $c\mathbf{X}$ is a position vector in the *opposite* direction as $\mathbf{X}$ and its length is longer or shorter depending on the absolute value of c, $|c|$. We can reduce this to the case of scalar multiplication by -1, followed by a stretch or compression of $|c|$. If $c = -1$, $c\mathbf{X}$ has the same length as $\mathbf{X}$, but points in the opposite direction.

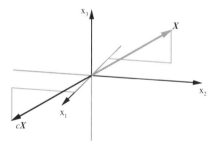

Figure 15.35: Reversal of a vector

If $c = 1$, the position vector remains unchanged since $c\mathbf{X} = 1\mathbf{X} = \mathbf{X}$

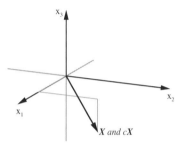

Figure 15.36: No change

What happens when $c = 0$?

Example 15.27.   *The Algebraic Length Computation for a Scalar Multiple*

If $\mathbf{Y} = c\mathbf{X}$, we have the following length computation for positive c,

$$|\mathbf{Y}| = \left|\begin{bmatrix} c\,x_1 \\ c\,x_2 \\ c\,x_3 \end{bmatrix}\right| = \sqrt{(c\,x_1)^2 + (c\,x_2)^2 + (c\,x_3)^2}$$

$$= \sqrt{c^2 x_1^2 + c^2 x_2^2 + c^2 x_3^2} = \sqrt{c^2\,(x_1^2 + x_2^2 + x_3^2)}$$

$$= c\,\sqrt{x_1^2 + x_2^2 + x_3^2}$$

$$|c\,\mathbf{X}| = c\,\sqrt{x_1^2 + x_2^2 + x_3^2} = c\,|\mathbf{X}|$$

This computation algebraically shows that the length of a positive scalar multiple is that multiple of the original length.

If c is negative, we get $|c\,\mathbf{X}| = |c|\,|\mathbf{X}|$, but not $|c\,\mathbf{X}| = c\,|\mathbf{X}|$. The direction of the vector is reversed, but the reversed vector still has a positive length.

Example 15.28. $\mathbf{Y} = 3\mathbf{X}$ *Geometrically*

Examine the position vector $\mathbf{X} = \begin{bmatrix} 1 \\ 4 \\ 2 \end{bmatrix}$ and the position vector $\mathbf{Y} = \begin{bmatrix} 3 \\ 12 \\ 6 \end{bmatrix}$ shown in Figure 15.37.

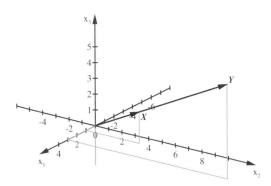

Figure 15.37: $\mathbf{X}$ and $\mathbf{Y}$

Position vector $\mathbf{Y} = \begin{bmatrix} 3 \\ 12 \\ 6 \end{bmatrix}$ can be rewritten as $\begin{bmatrix} 3 \cdot 1 \\ 3 \cdot 4 \\ 3 \cdot 2 \end{bmatrix}$ or $3 \begin{bmatrix} 1 \\ 4 \\ 2 \end{bmatrix}$. By the algebraic operation above, that is $\mathbf{Y} = 3\mathbf{X}$.

Example 15.29. $\mathbf{S} = \frac{1}{3}\mathbf{R}$ *Geometrically*

Examine the position vector $\mathbf{R} = \begin{bmatrix} 3 \\ 9 \\ 6 \end{bmatrix}$ and the position vector $\mathbf{S} = \begin{bmatrix} 1 \\ 3 \\ 2 \end{bmatrix}$.

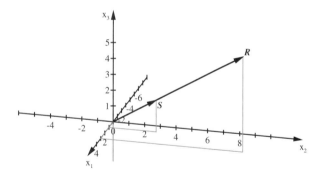

Figure 15.38: **R** and **S**

Position vector **S** is one third the length of position vector **R**. $\mathbf{S} = \begin{bmatrix} 1 \\ 3 \\ 2 \end{bmatrix}$ can be rewritten as

$\begin{bmatrix} \frac{1}{3} \cdot 3 \\ \frac{1}{3} \cdot 9 \\ \frac{1}{3} \cdot 6 \end{bmatrix}$ or $\frac{1}{3} \begin{bmatrix} 3 \\ 9 \\ 6 \end{bmatrix}$ which is $\frac{1}{3}\mathbf{R}$.

Example 15.30. *Reverse and Squash*

Consider the position vectors $\mathbf{X} = \begin{bmatrix} 2 \\ 4 \\ 6 \end{bmatrix}$ and $\mathbf{Y} = \begin{bmatrix} -1 \\ -2 \\ -3 \end{bmatrix}$ shown in Figure 15.39.

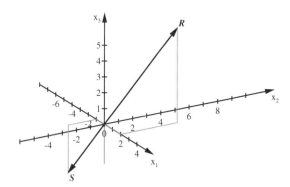

Figure 15.39: **X** and **Y**

Notice that position vector **Y** is one half the length of position vector **X** but in exactly the
opposite direction. $\mathbf{Y} = \begin{bmatrix} -1 \\ -2 \\ -3 \end{bmatrix}$ can be rewritten as $\begin{bmatrix} -\frac{1}{2} \cdot 2 \\ -\frac{1}{2} \cdot 4 \\ -\frac{1}{2} \cdot 6 \end{bmatrix}$ or $-\frac{1}{2} \begin{bmatrix} 2 \\ 4 \\ 6 \end{bmatrix}$ which is $-\frac{1}{2}\mathbf{X}$.

Example 15.31. *2-D Scalar Multiple*

Given $\mathbf{R} = \mathbf{R}(-3, 2)$ and $c = 3$, calculate and sketch $c\mathbf{R}$.
SOLUTION:

$$c\mathbf{R} = 3\begin{bmatrix} -3 \\ 2 \end{bmatrix} = \begin{bmatrix} 3 \cdot -3 \\ 3 \cdot 2 \end{bmatrix} = \begin{bmatrix} -9 \\ 6 \end{bmatrix}$$

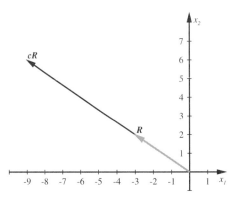

Figure 15.40: Solution

Example 15.32. *3-D Scalar Multiple*

Given $\mathbf{S} = \mathbf{S}(-5, 8, 2)$ and $c = 1/4$, calculate and sketch $c\mathbf{S}$.

SOLUTION:

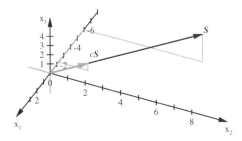

Figure 15.41: Solution

Example 15.33. *Unit Vectors*

Scalar multiplication can be used to find a vector of unit length that points in the same direction as a given vector. For example, the vector

$$\mathbf{Z} = \begin{bmatrix} -6 \\ -3 \\ -8 \end{bmatrix} \qquad \text{has length} \qquad |\mathbf{Z}| = \sqrt{6^2 + 3^2 + 8^2} = \sqrt{109}$$

We need to compress the vector by the factor $1/\sqrt{109}$ in order to make it unit length,

$$\mathbf{Z}_u = \frac{1}{\sqrt{109}} \cdot \begin{bmatrix} -6 \\ -3 \\ -8 \end{bmatrix} = \begin{bmatrix} \frac{-6}{\sqrt{109}} \\ \frac{-3}{\sqrt{109}} \\ \frac{-8}{\sqrt{109}} \end{bmatrix} \approx -\begin{bmatrix} 0.575 \\ 0.287 \\ 0.794 \end{bmatrix}$$

Exercise set 15.4

1. *In each example below a vector* **X** *and a scalar c are given.*
   Compute: (1) $\mathbf{Y} = c\mathbf{X}$,   (2) $|\mathbf{Y}|$ (3) $|\mathbf{X}|$ (4) $|c|\,|\mathbf{X}|$
   (5) *Sketch* **X**, **Y** *and a unit vector in the same direction as* **X**.

   a) $\mathbf{X} = \begin{bmatrix} 6 \\ 4 \end{bmatrix}$ $c = 2$

   b) $\mathbf{X} = \begin{bmatrix} -3 \\ 1 \end{bmatrix}$ $c = 3$

   c) $\mathbf{X} = \begin{bmatrix} -3 \\ 6 \end{bmatrix}$ $c = -\frac{1}{3}$

   d) $\mathbf{X} = \begin{bmatrix} 5 \\ 4 \end{bmatrix}$ $c = 1$

   e) $\mathbf{X} = \begin{bmatrix} 5 \\ 4 \end{bmatrix}$ $c = 0$

   f) $\mathbf{X} = \begin{bmatrix} 5 \\ 4 \end{bmatrix}$ $c = -1$

   g) $\mathbf{X} = \begin{bmatrix} 8 \\ 5 \\ 2 \end{bmatrix}$ $c = -2$

   h) $\mathbf{X} = \begin{bmatrix} 8 \\ 5 \\ 2 \end{bmatrix}$ $c = \frac{1}{2}$

   *Check your work with the program **ScalarMult**.*

2. *Compute the length of* $\mathbf{Y} = \begin{bmatrix} 8 \\ 4 \\ 2 \end{bmatrix}$ *and the length of* $\mathbf{X} = \begin{bmatrix} -4 \\ -2 \\ -1 \end{bmatrix}$. *Show that* $|\mathbf{Y}| \neq -2|\mathbf{X}|$.
   If $\mathbf{Y} = c\mathbf{X}$, *when is* $|\mathbf{Y}| \neq c|\mathbf{X}|$?

3. Parallel Position Vectors
   Use the geometric meaning of scalar multiplication to give an algebraic condition equivalent to two position vectors being parallel:

   X *is parallel to* **Y** *if and only if ???*

   Test your condition on the following pairs of vectors:
   a) $X(4,5)$; $Y(6,7)$ b) $X(4,5)$; $Y(8,10)$ c) $X(-3,4)$; $Y(6,-8)$
   d) $X(3,4,5)$; $Y(6,7,8)$ e) $X(3,4,5)$; $Y(6,8,10)$ f) $X(-3,4,-5)$; $Y(6,-8,10)$

4. A Line

   (a) *Describe geometrically the set of points that consists of the tips of all position vectors of the form* $\begin{bmatrix} 2c \\ 3c \end{bmatrix}$, *where c takes on all real number values. Sketch several values, $c = 1$, $c = 3/2$, $c = 2$, $c = 5/2$, $c = 1/2$, $c = -1/2$, $c = -1$, $c = -3/2 \cdots$ until you see the general pattern.*

   (b) *Repeat part (1) using the 3-D position vector* $\begin{bmatrix} 3c \\ 5c \\ 8c \end{bmatrix}$.

15.5   Differences and Displacements

This section shows how to treat arrows that do not start at the origin.

A position vector $\mathbf{X}$ in 2-space or in 3-space can be visualized geometrically by an arrow from the origin to the point $\mathbf{X}$. It is essential that we always treat vectors as starting from the origin in our algebraic computations, but we also want "free arrows." This section shows how they are related to position vectors by algebraic differences.

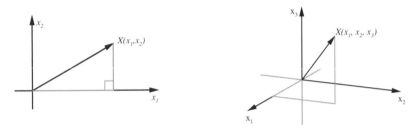

Figure 15.42: Two-dimensional and three-dimensional vectors

We also want to consider oriented displacements (or arrows) from any one point to another. We must be careful not to treat these as position vectors. These vectors are sometimes called "free vectors." We prefer to call them oriented displacements or displacement arrows. Consider any two points $\mathbf{S}$ and $\mathbf{T}$ in a 2-dimensional cartesian coordinate system. The oriented displacement from $\mathbf{S}$ to $\mathbf{T}$ can be visualized geometrically by an arrow from $\mathbf{S}$ to $\mathbf{T}$ as shown in Figure 15.43.

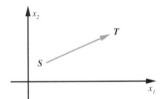

Figure 15.43: A 2-D displacement

This oriented displacement from $\mathbf{S}$ to $\mathbf{T}$ cannot be described using a position vector, because the tail of this arrow would not lie at the origin. If we first move the x_1x_2-coordinate axes such that the translated $x_1'x_2'$-coordinate axes are parallel to the corresponding x_1x_2-coordinate axes and point $\mathbf{S}$ is at the origin of the $x_1'x_2'$-coordinate system as shown below, then the components of the new arrow are a position vector in the new system shown in Figure 15.44.

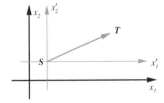

Figure 15.44: New coordinates

An ordinary position vector is associated with the oriented displacement from $\mathbf{S}$ to $\mathbf{T}$ because of the geometric 'tips to tails' interpretation of vector sum. If an ordinary unknown position vector $\mathbf{X}$ is added to $\mathbf{S}$ and the sum $\mathbf{S} + \mathbf{X}$ points from the origin to the tip of $\mathbf{T}$, then the 'tips to tails' law of vector addition says that the arrow with its tail at $\mathbf{S}$ the same length and parallel to $\mathbf{X}$ connects the tip of $\mathbf{S}$ and the tip of $\mathbf{T}$ as shown in Figure 15.45.

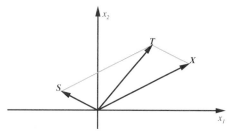

Figure 15.45: A position vector parallel to the displacement from $\mathbf{S}$ to $\mathbf{T}$

Algebraically, we have $\mathbf{S} + \mathbf{X} = \mathbf{T}$; so, if both $\mathbf{S}$ and $\mathbf{T}$ are known, we solve for $\mathbf{X}$. This is an important observation.

Procedure 15.6.   A position vector parallel to an oriented displacement from the tip of $\mathbf{S}$ to the tip of $\mathbf{T}$ is given by
$$\mathbf{X} = \mathbf{T} - \mathbf{S}$$

In other words, the position vector parallel to the (displacement) arrow pointing from the tip of $\mathbf{S}$ to the tip of $\mathbf{T}$ is given by the vector difference.

Example 15.34.   *The Oriented Displacement from $S(2,3)$ to $T(5,9)$*

Find the position vector $\begin{bmatrix} x_1 \\ x_2 \end{bmatrix}$ describing the oriented displacement from $S(2,3)$ to $T(5,9)$.

SOLUTION:

We begin by plotting the points $\mathbf{S}$ and $\mathbf{T}$ and drawing the arrow from $\mathbf{S}$ to $\mathbf{T}$.

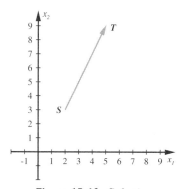

Figure 15.46: Solution

Next we translate the axes, so we can see the derivation of the formula $\mathbf{X} = \mathbf{T} - \mathbf{S}$ again.

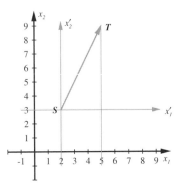

Figure 15.47: Solution with new axes at the tip of **S**

From the diagram, we see that the coordinates of $T(5, 9)$ in the $x_1' x_2'$-coordinate system are $X(3, 6)$ and, hence, the vector $\begin{bmatrix} 3 \\ 6 \end{bmatrix}$ is the desired position vector. Algebraically, $\begin{bmatrix} 5 \\ 9 \end{bmatrix} - \begin{bmatrix} 2 \\ 3 \end{bmatrix} = \begin{bmatrix} 3 \\ 6 \end{bmatrix}$.

Example 15.35.   *Vector Difference*

Given the position vectors $\mathbf{X} = \mathbf{X}(5, 3)$ and $\mathbf{Y} = \mathbf{Y}(2, -4)$, find and sketch a diagram of $\mathbf{Y} - \mathbf{X}$.

SOLUTION:

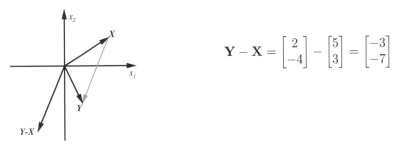

$$\mathbf{Y} - \mathbf{X} = \begin{bmatrix} 2 \\ -4 \end{bmatrix} - \begin{bmatrix} 5 \\ 3 \end{bmatrix} = \begin{bmatrix} -3 \\ -7 \end{bmatrix}$$

The same formula for displacement works in 3-D:

Procedure 15.7.   The 3-D position vector parallel to an oriented displacement from the tip of **S** to the tip of **T** is given by:

$$\mathbf{X} = \mathbf{T} - \mathbf{S}$$

We verify this in an example. An oriented displacement from point **S** to point **T** in a 3-dimensional cartesian coordinate system can be described with the tips to tails law of vector addition. Tips-to-tails tells us that a position vector **X** such that the sum **X** + **S** points to the tip of **T** satisfies **X** + **S** = **T**. We can draw this by sliding **X** parallel until its tail is at the tip of **S**. The moved arrow points from **S** to **T**. We can also describe this by drawing $\mathbf{X}$ as a position vector in a new coordinate system with its origin at the tip of $\mathbf{S}$. In this system $\mathbf{X}$ points to the tip of $\mathbf{T}$. Figure 15.48 shows this with the tips of $\mathbf{S}$ and $\mathbf{T}$ labeled by letters, but without position arrows for $\mathbf{S}$ and $\mathbf{T}$. The gray arrow is $\mathbf{X}$ as we show in the following discussion.

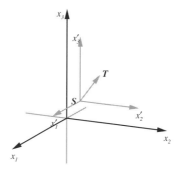

Figure 15.48: A 3-D displacement in new coordinates

The specific arrow along the oriented displacement from $S(2, 3, 4)$ to $T(5, 7, 9)$ is shown in regular coordinates in Figure 15.49.

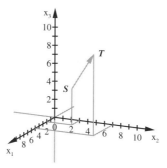

Figure 15.49: Displacement from $\mathbf{S}$ to $\mathbf{T}$

If we translate the axes, we see Figure 15.50 with $\mathbf{X}$ plotted in the new coordinate system. This is also the tips to tails interpretation of $\mathbf{S} + \mathbf{X} = \mathbf{T}$, so $\mathbf{X} = \mathbf{T} - \mathbf{S}$.

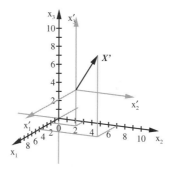

Figure 15.50: $\mathbf{X}$ in new coordinates

From the diagram, the measured coordinates of $\mathbf{X}$ in the $x_1' x_2' x_3'$-coordinate system are $(3, 4, 5)$ and, hence, $\begin{bmatrix} 3 \\ 4 \\ 5 \end{bmatrix}$ is the desired position vector. Verify this by your own observation of Figure 15.50.

We can also do this by vector algebra. We know that $\mathbf{X} + \mathbf{S} = \mathbf{T}$ from tips to tails vector addition, so $\mathbf{X} = \mathbf{T} - \mathbf{S} = \mathbf{T}(5, 7, 9) - \mathbf{S}(2, 3, 4) = \mathbf{X}(3, 4, 5)$.

Example 15.36.   *3-D Displacement*

Sketch the displacement arrow from the tip of $\mathbf{R} = \mathbf{R}(-3, 4, 1)$ to the tip of $\mathbf{S} = \mathbf{S}(2, 3, -2)$. Calculate and sketch a diagram of $\mathbf{S} - \mathbf{R}$.

SOLUTION:

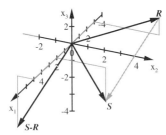

$$\mathbf{S} - \mathbf{R} = \begin{bmatrix} 2 \\ 3 \\ -2 \end{bmatrix} - \begin{bmatrix} -3 \\ 4 \\ 1 \end{bmatrix} = \begin{bmatrix} 5 \\ -1 \\ -3 \end{bmatrix}$$

Figure 15.51: Solution

Exercise set 15.5

1. Finish - Start

In Figure 15.52 below a displacement arrow (or oriented displacement) from the end of $\mathbf{R}$ *to the end of* $\mathbf{S}$ *is shown. Sketch the parallel position vector, and give an algebraic formula for it.*

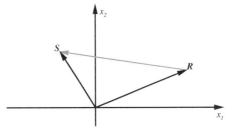

Figure 15.52: Displacement from $\mathbf{R}$ to $\mathbf{S}$

C

2. *For each of the following, use a diagram to find the position vector that describes the oriented displacement from* **S** *to* **T**. *Calculate the numerical vector displacement and find the distance from* **S** *to* **T**.
2-D:

a) $S(4,5)$; $T(6,9)$ b) $S(-4,6)$; $T(3,-5)$ c) $S(-3,-4)$; $T(5,-4)$
3-D:

a) $S(2,5,3)$; $T(7,10,12)$ b) $S(13,10,9)$; $T(4,6,2)$ c) $S(4,-5,8)$; $T(-2,-6,4)$
*Check your work with the program **DisplVect**.*

3. *Give a geometric condition so that the formula* $|\mathbf{T} - \mathbf{S}| = |\mathbf{T}| - |\mathbf{S}|$ *is true. Which of the vectors in the previous exercise satisfy your condition?*

15.6   Angles and Projections

This section shows how to find "direction" from cartesian coordinates. It also gives the useful and simple projection formula. Both of these formulas use the dot product.

The dot (or inner) product does not have an immediate geometric interpretation, but is used to compute angles. Algebraically, it is very simple.

Definition 15.8. *Dot Product Formula*
The dot product of position vectors **X** *and* **Y** *is the scalar quantity*

$$\langle \mathbf{X} \bullet \mathbf{Y} \rangle = x_1 y_1 + \cdots + x_d y_d$$

where $\mathbf{X} = \begin{bmatrix} x_1 \\ \vdots \\ x_d \end{bmatrix}$ *and* $\mathbf{Y} = \begin{bmatrix} y_1 \\ \vdots \\ y_d \end{bmatrix}$, *for any dimension d.*

Notice that the length is related to the dot product by

$$|\mathbf{X}| = \sqrt{\langle \mathbf{X} \bullet \mathbf{X} \rangle} = \sqrt{x_1 x_1 + \cdots + x_d x_d}$$

The dot product of two vectors is a scalar. The scalar product of a number and a vector is a vector. We need to be careful with the different kinds of operations on numbers and vectors.

Example 15.37. *2-D Dot Product*

Find the dot product of the position vectors $\begin{bmatrix} 4 \\ 3 \end{bmatrix}$ and $\begin{bmatrix} 7 \\ -11 \end{bmatrix}$.

SOLUTION:

$$\langle \begin{bmatrix} 4 \\ 3 \end{bmatrix} \bullet \begin{bmatrix} 7 \\ -11 \end{bmatrix} \rangle \quad = \quad 4 \cdot 7 + 3 \cdot (-11)$$
$$= \quad 28 - 33$$
$$= \quad -5$$

Example 15.38.   *3-D Dot Product*

Find the dot product of the position vectors $\begin{bmatrix} 2 \\ -3 \\ 5 \end{bmatrix}$ and $\begin{bmatrix} -4 \\ 3 \\ 6 \end{bmatrix}$.

SOLUTION:

$$\langle \begin{bmatrix} 2 \\ -3 \\ 5 \end{bmatrix} \bullet \begin{bmatrix} -4 \\ 3 \\ 6 \end{bmatrix} \rangle \quad = 2 \cdot (-4) + (-3) \cdot 3 + 5 \cdot 6$$
$$= \quad -8 - 9 + 30$$
$$= \quad 13$$

Given two position vectors $\mathbf{X}$ and $\mathbf{Y}$ with $\theta = \angle \langle \mathbf{X}, \mathbf{Y} \rangle$ the angle between them, the cosine of θ can be found by the formula:

Definition 15.9.   *The Formula for the Angle Between Two Vectors*

$$\mathrm{Cos}[\theta] = \frac{\langle \mathbf{X} \bullet \mathbf{Y} \rangle}{|\mathbf{X}| \, |\mathbf{Y}|}$$

In order to find the actual angle, we need to calculate the ArcCosine of the expression on the right. ArcCosine is difficult when computed by hand, but easy on your computer.

Example 15.39.   *Angles in 3-D*

If $\mathbf{X} = \begin{bmatrix} 3 \\ 5 \\ 6 \end{bmatrix}$, $\mathbf{Y} = \begin{bmatrix} 4 \\ -2 \\ 3 \end{bmatrix}$, and θ, find the angle between them.

SOLUTION:

$$\text{Cos}[\theta] = \frac{\left\langle \begin{bmatrix} 3 \\ 5 \\ 6 \end{bmatrix} \bullet \begin{bmatrix} 4 \\ -2 \\ 3 \end{bmatrix} \right\rangle}{\left| \begin{bmatrix} 3 \\ 5 \\ 6 \end{bmatrix} \right| \left| \begin{bmatrix} 4 \\ -2 \\ 3 \end{bmatrix} \right|}$$

$$= \frac{12 - 10 + 18}{\sqrt{9 + 25 + 36}\sqrt{16 + 4 + 9}}$$

$$= \frac{20}{\sqrt{2030}}$$

$$\theta = \text{ArcCos}[\frac{20}{\sqrt{2030}}]$$

$$\approx 1.11085 \quad (radians)$$

Vector projection may be visualized as one position vector casting a "shadow" on the other position vector under special lighting conditions.

Let $\mathbf{X}$ and $\mathbf{Y}$ be position vectors (2-dimensional or 3-dimensional). We are wish to compute the position vector formed by the "shadow" of $\mathbf{Y}$ cast onto $\mathbf{X}$ if the source of light is perpendicular to position vector $\mathbf{X}$ and is contained by the plane in which both $\mathbf{X}$ and $\mathbf{Y}$ lie.

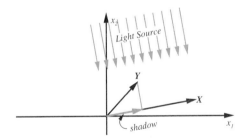

Figure 15.53: Vector projection

Notice that the "shadow" vector in Figure 15.53 is a scalar multiple of $\mathbf{X}$.

Procedure 15.10. Algebraically, the projection of $\mathbf{Y}$ onto $\mathbf{X}$ is

$$\text{Proj}_{\mathbf{X}}[\mathbf{Y}] = \frac{\langle \mathbf{Y} \bullet \mathbf{X} \rangle}{\langle \mathbf{X} \bullet \mathbf{X} \rangle} \mathbf{X}$$

Since the quotient of dot products is a scalar (or number), $c = \frac{\langle \mathbf{Y} \bullet \mathbf{X} \rangle}{\langle \mathbf{X} \bullet \mathbf{X} \rangle}$, the $\text{Proj}_{\mathbf{X}}[\mathbf{Y}]$ becomes $c\mathbf{X}$. Position vector projection is a special scalar multiplication. Scalar multiplication just produces a stretch, compression, or reversal of a vector, so the "shadow" lies along the direction of $\mathbf{X}$.

You can derive this formula using trigonometry and the dot product formula for the cosine of an angle. But whether you derive it or simply memorize it, this formula is worth remembering because it can be computed without square roots or ArcCosines. In other words, if you can reduce a geometric problem to computing a perpendicular projection, you have an easy computation.

Example 15.40.   *2-D Perpendicular Projection*

Given $\mathbf{X} = \mathbf{X}(-5, 2)$ and $\mathbf{Y} = \mathbf{Y}(-2, 6)$ calculate and sketch a diagram of the projection of $\mathbf{Y}$ onto $\mathbf{X}$.

SOLUTION:

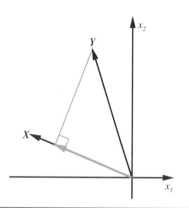

$$\text{Proj}_{\mathbf{X}}[\mathbf{Y}] = \frac{\begin{bmatrix} -5 \\ 2 \end{bmatrix} \bullet \begin{bmatrix} -2 \\ 6 \end{bmatrix}}{\begin{bmatrix} -5 \\ 2 \end{bmatrix} \bullet \begin{bmatrix} -5 \\ 2 \end{bmatrix}} \begin{bmatrix} -5 \\ 2 \end{bmatrix}$$

$$= \frac{22}{29} \begin{bmatrix} -5 \\ 2 \end{bmatrix}$$

$$= \begin{bmatrix} -\frac{110}{29} \\ \frac{44}{29} \end{bmatrix}$$

Example 15.41.   *3-D Perpendicular Projection*

Given $\mathbf{R} = \mathbf{R}(4, 6, -1)$ and $\mathbf{S} = \mathbf{S}(-2, 3, 1)$, calculate and sketch a diagram of the projection of $\mathbf{S}$ onto $\mathbf{R}$.

SOLUTION:

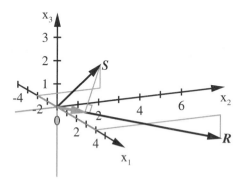

$$\text{Proj}_{\mathbf{R}}(\mathbf{S}) = \frac{\begin{bmatrix} 4 \\ 6 \\ -1 \end{bmatrix} \bullet \begin{bmatrix} -2 \\ 3 \\ 1 \end{bmatrix}}{\begin{bmatrix} 4 \\ 6 \\ -1 \end{bmatrix} \bullet \begin{bmatrix} 4 \\ 6 \\ -1 \end{bmatrix}} \begin{bmatrix} 4 \\ 6 \\ -1 \end{bmatrix}$$

$$= \frac{9}{53} \begin{bmatrix} 4 \\ 6 \\ -1 \end{bmatrix}$$

$$= \begin{bmatrix} 36/53 \\ 54/53 \\ -9/53 \end{bmatrix}$$

Exercise set 15.6

1. Your Own Condition for Perpendicularity

Give the simplest possible algebraic condition you can use to test whether two vectors are perpendicular. HINT: What is $\text{Cos}[\frac{\pi}{2}]$? *For example, are* $\begin{bmatrix} 1 \\ -3 \\ 2 \end{bmatrix}$ *and* $\begin{bmatrix} 2 \\ 2 \\ 1 \end{bmatrix}$ *perpendicular?*

Are $\begin{bmatrix} 1 \\ -3 \\ 2 \end{bmatrix}$ *and* $\begin{bmatrix} 1 \\ 2 \\ 1 \end{bmatrix}$? *What are the angles between them? Does the denominator matter?*

Your Condition for Perpendicularity: **A** is perpendicular to **B** if and only if ??

2. *Draw each pair of vectors* **X** *and* **Y** *below and compute* $\langle \mathbf{X} \bullet \mathbf{Y} \rangle$, $|\mathbf{X}|$, $|\mathbf{Y}|$, $\text{Cos}[\angle \langle A, B \rangle]$, *and* $\theta = \angle \langle \mathbf{X}, \mathbf{Y} \rangle$.

a) $\begin{bmatrix} -3 \\ -2 \end{bmatrix}, \begin{bmatrix} -8 \\ 12 \end{bmatrix}$ b) $\begin{bmatrix} 2 \\ 1 \end{bmatrix}, \begin{bmatrix} -4 \\ 5 \end{bmatrix}$ c) $\begin{bmatrix} 7 \\ -3 \end{bmatrix}, \begin{bmatrix} -2 \\ -8 \end{bmatrix}$ d) $\begin{bmatrix} 6 \\ -9 \end{bmatrix}, \begin{bmatrix} -3 \\ 2 \end{bmatrix}$

e) $\begin{bmatrix} 5 \\ 2 \\ -3 \end{bmatrix}, \begin{bmatrix} 2 \\ -2 \\ 2 \end{bmatrix}$ f) $\begin{bmatrix} 6 \\ -3 \\ -3 \end{bmatrix}, \begin{bmatrix} 2 \\ 9 \\ -5 \end{bmatrix}$ g) $\begin{bmatrix} 3 \\ 5 \\ 6 \end{bmatrix}, \begin{bmatrix} 4 \\ -2 \\ 3 \end{bmatrix}$ h) $\begin{bmatrix} 5 \\ 6 \\ 3 \end{bmatrix}, \begin{bmatrix} 1 \\ 3 \\ 7 \end{bmatrix}$

i) $\begin{bmatrix} 1 \\ 4 \\ 8 \end{bmatrix}, \begin{bmatrix} 8 \\ -4 \\ 1 \end{bmatrix}$ j) $\begin{bmatrix} 1 \\ 4 \\ 8 \end{bmatrix}, \begin{bmatrix} 4 \\ 7 \\ -4 \end{bmatrix}$ k) $\begin{bmatrix} 2 \\ 2 \\ 1 \end{bmatrix}, \begin{bmatrix} -2 \\ 1 \\ 2 \end{bmatrix}$ l) $\begin{bmatrix} 1 \\ -2 \\ 2 \end{bmatrix}, \begin{bmatrix} -2 \\ 1 \\ 2 \end{bmatrix}$

*Check your answers with the program **AnglPerp**.*

3. *Find the projection of the first vector,* **X**, *in the direction of the second,* **Y**, *for the pairs of vectors in the previous exercise.*

15.7   Cross Product in 3 Dimensions

A cross product contains geometric information, including 3-D orientation.

The cross product computation is a little complicated. Before you read the formula, think about how you might find a vector perpendicular to two given vectors, such as $\begin{bmatrix} 1 \\ 3 \\ 1 \end{bmatrix}$ and $\begin{bmatrix} -2 \\ -3 \\ 1 \end{bmatrix}$. Sketch a figure and convince yourself that there is a whole line of possibilities. Also convince yourself that the computation is not obvious. (HINT: Dot products do not help very much.)

Procedure 15.11.   Algebraic Cross Product

The 3-dimensional vector cross product is given by the computation

$$\mathbf{Z} = \mathbf{X} \times \mathbf{Y} = \begin{bmatrix} x_2 y_3 - x_3 y_2 \\ x_3 y_1 - x_1 y_3 \\ x_1 y_2 - x_2 y_1 \end{bmatrix}$$

The cross product computation may be easier to remember using the symbolic determinant:

$$\mathbf{Z} = \mathbf{X} \times \mathbf{Y} = \det \begin{vmatrix} i & j & k \\ x_1 & x_2 & x_3 \\ y_1 & y_2 & y_3 \end{vmatrix} = \begin{bmatrix} x_2 y_3 - x_3 y_2 \\ x_3 y_1 - x_1 y_3 \\ x_1 y_2 - x_2 y_1 \end{bmatrix}$$

where i, j, and k stand for the unit length positive axis vectors,

$$\mathbf{i} = \begin{bmatrix} 1 \\ 0 \\ 0 \end{bmatrix}, \qquad \mathbf{j} = \begin{bmatrix} 0 \\ 1 \\ 0 \end{bmatrix}, \qquad \mathbf{k} = \begin{bmatrix} 0 \\ 0 \\ 1 \end{bmatrix}$$

Example 15.42.   *Cross Product*

Compute the cross product $\mathbf{Z} = \mathbf{X} \times \mathbf{Y}$,

$$\mathbf{Z} = \begin{bmatrix} -8 \\ 10 \\ 2 \end{bmatrix} \times \begin{bmatrix} 2 \\ -4 \\ 5 \end{bmatrix}$$

SOLUTION:

$$\mathbf{Z} = \mathbf{X} \times \mathbf{Y}$$

$$= \det \begin{vmatrix} i & j & k \\ x_1 & x_2 & x_3 \\ y_1 & y_2 & y_3 \end{vmatrix} = \det \begin{vmatrix} i & j & k \\ -8 & 10 & 2 \\ 2 & -4 & 5 \end{vmatrix}$$

$$= \begin{bmatrix} x_2 y_3 - x_3 y_2 \\ x_3 y_1 - x_1 y_3 \\ x_1 y_2 - x_2 y_1 \end{bmatrix} = \begin{bmatrix} 10 \cdot 5 - 2 \cdot (-4) \\ 2 \cdot 2 - (-8) \cdot 5 \\ (-8) \cdot (-4) - 10 \cdot 2 \end{bmatrix}$$

$$= \begin{bmatrix} 50 + 8 \\ 4 + 40 \\ 32 - 20 \end{bmatrix} = \begin{bmatrix} 58 \\ 44 \\ 12 \end{bmatrix}$$

Theorem 15.12. *Geometric Cross Product*
   The cross product **Z** *of* **X** *and* **Y** *is the vector shown in Figure 15.54 with the following properties:*
   (a) **Z** *is perpendicular to both* **X** *and* **Y**.
   (b) *The length of* **Z** *equals the area of the parallelogram spanned by the edges of* **X** *and* **Y**.
   (c) *The triple* **X**, **Y**, **Z** *forms a hight-hand frame of vectors, that is, you can point your right thumb in the direction of* **X**, *your first finger in the direction of* **Y**, *and your second finger in the direction of* **Z**. *If you try this with your left hand, your second finger will point opposite the direction of* **Z**.

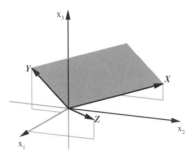

Figure 15.54: $\mathbf{Z} = \mathbf{X} \times \mathbf{Y}$

Exercise set 15.7

1. *Compute the vector (cross) product* $\mathbf{M} = \mathbf{P} \times \mathbf{D}$, *for the vectors given below. Find a vector perpendicular to BOTH* **P** *and* **D**. *Sketch the plane that contains* 0, **P** *and* **D**. *Sketch* **M**.

a) $\mathbf{P} = \begin{bmatrix} 1 \\ 3 \\ 1 \end{bmatrix}$ *and* $\mathbf{D} = \begin{bmatrix} -2 \\ -3 \\ 1 \end{bmatrix}$

b) $\mathbf{P} = \frac{1}{9}\begin{bmatrix} 1 \\ 4 \\ 8 \end{bmatrix}$ *and* $\mathbf{D} = \frac{1}{9}\begin{bmatrix} 8 \\ -4 \\ 1 \end{bmatrix}$

c) $\mathbf{P} = \frac{1}{3}\begin{bmatrix} 2 \\ 2 \\ 1 \end{bmatrix}$ *and* $\mathbf{D} = \frac{1}{3}\begin{bmatrix} -2 \\ 1 \\ 2 \end{bmatrix}$

d) $\mathbf{P} = \frac{1}{7}\begin{bmatrix} 2 \\ 3 \\ 6 \end{bmatrix}$ *and* $\mathbf{D} = \frac{1}{7}\begin{bmatrix} -6 \\ -2 \\ 3 \end{bmatrix}$

e) $\mathbf{P} = \frac{1}{11}\begin{bmatrix} 2 \\ 6 \\ 9 \end{bmatrix}$ *and* $\mathbf{D} = \frac{1}{11}\begin{bmatrix} 9 \\ -6 \\ 2 \end{bmatrix}$

f) $\mathbf{P} = \frac{1}{3}\begin{bmatrix} 2 \\ 2 \\ 1 \end{bmatrix}$ *and* $\mathbf{D} = \frac{1}{3}\begin{bmatrix} 1 \\ -2 \\ 2 \end{bmatrix}$

Check your work with the program ***CrossProd***.

Problem 15.1. Areas, Angles, and Cross Products ──────────▼

The area of a parallelogram with edges **X** *and* **Y** *of lengths* $|\mathbf{X}|$ *and* $|\mathbf{Y}|$ *can be computed by simple trigonometry. In Figure 15.55 below, a "height" or segment perpendicular to the side of* **X** *is shown.*

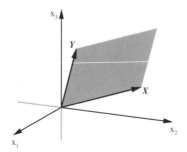

Figure 15.55: *Area by trig*

(a) *Show that the length of the "height" segment is* $h = |\mathbf{Y}| \operatorname{Sin}[\theta]$, *where* θ *is the angle between* **X** *and* **Y**.

(b) *Show that the area of the parallelogram is* $|\mathbf{X}| \cdot |\mathbf{Y}| \operatorname{Sin}[\theta]$. *Use trigonometry to find the height* $h = |\mathbf{Y}| \operatorname{Sin}[\theta]$ *and base* $b = |\mathbf{X}|$ *of the parallelogram.*

(c) *We also know that the area of the parallelogram is* $|\mathbf{X} \times \mathbf{Y}|$ *from the geometric definition of cross product. Show that*

$$\operatorname{Sin}[\theta] = \frac{|\mathbf{X} \times \mathbf{Y}|}{|\mathbf{X}| \cdot |\mathbf{Y}|}$$

(d) *Which is an easier computation, finding the angle between* $X(-6, 8, 2)$ *and* $Y(2, 3, 8)$ *by the dot product formula for cosine or using the cross product formula above for sine? (In both cases, use your calculator to find the arcsine or arccosine).*

Check your work with the program **CrossProd**.

──────────────────────────────────▲

15.8 Geometry and Algebra Lexicon

> *This section is a summary of the chapter. We want you to make your own "translation dictionary" or lexicon that will allow you to go back and forth between geometry and algebra.*

You should memorize the key algebraic computation of each subsection and draw a diagram to which it corresponds. Once you can translate these "words" that have counterparts in both the language of algebra and the language of geometry, all you need to do three dimensional analytical geometry is put your problem in a form to which your lexicon applies, make the translation, and proceed in the other language.

Recall that our formulas are for "position vectors" with their tails at the origin. When we want to draw an arrow with its tail at another point, "a displacement vector," we use the geometric meaning of vector addition or vector difference in making computations. (Use of sum or difference depends on which vectors are known and which you need to compute).

SCALAR MULTIPLICATION AND STRETCHING:
The scalar product $c\mathbf{X}$ is a vector that stretches, shrinks, or reverses $\mathbf{X}$, but does not change its direction.

$$c\,\mathbf{X} = c \begin{bmatrix} x_1 \\ x_2 \\ x_3 \end{bmatrix} = \begin{bmatrix} c \cdot x_1 \\ c \cdot x_2 \\ c \cdot x_3 \end{bmatrix}$$

Nonzero vectors $\mathbf{X}$ and $\mathbf{Y}$ are parallel if and only if there is a scalar c so that $c\mathbf{X} = \mathbf{Y}$.

VECTOR ADDITION AND TIPS-TO-TAILS:
The sum $\mathbf{X} + \mathbf{Y}$ is the position vector you reach by drawing a displacement vector parallel to $\mathbf{Y}$ but with its tail at the tip of $\mathbf{X}$.

$$\mathbf{X} + \mathbf{Y} = \begin{bmatrix} x_1 \\ x_2 \\ x_3 \end{bmatrix} + \begin{bmatrix} y_1 \\ y_2 \\ y_3 \end{bmatrix} = \begin{bmatrix} x_1 + y_1 \\ x_2 + y_2 \\ x_3 + y_3 \end{bmatrix}$$

VECTOR DIFFERENCE AND DISPLACEMENT:
The vector parallel to the displacement arrow pointing from the tip of $\mathbf{B}$ to the tip of $\mathbf{A}$ is $\mathbf{A} - \mathbf{B}$.

$$\mathbf{A} - \mathbf{B} = \begin{bmatrix} a_1 \\ a_2 \\ a_3 \end{bmatrix} - \begin{bmatrix} b_1 \\ b_2 \\ b_3 \end{bmatrix} = \begin{bmatrix} a_1 - b_1 \\ a_2 - b_2 \\ a_3 - b_3 \end{bmatrix}$$

LENGTH:

$$|\mathbf{X}| = \sqrt{x_1^2 + x_2^2 + x_3^2}$$

ANGLE:

$$\mathrm{Cos}[\theta] = \frac{\langle \mathbf{X} \bullet \mathbf{Y} \rangle}{|\mathbf{X}|\,|\mathbf{Y}|}$$

$$= \frac{x_1 \cdot y_1 + x_2 \cdot y_2 + x_3 \cdot y_3}{\sqrt{x_1^2 + x_2^2 + x_3^2} \cdot \sqrt{y_1^2 + y_2^2 + y_3^2}}$$

PERPENDICULARITY:
Non-zero vectors $\mathbf{X}$ and $\mathbf{Y}$ are perpendicular if and only if

$$\langle \mathbf{X} \bullet \mathbf{Y} \rangle = 0$$

$$x_1 \cdot y_1 + x_2 \cdot y_2 + x_3 \cdot y_3 = 0$$

PROJECTION:

The vector projection of $\mathbf{Y}$ in the direction of $\mathbf{X}$ is given by

$$\mathrm{Proj}_{\mathbf{X}}(\mathbf{Y}) = \frac{\langle \mathbf{Y} \bullet \mathbf{X} \rangle}{\langle \mathbf{X} \bullet \mathbf{X} \rangle} \mathbf{X}$$

$$= \frac{x_1 \cdot y_1 + x_2 \cdot y_2 + x_3 \cdot y_3}{x_1 \cdot x_1 + x_2 \cdot x_2 + x_3 \cdot x_3} \begin{bmatrix} x_1 \\ x_2 \\ x_3 \end{bmatrix}$$

CROSS PRODUCT:

The 3-D cross product is given by

$$\mathbf{Z} = \mathbf{X} \times \mathbf{Y} = \det \begin{vmatrix} \mathbf{i} & \mathbf{j} & \mathbf{k} \\ x_1 & x_2 & x_3 \\ y_1 & y_2 & y_3 \end{vmatrix} = \begin{bmatrix} x_2 y_3 - x_3 y_2 \\ x_3 y_1 - x_1 y_3 \\ x_1 y_2 - x_2 y_1 \end{bmatrix}$$

and $\mathbf{Z}$ is the vector with the properties that

(a) $\mathbf{Z}$ is perpendicular to both $\mathbf{X}$ and $\mathbf{Y}$.

(b) The length of $\mathbf{Z}$ equals the area of the parallelogram spanned by the edges of $\mathbf{X}$ and $\mathbf{Y}$.

(c) The frame $\mathbf{X}, \mathbf{Y}, \mathbf{Z}$ is right-hand oriented.

Problem 15.2. THE GEOMETRIC HALF OF THE LEXICON ──────▼

*Draw the geometric figure that corresponds to each of the formulas above. Compare your personal lexicon with the program **AlgGeoLx**.*

▲

15.8.1 Translation Exercises

Not every algebraic computation has a geometric interpretation, and not every geometric construction corresponds to a single simple formula in your lexicon. For example, if we want a vector of unit length in the same direction as a given vector $\mathbf{X}$, we use the length formula below to compute its length. Then, we use scalar multiplication to find the unit vector by stretching or shrinking the original vector by an appropriate amount to make its length 1. It is not necessary to have a separate formula for this idea if you understand both formulas separately. Here is some basic practice in using your lexicon of geometry and algebra.

────────(**Exercise set 15.8**)────────

1. Custom Uses of the Length Formula

Compute the length of the vectors

$$\mathbf{Y} = \begin{bmatrix} 4 \\ -4 \\ 2 \end{bmatrix} \quad and \quad \mathbf{Z} = \begin{bmatrix} 2 \\ -2 \\ 1 \end{bmatrix}$$

Find a scalar c so that $\mathbf{Z} = c\mathbf{Y}$.

Give your own personal formula for the unit length vector that points in the same direction as a given general vector, $\mathbf{X} = \begin{bmatrix} x_1 \\ x_2 \\ x_3 \end{bmatrix}$. *It is not essential that you add this formula to your personal lexicon, but you may if you wish. It is essential that you can find the unit vector. For example, test your formula on*

$$\begin{bmatrix} \frac{2}{3} \\ -\frac{1}{3} \\ \frac{2}{3} \end{bmatrix} \qquad \text{the unit vector in the direction of} \qquad \begin{bmatrix} 2 \\ -1 \\ 2 \end{bmatrix}$$

2. *Sketch* $\mathbf{U} + \mathbf{V} - \mathbf{W}$ *given the figure:*

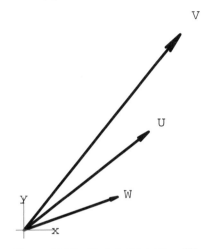

Figure 15.56: Sketch $\mathbf{U} + \mathbf{V} - \mathbf{W}$

3. *Sketch the vectors* $\mathbf{P} = \frac{\langle \mathbf{U} \bullet \mathbf{V} \rangle}{\langle \mathbf{U} \bullet \mathbf{U} \rangle} \mathbf{U}$ *and* $\mathbf{V} - \mathbf{P}$ *on the figure below:*

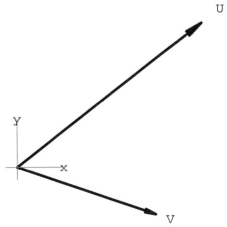

Figure 15.57: Projections

4. Circle

Describe geometrically the set of points that consists of the tips of all 2-D position vectors that have a length of five. Algebraically, this is the set of vectors $X(x_1, x_2)$ satisfying an equation in x_1 and x_2. What is this equation?

5. Sphere

Repeat Exercise 15.8.4 using 3-dimensional position vectors and give the equation for the three coordinates.

6. Dots in 2-D, Circle in 3-D

What is the intersection of a plane and a sphere? What is the intersection of a line and a circle?

(a) *Find and draw on the same coordinate system all 2-D position vectors of the form $\begin{bmatrix} x_1 \\ 7 \end{bmatrix}$ that also have length 25. The first condition is the equation $x_2 = 7$. What does it mean geometrically? What is the condition that the unknown vector $X(x_1, x_2)$ has length 25 in terms of algebra? What is the combined algebraic condition? (What are the simultaneous equations?)*

(b) *Find and draw on the same coordinate system all 3-D position vectors of the form $\begin{bmatrix} x_1 \\ 7 \\ x_3 \end{bmatrix}$ that also have length 25. The first condition is the equation $x_2 = 7$. What does it mean geometrically? What is the condition that the vectors have length 25 in terms of algebra? What is the geometric meaning of the equation $x_1^2 + x_2^2 + x_3^2 = 625$? What is the combined algebraic condition $x_2 = 7$ and $x_1^2 + x_2^2 + x_3^2 = 625$? What is the geometric set?*

(c) *Find and draw on the same coordinate system all 3-D position vectors of the form $\begin{bmatrix} x_1 \\ x_2 \\ 7 \end{bmatrix}$. What does it mean geometrically? What is the condition that the unknown vector $X(x_1, x_2, x_3)$ satisfies $x_1^2 + x_2^2 = 625$? (Careful, x_3 is not restricted). What is the combined algebraic condition $x_3 = 7$ and $x_1^2 + x_2^2 = 625$? What is the geometric set? Do these vectors all have the same length?*

7. Sliding the Sphere

(a) *Sketch the position vector $\mathbf{P} = \begin{bmatrix} 1 \\ -2 \\ 3 \end{bmatrix}$.*

(b) *Sketch and geometrically describe the set of all position vectors $\mathbf{X}$ that point to locations in space that are a distance 3 from the tip of $\mathbf{P}$.*

(c) *Sketch several vectors $\mathbf{V}_1$, $\mathbf{V}_2$, $\mathbf{V}_3$, ... of length 3 with their tail at the tip of $\mathbf{P}$.*

(d) *If you draw a vector, say $\mathbf{V}$ with its tail at the tip of $\mathbf{P}$, the position vector $\mathbf{X}$ that points from the origin to the tip of the (translated) $\mathbf{V}$ is given by $\mathbf{X} = ??$ (Consult your lexicon). Express $\mathbf{V}$ in terms of a formula with $\mathbf{X}$ and $\mathbf{P}$.*

(e) *Write the vector equation $|\mathbf{X} - \mathbf{P}| = 3$ in coordinates and show that it is equivalent to*

$$(x-1)^2 + (y+2)^2 + (z-3)^2 = 9$$

(f) *Express the equation* $|\mathbf{X} - \mathbf{P}| = 3$ *in English, "The set of unknown vectors* $\mathbf{X} = (x, y, z)$ *such that — from* $\mathbf{P}$ *is —."*

(g) *Sketch the set of all points* $\mathbf{X} = (x, y, z)$ *that satisfy*

$$x^2 - 2x + y^2 + 4y + z^2 - 6z + 5 = 0$$

Notice that it would be harder to do this problem in reverse order. We would write the last equation as a sum of squares and then interpret it geometrically using vector differences and lengths.

8. *Find the angle between two diagonals on adjacent faces of a cube.*

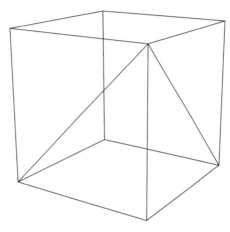

Figure 15.58: Diagonals on a cube

16

Parametric Curves

This chapter develops your skill at using position vectors to formu-
late geometric problems analytically. We begin with the parametric
straight line and move to parametric circles and curves.

The formulas summarized in the last section of Chapter 15 allow you to translate geometric ideas into algebra and algebraic ideas into geometry. You need to memorize the formulas along with their geometric meaning. The parametric line and implicit plane are explained in detail in sections below. These should be added to your lexicon of algebra and geometry when you master them.

16.1   The Vector Parametric Line

This section describes a line by a parametric vector equation.

> **Theorem 16.1.**   *Vector Form of the Parametric Line*
> *The line in the direction of the vector **D** passing through the tip of the vector **P***
> *is the set of all vectors **X** of the form **X** = **P** + t**D** for some real number t.*

You also need to be able to recognize this vector equation in its classical form:

$$x_1 = p_1 + d_1\, t$$
$$x_2 = p_2 + d_2\, t \qquad \text{for example}$$
$$x_3 = p_3 + d_3\, t$$

$$x = 2 - 5\,t$$
$$y = -3 + 2\,t$$
$$x = 1 + 7\,t$$

The explicit example at the right gives the vector equation:

$$\begin{bmatrix} x_1 \\ x_2 \\ x_3 \end{bmatrix} = \begin{bmatrix} 2 \\ -3 \\ 1 \end{bmatrix} + t \begin{bmatrix} -5 \\ 2 \\ 7 \end{bmatrix}$$

This is a line through the point **P**(2, −3, 1) in the direction of **D**(−5, 2, 7).

The vector equation describes a line because it is a combination of vector addition and scalar multiplication. Problem 16.1 uses the lexicon entries for scalar multiplication and vector addition to derive the parametric equation of a line and show you this connection in detail.

Problem 16.1. A LINE THROUGH **P** IN THE DIRECTION OF **D** ──────────────▼

$$\mathbf{P} = \begin{bmatrix} p_1 \\ p_2 \\ p_3 \end{bmatrix} = \begin{bmatrix} 1 \\ 3 \\ 1 \end{bmatrix} \qquad and \qquad \mathbf{D} = \begin{bmatrix} d_1 \\ d_2 \\ d_3 \end{bmatrix} = \begin{bmatrix} -2 \\ -3 \\ 1 \end{bmatrix}$$

(a) *Start a new sheet of paper. Let $t = 1$.*
    (1) *Draw a dot at the tip of the position vector **P**.*
    (2) *Draw the vector $t\mathbf{D}$ as a position vector and as a dotted displacement vector with its tail at the tip of **P**.*
    (3) *Compute $\mathbf{P} + t\mathbf{D}$ and plot it as a position vector.*
(b) *Start a new sheet of paper. Let $t = \frac{1}{2}$.*
    (1) *Draw a dot at the tip of the position vector **P**.*
    (2) *Draw the vector $t\mathbf{D}$ as a position vector and as a dotted displacement vector with its tail at the tip of **P**.*
    (3) *Compute $\mathbf{P} + t\mathbf{D}$ and plot it as a position vector.*
(c) *Start a new sheet of paper. Let $t = 3$.*
    (1) *Draw a dot at the tip of the position vector **P**.*
    (2) *Draw the vector $t\mathbf{D}$ as a position vector and as a dotted displacement vector with its tail at the tip of **P**.*
    (3) *Compute $\mathbf{P} + t\mathbf{D}$ and plot it as a position vector.*
(d) *Start a new sheet of paper. Let $t = -1$.*
    (1) *Draw a dot at the tip of the position vector **P**.*
    (2) *Draw the vector $t\mathbf{D}$ as a position vector and as a dotted displacement vector with its tail at the tip of **P**.*
    (3) *Compute $\mathbf{P} + t\mathbf{D}$ and plot it as a position vector.*
(e) *What is the set of tips of (position) position vectors of the form $X = \mathbf{P} + t\mathbf{D}$, for all real numbers t?*
(f) *Explain why the coordinate equations* $\begin{bmatrix} x_1 \\ x_2 \\ x_3 \end{bmatrix} = \begin{bmatrix} 1 \\ 3 \\ 1 \end{bmatrix} + t \begin{bmatrix} -2 \\ -3 \\ 1 \end{bmatrix}$ *or*

$$x = 1 - 2t$$
$$y = 3 - 3t$$
$$z = 1 + t$$

describe a parametric line that goes through the point $\mathbf{P} = \begin{bmatrix} 1 \\ 3 \\ 1 \end{bmatrix}$ and points parallel to

the vector $\mathbf{D} = \begin{bmatrix} -2 \\ -3 \\ 1 \end{bmatrix}$.

(g) *Draw this line by hand and check your work with the computer program **ParamLine**.*

──▲

Example 16.1.   *Plot the Parametric Equations*

$$x = 4 - 3t$$
$$y = 1 + t$$
$$z = 2t$$

We use the vector form

$$\mathbf{X} = \mathbf{P} + t\mathbf{D}$$

to plot it. The point $\mathbf{P} = \mathbf{P}(4, 1, 0)$ and the direction $\mathbf{D} = \mathbf{D}(-3, 1, 2)$. The vector form helps to make the plot because we simply draw $\mathbf{D}$ starting with its tail at the tip of $\mathbf{P}$ and then fill in the line through that arrow.

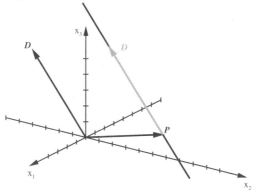

Figure 16.1: A line through $\mathbf{P}$ in direction $\mathbf{D}$

Not every geometric fact needs to have a formula. We know geometrically that "two points determine a line," but that does not mean we need to enter a formula corresponding to that fact in our lexicon. Instead, we can use the vector difference to find a direction and then use the parametric line to determine the line algebraically. The geometric fact that two points determine a line means that we can use the lexicon to find an algebraic representation of the line.

Example 16.2.   *Find the Parametric Equations of a Line Through the Points*

$$\mathbf{Q} = \begin{bmatrix} 2 \\ 3 \\ 4 \end{bmatrix} \qquad and \qquad \mathbf{R} = \begin{bmatrix} 1 \\ -2 \\ 3 \end{bmatrix}$$

You may want to sketch both vectors.

The geometric meaning of the vector difference is that $\mathbf{R} - \mathbf{Q}$ is a position vector parallel to the arrow pointing from $\mathbf{Q}$ to $\mathbf{R}$. Sketch the displacement vector pointing from $\mathbf{Q}$ to $\mathbf{R}$. We compute its position counterpart,

$$\mathbf{D} = \mathbf{R} - \mathbf{Q} = \begin{bmatrix} 1 \\ -2 \\ 3 \end{bmatrix} - \begin{bmatrix} 2 \\ 3 \\ 4 \end{bmatrix} = \begin{bmatrix} -1 \\ -5 \\ -1 \end{bmatrix}$$

Finally, we use one point and this direction vector to give equations for the line in the form $\mathbf{X} = \mathbf{P} + t\mathbf{D}$:

$$\begin{bmatrix} x_1 \\ x_2 \\ x_3 \end{bmatrix} = \begin{bmatrix} 1 \\ -2 \\ 3 \end{bmatrix} + t \begin{bmatrix} -1 \\ -5 \\ -1 \end{bmatrix}$$

There is more than one way to solve the previous exercise of determining the parametric equation of a line through two points. The vector from $\mathbf{Q}$ to $\mathbf{R}$ is one direction vector for the line, and the vector from $\mathbf{R}$ to $\mathbf{Q}$ is another. Of course, these two vectors are simply the negative of one another, but they are different. They both point along the line, but in opposite directions. If $\mathbf{D}$ is one vector in the direction of the line, any scalar multiple $c\mathbf{D}$ also points along the line. In addition, any point $\mathbf{P}$ that lies on the line can be used in the parametric form $\mathbf{X} = \mathbf{P} + t\mathbf{D}$.

Example 16.3. *Nonuniqueness of Parametric Equations*

We will show that both of the following sets of parametric equations go through both of the points $\mathbf{R} = \begin{bmatrix} 2 \\ 3 \\ 4 \end{bmatrix}$ and $\mathbf{Q} = \begin{bmatrix} 1 \\ -2 \\ 3 \end{bmatrix}$.

$$\begin{aligned} x &= 1 + t & \qquad\qquad x &= 2 - 2\,s \\ y &= -2 + 5\,t & y &= 3 - 10\,s \\ z &= 3 + t & z &= 4 - 2\,s \end{aligned}$$

In other words, we find values of $t = t_R$, t_Q, $s = s_R$, and s_Q so that each of the general vectors of the equations $X(x, y, z)$ matches both $\mathbf{R}$ and $\mathbf{Q}$.

Notice that the first equation is

$$\mathbf{X} = \mathbf{Q} + t(\mathbf{R} - \mathbf{Q})$$

so $\mathbf{X} = \mathbf{Q}$ when $t = 0$, $\mathbf{X} = \mathbf{R}$ when $t = 1$.

The second equation is

$$\mathbf{X} = \mathbf{R} + s(2(\mathbf{Q} - \mathbf{R}))$$

so $\mathbf{X} = \mathbf{R}$ when $s = 0$, $\mathbf{X} = \mathbf{Q}$ when $t = 1/2$. See the **ParamLine** program.

Exercise set 16.1

GENERAL INSTRUCTIONS:

Plot position vectors as black arrows (with their tails at the origin, 0). Plot displacement arrows ("free vectors") that have their tails at other places as dotted red arrows.

1. Warm Up

$$\mathbf{A} = \begin{bmatrix} a_1 \\ a_2 \\ a_3 \end{bmatrix} = \begin{bmatrix} 1 \\ 3 \\ 2 \end{bmatrix} \; and \; \mathbf{B} = \begin{bmatrix} b_1 \\ b_2 \\ b_3 \end{bmatrix} = \begin{bmatrix} -2 \\ 1 \\ 3 \end{bmatrix} \quad Plot \; A, \; B, \; \tfrac{1}{2}A, \; 3B.$$

(a) *Draw the displacement arrow parallel to $3B$ but with its tail at the tip of $\frac{1}{2}A$. Draw the displacement arrow parallel to $\frac{1}{2}A$ but with its tail at the tip of $3B$. Compute $3B + \frac{1}{2}A$ and plot it as a position vector.*

(b) *Draw the displacement arrow from the tip of B to the tip of A. Compute $A - B$. Plot $A - B$ as a position vector.*

2. Classical Parametric Equations

The parametric equations

$$x = 3 - 4t$$
$$y = 1 + t$$
$$z = 2$$

define a line. Use the vector form

$$\mathbf{X} = \mathbf{P} + t\mathbf{D}$$

*to plot it. What are $\mathbf{P}$ and $\mathbf{D}$? Why does the vector form help to make the plot? Find two nonparallel vectors $\mathbf{M}$ and $\mathbf{N}$, both perpendicular to the direction of this line. See the **ParamLine** program.*

3. Two Points Determine a Line

Find the parametric equations of a line through the points

$$\mathbf{Q} = \begin{bmatrix} -1 \\ 3 \\ 2 \end{bmatrix} \qquad and \qquad \mathbf{R} = \begin{bmatrix} 1 \\ 2 \\ 3 \end{bmatrix}$$

First, sketch both vectors.

Next, use the geometric meaning of the vector difference to find a direction vector for the line. Sketch the displacement vector pointing from $\mathbf{Q}$ to $\mathbf{R}$ and compute its position counterpart, $\mathbf{D}$.

*Finally, use one point and your direction vector to give equations for the line in the form $\mathbf{X} = \mathbf{P} + t\mathbf{D}$. See the **ParamLine** program.*

4. Nonuniqueness of Parametric Equations

Show that both of the following sets of parametric equations go through both of the points

$$\mathbf{R} = \begin{bmatrix} 1 \\ 2 \\ 3 \end{bmatrix} \; and \; \mathbf{Q} = \begin{bmatrix} 3 \\ -2 \\ 1 \end{bmatrix}.$$

$$
\begin{array}{ll}
x = 1 + 2\,t & \qquad x = 3 - s \\
y = 2 - 4\,t & \qquad y = -2 + 2\,s \\
z = 3 - 2\,t & \qquad z = 1 + s
\end{array}
$$

*In other words, find values of $t = t_R$, t_Q, $s = s_R$, and s_Q so that each of the general vectors of the equations $\mathbf{X}(x, y, z)$ match both $\mathbf{R}$ and $\mathbf{Q}$. See the **ParamLine** program.*

16.1.1   Distance from a Point to a Line

Sometimes we need to use the lexicon indirectly. We may want to find a certain formula and have no lexicon entry for it but rather have a lexicon entry for a closely associated vector. We compute the associated vector and use that to find the one we really want. One example of this is in finding parallel and perpendicular components of one vector relative to another. The next example shows you how this works in finding the distance from the origin to a line. The following problem tests your understanding of the principle.

Example 16.4.   *Distance from the Origin to a Line*

Use various parts of the algebra - geometry lexicon to find the distance from the origin to the parametric line $\mathbf{X} = \mathbf{P} + t\mathbf{D}$.

SOLUTION:

The idea of the following solution is to find position vectors $\mathbf{R}$ and $\mathbf{Q}$ such that

$$\mathbf{P} = \mathbf{R} + \mathbf{Q}$$

$\mathbf{R}$ is parallel to the line in direction $\mathbf{D}$.

$\mathbf{Q}$ is perpendicular to the line.

In other words, $\mathbf{R}$ and $\mathbf{Q}$ are parallel and perpendicular vector components of $\mathbf{P}$ shown in Figure 16.2.

1) Let $\mathbf{P} = \begin{bmatrix} p_1 \\ p_2 \\ p_3 \end{bmatrix} = \begin{bmatrix} 1 \\ 3 \\ 1 \end{bmatrix}$ and $\mathbf{D} = \begin{bmatrix} d_1 \\ d_2 \\ d_3 \end{bmatrix} = \begin{bmatrix} -2 \\ -3 \\ 1 \end{bmatrix}$. We sketch the line that points parallel to the direction of the arrow associated with $\mathbf{D}$ and that passes through the tip of $\mathbf{P}$. This is the set of position position vectors of the form $\mathbf{X} = \mathbf{P} + t\mathbf{D}$, a parametric line.

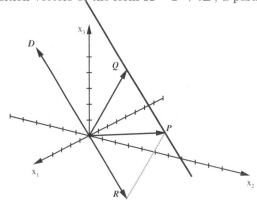

Figure 16.2: Distance to a line

2) We compute the vector $\mathbf{R}$ = the projection of $\mathbf{P}$ in the direction of $\mathbf{D}$, sketched above,

$$\mathbf{R} = \mathrm{Proj}_{\mathbf{D}}(\mathbf{P}) = \frac{\langle \mathbf{P} \bullet \mathbf{D} \rangle}{\langle \mathbf{D} \bullet \mathbf{D} \rangle}\,\mathbf{D} = \frac{\left\langle \begin{bmatrix} 1 \\ 3 \\ 1 \end{bmatrix} \bullet \begin{bmatrix} -2 \\ -3 \\ 1 \end{bmatrix} \right\rangle}{\left\langle \begin{bmatrix} -2 \\ -3 \\ 1 \end{bmatrix} \bullet \begin{bmatrix} -2 \\ -3 \\ 1 \end{bmatrix} \right\rangle} \begin{bmatrix} -2 \\ -3 \\ 1 \end{bmatrix} = -\frac{10}{14}\begin{bmatrix} -2 \\ -3 \\ +1 \end{bmatrix}$$

This is the vector component of $\mathbf{P}$ parallel to $\mathbf{D}$ and the line.

3) Now we compute the vector parallel to the displacement arrow that points from the tip of $\mathbf{R}$ to the tip of $\mathbf{P}$ as shown in Figure 16.2. The position vector parallel to this displacement is given by $\mathbf{Q} = \mathbf{P} - \mathbf{R}$ (by the lexicon entry for displacement vectors), also sketched above. (We took some poetic license and labeled both the position vector and displacement vector "Q").

This gives $\mathbf{Q}$ two important properties: $\mathbf{Q}$ points from the origin to the line, and $\mathbf{Q}$ is perpendicular to the line. Geometrically, the position vector $\mathbf{Q}$ points to the line because it is moved along the line through $\mathbf{D}$ to the origin.

Here is an algebraic proof that $\mathbf{Q}$ lies on the line:

WANT: t so $\mathbf{Q} = \mathbf{X} = \mathbf{P} + t\mathbf{D}$

KNOW: $\mathbf{Q} = \mathbf{P} - \mathbf{R} = \mathbf{P} - \frac{\mathbf{P} \bullet \mathbf{D}}{\mathbf{D} \bullet \mathbf{D}}\mathbf{D}$

TAKE: $t = -\frac{\mathbf{P} \bullet \mathbf{D}}{\mathbf{D} \bullet \mathbf{D}}$, then $\mathbf{Q} = \mathbf{P} + t\mathbf{D}$ and $\mathbf{Q}$ lies on the line.

4) $\mathbf{R}$ is the projection of $\mathbf{P}$ in the direction of $\mathbf{D}$, this means $\mathbf{Q}$ is perpendicular to the line. Now, we *prove* by computation that $\mathbf{Q}$ is perpendicular to $\mathbf{D}$:

$$\mathbf{Q} \bullet \mathbf{D} = (\mathbf{P} - \mathbf{R}) \bullet \mathbf{D}$$
$$= \mathbf{P} \bullet \mathbf{D} - \mathbf{R} \bullet \mathbf{D}$$
$$= \mathbf{P} \bullet \mathbf{D} - \frac{\mathbf{P} \bullet \mathbf{D}}{\mathbf{D} \bullet \mathbf{D}}\,\mathbf{D} \bullet \mathbf{D} = 0$$

5) Finally, the distance from the origin to the line is $|\mathbf{Q}|$, because the vector points from the origin to the line and is perpendicular to the line.

We could view these five steps as a big formula for the distance from the origin to a parametric line,

$$|\mathbf{Q}| = \left|\mathbf{P} - \frac{\mathbf{P} \bullet \mathbf{D}}{\mathbf{D} \bullet \mathbf{D}}\,\mathbf{D}\right| = \left\|\begin{bmatrix}1\\3\\1\end{bmatrix} - \frac{10}{14}\begin{bmatrix}2\\3\\-1\end{bmatrix}\right\| = \left\|\begin{bmatrix}-3/7\\6/7\\12/7\end{bmatrix}\right\| = 3\sqrt{\frac{3}{7}} \approx 1.964$$

We prefer not to memorize this messy formula but only to remember that we can project perpendicularly and subtract to find a vector perpendicular to the line. On the computer, we would not type this huge formula but would break the computation down into steps similar to the ones above. See the program **ParamLine** for the computer version of the solution we just found.

Problem 16.2. A CUSTOM PROGRAM FOR DISTANCE ▬▬▬▬▬▬▬▼

Given a general line in terms of the parametric equation $\mathbf{X} = \mathbf{P} + t\mathbf{D}$ and a point $\mathbf{Y}$ anywhere in space, use vector projection to find the distance from $\mathbf{Y}$ to the line. Rather than give a single formula, you may want to outline a procedure of several steps. Ideally, you should be able to use your procedure to write a brief computer program to compute the distance.

*HINT: Find a vector perpendicular to the line that points from the tip of $\mathbf{Y}$ to the line. Once you have this, just compute its length. Consider the line $\mathbf{Z} = (\mathbf{P} - \mathbf{Y}) + t\mathbf{D}$ or see the **ParamLine** program.*

16.2   Parametric Circles

This section recalls the definitions of radian measure, sine, and cosine, then uses them with vector operations to move in circles, loops, and spirals.

The radian measure of an angle θ is a length measured along a unit circle. Place the circle with its center at the apex of the angle. The two sides of the angle intercept the circle, and the radian measure of the angle is the distance between the sides as shown in Figure 16.3.

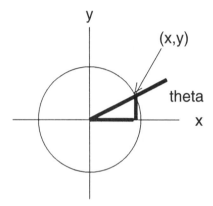

Figure 16.3: Radian measure

For the connection with the sine and cosine, we take one side of the angle to be the positive x-axis as shown below. The point where the other side of the angle meets the unit circle has coordinates

$$x = \mathrm{Cos}[\theta]$$
$$y = \mathrm{Sin}[\theta]$$

These equations follow from SOH-CAH-TOA. The right triangle with one leg on the x-axis and a vertical leg up to the intersection point has a hypotenuse of length 1 because that hypotenuse is a radius of the unit circle. Thus, as in Figure 16.4,

$$\mathrm{Cos}[\theta] = \frac{\text{adjacent}}{\text{hypotenuse}} = \frac{x}{1}$$
$$\mathrm{Sin}[\theta] = \frac{\text{opposite}}{\text{hypotenuse}} = \frac{y}{1}$$

A vector view of the pair of equations is that the vector

$$\mathbf{X}[\theta] = \begin{bmatrix} \mathrm{Cos}[\theta] \\ \mathrm{Sin}[\theta] \end{bmatrix}$$

points to the unit circle at the angle θ, a distance of θ measured along the circle. This is a vector parametric equation for a circle. In other words, if we plotted all vectors of this form without plotting θ, we would fill the unit circle.

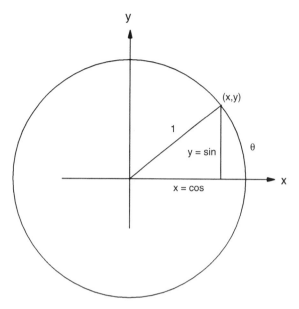

Figure 16.4: Radians, sine, and cosine

More information on sine and cosine as parametric equations for a circle is contained in Chapter 28.

Exercise set 16.2

1. Parametric Circles

(a) *Use scalar multiplication to show that the parametric vector equation of a circle of radius r centered at the origin is*

$$\mathbf{X}[\theta] = r \begin{bmatrix} \text{Cos}[\theta] \\ \text{Sin}[\theta] \end{bmatrix}$$

(b) *Use vector addition to show that the parametric vector equation of a circle of radius r centered at the point $\begin{bmatrix} a \\ b \end{bmatrix}$ is*

$$\mathbf{X}[\theta] = \begin{bmatrix} r\ \text{Cos}[\theta] + a \\ r\ \text{Sin}[\theta] + b \end{bmatrix}$$

but not

$$\mathbf{X}[\theta] = r \begin{bmatrix} \text{Cos}[\theta] + a \\ \text{Sin}[\theta] + b \end{bmatrix}$$

(c) *Verify that your derivations are correct by making computer animations of a point traveling around these circles using the program **Circles**.*

2. The Squashed Bug

A bug accidentally crawls onto a turning gear that revolves and meets the other gear. The bug is squashed... We want you to decide if its parts are ever reunited.

Suppose the small gear has radius r_s and the big gear has radius r_B.

(a) *Suppose the large wheel turns at the rate of one revolution per second. What is the speed of the small wheel? (HINT: Try some special cases such as $r_B = r_s$, $r_B = 2r_s$, etc., and then devise a general formula).*
Suppose we locate the large wheel at the origin, and time is t. Part of the bug lies on the big gear at the vector:

$$x_1 = r_B \cos[2\pi t]$$
$$x_2 = r_B \sin[2\pi t]$$

(b) *Place the small gear at the (x,y)-point $(r_B + r_s, 0)$ and assume that the bug is squashed at $t = 0$. Give equations for the position of the part of the bug on the small gear:*

$$y_1 = ??$$
$$y_2 = ??$$

(c) *If we let $\mathbf{X}[t] = \begin{bmatrix} x_1 \\ x_2 \end{bmatrix}$ and $\mathbf{Y}[t] = \begin{bmatrix} y_1 \\ y_2 \end{bmatrix}$, then the distance between the bug's parts is*

$$d[t] = |\mathbf{X}[t] - \mathbf{Y}[t]|$$

What is the maximum and the minimum of $d[t]$? In particular, is it ever zero or $2(r_B + r_s)$? (HINT: This can be plotted on the computer but is not a trivial max-min problem. The answer depends on the radii. See the program ClassicCrvs, and plot both curves as well as the function $d[t]$).

16.3   Polar Curves

Polar equations of curves are a special variation on the vector formula

$$\mathbf{X}[\theta] = r \begin{bmatrix} \cos[\theta] \\ \sin[\theta] \end{bmatrix}$$

where r is a function of θ. These are described in detail in the program **PolarCrvs**.

16.4   3-Dimensional Parametric Curves

3-D parametric curves are given by three coordinate functions.

A 3-D parametric curve is given by three coordinate functions, such as

$$\mathbf{X}[t] = \begin{bmatrix} x[t] \\ y[t] \\ z[t] \end{bmatrix}$$

Example 16.5. *The Plain Constant Pitch Helix*

We know that if we let

$$x[t] = \text{Cos}[t]$$
$$y[t] = \text{Sin}[t]$$

then the (x, y)-vector moves around a unit circle as t increases. Suppose that at the same time we let the z-component of a 3-D vector increase linearly:

$$\mathbf{X}[t] = \begin{bmatrix} \text{Cos}[t] \\ \text{Sin}[t] \\ k\,t \end{bmatrix}$$

Then z climbs as (x, y) goes round and round. This spiral curve is called a constant pitch helix. (The pitch is the rate of climb in z). It is shaped like the hand rail on a "circular" staircase or the thread of a bolt.

The program **Param3D** makes an animation of this spiral, partly shown in Figure 16.5.

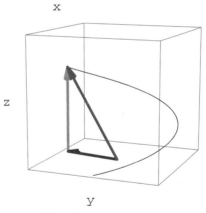

Figure 16.5: A constant pitch helix

Example 16.6. *An Increasing 3-D Spiral*

We know that if we let

$$x[t] = r[t]\ \text{Cos}[t]$$
$$y[t] = r[t]\ \text{Sin}[t]$$

then the (x, y)-vector lies on a circle of radius $r[t]$ at radian angle t. If we have $r[t]$ increasing as t increases, this curve will spiral outward in the x-y-plane. If, in addition, z increases linearly, we make a 3-D spiral shown in Figure 16.6.

$$\mathbf{X}[t] = \begin{bmatrix} \frac{t}{4\pi}\ \text{Cos}[t] \\ \frac{t}{4\pi}\ \text{Sin}[t] \\ \frac{t}{8\pi} \end{bmatrix}$$

x

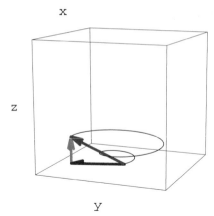

z

y

Figure 16.6: A 3-D spiral

Example 16.7. *A Curve on a Torus*

A torus is the donut-shaped surface given by all points that can be formed by adding a vector on a big circle to a vector on a small perpendicular circle centered on the large one as shown in Figure 16.7.

x

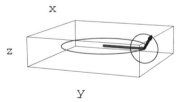

z

y

Figure 16.7: A small perpendicular circle revolving about a large circle

We can take the large circle in the x-y-plane

$$\mathbf{B}[\theta] = r_B \, \mathbf{U}[\theta]$$

where $\mathbf{U}$ is the unit vector

$$\mathbf{U}[\theta] = \begin{bmatrix} \mathrm{Cos}[\theta] \\ \mathrm{Sin}[\theta] \\ 0 \end{bmatrix}$$

The small circle perpendicular to the large circle at the point $\mathbf{B}[\theta]$ can be expressed as

$$\mathbf{S}[\phi] = r_s \left(\mathrm{Cos}[\phi] \, \mathbf{U}[\theta] + \mathrm{Sin}[\phi] \begin{bmatrix} 0 \\ 0 \\ 1 \end{bmatrix} \right)$$

A point on the torus is the sum of a vector pointing to the large circle plus the vector pointing from the center of the small circle to the surface of the torus.

$$\mathbf{X} = \mathbf{B}[\theta] + \mathbf{S}[\phi]$$

Writing this out in coordinates, we have

$$\mathbf{X} = r_B \begin{bmatrix} \text{Cos}[\theta] \\ \text{Sin}[\theta] \\ 0 \end{bmatrix} + r_s \left(\text{Cos}[\phi]\, \mathbf{U}[\theta] + \text{Sin}[\phi] \begin{bmatrix} 0 \\ 0 \\ 1 \end{bmatrix} \right)$$

$$= \begin{bmatrix} r_B\, \text{Cos}[\theta] + r_s\, \text{Cos}[\phi]\, \text{Cos}[\theta] \\ r_B\, \text{Sin}[\theta] + r_s\, \text{Cos}[\phi]\, \text{Sin}[\theta] \\ r_s\, Sin[\phi] \end{bmatrix}$$

$$= \begin{bmatrix} (r_B + r_s\, \text{Cos}[\phi])\, \text{Cos}[\theta] \\ (r_B + r_s\, \text{Cos}[\phi])\, \text{Sin}[\theta] \\ r_s\, \text{Sin}[\phi] \end{bmatrix}$$

Finally, suppose we decide to spin the little circle 4 times as fast as the vector moving around the large one

$$\mathbf{X}[t] = \begin{bmatrix} (r_B + r_s\, \text{Cos}[4t])\, \text{Cos}[t] \\ (r_B + r_s\, \text{Cos}[4t])\, \text{Sin}[t] \\ r_s\, \text{Sin}[4t] \end{bmatrix}$$

Figure 16.8: A curve winding 4 times about the torus

Exercise set 16.4

C

1. *Use the **Param3D** program to plot three turns of a constant pitch helix.*

C

2. *Use the **Param3D** program to plot*

$$\mathbf{X}[t] = \begin{bmatrix} \text{Cos}[t] \\ \text{Sin}[t]\, \text{Cos}[t] \\ \text{Cos}[t] \end{bmatrix}$$

C

3. *Use the **Param3D** program to plot* $\mathbf{Y}[t] = r[t]\, \mathbf{X}[t]$, *where*

$$\mathbf{X}[t] = \begin{bmatrix} \text{Cos}[t] \\ \text{Sin}[t] \\ \text{Cos}[t] \end{bmatrix}$$

and $r[t] = 1 - \text{Cos}[2t]$.

4. The Helicopter
The rotor on a helicopter has radius 10 feet and rotates 40 times per minute. The helicopter flies along a straight line at an elevation of 500 feet at 50 mph due east.

*The rotor turns in a horizontal plane (as the helicopter is coasting). Give parametric equations for the path of a point on the tip of the rotor with respect to a point on the ground. Plot the curve with help from the program **Param3D**.*

5. Bagel Curves

Parametric equations for a curve winding around a torus of large radius r_B and small radius r_s are

$$\mathbf{X}[t] == \begin{bmatrix} (r_B + r_s \, \text{Cos}[\phi]) \, \text{Cos}[\theta] \\ (r_B + r_s \, \text{Cos}[\phi]) \, \text{Sin}[\theta] \\ r_s \, \text{Sin}[\phi] \end{bmatrix}$$

where you specify both θ and ϕ in terms of t. Plot the curve that winds around the torus of big radius 3 and the little radius 1 and winds 3 times as you go around the large circle once.

Find the parametric equations of a curve that winds 4 times around the torus as you rotate 3 times around the large circle.

Also, see the program **ClassicCrvs**.

16.5   Tangents and Velocity Vectors

This section shows how to compute the tangent to a parametric curve.

Example 16.8.   *A Tangent to an Epicycloid*

Suppose that we have a parametric graph like the epicycloid

$$x = 4 \, \text{Cos}[\theta] + \text{Cos}[4\,\theta]$$
$$y = 4 \, \text{Sin}[\theta] + \text{Sin}[4\,\theta]$$

and want to find the tangent line to the graph at a point like $\theta = \pi/6$, where $4\theta = 2\pi/3$, so

$$x = 4 \, \frac{\sqrt{3}}{2} - \frac{1}{2} \approx 2.9641$$

$$y = 4 \, \frac{1}{2} + \frac{\sqrt{3}}{2} \approx 2.86603$$

We place parallel local coordinates (dx, dy) centered at this (x, y) point, $(x[\theta], y[\theta])$. The differentials of the original parametric equations

$$dx = (-4 \, \text{Sin}[\theta] - 4 \, \text{Sin}[4\theta])d\theta$$
$$dy = (4 \, \text{Cos}[\theta] + 4 \, \text{Cos}[4\theta])d\theta$$

at this specific $\theta = \pi/6$ give the parametric equations

$$dx = -(2 + 2\sqrt{3})d\theta \approx -5.46\ d\theta$$
$$dy = (2\sqrt{3} - 2)d\theta \approx 1.46\ d\theta$$

These equations are parametric equations for the tangent at $\pi/6$. They describe a line through the (dx, dy)-origin at $(x, y) = (x[\pi/6], y[\pi/6])$ pointing in the approximate direction

$$\begin{bmatrix} -5.46 \\ 1.46 \end{bmatrix}$$

because the vector form of the parametric line is

$$d\mathbf{X} = \mathbf{P} + d\theta \cdot \mathbf{D}$$

with components

$$\begin{bmatrix} dx \\ dy \end{bmatrix} = \begin{bmatrix} 0 \\ 0 \end{bmatrix} + d\theta \begin{bmatrix} -5.46 \\ 1.46 \end{bmatrix}$$

Once we know the tangent vector, we can sketch the tangent simply by putting the pen at the (x, y)-point of tangency and sketching the line through this vector as in Figure 16.9 (drawn with its tail at the tangency point.)

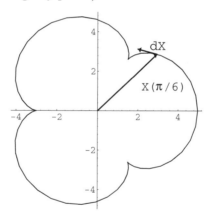

Figure 16.9: The tangent at $\theta = \pi/6$

If needed, the parametric equations in (x, y)-coordinates are obtained by replacing the local coordinates by their defining differences (after we compute the specific $\mathbf{X}'[t]$).

$$\begin{bmatrix} dx \\ dy \end{bmatrix} = \begin{bmatrix} 0 \\ 0 \end{bmatrix} + d\theta \begin{bmatrix} -(2 + 2\sqrt{3}) \\ (2\sqrt{3} - 2) \end{bmatrix}$$
$$\begin{bmatrix} x - x_0 \\ y - y_0 \end{bmatrix} = \begin{bmatrix} 0 \\ 0 \end{bmatrix} + d\theta \begin{bmatrix} -(2 + 2\sqrt{3}) \\ (2\sqrt{3} - 2) \end{bmatrix}$$
$$\begin{bmatrix} x \\ y \end{bmatrix} = \begin{bmatrix} x_0 \\ y_0 \end{bmatrix} + d\theta \begin{bmatrix} -(2 + 2\sqrt{3}) \\ (2\sqrt{3} - 2) \end{bmatrix}$$
$$\begin{bmatrix} x \\ y \end{bmatrix} = \begin{bmatrix} \frac{\sqrt{3}}{2} - \frac{1}{2} \\ 4\frac{1}{2} + \frac{\sqrt{3}}{2} \end{bmatrix} + d\theta \begin{bmatrix} -(2 + 2\sqrt{3}) \\ (2\sqrt{3} - 2) \end{bmatrix}$$

Since we do not plot the parameter, we can use either $d\theta$ or the original parameter. For example, in a computer program that also plots the nonlinear curve with θ, we might use the following.

$$\begin{bmatrix} x \\ y \end{bmatrix} = \begin{bmatrix} \frac{\sqrt{3}}{2} - \frac{1}{2} \\ 4\frac{1}{2} + \frac{\sqrt{3}}{2} \end{bmatrix} + (\theta - \pi/6) \begin{bmatrix} -(2 + 2\sqrt{3}) \\ (2\sqrt{3} - 2) \end{bmatrix}$$

In general

> **Theorem 16.2.**   *Parametric Tangents*
>
> *The family of differentials of equations defining parametric curves defines the parametric tangent line in local coordinates (where t is considered fixed):*
>
> | *Curve in (x, y)-coordinates* | *Tangent in (dx, dy)-coordinates at (x_0, y_0), t fixed* |
> |---|---|
> | $x = x[t] = t^2$ | $dx = x'[t]\,dt$ |
> | $y = y[t] = t$ | $dy = y'[t]\,dt$ |
>
> *Specifically, at $t = t_0$, the tangent vector is $\begin{bmatrix} x'[t_0] \\ y'[t_0] \end{bmatrix}$, so the tangent line can be plotted by drawing this vector with its tail at $\begin{bmatrix} x[t_0] \\ y[t_0] \end{bmatrix}$. In the original variables, the parametric tangent has equation*
>
> $$\begin{bmatrix} x \\ y \end{bmatrix} = \begin{bmatrix} x[t_0] \\ y[t_0] \end{bmatrix} + (t - t_0) \begin{bmatrix} x'[t_0] \\ y'[t_0] \end{bmatrix}$$

The three dimensional parametric tangent is given similarly by differentiating three x-y-z-equations.

16.5.1   Velocity Vectors

The parametric tangent vector can also be viewed as a velocity when the parameter is time t and the equations represent position (x, y, z). The reason that the parametric tangent is velocity comes from the increment equations defining derivatives with respect to t, as in Definition 5.5. A change in position over a small time increment is $\mathbf{X}[t + \delta t] - \mathbf{X}[t]$, and the time rate of change of position, called velocity, is

$$\frac{1}{\delta t}(\mathbf{X}[t + \delta t] - \mathbf{X}[t]) = \begin{bmatrix} (x[t + \delta t] - x[t])/\delta t \\ (y[t + \delta t] - y[t])/\delta t \\ (z[t + \delta t] - z[t])/\delta t \end{bmatrix} = \begin{bmatrix} x'[t] \\ y'[t] \\ z'[t] \end{bmatrix} + \begin{bmatrix} \varepsilon_x \\ \varepsilon_y \\ \varepsilon_z \end{bmatrix} \approx \begin{bmatrix} x'[t] \\ y'[t] \\ z'[t] \end{bmatrix}$$

This proves the following theorem.

Theorem 16.3.   *Velocity*

Suppose a position in 3-space is given as a function of time, t, by a vector function

$$\mathbf{X}[t] = \begin{bmatrix} x[t] \\ y[t] \\ z[t] \end{bmatrix}$$

Then the velocity at a particular time t is given by the derivative,

$$velocity\ vector = \frac{d\mathbf{X}}{dt} = \begin{bmatrix} \frac{dx}{dt} \\ \frac{dy}{dt} \\ \frac{dz}{dt} \end{bmatrix} = \mathbf{X}'[t] = \begin{bmatrix} x'[t] \\ y'[t] \\ z'[t] \end{bmatrix}$$

Exercise set 16.5

1. *Sketch the parametric curve and its tangent at the point indicated. Simply compute the proper vector and sketch it with its tail at the point of tangency.*

   (a) *Curve in (x, y)-coordinates*   *Tangent in (dx, dy)-coordinates at $t = \pi/3$*

$$x = x[t] = \text{Cos}[t] \qquad\qquad dx = x'[t]\, dt$$
$$y = y[t] = \text{Sin}[t] \qquad\qquad dy = y'[t]\, dt$$

   (b) *Curve in (x, y)-coordinates*   *Tangent in (dx, dy)-coordinates at $t = 2$*

$$x = x[t] = t^2 \qquad\qquad dx = x'[t]\, dt$$
$$y = y[t] = t \qquad\qquad dy = y'[t]\, dt$$

   (c) *Curve in (x, y)-coordinates*   *Tangent in (dx, dy)-coordinates at $t = \pi/6$*

$$x = x[t] = \text{Cos}[t] \qquad\qquad dx = x'[t]\, dt$$
$$y = y[t] = \text{Sin}[2t] \qquad\qquad dy = y'[t]\, dt$$

   (d) *Curve in (x, y)-coordinates*   *Tangent in (dx, dy)-coordinates at $t = \pi/6$*

$$x = x[t] = \sqrt{3}\,\text{Cos}[t] \qquad\qquad dx = x'[t]\, dt$$
$$y = y[t] = \text{Sin}[t] \qquad\qquad dy = y'[t]\, dt$$

   (e) *Curve in (x, y)-coordinates*   *Tangent in (dx, dy)-coordinates at $t = \text{Log}[2]$*

$$x = x[t] = e^{2t} \qquad\qquad dx = x'[t]\, dt$$
$$y = y[t] = e^{-t} \qquad\qquad dy = y'[t]\, dt$$

Here is a more dynamic test for your lexicon.

2. Skeet Shooting

A new skeet range opens near your house with helium-filled, gas-propelled clay targets that travel with constant velocity vectors (along straight lines with fixed speed). You decide to test your skill and stand at the origin of a coordinate system that has x pointing south, y pointing east, and z pointing up. Distances are measured in feet and times in seconds. Draw figures for all of the following descriptions and questions.

(a) *A target thrower is located at the tip of the vector* $\mathbf{T} = \begin{bmatrix} -20 \\ 15 \\ 10 \end{bmatrix}$. *How far away is it from you? Label the distance on your figure.*

(b) *The target velocity vector is* $\mathbf{V} = \begin{bmatrix} 8 \\ 1 \\ -1 \end{bmatrix}$. *How fast is it going?*

(c) *At what time t does the target hit the ground* $(\{\mathbf{X} = \begin{bmatrix} x \\ y \\ z \end{bmatrix} : z = 0\})$?

(d) *If your eye is 5 feet above the ground, what is the angle between the horizontal and your line of sight as a function of time?*

The next exercise combines calculus and geometry to follow a curved trajectory.

3. A Curved Trajectory

An old-fashioned 20-th century clay target with a chip on one side comes out of the thrower and traverses the path given by

$$\mathbf{X}[t]: \begin{aligned} x &= \text{Log}[1 + 8t] - 20 \\ y &= \text{Sin}[t] + 15 \\ z &= 10 - \frac{5}{2}t^2 \end{aligned}$$

(a) *When does this target hit the ground?*

(b) *How far from you is it when it hits the ground?*

(c) *Show that the target leaves the thrower at* $t = 0$ *with velocity* $\begin{bmatrix} 8 \\ 1 \\ 0 \end{bmatrix}$.

(d) *How fast is it going when it hits the ground?*

Problem 16.3. ────────────────────────────────▼

A simple case of velocity vectors is the vector equation $\mathbf{X} = \mathbf{P} + t\mathbf{D}$, *such as*

$$\begin{aligned} x &= 1 - 2t \\ y &= 3 - 1t \\ z &= 1 + 2t \end{aligned}$$

Suppose you are in space moving along a path described by these equations for time, t in seconds, where x, y, and z are measured in miles. Make sketches as you do the following computations.

Find your position at time zero, t = 0, and $\mathbf{X}_0 = ?$

Find your position one time unit later, t = 1, and $\mathbf{X}_1 = ?$

Find the vector displacement that you moved during this second. You should think of this as given in units of miles per second.

Find the length of the vector displacement. This represents your speed. Why?

Compute the velocity vector $\mathbf{X}'[t]$. *Why is this the same as your displacement in 1 second?*

A klingon starship moves along the path

$$x = 1 - 4\,t$$
$$y = 3 - 2\,t$$
$$z = 1 + 4\,t$$

with the same time units, t. How fast is it going? When are you together? Why will you never be together again?

The next problem is a parametric approach to finding the distance from a point to an ellipse. We solved this with implicit equations in the CD section on constrained max-min of Chapter 11.

Problem 16.4. DISTANCE TO A CURVE IN PARAMETRIC FORM

Find the points on the ellipse $\left(\frac{x}{2}\right)^2 + \left(\frac{y}{3}\right)^2 = 1$ *that are nearest and farthest from the point* $(1, 1)$ *as illustrated in figures of the CD section on constrained max-min of Chapter 11, but use the parametric equations*

$$x = 2\,\text{Cos}[\theta]$$
$$y = 3\,\text{Sin}[\theta]$$

HINTS: Write the distance, $\sqrt{(x-1)^2 + (y-1)^2}$, *from* $(1,1)$ *to a general* (x,y)-*point on the curve, in terms of the parameter* θ. *Then find the max and the min of the square of the distance* $D[\theta]$ *in terms of* θ. *Why is extremizing the square equivalent? Why is it easier?*

Many things vary as only a function of the distance to an object. The intensity of radiation is proportional to one over the square of the distance from the source. Hence, it falls off fast as you move away from the source. For example, the apparent brightness of a planet being observed by a space probe varies inversely as the square of the distance to the object

$$I = k\frac{1}{D}$$

where D is the square of the distance between the planet and observer.

Problem 16.5. ———————————————————————▼

Find the time t and the place (x, y) when a planet at $(a, b) = (0, 1)$ appears brightest for the following rocket trajectories. (The parabolic trajectories correspond to different primary missions).

    (a)$x = t - 1,$ $y = 0$
    (b)$x = t - 1,$ $y = 2x^2 = 2(t - 1)^2$
    (c)$x = t - 1,$ $y = \frac{1}{3}x^2 = \frac{1}{3}(t - 1)^2$

HINT: In coordinates, the distance from (a, b) to (x, y) is $\sqrt{(x - a)^2 + (y - b)^2}$.

16.6   Projects

16.6.1   Cycloids

A small wheel rolls around a large fixed wheel without slipping. What path does a point on the edge of the small wheel trace out as it goes?

You can answer this with vectors and an understanding of radian measure in the project on cycloids. The Spirograph toy makes such plots and the next figure is a sample path.

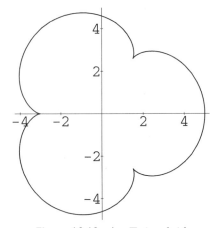

Figure 16.10: An Epicycloid

CHAPTER 17

Graphs in Several Variables

This chapter studies plotting in 3 dimensions and analyzes the case of linear functions in detail.

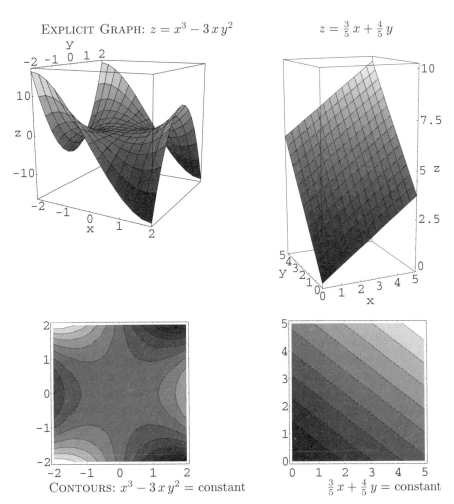

EXPLICIT GRAPH: $z = x^3 - 3xy^2$

$z = \frac{3}{5}x + \frac{4}{5}y$

CONTOURS: $x^3 - 3xy^2 = \text{constant}$

$\frac{3}{5}x + \frac{4}{5}y = \text{constant}$

Figure 17.1: Nonlinear and linear explicit surface plots and contour plots

There are three major types of graphs: explicit, implicit, and parametric. We studied

parametric lines and curves in Chapter 16, and whereas parametric surfaces are very useful
(especially in computer graphics), we do not include them in this chapter.

Explicit graphs are of the form

$$z = f[x] \qquad \text{or} \qquad z = f[x, y]$$

with all variables plotted. These are probably the most familiar to you, especially things
like the graphs of $y = x^2$ or $y = m\,x + b$. Now, we study graphs such as $z = 3x^2 + 5y^2$ and
$z = 3\,x - 5\,y + 3$.

An implicit curve is the set of (x, y)-points that satisfy an equation such as

$$3x^2 + 5y^2 = 7$$

where neither x nor y is separate (or explicitly given in terms of the other).

A "contour" is a curve of constant "height" or a "level curve." The contour graph of
the height of a mountain is called a "topographical map." If $z = h[x, y]$ gives the height z
above a base plane point (x, y), then the contour map is a plane filled with implicit curves
of constant height, $h[x, y] = c$, for various constants c. (For example, $3x^2 + 5y^2 = 7, 8, 9, \ldots$,
if $z = 3x^2 + 5y^2$.) Farmers use contour plowing to keep rain from eroding the land. It is the
same word for the same reason: The furrows are contours of constant height that therefore
do not run downhill.

The computer can easily make both explicit and implicit plots such as the ones shown in
Figure 17.1. One main goal of this chapter is to understand how these graphs are made, by
hand and with the computer. Another goal is to understand what they mean. A specific
main technical goal of the chapter is to connect the parameters m, n, and h in the general
explicit linear function

$$z = m\,x + n\,y + h$$

with both the explicit plot and the contour plot of the linear function.

Before we begin studying the details of linear graphs, you should plot some 3-D graphs
with the computer.

Exercise set 17.0

1. *Plot the functions below using the program **BasicGrfs3D**. You will recognize them if
you think about the curves you get by holding x fixed and varying y or holding y fixed
and varying x.*

a) $z = x^2 + y^2$ *b)* $z = (x/5)^2 + (y/3)^2$

c) $z = x\,y^3$ *d)* $z = x^2\,y^3$

e) $z = \text{Sin}[x] + \text{Cos}[y]$ *f)* $z = x \cdot \text{Cos}[2\pi y]$

g) $z = e^{-(x^2+y^2)}$ *h)* $z = e^{-(x^2+y^2)}\,\text{Sin}[x^2 + y^2]$

*Examine some of these surfaces with the program **SurfaceFlyBy**.*

17.1   The Expicit Plane in 3-D

The explicit plot of a linear function of two variables, $z = L[x, y]$, is a plane in 3-space. This section gives a first geometric interpretation of the algebraic parameters in $L[x, y] = m\,x + n\,y + h$.

The basic linear function of two variables is given by the explicit equation

$$z = m\,x + n\,y + h$$

or $z = L[x, y] = m\,x + n\,y + h$, for constants m, n, and h. For example, if $m = \frac{3}{5}$, $n = \frac{4}{5}$, and $h = 0$, then $z = \frac{3}{5}\,x + \frac{4}{5}\,y$.

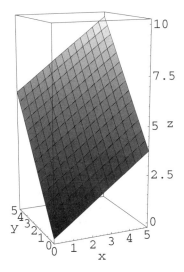

Figure 17.2: $z = \frac{3}{5}\,x + \frac{4}{5}\,y$

The 3-D plot of $z = \frac{3}{5}\,x + \frac{4}{5}\,y$ above consists of "slices" obtained by fixing either y or x and plotting z versus the variable one. For example, if we fix $y = 0$, then we are in the x-z-plane. The equation $z = \frac{3}{5}\,x + \frac{4}{5}\,y$ becomes

$$z = \frac{3}{5}\,x + 0$$

a line of slope 3/5 through $x = z = 0$ in the x-z-plane shown in Figure 17.3.

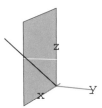

Figure 17.3: $z = \frac{3}{5}x$ in $y = 0$

If we fix $y = 1$, the equation $z = \frac{3}{5}x + \frac{4}{5}y$ becomes

$$z = \frac{3}{5}x + \frac{4}{5}$$

a line of slope $3/5$ through $z = 4/5$ when $x = 0$ in the plane parallel to the x-z-plane at $y = 1$ shown in Figure 17.4.

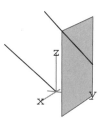

Figure 17.4: $z = \frac{3}{5}x + \frac{4}{5}$ in $y = 1$

Similarly, if we fix $x = c$, a constant, the equation $z = \frac{3}{5}x + \frac{4}{5}y$ becomes

$$z = \frac{4}{5}y + b$$

a line of slope $4/5$ through $z = b$ when $y = 0$ in the plane parallel to the y-z-plane at $x = c$ shown in Figure 17.5.

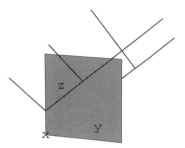

Figure 17.5: $z = \frac{y}{5}\,y + \frac{3}{5}$ in $x = 1$

The explicit graph of the linear equation

$$z = m\,x + n\,y + h$$

is a plane in three dimensions such that the intersection of this plane with a plane parallel to the x-z-plane (y fixed) is a line of slope m, and the intersection of this plane with a plane parallel to the y-z-plane (x fixed) is a line of slope n (see Figure 17.6). The plane intersects the z-axis at $z = h$ (when $x = y = 0$).

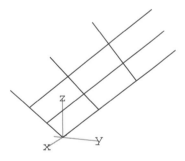

Figure 17.6: $z = \frac{3}{5}\,x + \frac{4}{5}\,y$

17.1.1 Gradient Vectors & Slopes

Although slope of a line $y = m\,x + b$ is the number m, the "slope" of a plane has both magnitude and direction. The explicit plane $z = m\,x + n\,y + h$ tilts upward with a certain steepness in a certain direction. This subsection shows how to measure "slope of a plane" with the gradient vector. The questions about a two dimensional plane lying in three dimensional space that we want to answer are as follows:

(a) In which x-y-direction does the plane tilt up most steeply?

(b) How steep is the plane in this direction?

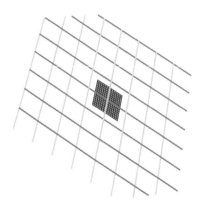

Figure 17.7: The steepest way up $z = -\frac{4}{5}x - \frac{3}{5}y$?

Definition 17.1. *The Linear Gradient Vector*
The gradient of the linear function

$$L[x, y] = m\,x + n\,y + h$$

is the 2-D vector $\mathbf{G} = \begin{bmatrix} m \\ n \end{bmatrix}.$

The slope of the line $y = L[x] = m\,x + b$ is a simple idea, but let us review the "function change" version of it. If we change x by Δx, then y changes by

$$\Delta y = L[x + \Delta x] - L[x] = m \cdot \Delta x$$

The ratio of these changes is the slope, $m = \Delta y / \Delta x$.

The constant m in the two-variable linear function $L[x, y] = m\,x + n\,y + h$ is the slope of the graph if we only move in the x-direction or only change x,

$$\Delta z = L[x + \Delta x, y] - L[x, y]$$
$$= (m \cdot (x + \Delta x) + n\,y + h) - (m\,x + n\,y + h) = m \cdot \Delta x$$

The constant n in the two-variable linear function is the slope of the graph if we only move in the y-direction or only change y,

$$\Delta z = L[x, y + \Delta y] - L[x, y]$$
$$= (m\,x + n \cdot (y + \Delta y) + h) - (m\,x + n\,y + h) = n \cdot \Delta y$$

The lines on the plane $z = m\,x + n\,y + h$ moving in the x-direction have slope m and the lines moving in the y-direction have slope n. This is shown (with a shortened vertical scale) in Figure 17.8. Notice that it is clear that neither the slope 3 in the x-direction nor the slope 4 in the y-direction is the steepest way up the plane.

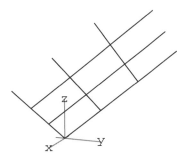

Figure 17.8: $z = 3x + 4y$

We want to change both x and y and compute the corresponding change in z. In order to make this a slope, suppose that we move a *unit* distance,

$$\mathbf{U} = \begin{bmatrix} \Delta x \\ \Delta y \end{bmatrix} \qquad \text{with} \qquad |\mathbf{U}| = \sqrt{\Delta x^2 + \Delta y^2} = 1$$

The change in z is

$$
\begin{aligned}
\Delta z &= L[x + \Delta x, y + \Delta y] - L[x, y] \\
&= (m \cdot (x + \Delta x)x + n \cdot (y + \Delta y) + h) - (m\,x + n\,y + h) \\
&= m \cdot \Delta x + n \cdot \Delta y \\
&= \begin{bmatrix} m \\ n \end{bmatrix} \bullet \begin{bmatrix} \Delta x \\ \Delta y \end{bmatrix} = \langle \mathbf{G} \bullet \mathbf{U} \rangle
\end{aligned}
$$

where $\mathbf{G}$ is the gradient vector of Definition 17.1. Since we chose $\mathbf{U}$ to have unit length, the slope of the plane in the direction of $\mathbf{U}$ is therefore $\langle \mathbf{G} \bullet \mathbf{U} \rangle$ (the z-change per unit moved.)

Combine this with the angle formula from your algebra-geometry lexicon to see that

$$
\begin{aligned}
\text{slope in direction } \mathbf{U} &= \langle \mathbf{G} \bullet \mathbf{U} \rangle \\
&= |\mathbf{G}| \cdot |\mathbf{U}| \, \mathrm{Cos}[\theta] \\
&= |\mathbf{G}| \cdot \mathrm{Cos}[\theta]
\end{aligned}
$$

since $|\mathbf{U}| = 1$. The cosine of the angle θ varies from $+1$ through 0 to -1 as

(a) The angle θ has $\mathbf{U}$ pointing in the same direction as $\mathbf{G}$ when $\theta = 0$ and $\mathrm{Cos}[0] = +1$.
(b) The angle θ has $\mathbf{U}$ is perpendicular to $\mathbf{G}$ when $\theta = \pm\pi/2$ and $\mathrm{Cos}[\pm\pi/2] = 0$.
(c) The angle θ has $\mathbf{U}$ pointing opposite $\mathbf{G}$ when $\theta = \pi$ and $\mathrm{Cos}[\pi] = -1$.

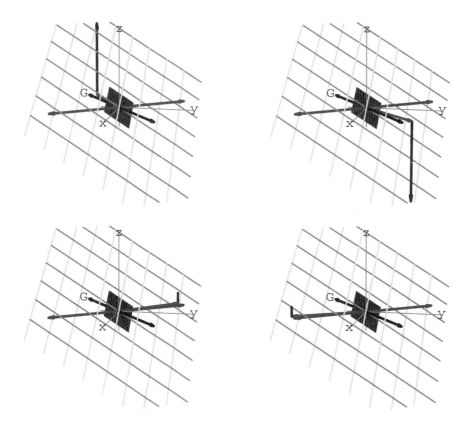

Figure 17.9: Four slopes on $z = -\frac{3}{5}x - \frac{4}{5}y$

Combining these observations, we have

Theorem 17.2. *The Gradient as a "Vector Slope" of an Explicit Plane*
Consider the explicit plane with equation

$$z = mx + ny + h \qquad and\ gradient \qquad \mathbf{G} = \begin{bmatrix} m \\ n \end{bmatrix}$$

(a) *The slope of a line in this plane is $\langle \mathbf{G} \bullet \mathbf{U} \rangle$ when the horizontal projection of the line points in the x-y unit vector direction $\mathbf{U}$.*

(b) *The steepest slope of a line in this plane is $|\mathbf{G}|$ when the horizontal projection of the line points in the x-y-direction of the gradient vector $\mathbf{G}$.*

(c) *The line in this plane above the line in the (x,y)-plane that is perpendicular to $\mathbf{G}$ is horizontal; that is, z is constant on the plane if we move x and y in a direction perpendicular to $\mathbf{G}$.*

The 2-D gradient vector

$$\mathbf{G} = \begin{bmatrix} m \\ n \end{bmatrix}$$

points in the x-y-direction of fastest increase of $z = mx + ny + h$, and we may plot $\mathbf{G}$ on the contour plot, where it is normal to the contour lines and tells how far apart those contours are spaced. We must learn to not confuse **G** with the normals to the explicit 3-D graph given by

$$\mathbf{N} = \begin{bmatrix} -m \\ -n \\ +1 \end{bmatrix} \quad \text{or} \quad -\mathbf{N} = \begin{bmatrix} +m \\ +n \\ -1 \end{bmatrix}$$

The main idea of differential calculus is that smooth functions microscopically "look" like linear functions. When we understand the linear case, we can proceed with calculus of several variables by applying the linear formulas to "small pieces" of the nonlinear graphs.

Exercise set 17.1

1. *Sketch the graph of $z = \frac{2}{3}x - \frac{3}{2}y + 2$ with paper and pencil. Check your work using the program **SurfSlices**.*

2. *Run the program **PlaneLines**, and observe the changes of a linear function for unit x-y-displacements in different directions.*

3. (a) *Which plane is steepest $z = 2x - 3y$ or $z = 3x + 2y$?*
       (b) *Which plane is steepest $z = 3x - 4y$ or $z = \frac{3}{5}x + \frac{4}{5}y$?*
       (c) *Verify your answers using the program **PlaneLines**.*

4. *Let $L[x,y] = 2x - 3y$. Let $\mathbf{U} = \begin{bmatrix} 3/5 \\ 4/5 \end{bmatrix}$.*

   (a) *Show that $|\mathbf{U}| = 1$*
   (b) *Plot the line on $z = L[x,y]$ pointing in the direction of **U** that goes through $(x,y,z) = (1, 2, -4)$. (That is, the line of intersection with the vertical plane that contains $(1, 2, 0)$ and a horizontal x-y-vector parallel to **U**).*
   (c) *The set of points*

$$x = 1 + t \cdot \frac{3}{5},$$

$$y = 2 + t \cdot \frac{4}{5}$$

$$z = L[1 + t \cdot \frac{3}{5}, 2 + t \cdot \frac{4}{5}]$$

   is the same as the line in part (2). Why?
   (d) *Write $z = L[1 + t \cdot \frac{3}{5}, 2 + t \cdot \frac{4}{5}]$ in the form*

$$z = \mu \cdot t + \beta$$

   *and show that the slope is $\mu = \langle \mathbf{G} \bullet \mathbf{U} \rangle$, where **G** is the gradient of $L[x,y]$. (For this step, it is best to write the substitution from the previous part in vector notation, $L[\mathbf{X}] = \langle \mathbf{G} \bullet \mathbf{X} \rangle$, so $L[\mathbf{P} + t\,\mathbf{U}] = \langle \mathbf{G} \bullet ? \rangle = ...)*

17.2   Vertical Slices and Chickenwire

This section builds explicit nonlinear surface plots out of slices made by holding one of the independent variables fixed and varying the other.

This section helps you make some hand-drawn 3-D graphs, linear, quadratic, and cubic. Once you understand what these graphs mean, the computer will be immensely helpful in 3-D graphing. It is also very helpful in visualizing, as you will see in the program **SurfaceFlyBy**.

Example 17.1.   *Vertical Slices of $z = x^2 + \frac{1}{2}y^2$*

We begin with the function $z = f[x,y] = x^2 + \frac{1}{2}y^2$ finding the vertical slices parallel to the $x - z$-plane. If we fix the value of y, say $y = -2$, then the function becomes $z = f[x,-2] = x^2 + 2$. This is a simple parabola and its graph is shown in the upper left figure plotted from $x = -2$ to $x = 3$ in Figure 17.10 below. The plot is drawn on the vertical (x,z)-plane where $y = -2$.

Next, we fix $y = -1$ and plot $z = f[x,-1] = x^2 + \frac{1}{2}$. This is also a parabola with the same basic shape as the first, but translated down by $\frac{3}{2}$. It is shown in the second figure of Figure 17.10 plotted in the vertical plane where $y = -1$ and x and z vary.

The next 4 figures of Figure 17.10 show the plots $z = x^2$ on the plane $y = 0$, $z = x^2 + \frac{1}{2}$ on the plane $y = 1$, $z = x^2 + 2$ on the plane $y = 2$, and $z = x^2 + \frac{9}{2}$ on the plane $y = 3$.

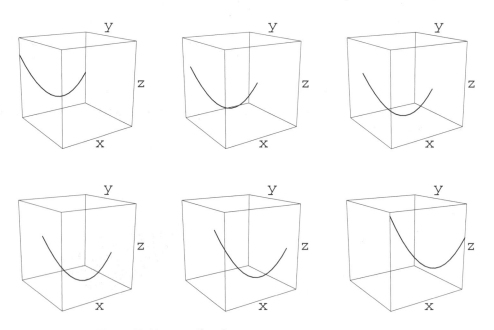

Figure 17.10: $z = f[x,a]$, $a = -2,-1,0,1,2,3$

Now, vertically slice the function $z = f[x,y] = x^2 + \frac{1}{2}y^2$ parallel to the y-z-plane. When we fix $x = -2$, the function becomes $z = f[-2,y] = 4 + \frac{1}{2}y^2$. This is a parabola and its graph is shown in the upper left figure of Figure 17.11 below plotted from $y = -2$ to $y = 3$ on the vertical (y, z)-plane where $x = -2$.

Next, we plot $z = f[-1, y] = 1 + \frac{1}{2}y^2$. This is also a parabola with the same basic shape as the first, but translated down by 3. It is plotted in Figure 17.11 in the vertical plane where $x = -1$ and y and z vary. The next 4 figures of Figure 17.11 show the plots $z = \frac{1}{2}y^2$ on the plane $y = 0$, $z = 1 + \frac{1}{2}y^2$ on the plane $x = 1$, $z = 4 + \frac{1}{2}y^2$ on the plane $x = 2$, and $z = 9 + \frac{1}{2}y^2$ on the plane $x = 3$.

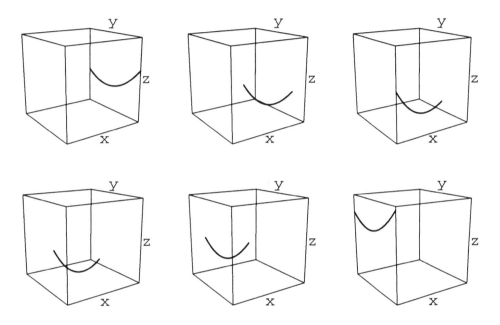

Figure 17.11: $z = f[a, y]$, $a = -2, -1, 0, 1, 2, 3$

When all these curves are shown together and the segments between them are shaded, we see the surface as in Figure 17.12.

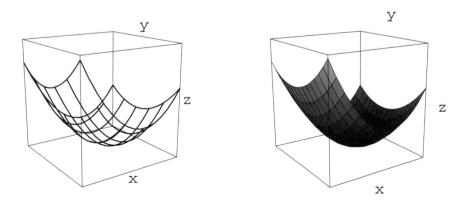

Figure 17.12: $z = x^2 + \frac{1}{2}y^2$

Exercise set 17.2

1. *For each of the functions $z = f[x, y]$ graphed below plot the curves:*

    (a) *$z = f[x, -2]$, $z = f[x, -1]$, $z = f[x, 0]$, $z = f[x, 1]$, $z = f[x, 2]$, $z = f[x, 3]$.*

    (b) *$z = f[-2, y]$, $z = f[-1, y]$, $z = f[0, y]$, $z = f[1, y]$, $z = f[2, y]$, $z = f[3, y]$.*

    (c) *Combine your plots from (1) and (2) above on a single 3-D graph, plotting $z = f[x, 2]$ in the vertical plane $y = 2$, and plotting $z = f[-1, y]$ in the vertical plane $x = -1$, etc.*

    (d) *Compare your sketches with the figures below and draw the slices with the program **SurfSlices**.*

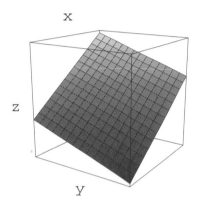

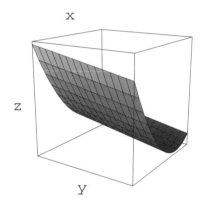

Figure 17.13: $z = -x - \frac{1}{2}y$ $z = x^2 - \frac{1}{2}y$

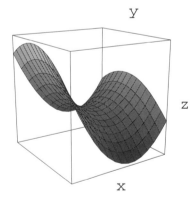

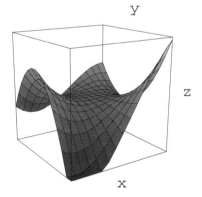

Figure 17.14: $z = x^2 - \frac{1}{2}y^2$ $z = x^3 - 3xy^2$

2. *Consider the formula*

$$T = M \operatorname{Sin}\left[\pi \frac{t}{6}\right] e^{-d/20}$$

Let $M = 15$ at first. Then later show the parameter M on your plots. (T is temperature in Celsius, t is time in years, and d is depth in meters for seasonably fluctuating sea water.)

(a) *Plot T vs. t treating d as a parameter. How do the graphs change as d increases? As the depth increases, do seasonal fluctuations become bigger or smaller?*

(b) *Plot T vs. d treating t as a parameter.*

Here are both families at once:

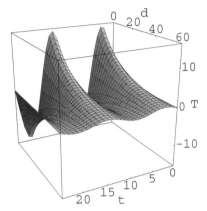

Figure 17.15: Variations with time and depth

17.3   Implicit Linear Equations

The algebra and geometry of an implicit linear equation are connected by "reading" the equation as,

"The set of all unknown vectors $\mathbf{X}$ whose displacements, $\mathbf{X} - \mathbf{R}$, from a fixed point $\mathbf{R}$ are perpendicular to a normal vector $\mathbf{N}$." $\Leftrightarrow$ $\mathbf{N} \bullet (\mathbf{X} - \mathbf{R}) = 0$

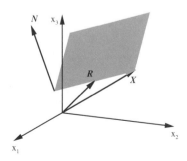

Figure 17.16: A plane normal to $\mathbf{N}$ through $\mathbf{R}$

Theorem 17.3. *Implicit Plane (or 2-D Implicit Line)*
Given known vectors $\mathbf{N}$ *and* $\mathbf{R}$, *the set of all (unknown) vectors* $\mathbf{X}$ *satisfying*

$$\langle \mathbf{N} \bullet (\mathbf{X} - \mathbf{R}) \rangle = 0$$

is the plane perpendicular to $\mathbf{N}$ *passing through the tip of* $\mathbf{R}$.
(If the vectors are 2-D, this is the line perpendicular to $\mathbf{N}$ *through the tip of* $\mathbf{R}$.)

You can see step by step why this theorem is true by working Problem 17.4. Recall (from your algebra - geometry lexicon) that two vectors are perpendicular if their dot product is zero. Also, the displacement from the tip of $\mathbf{R}$ to the tip of $\mathbf{X}$ is given by the vector difference, $\mathbf{X} - \mathbf{R}$. These two things "say" that the vector pointing from the tip of $\mathbf{R}$ to the tip of $\mathbf{X}$ is perpendicular to $\mathbf{N}$ exactly when $\langle \mathbf{N} \bullet (\mathbf{X} - \mathbf{R}) \rangle = 0$.

Example 17.2. *The Plane* $2(x - 3) - (y + 4) + z = 0$

The vectors $\mathbf{N}$ and $\mathbf{R}$ hidden in the equation $2(x - 3) - (y + 4) + z = 0$ are

$$\mathbf{N} = \begin{bmatrix} +2 \\ -1 \\ +1 \end{bmatrix} \quad \text{and} \quad \mathbf{R} = \begin{bmatrix} +3 \\ -4 \\ 0 \end{bmatrix}$$

as you can verify by substitution:

$$\langle \mathbf{N} \bullet (\mathbf{X} - \mathbf{R}) \rangle = \langle \begin{bmatrix} 2 \\ -1 \\ 1 \end{bmatrix} \bullet \left(\begin{bmatrix} x \\ y \\ z \end{bmatrix} - \begin{bmatrix} 3 \\ -4 \\ 0 \end{bmatrix} \right) \rangle = 2(x - 3) - (y + 4) + z$$

Once we know $\mathbf{N}$ and $\mathbf{R}$, we plot the plane by sketching a plane at the tip of $\mathbf{R}$ perpendicular to a translated $\mathbf{N}$ drawn at the tip of $\mathbf{R}$.

You should be able to recognize the "hidden" vectors if you are given an implicit linear equation in a classical form such as

$$n_1 x_1 + n_2 x_2 + n_3 x_3 = k \quad \text{or} \quad a x + b y + c z = k$$

For example,

$$x - 3 y + 5 z = 2$$

This is explained by the following theorem and example.

Theorem 17.4. *Another geometric formula for a plane in 3-D is*

$$\langle \mathbf{N} \bullet \mathbf{X} \rangle = k$$

where $\mathbf{N} = \begin{bmatrix} a \\ b \\ c \end{bmatrix}$ *is the (fixed) normal vector to the plane,* $\mathbf{X} = \begin{bmatrix} x \\ y \\ z \end{bmatrix}$ *is the unknown*
(variable) vector of coordinates, and k *is a constant satisfying*

$$k = |N| \times [\text{distance from the origin to the plane}]$$

Exercise 17.3.3 shows that k satisfies the distance formula given in the theorem above.

Example 17.3. *A Graph of* $x - 3\,y + 5\,z = 8$

One way to plot the plane of

$$x - 3\,y + 5\,z = 8$$

is first to notice that the vector $\mathbf{N} = \begin{bmatrix} 1 \\ -3 \\ 5 \end{bmatrix}$ is normal to the plane because

$$\langle \begin{bmatrix} 1 \\ -3 \\ 5 \end{bmatrix} \bullet \begin{bmatrix} x \\ y \\ z \end{bmatrix} \rangle = x - 3\,y + 5\,z$$

Next, measure out from the origin along the direction of $\mathbf{N} = \begin{bmatrix} 1 \\ -3 \\ 5 \end{bmatrix}$ a distance of

$$\frac{k}{|\mathbf{N}|} = \frac{8}{\sqrt{1^2 + 3^2 + 5^2}} = \frac{8}{\sqrt{35}}$$

This point is at the tip of the vector given by the distance times a unit vector:

$$\mathbf{R} = \frac{k}{|\mathbf{N}|} \cdot \frac{1}{|\mathbf{N}|} \cdot \mathbf{N} = \frac{k}{\langle \mathbf{N} \bullet \mathbf{N} \rangle} \cdot \mathbf{N} = \frac{8}{35} \begin{bmatrix} 1 \\ -3 \\ 5 \end{bmatrix}$$

Once we have the normal and a point on the plane, we sketch the plane through the tip of $\mathbf{R}$ perpendicular to $\mathbf{N}$, as shown in Figure 17.17.

Figure 17.17: Two ways to plot $x - 3\,y + 5\,z = 8$

Example 17.4. *A Simpler Approach to Graphing* $x - 3\,y + 5\,z = 8$

An easier way to find a point on the plane is simply to set $y = z = 0$ and solve for $x = 8$. Then, we know that the plane is normal to $\mathbf{N}$ above and goes through the tip of the vector

$$\mathbf{R}_x = \begin{bmatrix} 8 \\ 0 \\ 0 \end{bmatrix}$$

17.3.1   Explicit vs. Implicit Linear Equations

We may write the vector equation $\langle \mathbf{N} \bullet \mathbf{X} \rangle = k$ in components

$$a\,x + b\,y + c\,z = k$$

and then solve for z (if $c \neq 0$) to put it in explicit form:

$$z = m\,x + n\,y + h$$

Simple algebra gives us the connection between the parameters, the slope in the x-direction $m = -a/c$, the slope in the y-direction $n = -b/c$, and the z-intercept $h = k/c$.

If we start with the explicit formula, $z = m\,x + n\,y + h$, and want the normal vector, we rewrite the equation in the form $-m\,x - n\,y + z = h$. Then $a = -m$, $b = -n$, $c = 1$ and $k = h$ gives the parameters for the geometric form.

Notice that it is easy to plot a plane if we know its normal and one point on it, so remember this simple algebraic conversion:

$$z = m \cdot x + n \cdot y + h \quad \Leftrightarrow \quad \left\langle \begin{bmatrix} -m \\ -n \\ 1 \end{bmatrix} \bullet \begin{bmatrix} x \\ y \\ z \end{bmatrix} \right\rangle = h \quad \Leftrightarrow -m \cdot x - n \cdot y + z = h$$

$$z = -\frac{a}{c} \cdot x - \frac{b}{c} \cdot y + \frac{k}{c} \quad \Leftrightarrow \quad \left\langle \begin{bmatrix} a \\ b \\ c \end{bmatrix} \bullet \begin{bmatrix} x \\ y \\ z \end{bmatrix} \right\rangle = k \quad \Leftrightarrow a \cdot x + b \cdot y + c \cdot z = h$$

Which formula is best? Neither one. The vector equation gives us some useful geometric information, but the explicit equation is useful in other contexts.

Exercise set 17.3

You need to be able to recognize the geometry contained in old-fashioned equations.

1. The Classical Form of a Plane

   Consider parameters a, b, c, and k and unknowns x, y, and z. The graph of the general linear equation in 3 variables $ax + by + cz = k$ is a plane. Write this equation in the vector form $\mathbf{N} \bullet \mathbf{X} = k$.

   In terms of the given letters, what is the vector $\mathbf{N}$? What is the unknown vector $\mathbf{X}$? Take the special case $x + 3y + 2z = 6$. What is the vector normal to this plane? Find the three intercepts of this plane and the axes. (HINT: An unknown vector of the form $\mathbf{X} = \begin{bmatrix} x \\ 0 \\ 0 \end{bmatrix}$ lies on the x axis.) Sketch the plane and its (displacement) normal vector with its tail somewhere on the plane.

   How could you put this equation in the form $\langle \mathbf{N} \bullet (\mathbf{X} - \mathbf{R}) \rangle = 0$?

   Now, combine your knowledge of lines and planes.

2. *Find an equation of the line that contains the point $\begin{bmatrix} 1 \\ 3 \\ 2 \end{bmatrix}$ and is perpendicular to the plane $2x + 3y - 4z = 5$.*

3. Given the plane $\quad\langle \mathbf{N} \bullet \mathbf{X} \rangle = k \;\Leftrightarrow\; a\,x + b\,y + c\,z = k$
The Distance to the Origin is $\frac{k}{|\mathbf{N}|} = \frac{k}{\sqrt{a^2+b^2+c^2}}$

Suppose we are given a plane in the form $\langle \mathbf{N} \bullet \mathbf{X} \rangle = a\,x + b\,y + c\,z = k$. *We want you to prove that the distance from the origin to the plane is* $\frac{k}{|\mathbf{N}|}$. *Your job is to find a general formula, but here is an example to work with (and draw pictures of) as you discover why this formula is true:*

$$2\,x - y + 2\,z = 6 \qquad \Leftrightarrow \qquad \begin{bmatrix} 2 \\ -1 \\ 2 \end{bmatrix} \bullet \begin{bmatrix} x \\ y \\ z \end{bmatrix} = 6$$

Our first step in helping you find this equation asks you to find a vector $\mathbf{M}$ *parallel to* $\mathbf{N}$ *whose tip lies on the plane.*

(a) *If* $\mathbf{M}$ *is parallel to* $\mathbf{N}$, *then it must be a scalar multiple of* $\mathbf{N}$. *Why?*

(b) *Say,* $\mathbf{M} = h\,\mathbf{N}$ *and the tip of* $\mathbf{M}$ *lies on the plane. Solve for* h *in terms of* a, b, c, *and* k. *Show that your component equations are equivalent to*

$$h = \frac{k}{\langle \mathbf{N} \bullet \mathbf{N} \rangle}$$

In particular, find the value of h *for the sample above.*

(c) *Use* $\mathbf{M}$ *to find a formula for the distance from the origin to the plane. (HINT: Draw the figure of* $\mathbf{M}$ *pointing from the origin to the plane. Why is it perpendicular to the plane? Simplify and show that the length is* $\ell = k/|\mathbf{N}|$*)*

(d) *The equation* $\langle \mathbf{N} \bullet \mathbf{X} \rangle = a\,x + b\,y + c\,z = k$ *is not unique because we can multiply both sides by any non-zero constant without changing the solutions. For example, the following equations all describe the same plane:*

$$2\,x - y + 2\,z = 6 \qquad \Leftrightarrow \qquad \begin{bmatrix} 2 \\ -1 \\ 2 \end{bmatrix} \bullet \begin{bmatrix} x \\ y \\ z \end{bmatrix} = 6$$

$$\Leftrightarrow$$

$$\frac{2}{3}\,x - \frac{1}{3}\,y + \frac{2}{3}\,z = 2 \qquad \Leftrightarrow \qquad \begin{bmatrix} \frac{2}{3} \\ -\frac{1}{3} \\ \frac{2}{3} \end{bmatrix} \bullet \begin{bmatrix} x \\ y \\ z \end{bmatrix} = 2$$

Apply your formula from above to find the distance from the origin to this plane using each of the formulas for the plane.

(e) *The unit length vector in the same direction as* $\mathbf{N}$ *is* $\mathbf{U} = \frac{1}{|\mathbf{N}|}\,\mathbf{N}$. *Show that* $\ell = \frac{k}{|\mathbf{N}|}$ *makes the following equations equivalent:*

$$\langle \mathbf{N} \bullet \mathbf{X} \rangle = k \qquad \Leftrightarrow \qquad \langle \mathbf{U} \bullet \mathbf{X} \rangle = \ell$$

(f) *Which multiple of the unit vector* $\mathbf{U}$ *has its tip on the plane* $\langle \mathbf{U} \bullet \mathbf{X} \rangle = \ell$? *What is the length of this multiple of the vector?*

Compare your work to the program ***PlaneLines***.

The next problem uses your lexicon to extend the idea of Exercise 17.3.3. Vector projection is a very useful algebraic computation. In order to use it effectively, often we must compute an auxiliary vector.

Problem 17.1. THE DISTANCE FROM A POINT TO A PLANE ―――――――――――――▼
Given a plane in the form $\langle \mathbf{N} \bullet \mathbf{X} \rangle = k$ *and a point* $\mathbf{Q}$ *anywhere in space, use vector projection to describe a procedure to compute the distance from* $\mathbf{Q}$ *to the plane.*
      See the program ***PlaneLines***.

―――▲

Geometrically, three points determine a plane. The next exercise shows you how to use the lexicon to find the algebraic equation of the plane the three points determine. Solution of the exercise is an algebraic proof that three points really do determine a plane but it is a specific proof, because it gives you the specific equation.

Problem 17.2. THREE POINTS DETERMINE A PLANE ―――――――――――――――――▼

(a) *Plot the vectors*

$$\mathbf{R} = \begin{bmatrix} r_1 \\ r_2 \\ r_3 \end{bmatrix} = \begin{bmatrix} 3 \\ 0 \\ 0 \end{bmatrix}, \qquad \mathbf{S} = \begin{bmatrix} s_1 \\ s_2 \\ s_3 \end{bmatrix} = \begin{bmatrix} 0 \\ 2 \\ 0 \end{bmatrix} \quad and \quad \mathbf{T} = \begin{bmatrix} t_1 \\ t_2 \\ t_3 \end{bmatrix} = \begin{bmatrix} 0 \\ 0 \\ 1 \end{bmatrix}$$

   Also plot the displacement arrows $\mathbf{E}$ *from the tip of* $\mathbf{R}$ *to the tip of* $\mathbf{S}$ *and the displacement arrow* $\mathbf{F}$ *from* $\mathbf{R}$ *to* $\mathbf{T}$.
(b) *Compute the legal vectors* $\mathbf{E} = \mathbf{S} - \mathbf{R}$ *and* $\mathbf{F} = \mathbf{T} - \mathbf{R}$. *These are parallel to the arrows drawn in part 1 by our lexicon entry for vector difference.*
(c) *Compute and sketch a vector* $\mathbf{N}$ *that is perpendicular to both* $\mathbf{E}$ *and* $\mathbf{F}$.
(d) *Calculate* $\mathbf{N} \bullet \mathbf{R}$, $\mathbf{N} \bullet \mathbf{S}$, $\mathbf{N} \bullet \mathbf{T}$. *Explain algebraically why* $\mathbf{R}$, $\mathbf{S}$, *and* $\mathbf{T}$ *are not perpendicular to* $\mathbf{N}$. *Do they appear to be perpendicular on your plot?*
(e) *The equation* $\mathbf{N} \bullet (\mathbf{X} - \mathbf{R}) = 0$ *in the unknown vector* $\mathbf{X}$ *says, "*$\mathbf{X}$ *such that the displacement from* $\mathbf{R}$ *to* $\mathbf{X}$ *is perpendicular to* $\mathbf{N}$.*" Sketch this in a figure and explain why this is the equation for the plane determined by the tips of the three vectors* $\mathbf{R}$, $\mathbf{S}$, *and* $\mathbf{T}$.
(f) *The equations* $\mathbf{N} \bullet (\mathbf{X} - \mathbf{R}) = 0$, $\mathbf{N} \bullet (\mathbf{X} - \mathbf{S}) = 0$, *and* $\mathbf{N} \bullet (\mathbf{X} - \mathbf{T}) = 0$ *in the unknown vector* $\mathbf{X}$ *are all equations for this plane because each is an equation of a plane through one of* $\mathbf{R}$, $\mathbf{S}$, *or* $\mathbf{T}$ *and perpendicular to the same normal* $\mathbf{N}$. *In this form, the equations are different, but write them explicitly in components and simplify them to the form*

$$m\,x + n\,y + c\,z = k$$

   Are their simplified forms different? Why?
   See the program ***ThreePoints***.

―――▲

The next problem looks at the tilt of the explicit plane again.

Problem 17.3. STEEPNESS AND $|\mathbf{G}|$ ▬▬▬▬▬▬▬▬▬▬▬▬▬▬▬▬▬▼

An explicit plane through the origin in 3-D has an equation of the form

$$z = m\,x + n\,y$$

(a) *For example, $z = \sqrt{3}\,x + y$. Sketch this plane by x-y-slices, holding x fixed and varying y, holding y fixed and varying x, but plotting everything in 3-D. How do m and n appear on your plot?*

(b) *It is easier to plot by rewriting the equation in the form $\mathbf{N} \bullet \mathbf{X} = 0$, where $\mathbf{N} = \begin{bmatrix} -m \\ -n \\ 1 \end{bmatrix}$.*

Show that this is equivalent to the original equation. What is the geometric meaning of $\mathbf{N} \bullet \mathbf{X} = 0$? Sketch the 3-D vector $\mathbf{N}$ on your 3-D figure for this case. (Hint: Use your geometric-algebraic lexicon.)

(c) *Repeat the exercise for the equation $z = \sqrt{3}\,x/2 + y/2$; that is, sketch the plane perpendicular to $\begin{bmatrix} -\frac{\sqrt{3}}{2} \\ -\frac{1}{2} \\ 1 \end{bmatrix}$ through the origin. Do you need to plot slices?*

(d) *As the 2-D vector $\mathbf{G} = \begin{bmatrix} m \\ n \end{bmatrix}$ gets larger, what happens to the tilt of the 3-D vector $\mathbf{N} = \begin{bmatrix} -m \\ -n \\ 1 \end{bmatrix}$? (Is it more horizontal or more vertical?) For example, plot $\begin{bmatrix} -m \\ -n \\ 1 \end{bmatrix}$ and the plane perpendicular to it through the origin in case $m = n = 1/10$ and in the case $m = n = 10$.*

(e) *How is the steepness of the plane related to the length of $\mathbf{G}$?*

*See the program **LinearContours**.*

▬▬▬▬▬▬▬▬▬▬▬▬▬▬▬▬▬▬▬▬▬▬▬▬▬▬▬▬▬▬▲

The next problem shows that the implicit linear equation says, "The plane is the set of vectors whose displacements from a certain point $\mathbf{R}$ are perpendicular to a certain normal vector $\mathbf{N}$."

Problem 17.4. THE IMPLICIT EQUATION OF A PLANE ▬▬▬▬▬▬▬▬▬▬▬▼

We are given vectors $\mathbf{R}$ on a plane and $\mathbf{N}$ perpendicular (or normal) to it. We wish to find an equation in the unknown $\mathbf{X} = \begin{bmatrix} x_1 \\ x_2 \\ x_3 \end{bmatrix}$ that describes the plane. That is, so that the tips of all the $\mathbf{X}$ that satisfy the equation lie on the plane and completely fill it.

(a) *Take the example $\mathbf{N} = \begin{bmatrix} -6 \\ -4 \\ 3 \end{bmatrix}$ and $\mathbf{R} = \begin{bmatrix} -6 \\ 8 \\ 2 \end{bmatrix}$. Sketch the plane through the tip of $\mathbf{R}$ that is perpendicular to $\mathbf{N}$.*

(b) *Draw a generic unknown vector $\mathbf{X}$ whose tip lies somewhere on the plane.*

(c) *Draw the displacement arrow that points from the tip of $\mathbf{R}$ to the tip of $\mathbf{X}$. Show that the legal vector parallel to this arrow is $\mathbf{X} - \mathbf{R}$.*

(d) *Write the statement "$\mathbf{X} - \mathbf{R}$ is perpendicular to $\mathbf{N}$" as an algebraic formula.*

(e) *Show that the equation of this specific example is*

$$\mathbf{N} \bullet (\mathbf{X} - \mathbf{R}) = 0$$

$$\begin{bmatrix} -6 \\ -4 \\ 3 \end{bmatrix} \bullet \begin{bmatrix} x_1 + 6 \\ x_2 - 8 \\ x_3 - 2 \end{bmatrix} = 0$$

$$6\,x_1 + 4\,x_2 - 3\,x_3 = -10$$

*See the program **PlaneLines**.*

17.4   Gradient Vectors & Contour Plots

The gradient vector (Definition 17.1) is perpendicular to the contour lines of a linear function $L[x, y] = m\,x + n\,y + h$ and the density of the contour lines varies inversely with the magnitude of the gradient.

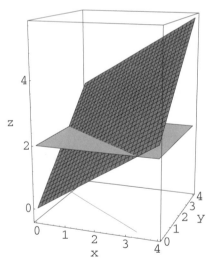

Figure 17.18: The contour of $z = \frac{3}{5}x + \frac{4}{5}y$ at $z = 2$

A contour map of an explicit surface $z = f[x, y]$ is the planar sketch of the curves that cut the surface at various fixed values of $z = k$, a constant. In the case of a linear function, $z = m\,x + n\,y + h$, the contours are parallel straight lines. The questions then are

(a) In which direction does the family of lines point?
(b) How far apart are the lines for a given spacing of the constant values of z?

Example 17.5.   *The Contour Plot of* $L[x, y] = 3\,x + 4\,y$

We wish to plot the collection of lines $3\,x + 4\,y = c$ for various constants c. Let us begin with $c = 0$. To plot $3\,x + 4\,y = 0$ we do *not* rewrite the equation in the form $y = m\,x + b$, but rather use Theorem 17.4 (and the examples right after it) in the 2-D case. The vector form of $3\,x + 4\,y = 0$ is

$$\begin{bmatrix} 3 \\ 4 \end{bmatrix} \bullet \begin{bmatrix} x \\ y \end{bmatrix} = 0$$

so the line consists of all vectors $\mathbf{X} = \begin{bmatrix} x \\ y \end{bmatrix}$ that are perpendicular to $\mathbf{N} = \begin{bmatrix} 3 \\ 4 \end{bmatrix}$. Simply sketch the vector $\mathbf{N}$ and draw the line through the origin perpendicular to it, as shown in Figure 17.19.

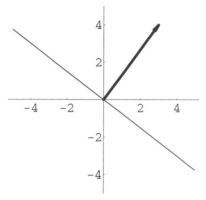

Figure 17.19: $3\,x + 4\,y = 0$

Next, let $c = 3$ and plot $3\,x + 4\,y = 3$. This is easy because the line is still perpendicular to $\mathbf{G}(3, 4)$ but now goes through $(1, 0)$, as shown in Figure 17.20.

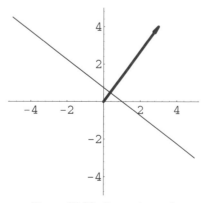

Figure 17.20: $3\,x + 4\,y = 3$

Next, let $c = -3$ and plot $3\,x + 4\,y = -3$. This is still perpendicular to $\mathbf{G}(3, 4)$ but now goes through $(-1, 0)$, as shown in Figure 17.21.

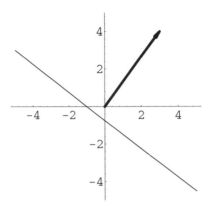

Figure 17.21: $3x + 4y = -3$

The contour plot consists of lines perpendicular to $\mathbf{G}(3,4)$, as shown in Figure 17.22.

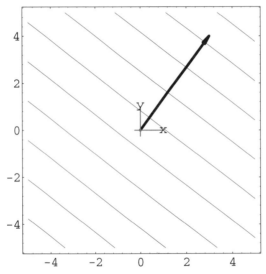

Figure 17.22: Contours of $L[x,y] = 3x + 4y$

In general, you can see that:

Theorem 17.5. *The contour plot of the linear function $L[x,y] = mx + ny$ consists of a family of lines perpendicular to the gradient vector*

$$\mathbf{G} = \begin{bmatrix} m \\ n \end{bmatrix}$$

17.4.1 The Gradient and Contour Direction

We want to change both x and y and compute the corresponding change in z. Consider the change,

$$\Delta \mathbf{X} = \begin{bmatrix} \Delta x \\ \Delta y \end{bmatrix} \qquad \text{with} \qquad |\Delta \mathbf{X}| = \sqrt{\Delta x^2 + \Delta y^2}$$

It may be helpful for you to think of the function as a function of a vector input, $L[x, y] = L[\mathbf{X}]$, where $\mathbf{X} = \begin{bmatrix} x \\ y \end{bmatrix}$. The change in z is

$$\begin{aligned} \Delta z &= L[\mathbf{X} + \Delta\mathbf{X}] - L[\mathbf{X}] \\ &= L[x + \Delta x, y + \Delta y] - L[x, y] \\ &= (m \cdot (x + \Delta x)x + n \cdot (y + \Delta y) + h) - (m\,x + n\,y + h) \\ &= m \cdot \Delta x + n \cdot \Delta y \\ &= \begin{bmatrix} m \\ n \end{bmatrix} \bullet \begin{bmatrix} \Delta x \\ \Delta y \end{bmatrix} = \langle \mathbf{G} \bullet \Delta\mathbf{X} \rangle = |\mathbf{G}| \cdot |\Delta\mathbf{X}| \cdot \mathrm{Cos}[\theta] \end{aligned}$$

where $\mathbf{G}$ is the gradient vector and θ is the angle between $\mathbf{G}$ and $\Delta\mathbf{X}$. Since the distance moved is the length $|\Delta\mathbf{X}|$, the rate of change of z in the direction of $\Delta\mathbf{X}$ is therefore $\langle \mathbf{G} \bullet \mathbf{U} \rangle / |\Delta\mathbf{X}|$ (the z-change per unit moved.) Using the angle formula for the dot product above, we have

$$\boxed{\text{The rate of change of } z \text{ per unit change in direction } \Delta\mathbf{X} = |\mathbf{G}| \cdot \mathrm{Cos}[\theta]}$$

In particular,

> **Theorem 17.6.** *The norm of the gradient $|\mathbf{G}| = \sqrt{m^2 + n^2}$ is the maximal rate of growth of the linear function $L[x, y] = m\,x + n\,y$. This rate is attained when we change x and y in the direction of $\mathbf{G}$.*

17.4.2 The Gradient and Contour Density

Roughly speaking, "The longer $\mathbf{G}$, the steeper the explicit plane graph and the denser the contour lines..." We want an algebraic connection between the closeness of the lines $m\,x + n\,y = k$; $k = 0, \pm 1, \pm 2, \cdots$, and the length of $\mathbf{G}$. (Section 17.1.1 gave the connection between the steepness of the explicit plane graph $z = m\,x + n\,y$ and the length of $\mathbf{G}$.) We show now that the distance between the line $\{\mathbf{X} : \langle \mathbf{G} \bullet \mathbf{X} \rangle = k_1\}$ and the second line $\{\mathbf{X} : \langle \mathbf{G} \bullet \mathbf{X} \rangle = k_2\}$ is $|k_2 - k_1|/|\mathbf{G}|$. We also need to understand this in terms of the function $L[\mathbf{X}]$ and its contour plot.

The distance between contours of $L[\mathbf{X}]$ is measured perpendicular to those lines or along the direction of $\mathbf{G}$. Let us assume that we have a point $\mathbf{X}$ on the first and $\mathbf{X}_2$ on the second and that $\Delta\mathbf{X} = \mathbf{X}_2 - \mathbf{X}$ is a multiple of $\mathbf{G}$,

$$L[\mathbf{X}_1] = k_1 \qquad \text{and} \qquad L[\mathbf{X} + \Delta\mathbf{X}] = k_2 = L[\mathbf{X} + t\mathbf{G}]$$

Now, we compute

$$\begin{aligned} k_2 - k_1 &= L[\mathbf{X} + \Delta\mathbf{X}] - L[\mathbf{X}] \\ &= \langle \mathbf{G} \bullet \Delta\mathbf{X} \rangle \\ &= \langle \mathbf{G} \bullet (t\mathbf{G}) \rangle \\ &= t\langle \mathbf{G} \bullet \mathbf{G} \rangle = t \cdot |\mathbf{G}|^2 \end{aligned}$$

and the distance moved is

$$|\Delta\mathbf{X}| = |t\mathbf{G}| = |t| \cdot |\mathbf{G}|$$

so the difference

$$|k_2 - k_1| = \text{distance moved} \cdot |\mathbf{G}|$$

Theorem 17.7. *The contour plot of the linear function*

$$L[x, y] = L[\mathbf{X}] = m\,x + n\,y + h$$

consists of the family of lines perpendicular to the vector $\mathbf{G} = \begin{bmatrix} m \\ n \end{bmatrix}$, *where the contour lines* $L[\mathbf{X}] = k_1$ *and* $L[\mathbf{X}] = k_2$ *are spaced apart by a distance*

$$\frac{|k_2 - k_1|}{|\mathbf{G}|}$$

Exercise set 17.4

1. *Make contour plots of the functions:*

1) $L[x, y] = 3\,x + 4\,y$
2) $L[x, y] = 3\,x - 4\,y$
3) $L[x, y] = 4\,x - 3\,y$
4) $L[x, y] = \frac{4}{5}\,x - \frac{3}{5}\,y$

2. *Sketch the lines:*

$$0 = 4\,x - 3\,y \qquad and \qquad 1 = 4\,x - 3\,y$$

How far apart are these lines? Sketch the lines:

$$0 = \frac{4}{5}\,x - \frac{3}{5}\,y \qquad and \qquad 1 = \frac{4}{5}\,x - \frac{3}{5}\,y$$

How far apart are these lines?

In the next problem, you should try to devise a geometric way to go from values for m, n, and h to the graph of the line $m\,x + n\,y = k$. At first, you may want to rewrite lines in the form $y = m\,x + b$, but it turns out to be better to work more directly from the vector $\mathbf{G}$. It is easy to plot the vector.

Problem 17.5. CONTOUR LINE DENSITY ⎯⎯⎯⎯⎯⎯⎯⎯⎯⎯⎯⎯⎯⎯▼
The purpose of this exercise is to have you work out the formulas needed to relate the steepness of an explicit linear graph $z = m\,x + n\,y$ and the density of the lines in its contour plot $m\,x + n\,y = h$. The secret of the whole exercise is in relating the length of the vector $G(m, n)$ to the steepness and density.

Consider the family of lines in 2-D given by

$$m\,x + n\,y = k \qquad for \quad k = 0, \pm 1, \pm 2, \cdots$$

We want you to develop a systematic geometric approach to this kind of plot, but you can use any methods you like in the first two parts. Check your later vector formulas on the first two special cases at each step of the exercise.

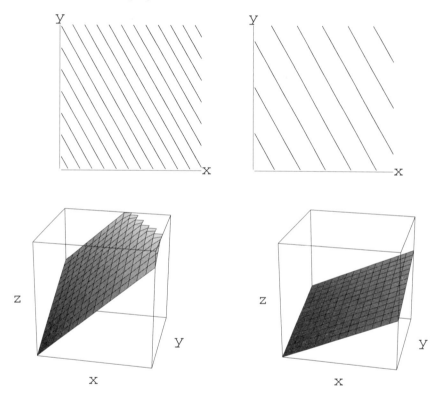

Figure 17.23: $z = \sqrt{3}\,x + y$ *and* $z = \sqrt{3}\,x/2 + y/2$

(a) *Plot the family of contour lines on a sheet of graph paper if $(m, n) = (\sqrt{3}, 1)$. Plot the vector $\mathbf{G} = \begin{bmatrix} m \\ n \end{bmatrix}$ on top of your family of lines. The next part of the exercise asks you to do the same thing for a shorter parallel vector. Remember the title of the problem.*

(b) *Plot the family of contour lines on a sheet of graph paper if $(m, n) = (\frac{\sqrt{3}}{2}, \frac{1}{2})$ Plot the vector $\mathbf{G} = \begin{bmatrix} m \\ n \end{bmatrix}$ on top of your family of lines.*

NOTE: $\mathrm{Cos}[\pi/6] = \sqrt{3}/2$ and $\mathrm{Sin}[\pi/6] = 1/2$, or the vectors from parts (a) and (b) are both inclined $30°$ above the x axis. The contour lines from parts (a) and (b) are both perpendicular to this direction, but the lines are closer together in one of them. Which has closer contour lines (for $k = 0, \pm 1, \pm 2, \cdots$), part (a) or part (b)?

Now consider the general geometric case of the family of lines $mx + ny = k$ for $k = 0, \pm 1, \pm 2, \cdots$ with unknown, but fixed m and n.

(c) Plot the vector $\mathbf{G} = \begin{bmatrix} m \\ n \end{bmatrix}$ and the line through the origin perpendicular to $\mathbf{G}$. Show that this is the line $\langle \mathbf{G} \bullet \mathbf{X} \rangle = mx + ny = 0$. (HINT: Use your geometric-algebraic lexicon.) Compare this procedure with the two test cases you did at the start of the exercise.

(d) Show that the 2-vector $\mathbf{G}$ is perpendicular to all the contour lines. Notice that

$$\langle \mathbf{G} \bullet \mathbf{X} \rangle = mx + ny = k; \qquad k = 0, \pm 1, \pm 2, \cdots$$

(e) What are all the vectors perpendicular to the lines $mx + ny = k$?

(f) What is the length of $\mathbf{G}$? (Compute the value in your two test cases.)

(g) Which unknown vector of the form $\mathbf{X} = t\mathbf{G}$ for some real t lies on the line $\mathbf{G} \bullet \mathbf{X} = 1$? You should treat the components of $\mathbf{G}$ as known parameters and solve for t. Sketch this vector $\mathbf{X}$ on your two test cases for $\mathbf{G}$ (parts (a) and (b) above).

(h) Show that the distance between the lines $\langle \mathbf{G} \bullet \mathbf{X} \rangle = 0$ and $\langle \mathbf{G} \bullet \mathbf{X} \rangle = +1$ is $1/|\mathbf{G}|$. See the program **LinearContours**.

▲

17.5   Horizontal Slices and Contours

The contour plot of a function $f[x, y]$ is a collection of implicit curves $f[x, y] = c$ for various values of the constant c.

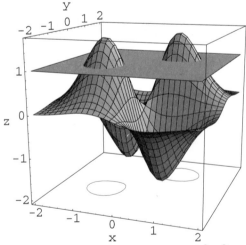

Figure 17.24: The contour of $z = 9xye^{-(x^2+y^2)}$ at $z = 1$

If we slice a surface $z = f[x, y]$ horizontally, we hold the value of z fixed. Consider the example from Example 17.1, $z = x^2 + \frac{1}{2}y^2$.

Example 17.6.   *Horizontal Slices of $z = x^2 + \frac{1}{2}y^2$*

If we let $z = 0$ we have the equation $0 = x^2 + \frac{1}{2}y^2$. This has only the solution $(x, y) = (0, 0)$. This is the single point at the very bottom of the surface.

If we let $z = 1$, the intersection of the plane with equation $z = 1$ and the surface $z = x^2 + \frac{1}{2}y^2$ has the equation $1 = x^2 + \frac{1}{2}y^2$. This is an ellipse.

Similarly, the intersections of the planes $z = c$, constant, are larger ellipses of similar shapes. The contour plot of the function $z = x^2 + \frac{1}{2}y^2$ is the family of these curves shown in Figure 17.25.

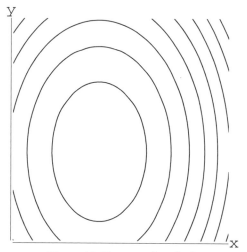

Figure 17.25: $z = x^2 + \frac{1}{2}y^2$

Exercise set 17.5

1. *Find the equations of the contours in the following examples for the z-values $z = -1$, $z = 0$, $z = 1$, $z = 2$. The plots are shown for $-2 \le x, y \le 3$ (even though the computer puts lines at the left and bottom).*

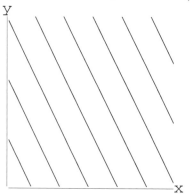

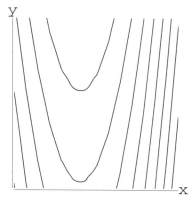

Figure 17.26: $z = -x - \frac{1}{2}y$ $\qquad\qquad\qquad$ $z = x^2 - \frac{1}{2}y$

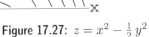

Figure 17.27: $z = x^2 - \frac{1}{2}y^2$ $\qquad$ $z = x^3 - 3xy^2$

2. *Make contour plots of the functions from Exercise 17.0.1 using the program Con-*
tourSlices and compare them to your surface plots.

The functions $z = f[x, y] = f[\mathbf{X}]$ given below are based on simple formulas, yet graphing
them by hand would be difficult because it is not at all easy to see what slices look like
when x, y, or z is fixed. The computer makes graphing them easy.

a) $\quad z = xy^2 + x^3 y$ $\qquad\qquad$ b) $\quad z = \text{Sin}^3(x) + \text{Cos}^4(y)$

c) $\quad z = \text{Sin}(xy)\,\text{Cos}(xy)$ $\qquad\qquad$ d) $\quad z = \sqrt{1 - x^2 - y^2} \cdot \text{Cos}(2\pi xy)$

e) $\quad z = \sqrt{x^2 + \text{Sin}(y^2)}$ $\qquad\qquad$ f) $\quad z = \dfrac{1}{\sqrt{x^2+y^2}}$

g) $\quad z = (2x^2 + 3y^2)e^{-(x^2+y^2)}$ $\qquad\qquad$ h) $\quad z = 2xye^{-(x^2/2+y^2/8)}$

i) $\quad z = \text{ArcTan}[x^2 y^3]$ $\qquad\qquad$ j) $\quad z = \text{Log}[x^2 + y^4]$

Problem 17.6. COMPARISON OF EXPLICIT AND CONTOUR PLOTS ⎯⎯⎯⎯⎯▼
Use the the computer program ExplicitSurfaces and the the computer program Contour-
Plots to plot all of the functions above. Compare the surface plots with their topographical
maps. (We have hidden one wrong contour plot in the program to keep you on your toes.)
▲

Save this work because we will return to these examples in the Chapter 18.

17.6   Explicit, Implicit, & Parametric

Before starting the new parts of calculus in several variables in Chapter 18,
we pause to review cases of the tangent that we have already seen.

Whenever we have an explicit formula, the differential is an explicit formula for the associated tangent. Whenever we have an implicit formula, the differential is an implicit formula for the tangent. Whenever we have a parametric formula, the differential is a parametric formula for the tangent. In other words, calculus always gives the tangent formula as the same type of graph, explicit, implicit, or parametric - but the differential is in local coordinates, (dx, dy) and dt, rather than (x, y) and t.

Type	Equation Nonlinear in x, y	Tangent Linear in dx, dy
Explicit 2-D	$y = f[x]$	$dy = m\, dx$
	$y = \sqrt{25 - x^2}$	$dy = -(4/3)dx$
Implicit 2-D	$F[x, y] = c$	$m\, dx + n\, dy = 0$
	$x^2 + y^2 = 25$	$4\, dx + 3\, dy = 0$
Parametric 2-D	$x = x[t]$	$dx = m\, dt$
	$y = y[t]$	$dy = n\, dt$
	$x = 5\,\mathrm{Cos}[t]$	$dx = -3\, dt$
	$y = 5\,\mathrm{Sin}[t]$	$dy = 4\, dt$

The main idea, however, is that the differential is the formula for the tangent. When we use calculus, there is a symbolic step between the specific tangent line and the nonlinear formula. (In addition, if we wish to express the tangent in x, y-coordinates, then there is another step for that.)

17.6.1 Explicit Graphs and Tangents - 2D

The graph of an explicit function is the set of (x, y)-coordinates that satisfy $y = f[x]$, for example, $y = \sqrt{25 - x^2}$ is shown in Figure 17.28. Its tangent is the explicit line $dy = m\, dx$ where $m = f'[x] = -x/\sqrt{25 - x^2}$ and x is considered fixed in the (dx, dy)-coordinates. For example, this is shown for $x = 4$ in Figure 17.28, where $y = 3$ and $-x/\sqrt{25 - x^2} = -4/3$.

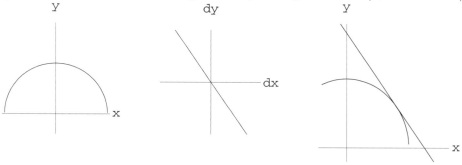

Figure 17.28: $y = \sqrt{25 - x^2}$ and $dy = -4\, dx/3 \iff y - 3 = -4(x - 4)/3$

The graph of an explicit function of two variables is the set of (x, y, z)-coordinates that satisfy $z = f[x, y]$; for example, $z = \text{Cos}[xy]$ is shown in Figure 17.29. The tangent plane is given in local coordinates at the point of tangency by an explicit linear equation in two variables, $dz = m\,dx + n\,dy$. In Chapter 18 we will see how to use rules of differentiation in order to find the values of m and n.

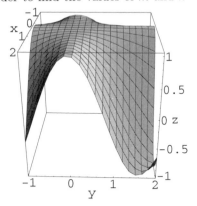
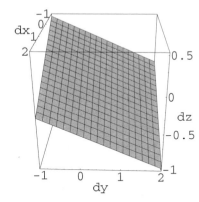

Figure 17.29: $y = \text{Cos}[xy]$ and $dz = -0.265\,dx - 0.229\,dy$

17.6.2 Implicit Equations

An equation such as the formula for the circle of radius 5,

$$x^2 + y^2 = 5^2$$

does not give an explicit way to compute y from x, but it is implicit in the formula that only certain values of y work for a given x. In general, an implicit formula $f[x, y] = c$ is associated with the set of (x, y)-coordinates that satisfy the equation.

An implicit linear equation in the two variables dx and dy has the form

$$m\,dx + n\,dy = k$$

for constants m, n, and k.

Implicit differentiation of the circle example gives

$$x^2 + y^2 = 25 \qquad \Leftrightarrow \qquad 2x\,dx + 2y\,dy = 0$$

When $(x, y) = (4, 3)$, we have the particular implicit tangent $4\,dx + 3\,dy = 0$ with normal vector $(4, 3)$. In other words, the tangent is perpendicular to the radius vector, which is clear geometrically without calculus. If we write this in terms of the original (x, y)-coordinates, we replace dx by $(x-4)$ and replace dy by $(y-3)$, obtaining the line $4 \cdot (x-4) + 3 \cdot (y-3) = 0$. In vector notation, this is the line perpendicular to $(4, 3)$ through the point $(4, 3)$ shown in Figure 17.30.

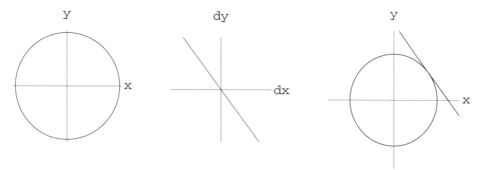

Figure 17.30: $x^2 + y^2 = 5$ and $4\,dx + 3\,dy = 0$

The contour plot of a function $f[x, y]$ is a collection of 2-D implicit curves, $f[x, y] = c$, for constants c. In the next chapter, we will use the linear gradient vector $\mathbf{G} = G(m, n)$ for the implicit tangent line $m\,dx + n\,dy = 0$ to understand the gradient of a nonlinear function.

17.6.3 Parametric Curves

A parametric circle of radius 5 is given by the following equations, shown with their symbolic differentials and the differentials at a specific point.

$$x = 5\,\mathrm{Cos}[t] \qquad dx = -5\,\mathrm{Sin}[t]\,dt \qquad dx = -3\,dt$$
$$y = 5\,\mathrm{Sin}[t] \qquad dy = 5\,\mathrm{Cos}[t]\,dt \qquad dy = 4\,dt$$

If we take $t = \mathrm{ArcTan}[3/4]$, then $\mathrm{Sin}[t] = 3/5$ and $\mathrm{Cos}[t] = 4/5$, so $x = 4$, $y = 3$, $dx = -3\,dt$, and $dy = 4\,dt$. The specific differentials are parametric equations for the tangent line to the circle at $(4, 3) = (x, y)$. They say in vector terms that the direction of the tangent line is the vector $(-3, 4)$. This can be viewed as the velocity vector if t represents time and x and y are positions. If we want to express the tangent in the same (x, y)-coordinates, we would replace dx by $(x - 4)$ and dy by $(y - 3)$. The parametric vector equations for the line in direction $(-3, 4)$ through the point $(4, 3)$ are $x = 4 - 3\,dt$ and $dy = 3 + 4\,dt$.

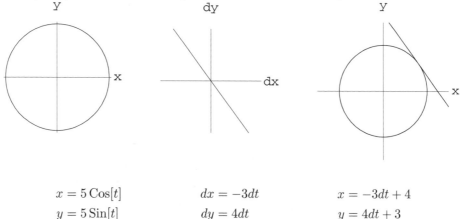

$$x = 5\,\mathrm{Cos}[t] \qquad dx = -3dt \qquad x = -3dt + 4$$
$$y = 5\,\mathrm{Sin}[t] \qquad dy = 4dt \qquad y = 4dt + 3$$

It is not necessary to replace the parameter dt, since it is not plotted.

Exercise set 17.6

1. *Use the computer to plot the curve* $y = \frac{3}{4}\sqrt{25 - x^2}$. *Find the symbolic differential* $dy = f'[x]\, dx$, *and use it to find the equation of the tangent to this curve when* $x = 4$. *Finally, use the computer to plot both curves.*

2. *Use the computer to plot the curve* $\frac{x^2}{16} + \frac{y^2}{9} = \frac{25}{16}$. *Find the symbolic differential* $m\, dx + n\, dy = 0$, *and use it to find the equation of the tangent to this curve when* $x = 4$. *Finally, use the computer to plot both curves.*

3. *Use the computer to plot the parametric curve* $x = 5\, \mathrm{Cos}[t]$, $y = \frac{15}{4}\, \mathrm{Sin}[t]$. *Find the symbolic differentials* $dx = a\, dt$ *and* $dy = b\, dt$, *and use them to find the equation of the tangent to this curve when* $t = \mathrm{ArcTan}[4/3]$. *Finally, use the computer to plot both curves.*

17.7   Review of Lines, Curves, & Planes

Expand your own 'lexicon' of geometric-algebraic translations from the one you made in Chapter 17 to include the algebra-geometry links of this chapter. This should be your personal list of fundamental facts that you refer to when translating between algebra and geometry. Test the usefulness of your lexicon by consciously using it when you solve the following review problems.

Exercise set 17.7

1. Lines and Planes

   (a) *A line through the point* $\begin{bmatrix} 1 \\ 2 \\ 3 \end{bmatrix}$ *intersects the plane* $3\,x + 2\,y + z = 10$ *at right angles.*
       Give equations for this line.

   (b) *At what point does this line intersect the plane? Give your reason.*

   (c) *Is the vector* $\begin{bmatrix} 1 \\ 2 \\ 3 \end{bmatrix}$ *perpendicular to the line? Give your reason.*

   (d) *Is the vector* $\begin{bmatrix} 1 \\ 2 \\ 3 \end{bmatrix}$ *perpendicular to the plane? Give your reason.*

2. *Write the vector* $\mathbf{A}$ *as a linear combination* $\mathbf{A} = b\mathbf{B} + c\mathbf{C}$; *that is, find* b *and* c.
   $\mathbf{A} = \begin{bmatrix} 2 \\ -1 \\ -1 \end{bmatrix}$, $\mathbf{B} = \begin{bmatrix} 0 \\ 2 \\ 1 \end{bmatrix}$, $\mathbf{C} = \begin{bmatrix} -10 \\ 3 \\ 4 \end{bmatrix}$ *Draw the parallelogram with a vertex at the origin,*
   one side the arrow $b\mathbf{B}$ *and another side the arrow* $c\mathbf{C}$. *Show the vector* $\mathbf{A}$ *on your figure.*

3. *Sketch and find the area a of the triangle with vertices*

$$\mathbf{A} = \begin{bmatrix} 1 \\ 1 \\ 1 \end{bmatrix} \quad , \quad \mathbf{B} = \begin{bmatrix} 2 \\ 3 \\ 5 \end{bmatrix} \quad , \quad \mathbf{C} = \begin{bmatrix} -1 \\ 3 \\ 1 \end{bmatrix}$$

4. *Show that the tips of the vectors* $\mathbf{A} = \begin{bmatrix} 3 \\ 2 \\ 0 \end{bmatrix}$, $\mathbf{B} = \begin{bmatrix} 1 \\ 1 \\ -1 \end{bmatrix}$, $\mathbf{C} = \begin{bmatrix} 5 \\ 3 \\ 1 \end{bmatrix}$ *are colinear. Find*
an equation for the line on which they lie and show that all three satisfy the equation.

5. *Show that the tips of the vectors* $\mathbf{A} = \begin{bmatrix} 0 \\ 2 \\ 1 \end{bmatrix}$, $\mathbf{B} = \begin{bmatrix} 1 \\ -1 \\ -1 \end{bmatrix}$, $\mathbf{C} = \begin{bmatrix} 13 \\ -1 \\ 5 \end{bmatrix}$, $\mathbf{D} = \begin{bmatrix} 14 \\ 2 \\ 8 \end{bmatrix}$ *are*
coplanar. Find the equation of the plane of three, and show that all four satisfy that
equation.

6. Normal and Tangential Components

   (a) *Suppose we have a plane given by*

$$z = m\,x + n\,y$$

   For example, consider $m = 3$ *and* $n = 2$. *Give a 3-D vector* $\mathbf{N}$ *normal (or*
   perpendicular) to the plane with an upward z-component.

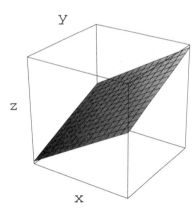

Figure 17.31: $z = 3\,x + 2\,y$

   (b) *Also suppose that a force* $\mathbf{F}$ *acts on the plane. For example, consider* $\mathbf{F} = \begin{bmatrix} 1 \\ 2 \\ 3 \end{bmatrix}$.

   Compute the projection of $\mathbf{F}$ *in the direction of the normal* $\mathbf{N}$.

$$\mathbf{F}_N = \mathrm{Proj}_{\mathbf{N}}[\mathbf{F}]$$

   What is the magnitude of the force acting against the plane?

   (c) *Sketch* $\mathbf{N}$, $\mathbf{F}$, $\mathbf{F}_N$, *and* $\mathbf{F} - \mathbf{F}_N$ *on Figure 17.31. How much force acts along or*
   tangent to the plane?

7. *Find the angle at which the line*

$$x = 1 - t$$
$$y = 2t - 3$$
$$z = 5$$

intersects the plane $z = 3x + 2y + 7$.

Now use calculus and geometry together.

8. Curves and Surfaces

(a) *The curve given parametrically by*

$$\mathbf{X}[t]: \qquad \begin{aligned} x &= t \\ y &= 2\sqrt{t} \\ z &= e^{-t} \end{aligned}$$

intersects the explicit surface

$$z = e^{x-y} \, \mathrm{Cos}[\pi x \, y]$$

at the point $\mathbf{P} = P(1, 2, 1/e)$. *Verify this.*

(b) *The velocity vector tangent to the curve at the time of intersection is* $\begin{bmatrix} 1 \\ 1 \\ -1/e \end{bmatrix}$.

Verify this.

(c) *The equation of the plane tangent to the surface at the point of intersection is* $dz = \frac{1}{e} \, dx - \frac{1}{e} \, dy$ *in local coordinates at* $\mathbf{P} = P(1, 2, 1/e)$. *(We will verify this in Chapter 18. Local coordinates mean* $dx = x - 1$, $dy = y - 2$ *and* $dz = z - 1/e$.*)*

Show that the vector $\begin{bmatrix} 1/e \\ -1/e \\ -1 \end{bmatrix}$ *is perpendicular to the plane.*

(d) *Find the angle that the curve makes with the surface at this point of intersection.*

9. *Show that the lines below never intersect by trying to solve for values of the parameters that would make* x, y, *and* z *equal.*

$x = 1 - t$	$x = s - 1$
$y = 2t - 3$	$y = 3 - 2s$
$z = 5$	$z = 1 + s$

Find the distance between these lines.

The parameter in a parametric line can represent time. Two airplanes flying along intersecting lines only collide if the point of intersection occurs at the same time.

10. Same Place vs. Same Time

(a) *Airplane* $\mathbf{X}$ *leaves the runway position* $(0, 0, 0)$ *at time* $t = 0$ *with a velocity of* $U(300, 450, 15)$. *What is the vector parametric equation in terms of* t *for the line along which it travels?* $\mathbf{X} =$?

(b) *Airplane* **Y**, *flying nearby, is at position* $P(100, 0, 5)$ *at time* $t = 0$ *when the first plane takes off. It is traveling with a velocity of* $V(0, 500, 0)$. *What is the vector parametric equation in terms of* t *for the line along which it travels?* **Y** =?

(c) *Show that the paths of* **X** *and* **Y** *cross but that the planes do not collide.*

(d) *Airplane* **Z**, *flying nearby, is at position* $Q(40, 44, 2)$ *at time* $t = 0$ *when the first plane takes off. It is traveling with a velocity of* $W(0, 120, 0)$. *What is the vector parametric equation in terms of* t *for the line along which it travels?* **Z** =?

(e) *Show that* **X** *and* **Z** *collide if they both maintain their courses.*

The flight deck of an aircraft carrier is inclined $10°$ off the axis of the ship. If you proceed with velocity vector **V** in still air, the wind will not blow straight down the flight deck because it gives the apparent wind of $-$**V**. However, if there is a real wind blowing with velocity **W**, then we can set a course so that the apparent wind does blow straight down the flight deck.

Problem 17.7. ⎯⎯⎯⎯⎯⎯⎯⎯⎯⎯⎯⎯⎯⎯⎯⎯⎯⎯▼

Find the heading (unit vector in the direction of your velocity **V***) and speed* $(|$**V**$|)$ *to proceed in a wind blowing at velocity* **W** *in order for the apparent wind to blow down the flight deck at 30 knots.*

HINTS: One approach is to choose coordinates so **V** $= V(0, v)$, *and wind is broken into normal and tangential components to the ship's axis,* **W** $= W(n, t)$. *The "wind" from the ship's motion is* $-$**V**, *so we want* **W** $-$ **V** *to be a 30-knot vector inclined* $10°$ *off the axis of the ship.*

What conditions does **W** *have to satisfy if the maximum speed of the ship is 30 knots?*

⎯⎯⎯⎯⎯⎯⎯⎯⎯⎯⎯⎯⎯⎯⎯⎯⎯⎯⎯⎯⎯⎯⎯⎯⎯⎯⎯⎯⎯⎯⎯⎯⎯⎯⎯▲

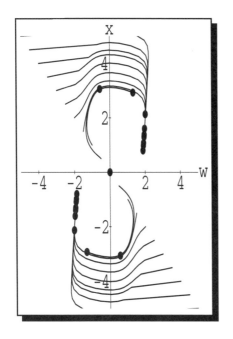

Part 4

Differentiation in Several Variables

Differentiation in Several Variables

This chapter uses the "microscope principle" to analyze nonlinear functions of several variables.

The graphical and symbolic theorems of one-variable differentiation say:

If we can compute the derivative $f'[x]$ of a function $f[x]$ using the rules of Chapter 6, then a sufficiently magnified view of the graph $y = f[x]$ appears linear. At high enough power, the error ε in Figure 18.1 is below the resolution of our microscope.

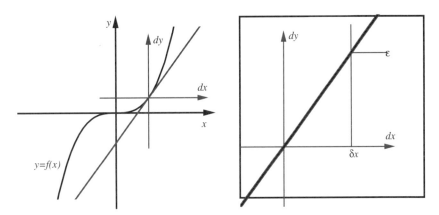

Figure 18.1: $y = f[x]$ and $dy = f'[x] \cdot dx$ in a powerful microscope

The condition of "tangency" is expressed symbolically by the formula that says the nonlinear change is a linear term plus something small compared to the change:

$$f[x + \delta x] - f[x] = f'[x] \cdot \delta x + \varepsilon \cdot \delta x$$

with $\varepsilon \approx 0$, whenever the perturbation is small, $\delta x \approx 0$, and x lies in an interval $[\alpha, \beta]$ where both $f[x]$ and $f'[x]$ are defined.

The microscopic error formula expresses the nonlinear change, $f[x + dx] - f[x]$, in terms of a "local variable" dx, with x fixed. The linear term in the local variable is called the differential,

$$dy = f'[x] \cdot dx \qquad \text{or} \qquad dy = m \cdot dx$$

in (dx, dy)-coordinates (with x fixed). We studied the geometric meaning of ε in the programs **Micro1D** or **Zoom**. See Chapter 6 and the Mathematical Background for details.

In this chapter we compare the change in a two variable nonlinear function with a linear function. You need to know some facts about linear functions. We review these facts from Chapter 17 using the variables (dx, dy, dz) because $z = f[x, y]$ will be needed for the nonlinear function. Here is a summary of the linear part of the last chapter:

Theorem 18.1. *The Linear Gradient Vector and Slope of a Plane*
The explicit plane in 3-space with equation

$$dz = m\,dx + n\,dy \quad \text{has gradient} \quad \mathbf{G} = \begin{bmatrix} m \\ n \end{bmatrix} \quad \text{and normal vector} \quad \mathbf{N} = \begin{bmatrix} -m \\ -n \\ +1 \end{bmatrix}$$

The slope of a line in this plane is $\langle \mathbf{G} \bullet \mathbf{U} \rangle$ when the horizontal projection of the line points in the x-y unit vector direction $\mathbf{U}$. The steepest slope of a line in this plane is $|\mathbf{G}|$ when the horizontal projection of the line points in the x-y-direction of the gradient vector $\mathbf{G}$.

The statement in the theorem about steepest slope can also be expressed as, "The norm of the gradient $|\mathbf{G}| = \sqrt{m^2 + n^2}$ is the maximal rate of growth of the function, and this rate is attained when the change vector $\mathbf{dX} = \begin{bmatrix} dx \\ dy \end{bmatrix}$ is in the direction of $\mathbf{G}$, $\mathbf{dX} = c\mathbf{G}$."

Theorem 18.2. *The Linear Gradient and Contours*
The contour plot in 2-space of the linear function $L[dx, dy] = m\,dx + n\,dy$ consists of a family of lines perpendicular to the gradient vector $\mathbf{G} = \begin{bmatrix} m \\ n \end{bmatrix}$ with the contours $L[dx, dy] = k_1$ and $L[dx, dy] = k_2$ spaced a distance $\frac{|k_2 - k_1|}{|\mathbf{G}|}$ apart.

18.1   Partial and Total Derivatives

The definition of the (total) derivative of a function of two real variables says a change in a smooth function is a two variable linear term plus an error that is small with respect to the change. The coefficients of the linear term are the partial derivatives, $\frac{\partial f}{\partial x}$ and $\frac{\partial f}{\partial y}$.

The computer programs **Micro3D** and **Zoom3D** make animations of both explicit graphs and contour graphs with their magnifications. This lets you see the linear approximation under magnification. The "movies" are great. Try them yourself.

Definition 18.3. *Partial Derivatives*
Let $z = f[x, y] = f[\mathbf{X}]$ be a function of two real variables. The derivative of the function $F[x] = f[x, y]$ (treating y as a constant) is called the partial derivative of f with respect to x, $F'[x] = \frac{\partial f}{\partial x}[x, y]$, and the derivative of the function $H[y] = f[x, y]$, (treating x as a constant) is called the partial derivative of f with respect to y, $H'[y] = \frac{\partial f}{\partial y}[x, y]$.

The partial derivatives, $\frac{\partial f}{\partial x}$ and $\frac{\partial f}{\partial y}$ are computed using the rules we learned in one variable by treating y as a constant or parameter when differentiating with respect to x and treating x as a parameter when differentiating with respect to y.

Example 18.1. *Partial Derivatives*

For example, if

$$z = f[x, y] = x^3 + y^5$$

then

$$\frac{\partial f}{\partial x}[x, y] = 3\, x^2$$

just treating the $3\,y^2$ term as a constant when differentiating with respect to x. Also,

$$\frac{\partial f}{\partial y}[x, y] = 5\; y^4$$

treating the x^2 term as a constant when differentiating with respect to y. We compute many examples in the next section.

Computation of partial derivatives is an extension of what we have learned in one-variable calculus that only requires us to apply those formulas with different letters.

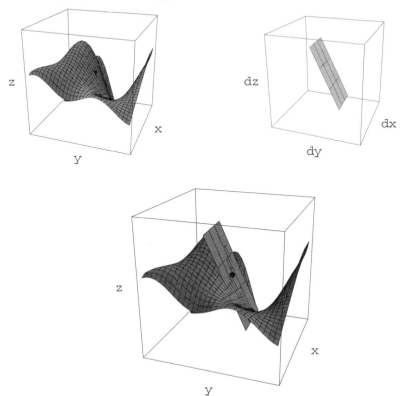

Figure 18.2: The surface and tangent

Procedure 18.4. Tangents in Two Variables

Suppose we are given a function $z = f[x, y]$ and we fix a point (x, y, z), with $z = f[x, y]$. Define parallel local coordinates (dx, dy, dz) with an origin at the fixed (x, y, z). The linear function in local (dx, dy, dz)-coordinates given by

$$dz = m\, dx + n\, dy$$

is the graph of the plane tangent to the surface graph of the nonlinear function

$$z = f[x, y]$$

at (x, y, z), shown in Figure 18.2, provided

$$m = \frac{\partial f}{\partial x}[x, y] \qquad \text{and} \qquad n = \frac{\partial f}{\partial y}[x, y]$$

and the partial derivatives can be computed by the rules of Chapter 6, yielding formulas valid in a rectangle around (x, y).

The condition of "tangency" is the approximation

$$f[x + \delta x, y + \delta y] - f[x, y] = \frac{\partial f}{\partial x}[x, y] \cdot \delta x + \frac{\partial f}{\partial y}[x, y] \cdot \delta y + \varepsilon \cdot \sqrt{\delta x^2 + \delta y^2}$$

where the magnified error $\varepsilon \approx 0$ is small when the changes are small, $\delta x \approx 0$ and $\delta y \approx 0$.

We may think of a function $f[x, y]$ as a function $f[\mathbf{X}]$ of a 2-D vector input. Let $\mathbf{X} = \begin{bmatrix} x \\ y \end{bmatrix}$ and $\delta \mathbf{X} = \begin{bmatrix} \delta x \\ \delta y \end{bmatrix}$, so the approximation formula above becomes

$$f[\mathbf{X} + \delta \mathbf{X}] - f[\mathbf{X}] = \langle \mathbf{G} \bullet \delta \mathbf{X} \rangle + \varepsilon \, |\delta \mathbf{X}|$$

with $\mathbf{G} = \begin{bmatrix} m \\ n \end{bmatrix} = \begin{bmatrix} \frac{\partial f}{\partial x}[\mathbf{X}] \\ \frac{\partial f}{\partial y}[\mathbf{X}] \end{bmatrix}$.

Definition 18.5. *Gradient*

The gradient of f at $\mathbf{X}$ is the vector $\mathbf{G} = \nabla f[\mathbf{X}] = \begin{bmatrix} \frac{\partial f}{\partial x}[x, y] \\ \frac{\partial f}{\partial y}[x, y] \end{bmatrix}$

The vector ∇f is pronounced "nabla f" (or sometimes "grad f"). The gradient vector, $\nabla f[\mathbf{X}]$ tells us the direction to move in the $\mathbf{X} = \begin{bmatrix} x \\ y \end{bmatrix}$-plane in order to make f increase fastest. This is only the geometrical interpretation; physically, it can be interpreted as the direction opposite heat flow, as an electrical or magnetic field, and in many other ways. We will see that its vector nature makes it quite useful.

The term

$$L[\delta x, \delta y] = \langle \mathbf{G} \bullet \delta \mathbf{X} \rangle = \frac{\partial f}{\partial x}[x, y] \cdot \delta x + \frac{\partial f}{\partial y}[x, y] \cdot \delta y$$

from the tangent approximation formula is a linear function of the variables δx and δy when x and y are fixed. If $f[x, y]$ is linear, recall that

$$f[x + \delta x, y + \delta y] - f[x, y] = f[x + \delta x - x, y + \delta y - y] = f[\delta x, \delta y] = L[\delta x, \delta y]$$

The term $\varepsilon \, |\delta \mathbf{X}| = \varepsilon \cdot \sqrt{\delta x^2 + \delta y^2}$ in the nonlinear approximation formula represents the deviation from linear change near $\mathbf{X}$.

Definition 18.6. *Total Differential and Derivative*

Consider a function $z = f[x, y]$ whose partial derivatives can be computed with the rules of Chapter 6 one variable at a time. The total differential of f at (x, y, z) is the linear equation in (dx, dy, dz):

$$dz = \frac{\partial f}{\partial x}[x, y] \cdot dx + \frac{\partial f}{\partial y}[x, y] \cdot dy$$

The linear function in the local variables (dx, dy),

$$L[dx, dy] = \frac{\partial f}{\partial x}[x, y] \cdot dx + \frac{\partial f}{\partial y}[x, y] \cdot dy$$

is the total derivative of f at the fixed input vector $\mathbf{X} = (x, y)$.

We actually used total differentials earlier in the course in implicit differentiation. For example, if

$$z = f[x, y] = x^3 + y^5$$

then

$$dz = 3\,x^2\,dx + 5\,y^4\,dy$$

We will compute more examples in the next section.

The total differential can be expressed by the dot product of the gradient and the perturbation vector,

$$dz = \langle \nabla f[\mathbf{X}] \bullet d\mathbf{X} \rangle$$

The vector form of the 3-D microscope approximation helps us to interpret the error term in the manner of the 2-D microscope. Magnification makes the length of the perturbation vector $\delta \mathbf{X} = (\delta x, \delta y)$ given by $|\delta \mathbf{X}| = \sqrt{\delta x^2 + \delta y^2}$ appear unit size. The condition $\varepsilon \approx 0$ means that the difference between the linear and the nonlinear graphs appears small after magnification.

Theorem 18.7. *The Increment Principle*

The function $f[\mathbf{X}]$ is smooth if and only if sufficiently magnified views of its explicit surface graph appear indistinguishable from linear planes; that is, $\varepsilon \approx 0$ when $|\delta \mathbf{X}| \approx 0$ in the microscopic approximation, $f[\mathbf{X} + \delta \mathbf{X}] - f[\mathbf{X}] = L[\delta \mathbf{X}] + \varepsilon \, |\delta \mathbf{X}|$.

When we magnify by $1/|\delta \mathbf{X}|$ to make the change appear to be unit length, we see ε as the magnified error (the actual error is $\varepsilon \cdot \sqrt{\delta x^2 + \delta y^2}$). When ε is small enough, we do not notice it. It is clear why we call this local linearity - under a microscope, or locally, things appear linear.

The beauty of the symbolic rules of calculus is the following rigorous version of Procedure 18.4.

Theorem 18.8. *Defined Formulas Imply Approximation*

Suppose that $z = f[x, y]$ is given by formulas and that the partial derivatives $\frac{\partial f}{\partial x}[x, y]$ and $\frac{\partial f}{\partial y}[x, y]$ can be computed using the rules of Chapter 6 (Specific Functions, Superposition Rule, Product Rule, Chain Rule) holding one variable at a time fixed. If the resulting three formulas $f[x, y]$, $\frac{\partial f}{\partial x}[x, y]$, $\frac{\partial f}{\partial y}[x, y]$, are all defined in a compact box, $\alpha \leq x \leq \beta$, $\gamma \leq y \leq \eta$, then

$$f[x + \delta x, y + \delta y] - f[x, y] = \frac{\partial f}{\partial x}[x, y] \cdot \delta x + \frac{\partial f}{\partial y}[x, y] \cdot \delta y + \varepsilon \cdot \sqrt{\delta x^2 + \delta y^2}$$

with ε uniformly small in the x-box for sufficiently small Δx.

We also formulate the geometric meaning of magnifications of the contour graphs later in this chapter and in the **Micro3D** program.

Exercise set 18.1

---

1. *Given that the function $z = f[x, y] = x^2\, y^3$ has partial derivatives, $\frac{\partial f}{\partial x}[x, y] = 2\, x\, y^3$ and $\frac{\partial f}{\partial y}[x, y] = 3\, x^2\, y^2$, give the gradient vector at a general point (x, y),*

$$\nabla f[\mathbf{X}] = \begin{bmatrix} ? \\ ?? \end{bmatrix}$$

Give the gradient at the specific point $(x, y) = (1, 1)$:

$$\nabla f[\mathbf{X}] = \begin{bmatrix} ? \\ ?? \end{bmatrix}$$

2. *Given that the function $z = f[x, y] = x^2\, y^3$ has partial derivatives, $\frac{\partial f}{\partial x}[x, y] = 2\, x\, y^3$ and $\frac{\partial f}{\partial y}[x, y] = 3\, x^2\, y^2$, give the total differential at a general point (x, y),*

$$dz = ???$$

Give the total differential at the specific point $(x, y) = (1, 1)$:

$$dz = ???$$

18.2   Partial Differentiation Examples

This section shows how to compute partial and total derivatives. Computation of $\frac{\partial f}{\partial x}$ and $\frac{\partial f}{\partial y}$ is only an extension of what we learned in Chapter 6 for one variable to different letters. The new difficulty that arises is in understanding where the formulas are valid.

Treating all letters equally is not easy because we all get used to "special" letters - especially x for the independent variable. We must free ourselves of this prejudice.

Example 18.2. $z = f[x, y] = x^2 + y^2/3$

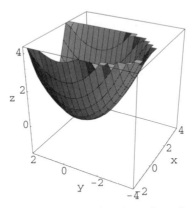

Figure 18.3: $z = f[x, y] = x^2 + y^2/3$

Then

$$\frac{\partial f}{\partial x}[x, y] = 2\,x$$

just treating the $3\,y^2$ term as a constant when differentiating with respect to x. Also

$$\frac{\partial f}{\partial y}[x, y] = 2\,y/3 = \frac{2}{3}\,y$$

treating the x^2 term as a constant when differentiating with respect to y. Putting these computations together, we obtain the total differential

$$dz = \frac{\partial f}{\partial x}\,dx + \frac{\partial f}{\partial y}\,dy = 2\,x\,dx + \frac{2}{3}\,y\,dy$$

The total differential for the magnified view at $(x, y) = (1, -1)$ is

$$dz = 2\,dx - \frac{2}{3}\,dy$$

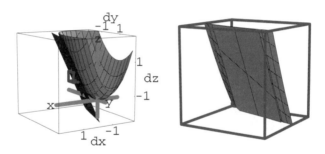

Figure 18.4: Zooming toward $(1, -1)$ on the surface $z = x^2 + 3\,y^2$

Example 18.3. $z = f[x, y] = x^2 y^3$

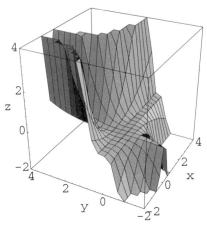

Figure 18.5: $z = f[x, y] = x^2 y^3$

Then

$$\frac{\partial f}{\partial x}[x, y] = 2xy^3$$

just treating the y^3 term as a constant when differentiating with respect to x. Also

$$\frac{\partial f}{\partial y}[x, y] = 3x^2 y^2$$

treating the x^2 term as a constant when differentiating with respect to y. Putting these computations together, we obtain the total differential

$$dz = \frac{\partial f}{\partial x}\, dx + \frac{\partial f}{\partial y}\, dy = 2xy^3\, dx + 3x^2 y^2\, dy$$

Now, consider the graph near $(1, 1)$ and a magnified view. The total differential for the magnified view has $(x, y) = (1, 1)$:

$$dz = 2\, dx + 3\, dy$$

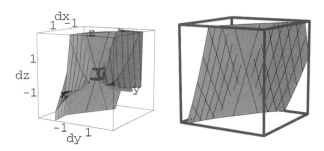

Figure 18.6: Zooming toward $(1, 1)$ on the surface $z = x^2 y^3$

Example 18.4. $z = f[x, y] = \mathrm{Cos}[x\ y]$

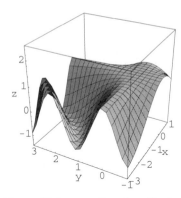

Figure 18.7: $z = f[x, y] = \mathrm{Cos}[xy]$

Then

$$\frac{\partial f}{\partial x}[x, y] = -y \cdot \mathrm{Sin}[x\ y]$$

treating the y term as a constant when differentiating with respect to x. This is the Chain Rule, $z = \mathrm{Cos}[u]$ and $u = xy$, so $\frac{\partial z}{\partial u} = -\mathrm{Sin}[u]$ and $\frac{\partial u}{\partial x} = y$. The Chain Rule yields

$$\frac{\partial z}{\partial x} = \frac{\partial z}{\partial u} \cdot \frac{\partial u}{\partial x} = -\mathrm{Sin}[u] \cdot y = -y \cdot \mathrm{Sin}[x\ y]$$

Also,

$$\frac{\partial f}{\partial y}[x, y] = -x \cdot \mathrm{Sin}[x\ y]$$

treating the x term as a constant when differentiating with respect to y. Putting these computations together, we obtain the total differential

$$dz = \frac{\partial f}{\partial x}\ dx + \frac{\partial f}{\partial y}\ dy = -y \cdot \mathrm{Sin}[x\ y]\ dx - x \cdot \mathrm{Sin}[x\ y]\ dy$$

The total differential at $(x, y) = (-1, 1)$ is

$$dz = \mathrm{Sin}[1]\ dx - \mathrm{Sin}[1]\ dy \approx 0.8415\ dx - 0.8415\ dy$$

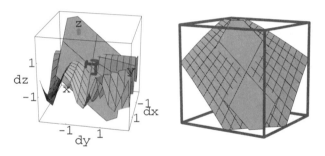

Figure 18.8: Zooming toward $(-1, 1)$ on the surface $z = \mathrm{Cos}[x\ y]$

Example 18.5. $z = f[x, y] = \sqrt{y^2 - x^3}$

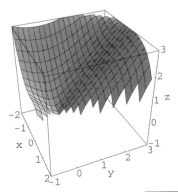

Figure 18.9: $z = f[x, y] = \sqrt{y^2 - x^3}$

The function $f[x, y]$ is defined when $y^2 - x^3 \geq 0$. We find the partial derivatives by using the Chain Rule.

$$z = u^{\frac{1}{2}} \qquad\qquad\qquad u = y^2 - x^3$$

$$\frac{\partial z}{\partial u} = \frac{1}{2} u^{-\frac{1}{2}} \qquad\qquad\qquad \frac{\partial u}{\partial x} = -3x^2$$

and

$$\frac{\partial u}{\partial y} = 2y$$

so

$$\frac{\partial z}{\partial x} = \frac{\partial z}{\partial u}\frac{\partial u}{\partial x} = \frac{1}{2} u^{-\frac{1}{2}} \cdot -3x^2 \qquad\qquad \frac{\partial z}{\partial y} = \frac{\partial z}{\partial u}\frac{\partial u}{\partial y} = \frac{1}{2} u^{-\frac{1}{2}} \cdot 2y$$

$$\frac{\partial z}{\partial x} = \frac{-3x^2}{2\sqrt{y^2 - x^3}} \qquad\qquad\qquad \frac{\partial z}{\partial y} = \frac{y}{\sqrt{y^2 - x^3}}$$

This makes the total differential

$$dz = \frac{\partial f}{\partial x}\, dx + \frac{\partial f}{\partial y}\, dy = \frac{-3x^2}{2\sqrt{y^2 - x^3}}\, dx + \frac{y}{\sqrt{y^2 - x^3}}\, dy$$

but this formula is not valid on the curve $y^2 = x^3$ nor in the region where the original function was undefined, $y^2 < x^3$. This is the region in the first quadrant below the explicit curve $y = x^{3/2}$, shown in Figure 18.10, plus all of the fourth quadrant. (You should sketch it in the x-y-plane of the 3-D picture.)

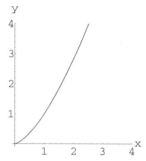

Figure 18.10: $y^2 = x^3$

Now, consider the graph near $(0,1)$ and a magnified view. The total differential for the magnified view is

$$dz = 0 \ dx + 1 \ dy$$

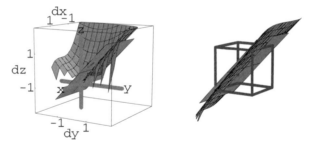

Figure 18.11: Zooming toward $(0,1)$ on the surface $z = \sqrt{y^2 - x^3}$

Example 18.6. $z = \mathrm{Tan}[(x^2 + y^2)]$

We may express the function as

$$f[x, y] = \mathrm{Tan}[(x^2 + y^2)] = \frac{\mathrm{Sin}[(x^2 + y^2)]}{\mathrm{Cos}[(x^2 + y^2)]} = \mathrm{Sin}[(x^2 + y^2)] \cdot (\mathrm{Cos}[(x^2 + y^2)])^{-1}$$

This function is undefined if cosine is zero - in other words, on the circles $x^2 + y^2 = \pi\frac{1}{2}$, $= \pi\frac{3}{2}$, $= \pi\frac{5}{2}$, ...

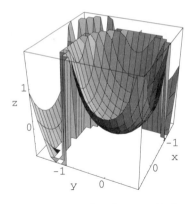

Figure 18.12: $z = f[x, y] = \text{Tan}[(x^2 + y^2)]$

The partial derivative computations require another use of the Chain Rule along with the Product Rule as follows:

$$z = \text{Sin}[u][\text{Cos}[u]]^{-1} \qquad\qquad\qquad\qquad u = (x^2 + y^2)$$

so

$$\frac{\partial z}{\partial u} = \text{Cos}[u][\text{Cos}[u]]^{-1} + \text{Sin}[u] \cdot (-1)[\text{Cos}[u]]^{-2} \cdot (-\text{Sin}[u])$$

$$= \frac{\text{Cos}^2(u) + \text{Sin}^2(u)}{\text{Cos}^2(u)} = \frac{1}{\text{Cos}^2(u)}$$

and

$$\frac{\partial u}{\partial x} = 2x \qquad\qquad\qquad\qquad\qquad\qquad \frac{\partial u}{\partial y} = 2y$$

so

$$\frac{\partial z}{\partial x} = \frac{\partial z}{\partial u}\frac{\partial u}{\partial x} = \frac{2x}{\text{Cos}^2(u)} \qquad\qquad \frac{\partial u}{\partial y} = \frac{\partial z}{\partial u}\frac{\partial u}{\partial y} = \frac{2y}{\text{Cos}^2(u)}$$

$$\frac{\partial z}{\partial x} = \frac{2x}{\text{Cos}^2[(x^2 + y^2)]} \qquad\qquad \frac{\partial u}{\partial y} = \frac{2y}{\text{Cos}^2[(x^2 + y^2)]}$$

The total derivative is thus

$$dz = \frac{\partial f}{\partial x}\,dx + \frac{\partial f}{\partial y}\,dy = \frac{2x}{\text{Cos}^2[(x^2 + y^2)]}\,dx + \frac{2y}{\text{Cos}^2[(x^2 + y^2)]}\,dy$$

which is valid for the same inputs as the function, $x^2 + y^2 \neq \pi(\frac{1}{2} + k)$. (Note that part of the computation above shows $\frac{d\,\text{Tan}[u]}{du} = \frac{1}{\text{Cos}^2(u)}$. If you already knew that, you could have shortened the calculation.)

Now, consider the graph near $(-1/2, 1/2)$ and a magnified view. The total differential for the magnified view is

$$dz \approx -1.298\,dx + 1.298\,dy$$

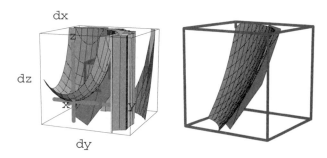

Figure 18.13: Zooming toward $(-1/2, 1/2)$ on the surface $z = \text{Tan}[x^2 + y^2]$

Exercise set 18.2

1. *The functions* $z = f[x, y]$ *or* $z = f[\mathbf{X}]$ *are given below. In each case, compute* $\frac{\partial f}{\partial x}$, $\frac{\partial f}{\partial y}$, $\nabla f[\mathbf{X}]$, *and the total differential. For which values of the vector* $\mathbf{X}$ *are your answers valid?*

a) $z = xy^2 + x^3 y$

b) $z = \text{Sin}^3(x) + \text{Cos}^4(y)$

c) $z = \text{Sin}[x\, y]\, \text{Cos}[x\, y]$

d) $z = \sqrt{1 - x^2 - y^2} \cdot \text{Cos}(2\pi xy)$

e) $z = \sqrt{x^2 + \text{Sin}(y^2)}$

f) $z = \frac{1}{\sqrt{x^2 + y^2}}$

g) $z = (2x^2 + 3y^2) e^{-(x^2 + y^2)}$

h) $z = 2xy e^{-(x^2/2 + y^2/8)}$

i) $z = \text{ArcTan}[x^2 y^3]$

j) $z = \text{Log}[x^2 + y^4]$

2. *Use the program* **PartialD** *to check your computations in the previous exercise.*

3. *For each of the functions above, find one* X *or* (x, y) *point where the graph has a horizontal tangent plane, if there is one. You may use your the computer homework from Exercise 17.6 to help see where to look for the solution of the appropriate condition on the partial derivatives you computed above.*

4. R *un the* **Micro3D** *and* **Zoom3D** *programs. They have a built-in functions and produce microscopic "zooming" animations in explicit surfaces as well as contour plots.*
Modify the function in the **Micro3D** *and* **Zoom3D** *programs to show microscopic views of some of the functions of the previous exercise. We especially recommend that you magnify* $z = \sqrt{x^2 + \text{Sin}(y^2)}$ *at* $(x, y) = (0, 0)$ *and explain what you see.*

5. Differentiation with Different Letters
Calculate the partial derivatives of the following:

a) $f[x, y] = x^2 + y^2$

b) $f[x, y] = \left(\frac{x}{5}\right)^2 + \left(\frac{y}{3}\right)^2$

c) $g[u, v] = \left(\frac{u}{5}\right)^2 + \left(\frac{v}{3}\right)^2$

d) $h[x, y] = x^2\, y^3$

e) $k[x, y] = \text{Sin}[x^2 + y^2]$

f) $h[u, v] = \text{Log}[u^2 + v^2]$

g) $f[x, y] = \text{Sin}[x] + \text{Cos}[y]$

h) $g[u, v] = \text{Sin}[u\, v]$

i) $h[u, v] = x\, \text{Tan}[2\pi y]$

Check your work with the program **PartialD**.

18.3   Differential Approximations

This section uses the differential to analyze measurement errors.

The microscope (or increment) equation in two variables is

$$f[x + dx, y + dy] - f[x, y] = \frac{\partial f}{\partial x}[x, y] \cdot dx + \frac{\partial f}{\partial y}[x, y] \cdot dy + \varepsilon \left| \begin{bmatrix} dx \\ dy \end{bmatrix} \right|$$

where $\varepsilon \approx 0$ whenever $dx \approx 0$ and $dy \approx 0$. This section uses the approximation obtained by dropping the error term,

$$f[x + dx, y + dy] - f[x, y] \simeq \frac{\partial f}{\partial x}[x, y] \cdot dx + \frac{\partial f}{\partial y}[x, y] \cdot dy$$

Notice that we wrote $\simeq$ rather than $\approx$ for our change approximation. We are trying to remind you that the error term $\varepsilon \left| \begin{bmatrix} dx \\ dy \end{bmatrix} \right|$ is *very* small, because it is a product of two small terms.

Example 18.7.   *Application of the Approximation to Gas Volume*

This example uses the differential to estimate the error in measuring the amount of gas in a container. We use the ideal gas law

$$PV = nRT$$

where P is the pressure in dynes/cm^2, V is the volume (in cm^3) taken up by the gas, T is the absolute temperature in degrees Kelvin, n is the amount of gas in moles, and R is the constant 8.3136×10^7 (for these units). At a fixed temperature $T = 294°K = 21°C$, the amount of gas is

$$n = n(P, V) = \frac{1}{RT} PV = kPV$$

where $k = 4.0913 \times 10^{-11}$.

We store the gas in a cylindrical container of measured radius 10 cm and height 1500 cm at 10 times atmospheric pressure. (One atmosphere $= 1.01325 \times 10^6$ dynes/cm^2, so our measured pressure is 1.0133×10^7.) Nominally, the number of moles of gas is

$$n = kPV = 4.0913 \times 10^{-11} \times 1.0133 \times 10^7 \times V$$

where

$$V = \pi r^2 h = 3.1416 \times 10^2 \times 1500$$

making

$$n = 195.36$$

If our measurements of P, r, and h are each accurate to at least 1%, what is the possible error in our computation of the amount of gas? We will use the approximation

$$n(P + dP, V + dV) - n(P, V) \simeq dn = \frac{\partial n}{\partial P} dP + \frac{\partial n}{\partial V} dV$$

obtained from the microscope equation. The differential of n is

$$dn = k(V dP + P dV)$$
$$= 4.0913 \times 10^{-11}(4.7124 \times 10^5 dP + 1.0133 \times 10^7 dV)$$

at the measured pressure and volume, whereas because $V = \pi r^2 h$

$$dV = 2\pi r h dr + \pi r^2 dh = 9.4248 \times 10^4 dr + 314.16 \ dh$$

One percent errors mean $|dP| \leq 1.0133 \times 10^5$, $|dr| \leq .1$, and $|dh| \leq 15$, so the largest

$$dV = 9.4248 \times 10^4 \times .1 + 314.16 \times 15 = 14137.$$

and

$$dn = 4.0913 \times 10^{-11}(4.7124 \times 10^5 \times 1.0133 \times 10^5 + 1.0133 \times 10^7 \times 1.4137 \times 10^4)$$
$$= 7.8144$$

or a 4% error.

All of this messy arithmetic is best done by first entering the symbolic expression

$$dV = k\,(V\ dP + P\ dV)$$

into the computer and then assigning values to the variables. This approach is clear and reliable. In addition, the symbolic expression can give us deeper insight into the basic measurement problem.

We can rewrite the differential for the amount of gas above in a "percentage form." The change in the amount (in moles) is dn and the amount itself is n, so the fraction dn/n is the relative change or $\frac{dn}{n} \times 100\%$ is the percent change. Similarly, dV/V and dP/P are the relative changes in volume and pressure. Algebra yields a simple formula,

$$dn = \frac{1}{RT}(V dP + P dV)$$
$$\frac{dn}{n} = \frac{RT}{PV} \frac{1}{RT}(V dP + P dV)$$
$$\frac{dn}{n} = \frac{dP}{P} + \frac{dV}{V}$$

To be sure of signs, perhaps we should summarize with the error inequality that says the relative error in amount is no more than the sum of the relative errors in pressure and volume,

$$|\frac{dn}{n}| \leq |\frac{dP}{P}| + |\frac{dV}{V}|$$

The relative volume error becomes

$$dV = 2\pi r \, , h \, dr + \pi r^2 \, dh$$

$$\frac{dV}{V} = \frac{2\pi r h}{\pi r^2 h} \, dr + \frac{\pi r^2}{\pi r^2 h} \, dh$$

$$\frac{dV}{V} = 2\frac{dr}{r} + \frac{dh}{h}$$

and we can see our 4% error in the formula

$$\frac{dn}{n} = \frac{dP}{P} + \frac{dV}{V}$$

$$\frac{dn}{n} = \frac{dP}{P} + 2\frac{dr}{r} + \frac{dh}{h}$$

when each single relative error can be as much as 1% and of any sign.

Exercise set 18.3

1. *A circular cone of radius r and height h has volume $V = \frac{1}{3}\pi r^2 h$.*
   (a) *Use the differential to show that the relative measurement of the radius is twice as important as the relative measurement of the height,*

$$\frac{dV}{V} = 2\frac{dr}{r} + \frac{dh}{h}$$

   (b) *Find the volume of a cone of radius 10 inches and height 20 inches.*
   (c) *If the radius is actually between 9.95 and 10.05 inches and the height is actually between 19.9 and 20.1 inches, find the actual range of values for the volume and give a prediction of the accuracy of the volume as a percent of volume.*
   (d) *Use the differential to predict the relative accuracy of the volume and compare this with your previous answer. (These only agree "in the limit" as r and h tend to zero.)*

2. *We wish to measure the gravitational constant of a moon with no atmosphere by dropping an object and timing its fall. Galileo's Law tells us the acceleration g is constant; so, if we observe a fall of a distance s in a time t, we have the relation $s = \frac{1}{2}gt^2$.*
   (a) *Suppose we drop an object 1 m and it takes 1 sec to fall. What is g?*
   (b) *Find a differential approximation for the relative error in predicting g by measuring s and t and use it to estimate the accuracy of g if both s and t are accurate to 1%.*
   (c) *Which is the more sensitive measurement, time or distance fallen?*

3. *Small oscillations of a pendulum may be approximated by a linear differential equation, yielding the equation for the time period of a complete oscillation, T, in terms of the length, L, and the gravitational constant, g:*

$$T = 2\pi \sqrt{\frac{L}{g}}$$

(a) *Use this formula to find a relative error formula between these quantities,*

$$\frac{dT}{T} = \frac{1}{2}\left(\frac{dL}{L} - \frac{dg}{g}\right)$$

(b) *We use a pendulum to measure the gravitational constant on the moon. A 1.0 meter pendulum swings through a full oscillation in 5.0 seconds. What is the nominal value of g?*

(c) *If the length and time period are both accurate to two significant figures, how accurate is your measurement of the moon's gravitational constant?*

NOTE: The formula $T = 2\pi\sqrt{L/g}$ is entirely false for large oscillations. See the Pendulum Project in the accompanying book of Scientific Projects.

4. *The total resistance, R, of two resistors, R_1, R_2, connected in parallel satisfies*

$$\frac{1}{R} = \frac{1}{R_1} + \frac{1}{R_2}$$

(a) *Show that*

$$\frac{dR}{R^2} = \frac{dR_1}{R_1^2} + \frac{dR_2}{R_2^2}$$

(b) *If 10 k ohm and 15 k ohm resistors are connected in parallel, show that the total resistance is 6 k ohm.*

(c) *If the measured accuracy of each of the resistors in part (b) is 5%, use differentials to predict the relative (%) accuracy of your 6 k total (in the worst case)?*

You are interested in the accuracy of your speedometer and perform the following experiment on an isolated stretch of flat straight interstate highway.

Problem 18.1. CHECK YOUR SPEEDOMETER ⎯⎯⎯⎯⎯⎯⎯⎯⎯⎯⎯⎯⎯▼

You drive at constant speed with your speedometer reading 60 mph. By your quartz watch, you cross between two consecutive mile markers in 57 seconds. We know "distance equals rate times time" or $D = RT$, so $R = D/T$ when time is in hours and distance is in miles.

(a) *Express R in miles per hour as a function of t in seconds and D in miles.*

(b) *Compute the total differential $dR = \frac{\partial R}{\partial t}dt + \frac{\partial R}{\partial D}dD$ symbolically and for the case where $D = 1$ and $t = 60$ corresponding to exactly 60 mph.*

(c) *Use the approximation from part (b) to estimate the error in your speedometer (when you travel 1 mile in 57 seconds, with the speedometer reading 60 mph, so $dt = -3$ and $dD = 0$.) Also, use your calculator (or the computer) to compute the error exactly (to the accuracy of your calculator).*

(d) *Suppose that there is as much as $\frac{1}{4}$ of 1% error in the placement of the mile markers and as much as 1% error in your time measurement of $t = 57$. What is the approximate absolute range of error of your speedometer?*

(e) *What other inaccuracies affect this experiment?*

(f) *What do all these parts mean together???*

18.4   Geometry of the Differential

The differential is used to compute a vector normal to a surface.

The Increment Principle 18.7 tells us that a small piece of a smooth graph $z = f[x, y]$ looks like the linear graph $dz = m\,dx + n\,dy$. This means that the vector

$$\begin{bmatrix} -m \\ -n \\ +1 \end{bmatrix} = \begin{bmatrix} -\frac{\partial f}{\partial x}[x, y] \\ -\frac{\partial f}{\partial y}[x, y] \\ +1 \end{bmatrix}$$

is normal to the surface $z = f[x, y]$ at (x, y, z) because we know

Theorem 18.9.   *The Linear Normal*
     The explicit graph of the linear equation

$$dz = m\,dx + n\,dy$$

is a plane in $\begin{bmatrix} dx \\ dy \\ dz \end{bmatrix}$ *-coordinates through* $\begin{bmatrix} dx \\ dy \\ dz \end{bmatrix} = \begin{bmatrix} 0 \\ 0 \\ 0 \end{bmatrix}$ *perpendicular to* $\begin{bmatrix} -m \\ -n \\ +1 \end{bmatrix}$.

Example 18.8.   *A Normal to* $z = 9xy\,e^{-(x^2+y^2)}$ *at* $(x, y) = (1/2, 1)$

The total differential of

$$z = 9xy\,e^{-(x^2+y^2)}$$

is

$$dz = 9\,e^{-(x^2+y^2)}\left((y - 2x^2 y)\,dx + (x - 2xy^2)\,dy \right)$$

at $(x, y) = (1/2, 1)$

$$dz = \frac{9}{2}\,e^{-5/4}\,(dx - dy) \approx 1.28927(dx - dy)$$

so the vector normal to the surface at $(1/2, 1)$ is

$$\begin{bmatrix} -1.28927 \\ +1.28927 \\ 1 \end{bmatrix}$$

as shown in Figure 18.14.

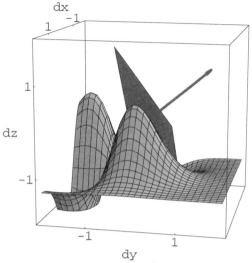

Figure 18.14: $z = 9xy\, e^{-(x^2+y^2)}$ at $(x, y) = (1/2, 1)$

We can reduce this to a procedure.

Procedure 18.10.

To find a vector normal to the explicit surface $z = f[x, y]$ at a specific point (x, y):

(a) Calculate the symbolic total differential $dz = \frac{\partial f}{\partial x}\, dx + \frac{\partial f}{\partial y}\, dy$.

(b) Calculate the specific numerical values of the terms m and n at the given (x, y) in the expression $dz = m\, dx + n\, dy$.

(c) Zoom in in your mind and recall that the normal to $dz = m\, dx + n\, dy$ is
$$\mathbf{N} = \begin{bmatrix} -m \\ -n \\ +1 \end{bmatrix}.$$

(d) Zoom back out and notice that $\mathbf{N}$ is normal to both the surface and plane.

18.4.1   The Tangent Plane in (x, y, z)-Coordinates

The equation of the tangent plane of the explicit surface $z = f[x, y]$, at (x_0, y_0) is

$$dz = m\, dx + n\, dy$$

where $m = \frac{\partial f}{\partial x}[x_0, y_0]$ and $n = \frac{\partial f}{\partial y}[x_0, y_0]$ and the local coordinates satisfy $dx = (x - x_0)$, $dy = (y - y_0)$, $dz = (z - z_0)$ with $z_0 = f[x_0, y_0]$. We can sketch the tangent plane from this information simply by putting the parallel (dx, dy, dz) coordinates at the point of tangency. However, sometimes we may want the equation of the tangent plane in the original coordinates. In that case, we need to be careful about what is fixed and what varies.

Example 18.9.   *The Tangent Plane to $z = x^2 y^3$ at $(x_0, y_0) = (2, 1)$*

Suppose we want the tangent to the explicit surface $z = x^2\, y^3$ at $(x_0, y_0, z_0) = (2, 1, 4)$. The symbolic differential is

$$dz = 2x\, y^3\, dx + 3x^2\, y^2\, dy$$

and at the specific point $(x, y, z) = (x_0, y_0) = (2, 1)$,

$$dz = 4\, dx + 12\, dy$$

Local coordinates at $(2, 1, 4)$ are $dx = x - 2$, $dy = y - 1$ and $dz = z - 4$, so the specific differential can be written:

$$z - 4 = 4(x - 2) + 12(y - 1) \qquad \Leftrightarrow \qquad z = 4x + 12y - 16$$

Here is the procedure.

Procedure 18.11.

To find the (x, y, z)-equation of the tangent to the explicit surface $z = f[x, y]$ at a specific point (x_0, y_0):

(a)   Calculate the symbolic total differential $dz = \frac{\partial f}{\partial x}\, dx + \frac{\partial f}{\partial y}\, dy$.

(b)   Calculate the specific numerical values of the terms m and n in the expression $dz = m\, dx + n\, dy$ at the given (x_0, y_0).

(c)   Calculate the value $z_0 = f[x_0, y_0]$.

(d)   Replace $dx \to (x - x_0)$, $dy \to (y - y_0)$ and $dz \to (z - z_0)$ in $dz = m\, dx + n\, dy$ and simplify.

This procedure is a step-by-step explanation of the tangent formula:

$$(z - z_0) = \frac{\partial f}{\partial x}[x_0, y_0]\, (x - x_0) + \frac{\partial f}{\partial y}[x_0, y_0]\, (y - y_0)$$

Exercise set 18.4

1. *Find a vector normal to the surface $z = x^2 + y^2/3$ at the points:*

a) $(x, y) = (0, 0)$ b) $(x, y) = (1, 1)$ c) $(x, y) = (-1, 0)$

2. R *un the program* **SurfNorm** *to sketch the results of the previous exercise.*

3. *Give the (x, y, z) equation of the tangent to $z = 9 - x^2 - y^2$ at $(x, y) = (1, -2)$. Check your work with the* **SurfNorm** *program.*

4. *What is wrong with the following computation of the equation for the "tangent plane" of $z = x^2 + y^3$ at $(x, y) = (-2, 3)$:*

$$z = x^2 + y^3$$
$$dz = 2x\, dx + 3y^2\, dy$$
$$(z - 31) = 2x\, (x + 2) + 3y^2\, (y - 3)$$
$$z = 31 + 4x + 2x^2 - 9y^2 + 3y^3$$

18.5   The Meaning of the Gradient

The gradient points the way toward the fastest change in a quantity. Heat flows from hottest to coldest, so it goes in the direction of the gradient.

Now, we turn our attention to a direct interpretation of the gradient vector $\nabla f[x]$. The first idea is to use the approximation

$$f[\mathbf{X} + \mathbf{dX}] - f[\mathbf{X}] \simeq \langle \nabla f[\mathbf{X}] \bullet \mathbf{dX} \rangle$$

to see:

Theorem 18.12.   *The Nonlinear Gradient*
The instantaneous rate of change of a nonlinear function $f[x, y] = f[\mathbf{X}]$ in the direction of a change $\mathbf{dX}$ is

$$\frac{\langle \nabla f[\mathbf{X}] \bullet \mathbf{dX} \rangle}{|\mathbf{dX}|} = \frac{\text{the approximate amount changed}}{\text{distance moved}}$$

In particular, the fastest rate of change is $|\nabla f[\mathbf{X}]|$ when we move in the direction of $\nabla f[\mathbf{X}]$.

The Increment Principle 18.7 tells us that if we view a very small portion of the graph, we cannot distinguish the nonlinear graph from the differential (in local coordinates). Imagine the magnified graph shown with its contour graph sketched in the base plane. This is shown in Figure 18.15 with the 2-D gradient vector pointing normal to the contour lines.

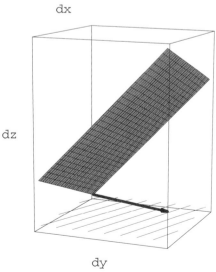

Figure 18.15: $z = x + 3y$

Theorem 18.13. *The Linear Gradient*
  The contour graph of the linear equation

$$dz = m \ dx + n \ dy$$

consists of lines in local $\begin{bmatrix} dx \\ dy \end{bmatrix}$ *-coordinates that are perpendicular to the gradient*

vector $\mathbf{G} = \begin{bmatrix} m \\ n \end{bmatrix}$. *In addition,* $\mathbf{G}$ *points toward lines with higher values of* $dz = k$ *and the rate of change of* dz *with respect to movement in this direction is* $|\mathbf{G}| = \sqrt{m^2 + n^2}$.

We can simply sketch the contour graph by drawing $\mathbf{G}$ and filling in perpendicular lines.

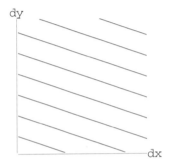

Figure 18.16: $x + 3y =$constant

In the computer **Micro3D** program, you can see that small magnified parts the contour graph of $z = f[\mathbf{X}]$ are indistinguishable from the corresponding contour graph of the linear differential approximation $dz = \nabla f[\mathbf{X}] \bullet d\mathbf{X}$. The result above when applied to the total differential says that $\nabla f[\mathbf{X}]$ points in the direction of fastest increase on the contour graph of the nonlinear function.

Another way to see this is the formula for the angle between two vectors.

$$\mathrm{Cos}[\theta] = \frac{\mathbf{A} \bullet \mathbf{B}}{|\mathbf{A}| \cdot |\mathbf{B}|} \qquad \text{or} \qquad \mathbf{A} \bullet \mathbf{B} = |\mathbf{A}| \cdot |\mathbf{B}| \, \mathrm{Cos}[\theta]$$

We think of moving in various directions $d\mathbf{X}$ but all of the same magnitude, $|d\mathbf{X}|$. In the linear case these perturbations could all be unit size, but they must be small to treat $f[\mathbf{X} + d\mathbf{X}] - f[\mathbf{X}]$ as near dz relative to $|d\mathbf{X}|$. It is clear that $\mathrm{Cos}[\theta]$ lies between -1 and $+1$, while $\nabla f[\mathbf{X}]$ and $|d\mathbf{X}|$ are fixed, so

$$\nabla f[\mathbf{X}] \bullet d\mathbf{X} = |\nabla f[\mathbf{X}]| \cdot |d\mathbf{X}| \, \mathrm{Cos}[\theta]$$

is largest when $\theta = 0$ [$\mathrm{Cos}[0] = 1$], the dot product is zero when $\theta = \frac{\pi}{2}$ or $\theta = -\frac{\pi}{2}$, and the dot product is smallest when $\theta = \pi$ [$\mathrm{Cos}[\pi] = -1$]. This has a simple meaning: Locally near X, $f[\mathbf{X}]$ increases fastest in the direction of $\nabla f[\mathbf{X}]$, is unchanged (up to linear approximation) in the directions perpendicular to $\nabla f[\mathbf{X}]$, and decreases fastest in the direction $-\nabla f[\mathbf{X}]$.

Example 18.10. *Heat Flow and the Gradient ∇f of a nonlinear $f[\mathbf{X}]$*

Heat flows from hot to cold, not just hot to cold, but hottest to coldest. This means it moves in the direction that makes the temperature function decrease at the fastest rate. This is exactly the description of the gradient we just gave - heat moves opposite the gradient.

Let $T[x, y]$ be the temperature at a point in the (x, y)-plane. Heat will flow in the direction of the vector $-\nabla T[\mathbf{X}]$.

For example, suppose we have a heated square plate described by coordinates $0 \leq x \leq 1$ and $0 \leq y \leq 1$. We heat the plate so that the initial temperature distribution is given by

$$T[x, y] = xy(1 - x)(1 - y)$$

Notice that $T = 0$ all around the edge of the square plate and $T > 0$ inside the plate. Suppose we hold the edges at zero temperature. Heat will flow toward the edges to cool the interior. It flows in the direction of $-\nabla T[\mathbf{X}]$. At $(x, y) = (\frac{1}{4}, \frac{1}{3})$, we have

$$\nabla T[\mathbf{X}] = \begin{bmatrix} \frac{\partial T}{\partial x} \\ \frac{\partial T}{\partial y} \end{bmatrix} = \begin{bmatrix} (1 - 2x)y(1 - y) \\ (1 - 2y)x(1 - x) \end{bmatrix}$$

$$\nabla T(\begin{bmatrix} \frac{1}{4} \\ \frac{1}{3} \end{bmatrix}) = \begin{bmatrix} \frac{1}{9} \\ \frac{1}{16} \end{bmatrix}$$

so heat flows in the opposite direction, $- \begin{bmatrix} \frac{1}{9} \\ \frac{1}{16} \end{bmatrix}$. We know that a microscopic view of the contour graph shown in Figure 18.17 is the same as the contour graph of the linear equation

$$\frac{dx}{9} + \frac{dy}{16} = \text{constant}$$

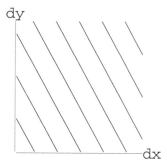

dy

dx

Figure 18.17: Microscopic isotherms normal to $-\nabla T(\frac{1}{9}, \frac{1}{16})$ near $(x, y) = (\frac{1}{4}, \frac{1}{3})$

If we want the direction of ∇T as an angle, rather than a vector, we could use the formula

$$\text{Tan}[\theta] = \frac{y}{x}$$

so $\theta = \text{ArcTan}(9/16) - \pi \approx -2.6292(\text{radians} \approx -151°)$. Actually, a better way to give a direction is with a unit length vector in the same direction. Our gradient has length $\sqrt{\frac{1}{9}^2 + \frac{1}{16}^2} = \frac{\sqrt{337}}{144}$. The vector

$$\frac{-144}{\sqrt{337}} \begin{bmatrix} \frac{1}{9} \\ \frac{1}{16} \end{bmatrix} = \begin{bmatrix} \text{Cos}[\theta] \\ \text{Sin}[\theta] \end{bmatrix}$$

is the unit vector for the same angle θ. This is "better" in the sense that it generalizes more easily to three (and more) dimensions. Any of the answers is sufficient - the vector, the unit vector in the same direction, or the angle θ. A contour plot of this temperature distribution is shown in Figure 18.18. The contours here are points of equal temperature called "isotherms."

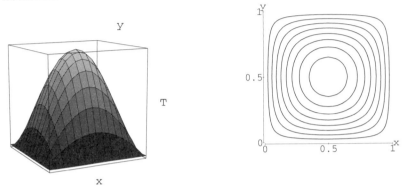

Figure 18.18: Graphs of explicit temperatures and isotherms

It is natural to ask now for the equations of the curves along which heat flows. Suppose heat moves in time along a parametric curve $(x[t], y[t])$. The tangent (or velocity) to this curve must be proportional to the negative of the gradient,

$$\frac{\mathbf{dX}}{dt} = -k\nabla T$$

$$\frac{dx}{dt} = -k\,y(1-y)(1-2\,x)$$

$$\frac{dy}{dt} = -k\,x(1-x)(1-2\,y)$$

so the slope of the low to high temperature curves is

$$\frac{dy}{dx} = \frac{x(1-x)(1-2\,y)}{y(1-y)(1-2\,x)}$$

We separate like variables and obtain

$$\frac{y(1-y)}{1-2\,y}\,dy = \frac{x(1-x)}{1-2\,x}\,dx$$

We can find antiderivatives of both sides by using long division and standard tricks from integration. However, the computer gives us

$$\int \frac{y(1-y)}{1-2\,y}\,dy = \int \frac{x(1-x)}{1-2\,x}\,dx$$

$$\frac{y^2}{4} - \frac{y}{4} - \frac{1}{8}\,\mathrm{Log}[2\,y-1] = \frac{x^2}{4} - \frac{x}{4} - \frac{1}{8}\,\mathrm{Log}[2\,x-1] + c$$

for an unknown constant c. We do not know a simple way to solve for y, but we can simply make the contour plot of

$$\frac{y^2}{4} - \frac{y}{4} - \frac{1}{8}\,\text{Log}[2\,y - 1] - (\frac{x^2}{4} - \frac{x}{4} - \frac{1}{8}\,\text{Log}[2\,x - 1])$$

being careful to avoid the converging flow lines at $(x, y) = (0.5, 0.5)$. This plot shows the (x, y)-curves at various values of the constant c. The graph of the flow lines and the graph of the isotherms are shown in Figure 18.19.

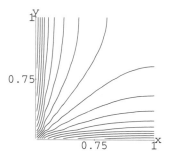
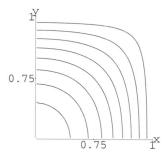

Figure 18.19: Flow lines and isotherms

Exercise set 18.5

1. *Find the fastest rate of change of $f[x, y] = x^2 + y^2/3$ at the points:*
a) $(x, y) = (0, 0)$ b) $(x, y) = (1, 1)$ c) $(x, y) = (-1, 0)$

2. U *se the program* **Grad** *to verify and graph your results from the previous exercise.*

3. *1) Suppose we make a small change in* $\mathbf{X}$ *parallel to* $\mathbf{G} = \nabla f[\mathbf{X}]$. *Show that the rate of change of $z = f[\mathbf{X}]$ is approximately $|\nabla f[\mathbf{X}]|$.*

2) Suppose we make a small change in $\mathbf{X}$ *parallel to* $\begin{bmatrix} -\frac{\partial f}{\partial y}[x, y] \\ \frac{\partial f}{\partial x}[x, y] \end{bmatrix}$. *Show that the rate of change of $z = f[\mathbf{X}]$ is approximately zero.*

4. Integration Review
Show that $\int \frac{x(1-x)}{1-2\,x}\,dx = \frac{x^2}{4} - \frac{x}{4} - \frac{1}{8}\,\text{Log}[2\,x - 1] + c$ *with the following hints:*
Use long division to show that $\frac{x(1-x)}{1-2\,x} = \frac{x}{2} - \frac{1}{4} - \frac{1}{4(2\,x-1)}$.
Use a change of variables, $u = 4(2\,x - 1)$, $du = 8\,dx$ to show that

$$\int \frac{1}{4(2\,x - 1)}\,dx = \frac{1}{8}\,\text{Log}[2\,x - 1]$$

We conclude this section with an exercise about the steepest path down the mountain.

Problem 18.2. ──▼

The Perfecto Ski Bowl has perfectly manicured slopes in the shape

$$a = x^2 + 2y^2$$

where a is the altitude, x measures east, and y measures north for $0 \leq x, y \leq 10$, all measured in leagues.

*a) Sketch a contour map of Perfecto and sketch the path of a downhill racer who starts at $(x, y) = (5, 10)$ always pointing her skis straight down the steepest direction. (Draw some gradient vectors and sketch in a path. Check your work with the program **ContourPlot**.)*

b) Write the specific vector equation

$$\frac{d\mathbf{X}}{dt} = -k\nabla a$$

for some unknown constant k and the specific gradient of a. What does this equation represent?

c) Solve the equation above for the unknown constant k, obtaining $-k = \frac{dx}{2x} = \frac{dy}{4y}$ so

$$\frac{dy}{dx} = \frac{2y}{x}$$

*d) Show that the curves $y = cx^2$ satisfy the differential equation in part c and sketch these curves on your contour map for various values of c. (The computer could make a Contour-Plot of x^2/y or y/x^2, but the singularities cause technical problems.) Which value of c passes through $(5,10)$ in particular? (See the program **MxMnThy** in the Chapter 19 folder.)*

──▲

18.6   Implicit Differentiation (Again)

Implicit differentiation often simplifies problems. This brief section recalls the method in two variables and extends it to three.

Implicit differentiation amounts to treating all variables equally at the differentiation stage. In Examples 7.3 and 7.5, we computed the tangent line to a circle using the implicit equation for the circle and implicit differentiation. Here is the generalization to three dimensions.

Example 18.11. *Implicit Tangent to a Sphere*

In Problem 18.3, you will calculate the tangent plane to a sphere by explicit differentiation. The implicit equation of the sphere centered at zero with radius r is $x^2 + y^2 + z^2 = r^2$.

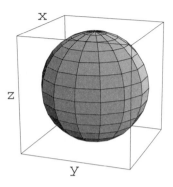

Figure 18.20: $x^2 + y^2 + z^2 = r^2$

Differentiating both sides with respect to each variable (including z) yields:

$$\frac{\partial(x^2 + y^2 + z^2)}{\partial x} = 2\,x, \qquad \frac{\partial(x^2 + y^2 + z^2)}{\partial y} = 2\,y, \qquad \frac{\partial(x^2 + y^2 + z^2)}{\partial z} = 2\,z$$

and $\frac{\partial r^2}{\partial -} = 0$ since r is constant. Now, we write the total differential in all the variables,

$$\frac{\partial(x^2 + y^2 + z^2)}{\partial x}\,dx + \frac{\partial(x^2 + y^2 + z^2)}{\partial y}\,dy + \frac{\partial(x^2 + y^2 + z^2)}{\partial z}\,dz = 0$$
$$2\,x\,dx + 2\,y\,dy + 2\,z\,dz = 0$$
$$x\,dx + y\,dy + z\,dz = 0$$
$$\left\langle \begin{bmatrix} x \\ y \\ z \end{bmatrix} \bullet \begin{bmatrix} dx \\ dy \\ dz \end{bmatrix} \right\rangle = 0$$
$$\langle \mathbf{X} \bullet d\mathbf{X} \rangle = 0$$

which says that $\mathbf{X}$ and $d\mathbf{X}$ are perpendicular.

Implicit Surface	Implicit Tangent Plane
$x^2 + y^2 + z^2 = r^2$	$x\,dx + y\,dy + z\,dz = 0$

Example 18.12. *A Specific Example*

The point $X(2, 1, 3)$ lies on a sphere centered at the origin of radius $\sqrt{14}$. The tangent plane passes through the tip of this vector and is perpendicular to it:

Implicit Surface	Implicit Tangent Plane
$x^2 + y^2 + z^2 = r^2$	$x\,dx + y\,dy + z\,dz = 0$

at $\quad X(2,1,3)$

| $2^2 + 1^2 + 3^2 = 14$ | $2\,dx + dy + 3\,dz = 0$ |

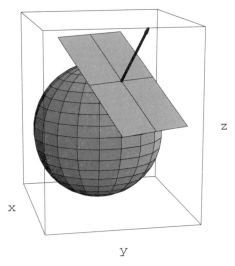

Figure 18.21: Tangent to $x^2 + y^2 + z^2 = 14$ at $X(2,1,3)$ with normal vector $N(2,1,3)$

In general, if we start with an implicit relationship among x, y, and z, with a function of all three equal to a constant, c, we differentiate both sides of the equation with respect to all of the variables and write the total differential as an implicit equation in dx, dy, and dz:

Implicit Surface	Implicit Tangent Plane
$g[x,y,z] = c$	$\dfrac{\partial g}{\partial x}\,dx + \dfrac{\partial g}{\partial y}\,dy + \dfrac{\partial g}{\partial z}\,dz = 0$

Exercise set 18.6

1. Tangent to an Ellipsoid

The ellipsoid $3\,x^2 + 2\,y^2 + z^2 = 20$ is shown in the CD section on implicit constraints of Chapter 19. Use implicit differentiation to show that the tangent plane to the ellipsoid at the point $X(x,y,z)$ is perpendicular to the vector

$$\mathbf{N} = \begin{bmatrix} 3\,x \\ 2\,y \\ z \end{bmatrix}$$

Use this normal to sketch the tangent plane at the points

$$\mathbf{X} = \begin{bmatrix} 1 \\ 1 \\ \sqrt{15} \end{bmatrix} \approx \begin{bmatrix} 1 \\ 1 \\ 3.87 \end{bmatrix} \qquad and \qquad \mathbf{X} = \begin{bmatrix} 2 \\ 2 \\ 0 \end{bmatrix}$$

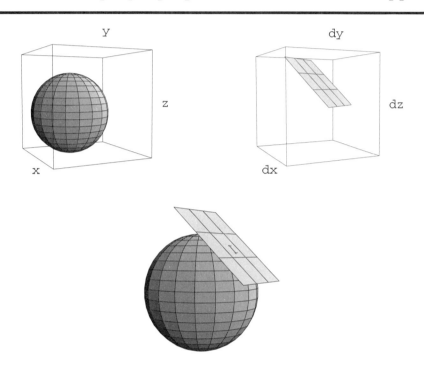

Figure 18.22: The tangent plane to a sphere

Problem 18.3. EXPLICIT TANGENT TO A SPHERE ───────────▼

It is geometrically clear that the tangent plane to a sphere is the plane perpendicular to a radius vector ending at the point of tangency. We computed this with calculus using implicit differentiation above and want you to use explicit formulas to compute it here.

The simplest implicit equation of a sphere of (fixed) radius r is

$$x^2 + y^2 + z^2 = r^2$$

because the vectors $\mathbf{X} = \begin{bmatrix} x \\ y \\ z \end{bmatrix}$ of length r are those satisfying $|\mathbf{X}| = r$ or

$$\sqrt{x^2 + y^2 + z^2} = r$$

Squaring both sides gives the simpler equation.

The vector $\mathbf{X}$ is the radius vector pointing from the origin to a point on the sphere.

1) Show that the vector $\mathbf{X} = \begin{bmatrix} 2 \\ 1 \\ 3 \end{bmatrix}$ *lies on the sphere of radius* $\sqrt{14}$ *and sketch.*

A general radius vector $\mathbf{X} = \begin{bmatrix} x \\ y \\ z \end{bmatrix}$ *may be considered as fixed somewhere on the sphere, whereas we consider other vectors in the parallel local* (dx, dy, dz)*-coordinate system centered at the tip of* $\mathbf{X}$*. We might call an unknown vector in the local system* $\mathbf{dX} = \begin{bmatrix} dx \\ dy \\ dz \end{bmatrix}$.

2) Sketch the (dx, dy, dz)*-axes at the tip of your radius vector from part (1) above.*

3) What is the geo-algebraic condition for $\mathbf{X}$ *and* $\mathbf{dX}$ *to be perpendicular? Show that this is the equation*

$$x \cdot dx + y \cdot dy + z \cdot dz = 0$$

when written in components. In particular, which vectors $\mathbf{dX}$ *are perpendicular to* $\mathbf{X} = \begin{bmatrix} 2 \\ 1 \\ 3 \end{bmatrix}$

on the sphere of radius $\sqrt{14}$*? (HINT:* $\mathbf{dX} = \begin{bmatrix} 2 \\ 2 \\ -2 \end{bmatrix}$ *is perpendicular, but* $\begin{bmatrix} -2 \\ 2 \\ 2 \end{bmatrix}$ *is not.) Add*

a sketch of the perpendicular displacement vector $\mathbf{dX} = \begin{bmatrix} 2 \\ 2 \\ -2 \end{bmatrix}$ *starting at the tip of* $\mathbf{X}$*.*

If we want to use explicit equations, we need to solve for z *and use the equation*

$$z = \sqrt{r^2 - x^2 - y^2}$$

for the top half of the sphere of radius r*. (The implicit equation does not need to be broken in half.)*

4) Use the Chain Rule to show that

$$\frac{\partial z}{\partial x} = \frac{-x}{\sqrt{r^2 - x^2 - y^2}} = -\frac{x}{z} \qquad \text{and} \qquad \frac{\partial z}{\partial x} = \frac{-y}{\sqrt{r^2 - x^2 - y^2}} = -\frac{y}{z}$$

so that the total differential of the explicit function from the surface $z = z[x, y]$ *above is*

$$dz = -\frac{x}{z} \cdot dx - \frac{y}{z} \cdot dy$$

which is algebraically equivalent to the equation for perpendicularity

$$x \cdot dx + y \cdot dy + z \cdot dz = 0$$

when $z > 0$*.*

5) The partial derivatives

$$\frac{\partial z}{\partial x} = \frac{-x}{\sqrt{r^2 - x^2 - y^2}} \qquad \text{and} \qquad \frac{\partial z}{\partial x} = \frac{-y}{\sqrt{r^2 - x^2 - y^2}}$$

are undefined on the circle $x^2 + y^2 = r^2$*. Why? Sketch this circle on your figure.*

6) What happens to the tangent to the sphere along the circle $x^2 + y^2 = r^2$? Is there no tangent, or is there something else going on that makes the partial derivatives undefined?

▲

18.7 Review Exercises

Calculus lets us "see" inside a powerful microscope without actually magnifying the nonlinear graph. We know that a smooth function "looks" like its tangent at high magnification and that the rules of calculus allow us to compute the equation of the tangent. In several variables, we can view the tangent as a plane touching an explicit surface or as a linear contour map approximating a small piece of a contour graph. Also, see the program **DiffReview** for a review of this chapter.

Exercise set 18.7

1. (a) *Sketch a set of (x, y, z)-axes and plot the point $P(1, 1, 2) = \mathbf{P} = \begin{bmatrix} 1 \\ 1 \\ 2 \end{bmatrix}$. Let x, y, and z run from -1 to 3.*
   (b) *The point $\mathbf{P}$ lies on the explicit surface $z = x^2 + y^3$. Verify this.*
   (c) *Add a pair of (dx, dy, dz)-axes centered at the (x, y, z)-point $P(1, 1, 2)$. How are these axes related to the (x, y, z)-axes?*
   (d) *Use rules of calculus to show that*

   $$z = x^2 + y^3 \quad \Rightarrow \quad dz = 2x\,dx + 3y^2\,dy$$

   (e) *Substitute $(x, y) = (1, 1)$ into your differential to show that*

   $$dz = 2\,dx + 3\,dy$$

   at the (x, y, z)-point $(1, 1, 2)$ or (dx, dy, dz)-point $(0, 0, 0)$.
   (f) *Plot the explicit plane $dz = 2\,dx + 3\,dy$ on your (dx, dy, dz)-axes.*
   (g) *What would you see if you looked at the graph of $z = x^2 + y^3$ under a very powerful microscope?*
   (h) *Use the computer program **Micro3D** or **DiffReview** to plot the function, its differential and make an animation of a microscope zooming in on the explicit graph at the (x, y, z)-point $(1, 1, 2)$.*
   (i) *Explain how the differential cell of the **Micro3D** program is actually solving parts (b) - (g) of this exercise.*

   Now, mentally zoom in contour plot mode.

2. (a) *Sketch a set of (x, y) axes and plot the point $P(1, 1) = \mathbf{P} = \begin{bmatrix} 1 \\ 1 \end{bmatrix}$. Let x and y run from -1 to 3.*
   (b) *The point $\mathbf{P}$ lies on the implicit curve $2 = x^2 + y^3$. Verify this.*
   (c) *Add a pair of (dx, dy)-axes centered at the (x, y)-point $P(1, 1)$. How are these axes related to the (x, y, z)-axes?*

(d) *Use rules of calculus to show that*

$$2 = x^2 + y^3 \qquad \Rightarrow \qquad 0 = 2x\ dx + 3\,y^2\ dy$$

(e) *Substitute $(x, y) = (1, 1)$ into your differential to show that*

$$0 = 2\ dx + 3\ dy$$

at the (x, y, z)-point $(1, 1)$ or (dx, dy)-point $(0, 0, 0)$.

(f) *Plot the implicit line $0 = 2\ dx + 3\ dy$ on your (dx, dy)-axes.*

(g) *What would you see if you looked at the contour graph of $z = x^2 + y^3$ under a very powerful microscope at $(1, 1)$?*

(h) *Use the computer program **Micro3D** to plot the function, its differential and to make an animation of a microscope zooming in on the contour graph at the (x, y, z)-point $(1, 1)$.*

(i) *Explain how the **Micro3D** program is actually solving parts (b) - (g) of this exercise.*

The next exercise reviews the three ways to define curves and the associated ways to define their tangents.

3. Implicit, Explicit, and Parametric Tangents

Consider the same ellipse shown below and given by the three types of equations:

Implicit:

$$\left(\frac{x}{3}\right)^2 + \left(\frac{y}{4}\right)^2 = 1$$

Explicit:

$$y = 4\sqrt{1 - \left(\frac{x}{3}\right)^2} \qquad or \qquad y = -4\sqrt{1 - \left(\frac{x}{3}\right)^2}$$

Parametric:

$$x = 3\ \text{Cos}[t]$$
$$y = 4\ \text{Sin}[t]$$

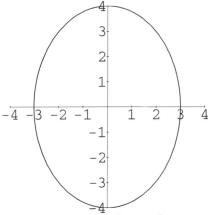

Figure 18.23: $\left(\frac{x}{3}\right)^2 + \left(\frac{y}{4}\right)^2 = 1$

(a) *Calculate the differential for the ellipse in each of the forms.*
    Implicit:

$$? \ dx + ? \ dy = 0$$

Explicit:

$$dy =? \ dx \qquad or \qquad dy =? \ dx$$

Parametric:

$$dx =? \ dt$$
$$dy =? \ dt$$

(b) *Use each of the forms to graph the tangent line to the ellipse at the point where* $x = 3/2$ *and* $y = -2\sqrt{3}$, $(t = -\pi/6)$.

Now, use projection of vectors to compute your tendency to slide down a slippery slope.

4. *In this exercise, we apply a force to a slippery slope and want to know how much acts perpendicular to the slope and how much acts tangent to it. The surface has equation* $z = x^3 - y^2$. *Draw figures to go with each part of the exercise.*

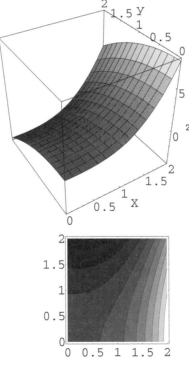

Figure 18.24: $z = x^3 - y^2$ *and* $x^3 - y^2 = constants$

(a) *What 2-D vector* **G** *is perpendicular to the level curve* $0 = x^3 - y^2$ *at the point*

$$\mathbf{X} = \begin{bmatrix} x \\ y \end{bmatrix} = \begin{bmatrix} 1 \\ 1 \end{bmatrix} ?$$

(b) *What 3-D vector* **N** *with upward z-component is perpendicular to the surface* $z = x^3 - y^2$ *at the point* $\mathbf{X} = \begin{bmatrix} x \\ y \\ z \end{bmatrix} = \begin{bmatrix} 1 \\ 1 \\ 0 \end{bmatrix}$?

(c) *A force* $\mathbf{F} = \begin{bmatrix} 1 \\ 2 \\ 3 \end{bmatrix}$ *is pushing up on the surface at the point* $\begin{bmatrix} x \\ y \\ z \end{bmatrix} = \begin{bmatrix} 1 \\ 1 \\ 0 \end{bmatrix}$. *What is the vector projection of* **F** *in the direction of the normal to the surface at this point? Call this normal force vector* $\mathbf{F}_n$.

(d) *Describe geometrically how the force* $\mathbf{F} - \mathbf{F}_n$ *acts at the point* $\begin{bmatrix} x \\ y \\ z \end{bmatrix} = \begin{bmatrix} 1 \\ 1 \\ 0 \end{bmatrix}$.

The next exercise concerns the approximation of a function of two variables and it uses the formulas

$$\frac{de^u}{du} = e^u \qquad \text{and} \qquad \frac{d\,\mathrm{Log}[v]}{dv} = \frac{1}{v}$$

To differentiate $z = x^y$, we first rewrite it using $x = e^{\mathrm{Log}[x]}$, so $z = [e^{\mathrm{Log}[x]}]^y = e^{y\,\mathrm{Log}[x]}$. Now, use the Chain Rule.

5. Continuity of Exponentiation

We are interested in the accuracy of the computation of e^π *when we use the rational approximations* $e \approx 2.718$ *and* $\pi \approx 3.142$. *We let* $x + dx = e$ *and* $y + dy = \pi$. *The decimal approximations are* x *and* y. *We know* $|dx| < 0.001$ *and* $|dy| < 0.001$. *Let*

$$z = x^y$$

and compute the total differential $dz =$?. *Use this to estimate the maximum error in the computation of* e^π *as* $2.718^{3.142}$?

6. The Mother of All Review Exercises

The program **DiffReview** *ends with a general review exercise based on plots.*

Several Variable Optimization - CD

This chapter applies the differential approximation to max-min problems in several variables. It appears only on the CD.

This chapter only on CD

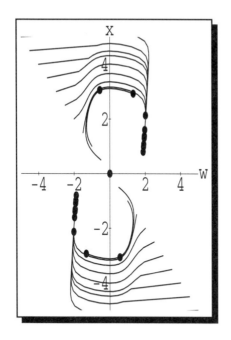

Part 5

Dynamics

Discrete Dynamical Systems - CD

*Dynamical systems are mathematical models of "how things move."
The motion of the bodies in the solar system comes to mind as a
physical dynamical system. The moons and planets move in complicated ways around the sun and one another, each exerting forces
on the others. Mathematical dynamical systems are broadly appli-
cable to other kinds of "dynamics." We study discrete dynamical
systems in this chapter. It appears only on the CD.*

This introduction tries to answer the questions: What is a practical mathematical meaning of "dynamical system?" What does "discrete" or "continuous" mean?

This chapter only on CD

Continuous Dynamical Systems in 1-D

This chapter begins a systematic study of the dynamic "movement" produced by following solutions of a differential equation. A differential equation describes "how things change," and, if we know where we start, we should be able to predict where we go - and how fast. The "systems" we want to study are initial value problems given by a differential equation and a starting position.

We have already seen many examples in which differential equations play a key role in describing change: the S-I-R epidemic, the serious law of cooling introduced in the silly canary story, Bugs Bunny's and Galileo's Laws of Gravity, air resistance for a falling object, and a mathematical definition for irrational exponents. We did not title this chapter "differential equations," because our study does not focus on tricks to "solve" equations; it primarily focuses on the dynamic "movement" produced by the differential equation.

One good analogy for a continuous dynamical system is the "flow" on the surface of a smoothly moving river. The differential equation corresponds to the velocity vector at each point on the surface, whereas the collection of paths of all the water particles constitutes the solution flow. We will find numerical, graphical, and symbolic descriptions of mathematical flows. This will shed new light even on the previous examples such as the S-I-R model, where we have seen the solutions but not analyzed them geometrically. Geometry is a powerful qualitative tool, answering the "where we go" questions. Numerics and symbolics answer the quantitative "how fast" questions.

The most important thing is to learn to "speak" the language of change by expressing how variables change one another. We solve linear equations and a few nonlinear ones explicitly, but we let the computer find the solution flows the rest of the time. This lets us concentrate on the most important issues as follows.

> Describing and Analyzing Continuous Change:
>   (a) What are the assumptions that go into describing change mathematically?
>   (b) What dynamic movement does the change law produce?

Responses to loudness of sounds, brightness of lights, and other broad range phenomena are often nonlinear in order to accommodate the range. The "Weber-Fechner law" even proposes a specific nonlinear psychological response law. These things are awkward to state in English, but really not very hard to understand once you express them mathematically. Here is an example:

Problem 21.1. SOMETIMES DIFF E Q s ARE BETTER THAN ENGLISH ⸺▼

The "Richter scale" of earthquake magnitudes is defined by the following statements:
  (a) *The lowest intensity setting of a seismograph is magnitude 0 at an energy level of 0.01754 kilowatt-hours of energy.*
  (b) *The rate of change of magnitude with respect to energy intensity is inversely proportional to intensity (or, in other words, is proportional to the reciprocal of intensity).*
    Identify the independent and dependent variables in the above statements and give them variable names. Using your variables, write the two statements as a mathematical initial value problem. Show that the solution to your initial value problem is given by the natural logarithm with some appropriate constants. You need more information to determine these constants.
    Here are some actual data:

| Energy Intensity | Magnitude | Earthquake |
millions of kw-h	Richter Scale	
8.8	5.8	Armenia 1988
197	6.7	California 1971
279	6.8	Armenia 1988
$7. \times 10^4$	8.4	Alaska 1964
1.4×10^5	8.6	San Francisco 1906
3.94×10^5	8.9	Japan 1933

Find the Richter magnitude as a function of the energy intensity for each of these. (They are not all exactly the same.) Use your formula from the 1964 Alaska quake (or an average) to compute the magnitude of an earthquake half as intense as the 1964 Alaska quake.

⸻▲

21.1   Exponential Growth and Decay

This section uses the general exponential function $x[t] = x_0\, e^{r\,t}$ and its description by a differential equation and initial value.

This section is self-contained but moves quickly because many of you studied Chapter 8 earlier. It may not be "obvious" at first that $x = e^{r\,t}$ satisfies the differential equation $\frac{dx}{dt} = r\,x$, but this is true and it means that once we know this we do not need to worry about how to solve this equation. Recall that, "If you know the starting value $x[0] = x_0$ and $\frac{dx}{dt} = r\,x$ for an unknown function $x[t]$, then the function must be $x[t] = x_0\, e^{r\,t}$. Conversely, the function $x[t] = x_0\, e^{r\,t}$ has $x[0] = x_0$ and satisfies this differential equation."

Theorem 21.1. *Exponential Functions as Solutions to an Initial Value Problem*
Given two constants x_0 and r, the following are equivalent statements about an
unknown function $x[t]$:

$$x[t] = x_0\, e^{r\,t} \qquad\qquad \Leftrightarrow \qquad\qquad \begin{cases} x[0] = x_0 \\ \dfrac{dx}{dt} = r\,x \end{cases}$$

To show that this function satisfies the differential equation, we differentiate with the Chain Rule,

$$x = x_0\, e^u \qquad\qquad\qquad\qquad u = r\,t$$

$$\frac{dx}{du} = x_0\, e^u \qquad\qquad\qquad\qquad \frac{du}{dt} = r$$

$$\frac{dx}{dt} = \frac{dx}{du}\frac{du}{dt} = x_0\, e^u\, r = r\, x_0\, e^{r\,t} = r\, x[t]$$

Uniqueness of the solution is mathematically important, and we discuss this issue in more detail in Section 21.3.

The description of exponential functions often arises in science from the change law rather than from the explicit formula. Here is a simple starting example:

Example 21.1. *Hourly Algae Growth*

Suppose we introduce algae into a large lake that has plenty of nutrient in the water. Each bit of algae gobbles up nutrient and itself produces new algae, so the more you have, the faster the total increases. If it takes a "mother" algae about one hour to have a "baby," and mothers are giving "birth" asynchronously, we might hypothesize:

"The rate of growth of algae per hour is equal to the amount present."

Let x denote the amount of algae (in grams) in the whole lake. Let t denote the elapsed time (in hours) measured from some starting time. The growth statement above is the differential equation,

$$\frac{dx}{dt} = x$$

because the instantaneous rate of growth is $\frac{dx}{dt}$ and that equals x. Another way to say this is that the rate of growth per mother, $\frac{1}{x}\frac{dx}{dt}$, is one baby per hour.

How much algae will be present after 3 hours? We cannot be expected to answer this question with only this information. Why? Because we do not know how much algae we started with. For example, if we start with no algae, we never get any, whereas if we start with lots, we get much more.

If we also give the initial information, $x[0] = 57$, Theorem 21.1 says

$$x[t] = 57\, e^{1\,t} \qquad\qquad \Leftrightarrow \qquad\qquad \begin{cases} x[0] = 57 \\ \dfrac{dx}{dt} = 1\,x \end{cases}$$

so $x[3] = 57e^3 \approx 1144$ gm.

The following examples cast exponential growth in a general form. This growth law is relatively simple because it is equivalent to saying $x[t] = x_0\, e^{r\,t}$. We want you to understand what the exponential growth laws say in terms of the parameters x_0 and r and to get some practice with the new terminology.

Example 21.2. *General Exponential Growth*

If a population has a constant per-capita fertility rate, independent of size, then its growth is governed by the law

$$\frac{dx}{dt} = r\,x \qquad \text{for some constant } r$$

because $\dfrac{1}{x}\dfrac{dx}{dt} = r$. The expression $\dfrac{1}{x}\dfrac{dx}{dt}$ is the rate of growth per unit of x, or per-capita growth rate. In picturesque language, this says "babies per mother in a unit of time equals r." The constant r is the "fertility rate" of each mother algae. In the hourly example above, we assumed that $r = 1$ - each mother has a baby each hour.

Let us consider a "practical" example of general exponential growth.

Example 21.3. *Exponential Mice*

In the spring, the growth of the mouse population in a field near my house is truly prolific. Let us suppose that it is governed by the law

$$\frac{dx}{dt} = r\,x \qquad \text{for some constant } r$$

where x represents the number of female mice per acre and t is time in days. We want to understand the meaning of the per-capita growth rate parameter, r.

Example 21.4. *Estimating r by One Day's Observation*

Monday morning before I went to work, there were $1{,}000$ (female) mice per acre; 8 hours later, there were $1{,}100$. On average, at this rate, how many babies is each mother having per day?

The change in the population while I was at work is $1100 - 1000 = 100$. There were 1000 mothers, so the average per-capita rate of growth Monday was

$$\frac{1100 - 1000}{1000} = \frac{1}{10} \quad \text{in 8 hours, or} \quad \frac{1}{10} \quad \text{babies per mother in one third of a day}$$

There is an average per-capita rate of growth of $\frac{3}{10}$ babies per mother each day. Since this is a fraction, let us re-express this rate as the average number of days between each mother mouse's birth of a baby,

$$\text{mother-days per baby mouse} = \frac{10}{3} \approx 3.333$$

Each mother mouse has a baby every three and a third days on the average.

Example 21.5. *Measuring r with the Symbolic Solution*

The exact symbolic solution to the initial value problem $x[0] = 1000$ and $\frac{dx}{dt} = r\,x$ is

$$x[t] = 1000\, e^{r\,t}$$

Eight hours later is $t = 1/3$ days, and we know $x[1/3] = 1000\, e^{r/3}$, so

$$1100 = 1000\, e^{r/3}$$

$$e^{r/3} = 1.1$$

$$\text{Log}[e^{r/3}] = \text{Log}[1.1]$$

$$\frac{r}{3} = \text{Log}[1.1] \approx 0.0953102$$

$$r = 3\,\text{Log}[1.1] \approx 0.285931$$

What is the instantaneous rate at which each female mouse is having babies? An answer of 0.285931 is not very satisfying. The reciprocal $1/0.285931 \approx 3.49735$ would have units of "mother-days per baby." Each mother has a baby every three and a half days on average. This is only slightly bigger than the average computation above. (It is larger because it includes old baby mice becoming mothers. As a rough approximation, $r \approx 0.285931 \sim 0.29 \sim 0.3$ and the reciprocals $3.5 \sim 3.3$.)

Exercise set 21.1

1. The Natural Exponential
   *What is the text's official mathematical definition of $x = e^t$? (HINT: Look it up in Chapter 8, and change variables. Also see the program **ExpEquns** from Chapter 8.)*

2. Exponential Growth
   Show that the function $x[t] = 7\,e^{3t}$ satisfies

   $$x[0] = 7$$

   $$\frac{dx}{dt} = 3\,x$$

   *In other words, $x[t] = 7\,e^{3t}$ satisfies this initial value problem. (Plug in $t = 0$ and verify the initial condition. Compute $\frac{dx}{dt}$. Compute $3\,x[t]$. Compare. See the program **ExpEquns** from Chapter 8.)*

3. *Find the exact symbolic solution to the initial value problem*

   $$x[0] = 57$$

   $$\frac{dx}{dt} = -23\,x$$

   *(See the program **ExpEquns** from Chapter 8.)*

4. *Find the exact symbolic solution to the general initial value problem*

   $$x[0] = x_0 \qquad x_0 \text{ a constant}$$

   $$\frac{dx}{dt} = r\,x \qquad r \text{ a constant}$$

   *(See the program **ExpEquns** from Chapter 8.)*

The way we want to study differential equations is to ask questions such as the following. How long does it take to grow a metric ton of algae for the algae growth model

$$x[0] = x_0$$

$$\frac{dx}{dt} = x$$

Mathematically, we want to find t_1 (depending on x_0), such that $x[t_1] = 1,000,000$.

5. *Show that the algae growth model*

$$x[0] = 57$$

$$\frac{dx}{dt} = x$$

*predicts $x[t_1] = 1,000,000$ when $t_1 = \text{Log}[\frac{10^6}{57}] = 6\ \text{Log}[10] - \text{Log}[57] \approx 9.77246$ (see the program **ExpEquns** from Chapter 8).*

6. *Show that the general algae growth model*

$$x[0] = x_0$$

$$\frac{dx}{dt} = r\,x$$

predicts $x[t_1] = 1,000,000$ when $t_1 = \text{Log}[\frac{10^6}{x_0}]/r.$

7. Doubling Time
The function $x[t]$ satisfies the differential equation $\frac{dx}{dt} = 3\,x$ and has $x[0] \neq 0$.
a) Find the time t_2 where $x[t_2] = 2\,x[0]$. Does this depend on the value of $x[0] \neq 0$?
b) If the time t_2 is where $x[t_2] = 2\,x[0]$, show that $x[t + t_2] = 2\,x[t]$ for any time t and explain this identity in words. (HINT: Use a functional identity of the exponential solution for the math. Use the word "doubles" in your verbal description.)

8. Half-Life
The amount x of a radioactive substance at time t satisfies $x[0] = x_0$ and $\frac{dx}{dt} = -r\,x$, for a positive constant r. Show that the half-life of the substance is $t_2 = \text{Log}[2]/r$; that is, the time t_2 makes $x[t_2] = x[0]/2$. Also, show that for any later time t, $x[t+t_2] = x[t]/2$.

9. Growth Law from Info
The spring mouse population in my neighbor's field doubles every 5 days. If they are growing at a constant per-capita rate, what is the constant of proportionality or "fertility" constant r? (HINT: $x[5] = x_0\,2$, $x[10] = x[5]\,2 = x_0\,2^2$, $x[15] = x[10]\,2 = x_0\,2^3$, $x[t] = x_0\,2^{t/5}$. AND $x[t]$ satisfies $\frac{dx}{dt} = r\,x$. Find r by differentiating or by solving $2^{t/5} = e^{r\,t}$.)

Radioactive substances decay at a rate proportional to the amount of the substance present. When a radioactive substance decays, it changes into something else - hence, is gone. Roughly speaking, each atom has an independent probability of emitting a particle; so, the more you have, the more decay you see.

Problem 21.2. EXPONENTIAL DECAY ──────────────────────────────▼

Let x equal the amount of a radioactive substance (in grams), and let t denote the time in years.

  (a) What expression about x represents decay? What does $-\frac{dx}{dt}$ represent?

  (b) Express the statement,

      "The rate of decay is proportional to the amount present."

      as a differential equation.

  (c) Show that the function $x[t] = x_0\, e^{-rt}$ has $x[0] = x_0$ and satisfies your differential equation for the correct choice of a constant r. How is the constant r referred to in the original statement above in quotes?

──▲

Problem 21.3. EXPONENTIAL GROWTH IS FAST ──────────────────────▼

The mass of algae in the pond near Chicago (Lake Michigan) doubles every 6 hours. Show that this is the same as the growth law:

$$\frac{dx}{dt} = r\, x \qquad with \qquad r = \frac{\text{Log}[2]}{6} \approx 0.115525$$

(HINT: Show that $x[t] = x_0\, 2^{t/6}$ satisfies the equation by differentiating.)

   Show that $x[t] = x_0\, e^{rt} = x_0\, 2^{t/6}$.

(HINT: Show that both satisfy the same initial value problem. Read Theorem 21.1.)

   On a nice summer day 1 gram of algae is introduced into the clear water. If this rate of growth persists, how long until all of Lake Michigan is completely full of algae? See the program **ExpGth** in the Chapter 28 folder. Also see Problem 28.2 and Section 8.3.

──▲

21.2   Logistic Growth Laws

A somewhat more realistic algae growth model has the "logistic" form - for example,

$$x[0] = x_0$$

$$\frac{dx}{dt} = x(1 - \frac{x}{3457.6})$$

This model will never produce a metric ton of algae, which will be clear once we understand simply what the growth law says.

First, we will see what this growth law says in English (which is not very good at describing change, but is more familiar). In the process of describing the growth, we will find a simple geometric way to show where x grows and where it declines. This analysis (or direction line) gives us an easy way to see the long-term consequences of this growth law. (Biologically, the law says that the species tends to the carrying capacity 3457.6.)

Exercise 21.2.6 and Example 21.14 give the exact symbolic solution to this problem. If you like, you can go to Section 21.4 now to learn how to find symbolic solutions to some nonlinear equations, including this one. (That section is independent of this one.) However, we will see that it is easier to understand the long-term growth without the symbolic solution in Exercise 21.2.6. The rather complicated formula can tell us how fast certain things happen, but geometry is more powerful and simpler in seeing where the formula tends to go.

Example 21.6. *Logistic Growth in English*

Since we are interested only in solutions when $x > 0$, we divide both sides by x obtaining the per-capita growth law

$$\frac{1}{x}\frac{dx}{dt} = 1 - \frac{x}{3457.6}$$

The left-hand side of this equation is the per-capita rate of growth, that is, the rate of adding new x per unit x, which is new algae babies per algae parent.

Notice that the per-capita growth in our model from the previous section is

$$\frac{1}{x}\frac{dx}{dt} = 1$$

This is the growth law that says each individual produces at unit rate no matter how large the population becomes. The per-capita growth rate in the logistic law changes with x, but when x is small the term $x/3457.6$ does not contribute much change; the two models "start off" the same.

So, how do we express this logistic growth law in English? "The per-capita growth of algae is equal to one minus the fraction of x over 3457.6." This is literally correct but misses the meaning. It is like a word-by-word translation from German to English; all the words are there, but the meaning is lost in a jumble of inappropriate order for the other language.

We can understand this growth law by describing how it works qualitatively. If $x \approx 0$, we have

$$\frac{1}{x}\frac{dx}{dt} = 1 - \frac{x}{3457.6} \approx 1$$

which is approximately our original model, $\frac{dx}{dt} = x$. However, as x grows larger, $1 - \frac{x}{3457.6}$ decreases until $x = 3457.6$, where the right-hand side equals zero. At zero growth, x does not change. If $x > 3457.6$, the right-hand side is negative, which makes the per-capita growth negative. Negative growth is the same thing as decline or decrease; so, if x starts at a value larger than 3457.6, it decreases toward this value. All of this shows

$$(\frac{dx}{dt} < 0 \quad \text{if} \quad x < 0)$$

$$\frac{dx}{dt} = 0 \quad \text{if} \quad x = 0$$

$$\frac{dx}{dt} > 0 \quad \text{if} \quad 0 < x < 3457.6$$

$$\frac{dx}{dt} = 0 \quad \text{if} \quad x = 3457.6$$

$$\frac{dx}{dt} < 0 \quad \text{if} \quad x > 3457.6$$

Example 21.7. *The Vertical Direction Line of Logistic Growth*

The best translation of this information is in the language of geometry. If we mark an
x-line with arrows that point in the direction of the change in x, the above information is
simply the figure

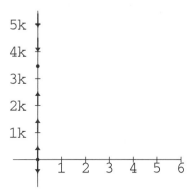

Figure 21.1: Directions of algae growth by value

From this geometric translation, we can see "where things go," namely, all the initial
value problems with $x_0 > 0$, tend toward a limit of 3457.6. Convince yourself of this! It's
easy. In fact, it is easier to use the direction line of this equation than it is to use the explicit
solution. Exercise 21.2.6 will convince you of this.

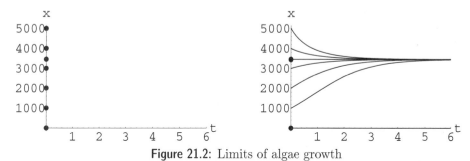

Figure 21.2: Limits of algae growth

Example 21.8. *S-I-S Diseases Are Logistic*

Logistic-like equations arise in other contexts. For example, the Scientific Project on
S-I-S diseases and endemic limits studies the system of equations

$$\frac{ds}{dt} = b\,i - a\,s\,i$$

$$\frac{di}{dt} = a\,s\,i - b\,i$$

where t is time in days, s is the fraction of the population that is susceptible (or well) and i
is the fraction of the population that is infectious (or sick). We know (in this context) that
the sum of the infective and susceptible fractions is one, $s + i = 1$. Substituting $i = 1 - s$

into the first equation gives

$$\frac{ds}{dt} = (b - a\,s)(1 - s)$$
$$= b(1 - \frac{a}{b} s)(1 - s)$$
$$= b(1 - c\,s)(1 - s)$$

where $\frac{a}{b} = c$, the contact number (see Chapter 2). This is just a logistic equation with the origin shifted. If $s = 1/c$ or if $s = 1$, then $\frac{ds}{dt} = 0$ and nothing changes. (Note: $s = 1$ means no one is sick, but $s = 1/c$ is the "endemic limit" when $c > 1$.)

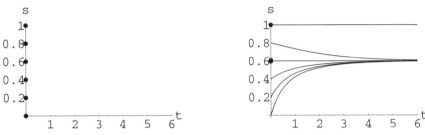

Figure 21.3: Limits of susceptibles, $c = 10/6$

21.2.1 General Terminology

A "differential equation" is a "change question" posed as follows:

GIVEN:

  (a) Where the continuous quantity x starts
  (b) How x changes with respect to t (a continuous variable like time)

FIND:

The function $x[t]$ for future times t
More specifically,

Definition 21.2. *One-Dimensional Initial Value Problem*
A one-dimensional initial value problem at starting time t_0 asks for an unknown function $x[t]$ for $t \geq t_0$ (either exactly or approximately) given:

WHERE WE START: $x[t_0] = a$ *for some constant a*

HOW WE CHANGE: $\frac{dx}{dt} = f[t, x]$ *for a given function of two variables, $f[\cdot, \cdot]$*

Technically, we seek a **function** $x[t]$ defined for $t \geq t_0$ with $x[t_0] = a$ and derivative $x'[t] = \frac{dx}{dt} = f[t, x[t]]$. Notice that the variable x stands for an unknown function in the differential equation but for a scalar variable in the expression $f[t, x]$. (This is confusing, but essential.) The solution function $x[t]$ is *not* the function $f[t, x]$. The function $f[., .]$ is the "growth law," not the solution $x[t]$.

Example 21.9. *A Growth Law Function*

The differential equation $\frac{dx}{dt} = t\,x$ has growth law

$$f[t,x] = t\,x$$

and the solution has the form $x[t] = x_0\,e^{t^2/2}$.

Definition 21.3. *Autonomous Differential Equation*
A differential equation is called autonomous if the "growth function" f does not depend on the independent variable, t. An autonomous equation may be written in the form
$$\frac{dx}{dt} = f[x]$$

Example 21.10. *Autonomous Growth Laws*

The differential equation $\frac{dx}{dt} = 3\,x$ has growth law

$$f[t,x] = 3\,x$$

This is autonomous, and the solution is $x[t] = x_0\,e^{3t}$.

The logistic differential equation above has growth function

$$f[t,x] = x(1 - \frac{x}{3457.6})$$

for the form in Definition 21.2. Notice that $f[t,x]$ does not depend on t in this case.

Similarly, the general logistic growth law $\frac{dx}{dt} = r\,x\,(1 - x/c)$ is given by the function

$$f[t,x] = r\,x\,(1 - x/c)$$

which does not depend on t.

Autonomy of the growth law is essential in the geometric analysis of growth by direction of change (because otherwise the arrows on our x-line would move in time).

Definition 21.4. *Direction Line*
An x-line with arrows pointing in the direction of x-change corresponding to an autonomous differential equation is called a one-dimensional direction field or direction line.

For the logistic example above we drew the direction line as follows:

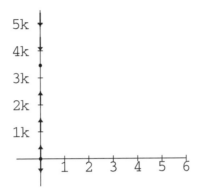

Figure 21.4: Directions of $\frac{dx}{dt} = x(1 - \frac{x}{3457.6})$ on the vertical axis

The t-axis is not needed for a direction line; we could have simply drawn the vertical x-axis and the directions of change.

Points on the direction line with zero x-derivative are stagnant; that is, if you start there, the flow never carries you off. A more official name for this is "equilibrium point."

Definition 21.5. *Equilibria, Attractors, and Repellers*
A point x_e on a direction line for the differential equation $\frac{dx}{dt} = f[x]$ with $f[x_e] = 0$ is called an equilibrium point. An equilibrium point is called a local attractor if the direction arrows point toward it on both sides. An equilibrium point is called a repeller if the direction arrows point away from it on both sides.

It is possible for an equilibrium point to be neither an attractor nor a repeller. If the direction field points toward the equilibrium point on one side and away from it on the other, then the equilibrium point is a "neither."

Exercise set 21.2

1. *Show that the function $x[t] = c e^{t^2}$, for a constant c, is a solution to the differential equation $\frac{dx}{dt} = 2tx$.*
   What is the initial value $x[0] = ?$
   Differentiate $x[t]$.
   Compute $2tx[t]$, and compare this with the previous result.
   Is this an autonomous differential equation?

2. *Write out the function $f[t, x]$ mentioned in Definition 21.2 for the equations:*

   a) $\frac{dx}{dt} = 3x$ b) $\frac{dx}{dt} = 3x(1 - x/5)$ c) $\frac{dx}{dt} = 3x(1 - xCos[t])$

   d) $\frac{dy}{dx} = 3x$ e) $\frac{dy}{dx} = 3y$ f) $\frac{dy}{dx} = 3xy$
   Solve these equations for your choice of initial conditions using the program AccDEsol.

3. Geometric Solution of $dx = r\,x\,dt$, r Constant
   Draw the direction line for the differential equation $\frac{dx}{dt} = r\,x$ and show that zero is the only equilibrium point and that it is a repeller when $r > 0$ and an attractor when $r < 0$.

4. *Show that the equilibrium point of*

$$\frac{dx}{dt} = x^2$$

is neither an attractor nor a repeller. (Which point is the equilibrium? Which way do the direction arrows point on the vertical x-axis?)

5. *Find and classify all the equilibrium points of the differential equation*

$$\frac{dx}{dt} = \text{Sin}[x]$$

*(This equation does not arise naturally but has a simple direction line. A similar-looking equation describes the pendulum.) Check your work with the program **Flow1D**.*

6. Directions Are Easier Than Formulas

The explicit solution to the initial value problem

$$x[0] = x_0$$
$$\frac{dx}{dt} = x(1 - \frac{x}{3457.6})$$

is given by the formula

$$x[t] = \frac{k\,e^t}{1 + \frac{k}{3457.6}e^t}$$

where $k = \frac{x_0}{1 - x_0/3457.6}$. Verify that this is the solution. (Ugh, subsitution, differentiating and showing $\frac{dx}{dt} = x(1 - x/c)$, is messy. You can use the computer to check your differentiation.) If $x_0 > 0$, show that $\lim_{t \to \infty} x[t] = 3457.6$.
*Run the program **Flow1D** to see the dynamic limit of population tending toward 3457.6.*

The formula in the previous exercise can answer quantitative questions such as "how fast" does the solution get somewhere, but the qualitative question of what limit the solution approaches is easier to see simply from the directions of change and a sketch.

7. Stagnant Points
Show that the solution to the initial value problem

$$x[0] = x_0$$
$$\frac{dx}{dt} = x(1 - \frac{x}{3457.6})$$

for $x_0 = 0$ is $x[t] = 0$ and for $x_0 = 3457.6$ is $x[t] = 3457.6$. Why are these the only constant solutions? Is it easier to answer this question using the differential equation or the explicit formula in the previous exercise? (What if you first had to find the explicit solution yourself?)

A typical natural environment can only support so many individuals of a particular species. This is referred to as the carrying capacity of the environment. The logistic algae model above had a carrying capacity of 3457.6.

Problem 21.4. GEOMETRIC ANALYSIS OF GENERAL LOGISTIC GROWTH ▬▬▬▬▬▬ ▼

A general "logistic growth" model is the initial value problem

$$x[0] = x_0$$
$$\frac{dx}{dt} = r\,x\left(1 - \frac{x}{c}\right)$$

for positive constants r and c. Draw a direction line corresponding to this family of initial value problems and explain why all positive solutions tend to the limit $x = c$. The constant c is called the "carrying capacity" because of this result. It represents the number of individuals that the "environment" can support.

How does the constant r affect the solutions? (Consider cases where r is very big and very small but positive.) Does r affect the direction line?

How could you measure the constants r and c in the field?

Show that mathematically $x = 0$ and $x = c$ are the only equilibrium points and that one is a repeller (although the negative directions are biologically meaningless) and the other an attractor. Attraction to the equilibrium $x = c$ is biologically important. Why? (What if the population gets temporarily depressed below c by disease?)

▲

Problem 21.5. THE ENDEMIC LIMIT OF S-I-S DISEASES ▬▬▬▬▬▬▬▬ ▼

This problem studies the S-I-S disease model $\frac{ds}{dt} = b(1 - c\,s)(1 - s)$ for constants $b > 0$ and $c > 1$. (You may want to begin with the case $b = 1$ and $c = 4/3$ as a warm up.)

(a) *Sketch a vertical s axis from $s = 0$ to $s = 1$ and show where the expression*

$$b(1 - c\,s)(1 - s)$$

   is zero.

(b) *Show that the function $s[t]$ satisfying the differential equation is constant, $s[t] = 1$ for all t, if $s[0] = 1$. (HINT: What is the derivative of a constant function?)*

(c) *Show that the function $s[t]$ satisfying the differential equation is constant, $s[t] = 1/c$ for all t, if $s[0] = 1/c$.*

(d) *Show that a function $s[t]$ satisfying the differential equation decreases if $1/c < s[t] < 1$.*

(e) *Show that a function $s[t]$ satisfying the differential equation increases if $0 < s[t] < 1/c$.*

(f) *Show that the value $s_e = 1$ is a repelling equilibrium point.*

(g) *Show that the value $s_e = 1/c$ is an attracting equilibrium point.*

Prove the endemic limit theorem that

$$\lim_{t \to \infty} s[t] = 1/c$$

for any solution to the differential equation with initial value $s[0]$ between zero and one.

*Run the program **Flow1D** to animate this limit.*

How does the parameter b affect the solutions? (You could experiment with the computer for a hint.)

▲

It is not very hard to find out if an equilibrium point is an attractor or repeller in one dimension. In two dimensions, we will see that this is much harder (and more interesting).

There we will use derivatives to find a criterion for attraction. The analogous idea works in one dimension. See if you can figure it out.

Problem 21.6. A CONDITION FOR ATTRACTION OR REPULSION WITH $f'[x_e]$ ───────────▼

Consider an autonomous differential equation $\frac{dx}{dt} = f[x]$ with an equilibrium point x_e. (Two prior exercises had $f[x] = x^2$ and $f[x] = \text{Sin}[x]$.) Show that if $f[x_e] = 0$ and $f'[x_e] < 0$, then x_e is an attractive equilibrium, whereas if $f[x_e] = 0$ and $f'[x_e] > 0$, then x_e is a repulsive equilibrium. What are the specific numerical values of $f'[x_e]$ for the equilibria of $f[x] = x^2$ and $f[x] = \text{Sin}[x]$ in Exercises 21.2.4 and 21.2.4? (HINT: Think of sketching the direction line near an attractor. You know $f[x_e] = 0$ and suppose $x[t]$ would increase for x slightly below x_e. What is the sign of $\frac{dx}{dt}$ when x has such a value? This is the sign of $f[x]$, too. What is the sign of $f[x]$ for x slightly above x_e? If $f[x]$ goes from positive to negative through $f[x_e] = 0$, what is the sign of $f'[x_e]$? What about the converse?)

▲

21.3   Some Helpful Theory

When we say that a model describes a situation, we mean that the mathematical solution is THE solution, not one of several possible behaviors. We give a theorem that says when we have one and only one solution. This theorem is easy to use - just differentiate and examine the result to know that the mathematical formulation of our problem determines THE outcome.

Euler's approximate solution of an initial value problem uses where we are and how we change to move ahead a small step. Once ahead, we know where we are and can compute how we change and move ahead another step. Our basic numerical method for solving initial value problems is quite simple provided you remember the increment equation (or microscope approximation) from Definition 5.5:

$$x[t + \delta t] = x[t] + \frac{dx}{dt}[t]\,\delta t + \varepsilon \delta t$$

where $\varepsilon \approx 0$ whenever $\delta t \approx 0$.

Even though we do not know a formula for the unknown function $x[t]$ in an initial value problem

$$x[0] = x_0$$
$$\frac{dx}{dt} = f[t, x]$$

we do know $x[0] = x_0$ and thus can compute $\frac{dx}{dt}[0] = f[0, x[0]]$. Substituting what we know into the right-hand side of the increment equation above, we obtain new information on the left

$$x[0 + \delta t] = x[0] + f[0, x[0]]\,\delta t + \varepsilon_1 \delta t$$
$$x[0 + \delta t] \approx x[0] + f[0, x[0]]\,\delta t$$

If we substitute this approximate value in the right-hand side of the increment approximation again, we obtain

$$x[t_1 + \delta t] \approx x[t_1] + f[t_1, x[t_1]] \, \delta t + \varepsilon \delta t$$
$$x[t_2] \approx x[t_1] + f[t_1, x[t_1]] \, \delta t$$

Euler's Method is tedious but easy to compute. We begin with the list $\{t_0, x_0\}$ and add $\{t_{n+1}, x_{n+1}\}$ to the list using the computations:

$$x_1 = x_0 + f[t_0, x_0] \, dt \qquad\qquad t_1 = t_0 + dt$$
$$x_2 = x_1 + f[t_1, x_1] \, dt \qquad\qquad t_2 = t_1 + dt$$
$$x_3 = x_2 + f[t_2, x_2] \, dt \qquad\qquad t_3 = t_2 + dt$$
$$x_4 = x_3 + f[t_3, x_3] \, dt \qquad\qquad \cdots$$

where $dt \approx 0$ is a small number.

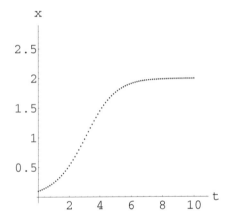

Figure 21.5: Euler solution of $dx = x(1 - x/2) \, dt$

Summarizing Euler's Method, we use the approximation $x[t+dt] - x[t] \approx dx$ and compute $dx = f[t, x] \, dt$. To help remember this, we sometimes write our differential equations using differentials (instead of derivatives),

$$x[t_0] = x_0$$
$$dx = f[t, x] \, dt$$

Further details on the amount of error are described in the Mathematical Background.

The orientation of this course and especially this chapter is to use calculus to describe change in concrete and scientific contexts. So, why do we need any theory? We know where we start and how we change. Can we not just use Euler's Method to see where we go - and how fast?

First of all, if we did write a mathematical model that happened not to have any solution, it would not be very useful in understanding science and the world. Euler's Method might turn out "approximations" that are not converging toward anything. That is not a very common occurrence; most models based on initial value problems have solutions. However, uniqueness of the solution turns out to be problematic.

Uniqueness has important consequences even in terms of sketching solutions. For example, the equilibrium point $x_e = c$ in the logistic equation $dx = r\,x\,(1-x/c)\,dt$ is an attractor; although positive solutions all tend toward $x = c$, they cannot get there in finite time.

Scientifically, when we say a model describes a situation we mean that the mathematical solution is *the* solution, not one of several possible behaviors. If our model mathematically has two or more solutions, then it does not determine the scientific outcome. We need a theorem that says we have solutions and that they determine the behavior completely. This will tell us that the mathematical formulation of our problem is complete.

Example 21.11. *Nonunique Solutions*

How could a single initial value problem have two outcomes? Unfortunately, the answer is, "It's easy." Consider

$$x[0] = 0$$
$$dx = 3\,x^{\frac{2}{3}}\,dt$$

The function $x_1[t] = 0$ is a solution. The function $x_2[t] = t^3$ also is a solution. We can even piece solutions together along the t-axis. For example, the function

$$x[t] = \begin{cases} (t + \frac{3}{2})^3, & \text{if } t < -\frac{3}{2} \\ 0, & \text{if } -\frac{3}{2} \le t \le \frac{1}{2} \\ (t - \frac{1}{2})^3, & \text{if } t > \frac{1}{2} \end{cases}$$

is a solution (shown below with flow lines). Having the dependent variable x in the expression for $\frac{dx}{dt}$ makes it quite different from the explicit formulas $\frac{dx}{dt} = f[t]$ with which we are most familiar.

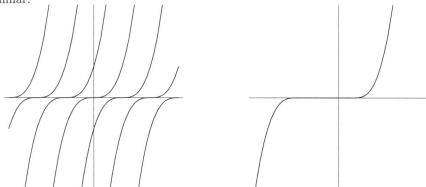

Figure 21.6: Tangent solution curves of $dx = 3\,x^{\frac{2}{3}}\,dt$

Fortunately, there is a simple criterion to use to determine whether an initial value problem has unique solutions. The theorem says that there is a solution and that there is only one.

Theorem 21.6. *Existence & Uniqueness for I. V. P.s*
   Suppose that the functions $f[t, x]$ and $\frac{\partial f}{\partial x}[t, x]$ are continuous in a rectangle around (t_0, x_0). Then, the initial value problem

$$x[t_0] = x_0$$

$$dx = f[t, x]\, dt$$

has a unique solution $x[t]$ defined for some small time interval $t_0 - \Delta < t < t_0 + \Delta$. In this case, Euler's Method converges to $x[t]$ on closed subintervals $[t_0, t_1]$, for $t_1 < t_0 + \Delta$.

This theorem is easy to use: Write the growth law function $f[t, x]$, and calculate its partials to be sure that the functions are continuous. (We know that smooth implies continuous.) The following example shows what can go wrong.

Example 21.12. *Lack of Smoothness*

The growth law function in the differential equation $dx = 3\, x^{\frac{2}{3}}\, dt$ is $f[t, x] = 3\, x^{\frac{2}{3}}$ with partial derivatives $\frac{\partial f}{\partial t}[t, x] = 0$ and

$$\frac{\partial f}{\partial x}[t, x] = 2x^{-\frac{1}{3}} = \frac{2}{x^{1/3}}$$

At the initial value $x[0] = 0$, this partial derivative is undefined; in fact, it has an infinite discontinuity. This means that Theorem 21.6 does not apply, and we saw in Example 21.11 that the I. V. P. starting at zero has nonunique solutions.

The solutions of an initial value problem can "explode to infinity" in finite time. This means that the Δ in the Existence and Uniqueness theorem may be limited.

Problem 21.7. Explosion in Finite Time ────────────────────────────▼

Show that the function $x[t] = \frac{1}{\frac{1}{x_0} - t}$ is a solution to the initial value problem

$$x[0] = x_0$$

$$dx = x^2\, dt$$

Also show that the function $f[t, x] = x^2$ has continuous partial derivatives, so it satisfies the hypotheses of the Existence and Uniqueness Theorem. At what time does x "escape to infinity"? What is the maximum Δ in the Existence and Uniqueness Theorem in this case?

──▲

21.3.1   Wiley Coyote's Law of Gravity

Galileo's Law of Gravity in a vacuum is a simple second-order differential equation,

$$\frac{d^2x}{dt^2} = g$$

that says, "Speed speeds up at a constant rate." We studied this in Chapter 10. It seemed plausible to the tourist in Exercise 4.2 that since, "the farther you go, the faster you fall," that the law of gravity might be,

$$\frac{dx}{dt} = r\,x$$

for some constant r. In Exercise 8.5, we learned that this leads to "Bugs Bunny's Law of Gravity: You don't fall if you don't look down." This silly remark means that the solution to $x[0] = 0$ and the differential equation is $x[t] = 0\,e^{r\,t} = 0$. The moral of Bug's Law is that a linear speed-up in speed is wrong. Next, we consider a nonlinear change that "gets faster the farther you go."

Nonuniqueness causes a different kind of problem in making good scientific statements.

Problem 21.8. WILEY COYOTE'S LAW OF GRAVITY: $\frac{dx}{dt} = \sqrt{2\,g\,x}$ ─────────▼

In a vacuum, an object released from rest and falling under gravity satisfies $x_g[t] = \frac{g}{2}\,t^2$, where x is the distance fallen, t is the time, and g is Galileo's universal gravitational constant (32 ft/sec^2 or 9.8 m/sec^2).

(a) *Show that $x_g[0] = 0$ and $\dfrac{d^2 x_g}{dt^2} = g$.*

(b) *Show that $x_g[0] = 0$ and $\dfrac{dx_g}{dt} = \sqrt{2\,g\,x_g}$. Conclude that Wiley Coyote's Law of Gravity*

$$\frac{dx}{dt} = \sqrt{2\,g\,x}$$

makes a correct prediction. (It does not make only one, however.)

(c) *Show that Bugs' solution $x_b[t] = 0$, for all t, also satisfies Wiley's Law.*

(d) *Show that Wiley's Law says, "You don't fall UNTIL you look down," in the sense that for any fixed $t_1 \geq 0$,*

$$x_w[t] = \begin{cases} 0, & \text{if } 0 \leq t < t_1 \\ \frac{g}{2}(t - t_1)^2, & \text{if } t \geq t_1 \end{cases}$$

is a solution to $x[0] = 0$ and $\frac{dx}{dt} = \sqrt{2\,g\,x}$.

(e) *Show that Wiley's Law does not satisfy the hypotheses of the Existence and Uniqueness Theorem 21.6 when $x_0 = 0$.*

───▲

───────────── **Exercise set 21.3** ─────────────

C

1. Computer and Exact Logistic Growth

Use Euler's Method to solve the initial value problem

$$x[0] = 1 \qquad \& \qquad dx = x(1 - x/2)\,dt$$

*by running the program **EulerApprox**. Verify that this equation satisfies the hypotheses of the Existence and Uniqueness Theorem 21.6. (What is $f[t, x]$? Is it differentiable?)*

2. *Show that the hypotheses of the Existence and Uniqueness Theorem 21.6 does not apply to the function $f[t,x] = 3x^{\frac{2}{3}}$ at a certain initial value. What is the function $f[t,x]$ of Theorem 21.6? Is the partial derivative $\frac{\partial f}{\partial x}$ continuous at all points? Show that the functions $x_1[t] = 0$ and $x_2[t] = t^3$ are both solutions with initial value 0. Are there other solutions?*

3. A Practical Consequence of Uniqueness
   The solutions of the logistic equation $dx = rx(1 - x/c)\,dt$ tend toward the carrying capacity,

$$\lim_{t \to \infty} x[t] = c$$

If $x[0] \neq c$, show that there is never a time t_1 such that $x[t_1] = c$. (HINT: Notice that $x_c[t] = c$ for all t, $-\infty < t < \infty$, is a solution. Suppose another solution satisfies $x_1[t_1] = c$, but $x_1[t]$ is not identically c. Can $x_c[t]$ and $x_1[t]$ both satisfy the differential equation and the initial condition $x[t_1] = c$?)

21.4   Separation of Variables

This section gives the solution method called "separation of variables."

There are many many old tricks for finding exact symbolic solutions of differential equations. We are only interested in two old ones and a new one:

  (a) Guessing that the answer might be an exponential and verifying that it is
  (b) "Separation of Variables"
  (c) Exact computer solutions

The first is used in linear equations, while the second works if we can write the differential equation with all the $x's$ on one side and all the $t's$ on the other. This is the form

$$g[x]\,dx = h[t]\,dt$$

To solve this differential equation, integrate both sides and solve for x. Keep the constants of indefinite integration so your expressions represent families of solutions to the differential equation without initial conditions. It does not always work, but it is so simple that it is helpful to know the idea.

Example 21.13. *Separation of Variables*

We begin with the differential equation

$$\frac{dx}{dt} = x \cdot t$$

which we rewrite as

$$\frac{dx}{x} = t\,dt$$

Now, indefinitely integrate both sides

$$\int \frac{dx}{x} = \int t \, dt$$

$$\text{Log}[x] = \frac{1}{2}t^2 + c$$

$$e^{\text{Log}[x]} = e^{\frac{1}{2}t^2 + c} = e^{\frac{1}{2}t^2} e^c$$

$$x = C \, e^{\frac{1}{2}t^2}$$

Note that we used the exponential functional identity $e^{\alpha+\beta} = e^\alpha \cdot e^\beta$. Also, since e^c is a constant, we may as well introduce a simpler name $C = e^c$. If we want $x[0] = 3$, we let $C = 3$. The function $x[t] = 3e^{\frac{1}{2}t^2}$ is the unique solution to that initial value problem by the Existence and Uniqueness Theorem 21.6, because $f[t, x] = t \cdot x$, so that $\frac{\partial f}{\partial x} = t$ and $\frac{\partial f}{\partial t} = x$ are defined and continuous everywhere.

The steps in the separation of variables can become technical. We could have trouble finding indefinite integrals, though the computer might help there. We could also have trouble solving for x. Here is an example where things still work out with considerable effort.

Example 21.14. *Integration and Solution Troubles for the Logistic Equation*

$$dx = r \, x \left(1 - \frac{x}{c}\right) dt$$

$$\frac{dx}{x\left(1 - \frac{x}{c}\right)} = r \, dt$$

$$\int \frac{dx}{x\left(1 - \frac{x}{c}\right)} = \int r \, dt = r \, t + k \qquad \text{for a constant } k$$

But how do we compute $\int \frac{dx}{x\left(1 - \frac{x}{c}\right)}$?

A trick from high school algebra says that we can write "partial fractions" as follows: Write the combined denominator in terms of separate denominators with unknown constants a and b in the numerator,

$$\frac{1}{x\left(1 - \frac{x}{c}\right)} = \frac{a}{x} + \frac{b}{1 - \frac{x}{c}}$$

$$= \frac{a\left(1 - \frac{x}{c}\right)}{x\left(1 - \frac{x}{c}\right)} + \frac{b \, x}{x\left(1 - \frac{x}{c}\right)}$$

$$= \frac{a - \frac{a}{c}x + b \, x}{x\left(1 - \frac{x}{c}\right)}$$

After expanding in terms of these unknowns, the numerators must match, so $a + (b - a/c) \, x =$

1. Hence, $a = 1$ and $b = \frac{1}{c}$, making

$$\int \frac{dx}{x(1 - \frac{x}{c})} = \int \frac{1}{x} + \frac{1}{c}\frac{1}{1 - \frac{x}{c}}\, dx$$

$$= \int \frac{1}{x}\, dx + \int \frac{1}{c}\frac{1}{1 - \frac{x}{c}}\, dx$$

$$= \text{Log}[x] - \int \frac{1}{u}\, du \qquad \text{where } u = 1 - \frac{x}{c}$$

$$= \text{Log}[x] - \text{Log}[1 - \frac{x}{c}]$$

$$= \text{Log}[\frac{x}{1 - \frac{x}{c}}]$$

Therefore, our separation of variables yields

$$\text{Log}[\frac{x}{1 - \frac{x}{c}}] = rt + k$$

$$e^{\text{Log}[\frac{x}{1 - \frac{x}{c}}]} = e^{rt+k} = e^{k}e^{rt}$$

$$\frac{x}{1 - \frac{x}{c}} = Ke^{rt}$$

Now, we solve for x,

$$x = (1 - \frac{x}{c})Ke^{rt}$$

$$x + \frac{K}{c}e^{rt}x = Ke^{rt}$$

$$x(1 + \frac{K}{c}e^{rt}) = Ke^{rt}$$

$$x = \frac{Ke^{rt}}{1 + \frac{K}{c}e^{rt}}$$

Whew!

Here is an outline of the method:

Procedure 21.7.   Separation of Variables

(a)   Given a differential equation $dx = f[t, x]\, dt$, separate the variables in the form

$$g[x]\, dx = h[t]\, dt$$

(b)   Indefinitely integrate both sides, keeping the "constant of integration" k

$$\int g[x]\, dx = \int h[t]\, dt$$

$$G[x] = H[t] + k$$

(c)   Solve the equation $G[x] = H[t] + k$ for x.

If you cannot do any one of these steps, use another method.

Exercise set 21.4

1. *Which of the following "solutions" of the differential equation make sense?*

$$\frac{dy}{dt} = t^2 \qquad\qquad\qquad \frac{dx}{dt} = x$$

$$\int dy = \int t^2 \, dt \qquad\qquad \int \frac{dx}{dt} = \int x$$

$$y = \frac{1}{3}t^3 + b \qquad\qquad x = \frac{1}{2}x^2 + c$$

Notice that the simple differential equation on the right has x on both sides of the equation unlike the differentiation formulas from earlier in the course. Compare and contrast the "solution" on the right with your completion of the following computation:

$$\frac{dx}{dt} = x$$

$$\frac{1}{x}\, dx = dt$$

$$\int \frac{1}{x}\, dx = \int dt$$

$$\vdots$$

2. Practice with Partial Fraction Integration
Solve for constants a and b so that

$$\frac{1}{(x-3)(x-2)} = \frac{a}{x-3} + \frac{b}{x-2}$$

If these constants are known, use properties of the integral to prove

$$\int \frac{1}{(x-3)(x-2)}\, dx = a \int \frac{1}{x-3}\, dx + b \int \frac{1}{x-2}\, dx$$

Combine the first two parts and use change of variables to show that

$$\int \frac{1}{(x-3)(x-2)}\, dx = \mathrm{Log}[\frac{x-3}{x-2}] + C$$

You can integrate with the computer.

3. Symbolic Solutions and a Check
Use separation of variables to solve the initial value problems:

(a)	$x[0] = 1$	$dx = x\,t\,dt$	
(b)	$x[0] = 0$	$dx = e^{-t-x}\,dt$	
(c)	$x(1) = 0$	$dx = \frac{1}{t}\,dt$	
(d)	$x[0] = 0$	$dx = 2t\,dt$	
(e)	$x[0] = 2$	$dx = x^2\,dt$	
(f)	$x[0] = 0$	$dx = 2x\,dt$	
(g)	$x[0] = 0$	$dx = 3\,x^{2/3}\,dt$	*Which solution is Euler's?*

*Use the program **EULER&exact** along with your solution to compare the Euler ap-proximations to the exact solution of part (2). (This comparison is already done for you for part (1) and part (3) as well as for the logistic equation.)*

Euler's Method is not very accurate. In some examples, the accumulating errors make small enough increment computations infeasible either because there are too many compu-tations or because machine arithmetic errors enter into the accuracy of the computations.

You will also notice computational difficulties in the growth of solutions. Of course, true solutions can "explode" to infinity in finite time and, in that case, Euler's Method will grow very large. However, we know that exponential functions grow very rapidly (see the NoteBook **ExpGth**).

4. Accurate Solutions to Differential Equations
   *Use the program **DEsoln** to solve one of the differential equations above in which Euler's Method seemed rather inaccurate.*

Now apply the method to an interesting application.

Problem 21.9. PARTIAL FRACTION INTEGRATION AND THE S-I-S MODEL ⎯⎯⎯⎯▼

Use separation of variables to find symbolic solutions of the S-I-S differential equation,

$$\frac{ds}{dt} = b(1 - cs)(1 - s)$$

Here is some help: Let c be a constant and s be a variable. Show that

$$\frac{1}{(1-s)(1-cs)} = h\left(\frac{c}{1-cs} - \frac{1}{1-s}\right)$$

for $h = 1/(c-1)$. Use this to rewrite the S-I-S differential equation in the form

$$\left(\frac{c}{1-cs} - \frac{1}{1-s}\right) ds = r \, dt, \qquad \text{where } r = b(c-1)$$

Show that

$$\int \frac{c}{1-cs} \, ds = -\operatorname{Log}[1-cs] + k_1 \qquad and \qquad -\int \frac{1}{1-s} \, ds = \operatorname{Log}[1-s] + k_2$$

so

$$\int \left(\frac{c}{1-cs} - \frac{1}{1-s}\right) ds = ? + k_3$$

for constants k_j.
   Consider the separated differential equation, and show that

$$\operatorname{Log}[\frac{1-s}{1-cs}] = rt + k$$

$$\frac{1-s}{1-cs} = e^{rt+k} = e^k \cdot e^{rt} = K e^{rt}$$

where $r = b(c-1)$ and K is constant.

Solve the equation

$$\frac{1-s}{1-cs} = E$$

for s in terms of c and E, and use your solution to show that the solution to the S-I-S equation has the form

$$s[t] = \frac{1 - k\,e^{-r\,t}}{c - k\,e^{-r\,t}}$$

Now, solve the initial value problem by finding K so that

$$s_0 = \frac{1-k}{c-k}$$

(HINT: $k = (c\,s_0 - 1)/(s_0 - 1)$; you can use Mathematica.)
Finally, use your analytical solution to prove that

$$\lim_{t\to\infty} s[t] = \frac{1}{c}, \qquad for\ 0 < s_0 < 1, \qquad when\ c > 1$$
$$\lim_{t\to\infty} s[t] = 1, \qquad for\ 0 < s_0 < 1, \qquad when\ 0 < c < 1$$

(Don't forget to compute the sign of r in each case.)

Compare the solution to the previous problem with the geometric solution of the "endemic limit" in Exercise 21.5.

21.5 Projects
21.5.1 Logistic Growth with Hunting

A field of logistically growing mice is visited by Voodoo the barn cat. This project examines at what happens to the population.

21.5.2 The Tractrix

A tractor pulls a log through the mud, but the tractor stays on the road and the log starts off the road. Describe the track made by the log.

21.5.3 The Isochrone

A bead slides down a wire that is shaped in such a way that the bead moves the same vertical distance in the "same time." Describe the wire.

21.5.4 The Catenary

A chain made of small links (so it cannot support bending forces) hangs between two poles. Describe the shape.

21.5.5 Linear Nonconstant Coefficient Problems

This project shows how $dx/dt = r[t]x[t]$ is solved by assuming $x[t] = \text{Exp}[R[t]]$ and making a substitution to find $R[t]$.

Continuous Dynamical Systems in 2-D

The conceptual idea of an initial value problem extends easily to vector equations,

$$\begin{cases} \mathbf{X}[0] = \mathbf{X}_0 \\ d\mathbf{X} = \mathbf{F}[t, \mathbf{X}] \, dt \end{cases}$$

The vector formula $d\mathbf{X} = \mathbf{F}[t, \mathbf{X}] \, dt$ should be thought of as the equation telling you an increment $d\mathbf{X} \approx \mathbf{X}[t + dt] - \mathbf{X}[t]$ that you would move in an "instant" dt from the given information t, $x[t]$, and $y[t]$. This is simply Euler's Method in vectors.

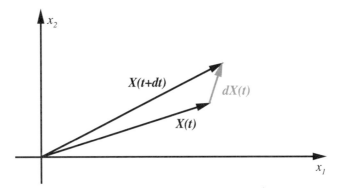

Figure 22.1: Increment of a vector function of t

The basic theory is similar, but the geometry of solutions is much richer. We state the 2-dimensional versions of the theory first and then examine some mathematical examples.

Phase plane analysis gives the whole story geometrically:

    (a) What are the assumptions that go into describing change mathematically?
    (b) How does the phase plane or flow describe the dynamic movement?

22.1 Basic Theory in 2-D

This section lists the 2-D versions of the theory results of the Chapter 21.

A 2-dimensional initial value problem

$$\begin{cases} \mathbf{X}[0] = \mathbf{X}_0 \\ d\mathbf{X} = \mathbf{F}[\mathbf{X}, t]\, dt \end{cases}$$

looks like

$$\begin{cases} \begin{bmatrix} x[0] \\ y[0] \end{bmatrix} = \begin{bmatrix} x_0 \\ y_0 \end{bmatrix} \\ \begin{bmatrix} dx \\ dy \end{bmatrix} = \begin{bmatrix} f[t, x, y] \\ g[t, x, y] \end{bmatrix} dt \end{cases}$$

in 2-dimensional components, or as given in the next theorem without brackets.

Theorem 22.1. *Vector Existence & Uniqueness for I. V. P.s*
Suppose that the functions $f[t, x, y]$, $g[t, x, y]$, $\frac{\partial f}{\partial x}[t, x, y]$, $\frac{\partial f}{\partial y}[t, x, y]$, $\frac{\partial g}{\partial x}[t, x, y]$, and $\frac{\partial g}{\partial y}[t, x, y]$ are continuous in a box around (t_0, x_0, y_0). Then, the initial value problem

$$\begin{cases} x[0] = x_0 \\ y[0] = y_0 \\ dx = f[t, x, y]\, dt \\ dy = g[t, x, y]\, dt \end{cases}$$

has a unique vector solution with component functions $x[t]$ and $y[t]$ defined for some time interval $t_0 - \Delta < t < t_0 + \Delta$. In this case, Euler's Method converges to $x[t]$ and $y[t]$ on closed subintervals $[t_0, t_1]$, for $t_1 < t_0 + \Delta$.

Example 22.1. *Nonautonomous Equations*

The differential equations

$$dx = t(1 - \frac{x}{7} - \frac{y}{7})\, dt$$
$$dy = t(1 - \frac{x}{10} - \frac{y}{5})\, dt$$

have growth functions:

$$f[t, x, y] = t(1 - \frac{x}{7} - \frac{y}{7}) \quad \text{with} \quad \frac{\partial f}{\partial x} = -\frac{t}{7} \quad \text{and} \quad \frac{\partial f}{\partial y} = -\frac{t}{7}$$

$$g[t, x, y] = t(1 - \frac{x}{10} - \frac{y}{5}) \qquad\qquad \frac{\partial g}{\partial x} = -\frac{t}{10} \qquad\qquad \frac{\partial g}{\partial x} = -\frac{t}{5}$$

These functions and partial derivatives are continuous everywhere, so every initial condition leads to a unique solution for at least some time interval.

Geometric analysis of solutions by direction of change only makes sense when the direction of change does not vary with time.

Definition 22.2. *Autonomous Differential Equation*
A vector differential equation of the form

$$\frac{d\mathbf{X}}{dt} = \mathbf{F}[\mathbf{X}]$$

where the function $\mathbf{F}$ *does NOT depend on the independent variable,* t, *is called autonomous. In components, this becomes*

$$dx = f[x, y]\, dt$$
$$dy = g[x, y]\, dt$$

where f *and* g *depend on* x *and* y *but not on* t.

Example 22.2. *Competition Equations*

The differential equations

$$dx = x\left(1 - \frac{x}{7} - \frac{y}{7}\right) dt$$
$$dy = y\left(1 - \frac{x}{10} - \frac{y}{5}\right) dt$$

represent "competition" between x and y. The growth functions are

$$f[x, y] = x\left(1 - \frac{x}{7} - \frac{y}{7}\right) \quad \text{with} \quad \frac{\partial f}{\partial x} = \left(1 - \frac{2x}{7} - \frac{y}{7}\right) \quad \text{and} \quad \frac{\partial f}{\partial y} = -\frac{x}{7}$$

$$g[x, y] = y\left(1 - \frac{x}{10} - \frac{y}{5}\right) \qquad \frac{\partial g}{\partial x} = -\frac{y}{10} \qquad\qquad\qquad \frac{\partial g}{\partial x} = \left(1 - \frac{x}{10} - \frac{2y}{5}\right)$$

and

$$\mathbf{F}[\mathbf{X}] = \begin{bmatrix} r\, x\left(1 - \frac{x}{7} - \frac{y}{7}\right) \\ s\, y\left(1 - \frac{x}{10} - \frac{y}{5}\right) \end{bmatrix}$$

These functions do not depend on t, so the system is autonomous. In addition, the functions and partials are continuous everywhere, so all initial conditions lead to unique solutions for at least a short time interval.

Autonomy is essential to the analysis of growth by just direction of change because otherwise the arrows on our x-y-plane would move in time at the same x-y position. The next section is devoted to analyzing these directions. We record some general terminology now:

Definition 22.3. *Direction Fields*
An x-y-*plane with arrows pointing in the direction of* x-y-*change corresponding to an autonomous differential equation is called a "two-dimensional direction field."*

The solution curves $x[t]$, $y[t]$, plotted parametrically (without t) pass through the direction field tangent to the arrows at every point.

Points on the direction field with both component derivatives zero are "in equilibrium;" that is, if you start there, the flow never carries you off, and the constant function $(x[t], y[t]) = (x_e, y_e)$ for all t is a solution.

Definition 22.4. *Equilibria, Attractors, and Repellers*
   A point (x_e, y_e) is called an equilibrium point for the vector differential equation

$$dx = f[x, y] \, dt$$
$$dy = g[x, y] \, dt$$

if $f[x_e, y_e] = g[x_e, y_e] = 0$. An equilibrium point is called a local attractor if solutions that start from nearby initial conditions tend toward the point. An equilibrium point is called a repeller if solutions that start from nearby initial conditions tend away from the point.

An equilibrium point does not have to be either an attractor or a repeller; some nearby points may move in while others move out.

Exercise set 22.1

1. *Give the functions $f[t, x, y]$ and $g[t, x, y]$ referred to in Theorem 22.1 for the differential equations below. Calculate $\frac{\partial f}{\partial x}$, $\frac{\partial f}{\partial y}$, $\frac{\partial g}{\partial x}$, $\frac{\partial g}{\partial y}$.*

$$dx = t(1 - \frac{x}{3} - \frac{y}{4}) \, dt$$
$$dy = t(1 - \frac{x}{7} - \frac{y}{5}) \, dt$$

and

$$dx = y(1 - \frac{x}{3} - \frac{y}{4}) \, dt$$
$$dy = x(1 - \frac{x}{7} - \frac{y}{5}) \, dt$$

Which of these systems is autonomous?

22.2   Geometric Solution in 2-D

Our first geometric analysis is of the equations

$$dx = r \, x(1 - \frac{x}{7} - \frac{y}{7}) \, dt$$
$$dy = s \, y(1 - \frac{x}{10} - \frac{y}{5}) \, dt$$

for positive constants r and s.

We can think of these equations as describing an interaction between two species. Notice that if $y[t] = 0$ for any time, then $dy = 0$, so y remains zero for all future time. In this case, the first equation becomes $dx = r \, x(1 - x/7) \, dt$, which is a logistic growth equation like the ones we studied above. The points $(0,0)$ and $(7,0)$ are equilibria, with $(0,0)$ repelling along the x-axis and $(7,0)$ attracting along the x-axis. Similarly, if $x = 0$, then $dx = 0$, so x does not change and the equation $dy = s\,y(1 - y/5)\,dt$ is another logistic growth equation with $(0,0)$ repelling along the y-axis and $(0,5)$ attracting along the y-axis.

Example 22.3. *Competition*

The important new ingredient we need to study is the interaction (or coupling) between x and y. First, the change is strictly vertical if x does not change or $dx = 0$. This is the equation

$$r\,x(1 - \frac{x}{7} - \frac{y}{7}) = 0$$

We have already discussed the solution $x = 0$. This is when we are on the y-axis and have the dynamics of $dy = s\,y(1 - y/5)\,dt$. The other solution is the line shown in Figure 22.2,

$$(1 - \frac{x}{7} - \frac{y}{7}) = 0$$

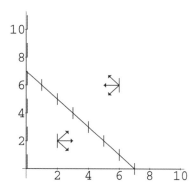

Figure 22.2: Vertical change, $dx = r\,x(1 - \frac{x}{7} - \frac{y}{7})\,dt = 0$

To the left of this line in the first quadrant (where x and y are positive), the value of dx is positive, so the vector

$$dX = \begin{bmatrix} dx \\ dy \end{bmatrix}$$

points "easterly." It may point due east, northeast, or southeast; but it cannot have a westerly component.

To the right of the line $(1 - \frac{x}{7} - \frac{y}{7}) = 0$, the value of dx is negative, so the vector dX points "westerly."

The change is strictly horizontal if y does not change or $dy = 0$. This is the equation

$$s\,y(1 - \frac{x}{10} - \frac{y}{5}) = 0$$

We have already discussed the solution $y = 0$. This is when we are on the x-axis and have the dynamics of $dx = r\,x(1 - x/7)\,dt$. The other solution is the line shown in Figure 22.3,

$$(1 - \frac{x}{10} - \frac{y}{5}) = 0$$

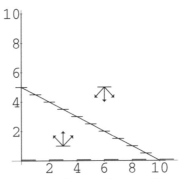

Figure 22.3: Horizontal change, $dy = s\,y(1 - \frac{x}{10} - \frac{y}{5})\,dt = 0$

In the first quadrant below this line, the value of dy is positive, so the vector $d\mathbf{X}$ points "northerly."

Above the line $(1 - \frac{x}{10} - \frac{y}{5}) = 0$, the value of dy is negative and the vector $d\mathbf{X}$ points "southerly."

All the preceding direction information is summarized in Figure 22.4. A more detailed computer calculation of directions for this example is shown to the right of the compass headings.

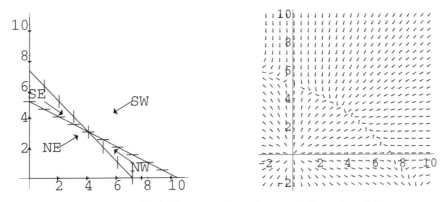

Figure 22.4: Compass directions and direction field
for
$$dx = r\,x(1 - \frac{x}{7} - \frac{y}{7})\,dt$$
$$dy = s\,y(1 - \frac{x}{10} - \frac{y}{5})\,dt$$

You can draw direction fields with the program **DirField**. Either of these figures can be used to sketch solutions of this vector differential equation. Put your pencil down on the compass direction chart and move in the directions indicated or on the computer-drawn direction field and move keeping your motion tangent to the little arrows.

The exact solution curves $(x[t], y[t])$, drawn parametrically, must pass through points in the x-y plane so that their tangents point in the directions of the direction field. Figure 22.5 shows the last frame of a computer animation of many solutions dynamically moving under the change law described by these differential equations. It was created with the program **Flow2D**. The heavy dots are the current positions of points in the solution.

Notice that the point of intersection of the lines

$$1 - \frac{x}{7} - \frac{y}{7} = 0$$

$$1 - \frac{x}{10} - \frac{y}{5} = 0$$

is $(x, y) = (4, 3)$, so that there are equilibrium points at $(4, 3)$, $(7, 0)$, $(0, 5)$, and $(0, 0)$. You can simply look at the solution flow to see that $(4, 3)$ is a local attractor, that $(0, 0)$ is a local repeller, and that the other two attract in some directions and repel in others. We called this example "competition" because, although the two "species" x and y each tend to diminish the other (with the terms $\frac{x}{10}$ and $\frac{y}{7}$), they do balance this competition at the point $(4, 3)$.

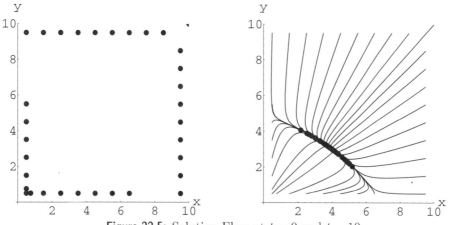

Figure 22.5: Solution Flow at $t = 0$ and $t = 10$

$$dx = r\,x(1 - \tfrac{x}{7} - \tfrac{y}{7})\,dt$$
$$dy = s\,y(1 - \tfrac{x}{10} - \tfrac{y}{5})\,dt$$

Our next example has very similar-looking equations but very different dynamics.

Example 22.4. *Fierce Competition*

This dynamical system is

$$dx = r\,x(1 - \frac{x}{10} - \frac{y}{5})\,dt$$

$$dy = s\,y(1 - \frac{x}{7} - \frac{y}{7})\,dt$$

Notice that if $y[t] = 0$ for any time, then $dy = 0$, so y remains zero for all future time. In this case, the first equation becomes $dx = r\,x(1 - x/10)\,dt$, which is a logistic growth equation like the ones we studied above. The points $(0, 0)$ and $(10, 0)$ are equilibria, with $(0, 0)$ repelling along the x-axis and $(10, 0)$ attracting along the x-axis. Similarly, if $x = 0$, then $dx = 0$, so x does not change and the equation $dy = s\,y(1 - y/7)\,dt$ is another logistic growth equation with $(0, 0)$ repelling along the y-axis and $(0, 7)$ attracting along the y-axis.

The change vector $d\mathbf{X}$ is strictly vertical if x does not change or $dx = 0$. This is the equation

$$r\,x(1 - \frac{x}{10} - \frac{y}{5}) = 0$$

We have already discussed the solution $x = 0$. The other solution is the line shown in Figure 22.6,

$$(1 - \frac{x}{10} - \frac{y}{5}) = 0$$

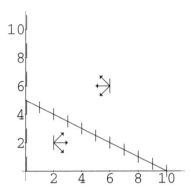

Figure 22.6: Vertical change, $dx = r\,x(1 - \frac{x}{10} - \frac{y}{5})\,dt = 0$

To the left of this line in the first quadrant (where x and y are positive), the value of $d\mathbf{X}$ points "easterly." It may point due east, northeast, or southeast, but it cannot have a westerly component.

To the right of the line $(1 - \frac{x}{10} - \frac{y}{5}) = 0$, the value of dx is negative, so the vector $d\mathbf{X}$ points "westerly."

The change is strictly horizontal if y does not change or $dy = 0$. This is the equation

$$s\,y(1 - \frac{x}{7} - \frac{y}{7}) = 0$$

We have already discussed the solution $y = 0$. This is when we are on the x-axis and have the dynamics of $dx = r\,x(1 - x/10)\,dt$. The other solution is the line shown in Figure 22.7,

$$(1 - \frac{x}{7} - \frac{y}{7}) = 0$$

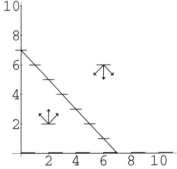

Figure 22.7: Horizontal change, $dy = s\,y(1 - \frac{x}{7} - \frac{y}{7})\,dt = 0$

In the first quadrant below this line, the value of dy is positive, so the vector $d\mathbf{X}$ points "northerly."

Above the line $(1 - \frac{x}{10} - \frac{y}{5}) = 0$, the value of dy is negative and the vector $d\mathbf{X}$ points "southerly."

All the preceding direction information is summarized in Figure 22.8. A more detailed computer calculation of directions for this example is shown to the right of the compass headings.

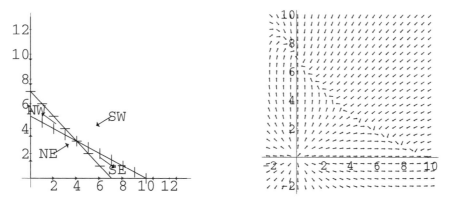

Figure 22.8: Compass directions and direction field

for
$$dx = r\,x(1 - \tfrac{x}{10} - \tfrac{y}{5})\,dt$$
$$dy = s\,y(1 - \tfrac{x}{7} - \tfrac{y}{7})\,dt$$

Use either the compass direction chart or the direction field to sketch several solutions of the differential equations now - just move your pen along the figure staying tangent to the arrows.

The true solution curves $(x[t], y[t])$, drawn parametrically in Figure 22.9, must pass through points in the x-y-plane so that their tangents point in the directions of the direction field.

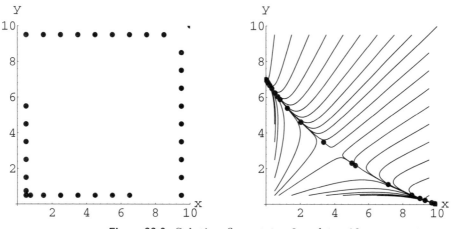

Figure 22.9: Solution flow at $t = 0$ and $t = 10$
$$dx = r\,x(1 - \tfrac{x}{10} - \tfrac{y}{5})\,dt$$
$$dy = s\,y(1 - \tfrac{x}{7} - \tfrac{y}{7})\,dt$$

Notice that the point of intersection of the lines

$$1 - \frac{x}{7} - \frac{y}{7} = 0$$

$$1 - \frac{x}{10} - \frac{y}{5} = 0$$

is still $(x, y) = (4, 3)$, so that there are equilibrium points at $(4, 3)$, $(10, 0)$, $(0, 7)$, and $(0, 0)$. This time the central equilibrium $(4, 3)$ is not attracting. Most solutions that begin in the first quadrant end up either at $(10, 0)$ or at $(0, 7)$. This is ferocious competition in which either x or y wins, killing their whole competing species.

Exercise set 22.2

1. Direction Fields & Flows

Geometrically analyze the differential equations

$$dx = 2x(1 - \frac{x}{4} - \frac{y}{4})\, dt$$

$$dy = y(1 - \frac{x}{3} - \frac{y}{6})\, dt$$

Do the calculations both by hand, as in the above discussion, and check your work with the example program **CowSheep** *example in the* **Flow2D** *program.*

2. *Geometrically analyze the differential equations*

$$dx = x(1 - \frac{x}{8} - \frac{y}{6})\, dt$$

$$dy = y(1 - \frac{x}{6} - \frac{y}{9})\, dt$$

3. *Geometrically analyze the differential equations*

$$dx = x(1 - \frac{x}{6} - \frac{y}{9})\, dt$$

$$dy = y(1 - \frac{x}{8} - \frac{y}{6})\, dt$$

22.3   Flows vs. Explicit Solutions

A rough sketch of the explicit x vs. t and y vs. t plots of the solution of a vector differential equation can be made from the flow.

The last parametric frame of the flow for the fierce competition equations, Figure 22.9, shows a "big picture" that is hard to aquire from individual explicit solutions. However, we can also give a rough sketch of the time-solutions using this picture.

The (vector) solution starting at $(2, 2)$ passes near the central equilibrium but then goes on toward the $(0, 7)$ equilibrium. As explicit solutions, the components look like Figure 22.10.

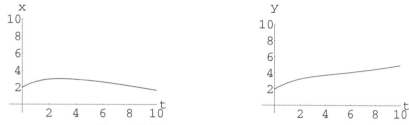

Figure 22.10: Explicit Solutions Starting at (2,2)

The solution starting at $(4, 2)$ tends to the $(10, 0)$ equilibrium and the components look explicitly as shown in Figure 22.11.

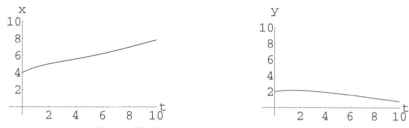

Figure 22.11: Explicit Solutions Starting at (4,2)

The solution starting at $(4, 6)$ tends to the $(0, 7)$ equilibrium and the components look explicitly like Figure 22.12.

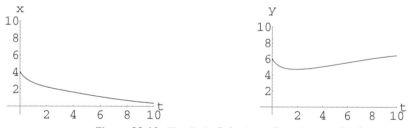

Figure 22.12: Explicit Solutions Starting at (4,6)

The "phase plane" of complete parametric curves gives us a simple geometric way to see how the whole dynamical system works. The static picture loses the "how fast" information, but the computer flow animation captures even that. The next section shows what phase plane analysis adds to your understanding of the S-I-R epidemic model.

Exercise set 22.3

1. Sketching Explicit Solutions

Use the solution flow in Figure 22.9 to sketch the explicit solution $(x[t], y[t])$ to these equations that starts at $(x_0, y_0) = (4, 8)$. Notice that $x[t]$ is decreasing, while $y[t]$ first decreases and then increases again. Check your work with the program SolnViews.

2. *Use the solution flow in Figure 22.5 to sketch the explicit solution* $(x[t], y[t])$ *to the competition equations of Example 22.3. Plot component time-solutions that start at the points*

$$(x_0, y_0) = (4, 8), (4, 6), (4, 1) \, and \, (1, 1)$$

and

$$(x_0, y_0) = (1, 2), (4, 2), (8, 2) \, and \, (8, 8)$$

*(You may check your work with the program **SolnViews**.)*

22.4   Flow Analysis of Applied Models

Now we apply our knowledge to some applications.

22.4.1   Predator-Prey Interactions

The first problem in this section explores a very simple model of a predator and prey species. Rabbits eat an unlimited supply of grass, while foxes eat rabbits, but find them in proportion to their numbers.

Problem 22.1. PREDATOR-PREY INTERACTION ⎯⎯⎯⎯⎯⎯⎯⎯⎯⎯⎯⎯⎯▼

The Lotka-Volterra equations describe a simple predator-prey system. The equations are

$$dx = (p - m\,y)x \; dt$$
$$dy = (-q + n\,x)y \; dt$$

One of these variables represents the density of rabbits, and the other represents the density of foxes in an environment with unlimited space and grass. Analyze the cases in which just one species is absent $(x = 0$ or $y = 0)$ as well as how one species affects the other. Notice that bunnies, $x > 0$, prosper in the absence of foxes, $y = 0$, but that foxes starve, $y > 0$, when bunnies are absent, $x = 0$. How do these facts appear in the model?

Sketch the lines where the vector field is horizontal. Which population has leveled off along these lines?

Sketch the lines where the vector field is vertical. Which population has leveled off along these lines?

Make a compass heading chart and a direction field by hand, and use them to sketch some parametric solutions of the equation.

Show that foxes and rabbits are in equilibrium when $y = p/m$ and $x = q/n$. Biologically, at this level, rabbits just replace their eaten babies.

*Check your work with the example **FoxRabbit** in the **Flow2D** program when $p = 0.1$, $q = 0.04$, $m = 0.005$, $n = 0.00004$.*

After watching the flow animation, you should sketch the solutions as functions of time, x vs. t, and y vs. t.

*Use **SolnViews** to solve the equations with the initial conditions (1000,5), (1000,20). Compare these solutions to your sketches.*

Notice that reducing the fox population below its equilibrium value in an effort to save baby bunnies works initially, causing an explosion in the rabbit population. What happens after that? Is the extermination effort beneficial to the bunnies in the long-term?

22.4.2   The Geometry of S-I-R Epidemics

Recall the differential equations from Chapter 2 that describe an S-I-R epidemic:

$$\frac{ds}{dt} = -a\,s\,i$$

(S-I-R DEs)

$$\frac{di}{dt} = a\,s\,i - b\,i$$

The variables are as follows:

$$t = \text{time measured in days continuously from } t = 0 \text{ at the start of the epidemic}$$

$$s = \text{the fraction of the population that is susceptible}$$

$$i = \text{the fraction of the population that is infected}$$

$$r = \text{the fraction of the population that is removed}$$

with n equal to the (fixed) size of the total population and $r = 1 - s - i$.

The parameter a is the rate at which members of the population mix in a manner sufficiently close to transmit the particular disease. When the contact involves a susceptible and infected person, this results in a new infection. This means that the number of new infected people per day is a times the number of infectious people $(i \times n)$ times the fraction of contacted people who are still susceptible, $s = \frac{\text{number susceptible}}{n}$:

$$\text{number of new infected per day} = a\,s\,i \times n$$

The fractional rate of new cases is $a\,s\,i$, because we divide by n to get the fraction of the population.

The parameter b is,

$$b = 1/(\text{the number of days infectious})$$

If the population is large and $i \cdot n$ people are infectious, we expect that, on average, the fraction $1/(\text{the number of days infectious})$ of those people will recover tomorrow. For example, if each person is infectious for 11 days, on average, 1/11th of the sick people got the disease 10 days ago and will recover tomorrow. This means that $b \cdot i \cdot n$ is the daily rate of recovery, and dividing by n gives the daily fraction of recovery, $b \cdot i$.

Inasmuch as new cases are removed from the susceptible fraction,

$$\frac{ds}{dt} = -a\,s\,i$$

New cases enter the infectious fraction and recoveries leave, so

$$\frac{di}{dt} = a\,s\,i - b\,i$$

The ratio

$$c = \frac{a}{b}$$

is called the "contact number" and intuitively represents the average number of contacts each infected person has over the course of the disease. (In Chapter 2, c was measured by the formula $c = (\text{Log}[s[\infty]] - \text{Log}[s[0]])/(s[\infty] - s[0])$ that came from simple tricks of calculus.) Typical values of a and b for rubella are

$$b = 1/11$$
$$c = 6.8$$
$$a = bc = 6.8/11$$

You found explicit computer solutions to these differential equations in Chapter 2, and you can recompute these with **AccAccDEsoln** if they are lost. However, the explicit solutions are not the best way to understand some aspects of the epidemic model. For example, what is a formula for

$$\lim_{t \to \infty} s[t] = ?$$

Or, at least, how does the number of people who escape infection during the epidemic depend on the number of people who are susceptible at the start of the epidemic? These questions become easier to understand with a hand-drawn sketch of the flow of the S-I-R equations.

Problem 22.2. ESCAPING INFECTION FOR INDIVIDUALS AND THE HERD ──────▼
Make a compass direction chart and hand sketch of parametric solutions to the S-I-R epi-demic differential equations. Analyze your flow and explain whether you would be more likely to escape infection in a rubella epidemic if it began with many more or relatively few people, besides you, susceptible.

*Modify the **Flow2D** program to animate the flow of an epidemic and compare the movie with your hand sketch. (It is helpful in a hand sketch to exaggerate the parameters, for example, using $c = 3$.)*

*Where does $\lim_{t \to \infty} s[t]$ show up on your phase plane and in the computer movie of the flow? How does this limit change when $s[0]$ is nearer to 1? Use the program **AccDEsoln** to solve the cases with 60% or with 90% of the population initially susceptible. Start both cases with 1% infected. Compare the explicit solution plots with the **Flow2D** plots.*

What is the significance of the line $s = 1/c$ on your phase diagram? In particular, how many people in a population must be immune in order to ensure that the disease decreases? (In other words, which portion of your compass heading chart has i decreasing, or equiva-lently, $\frac{di}{dt}$ negative?) Answer this in terms of parameters and s alone. Would you have an expanding epidemic in a population with $s < 1/c$?

──▲

22.4.3 S-I-R Epidemics with Births and Deaths

The previous model of an epidemic would not be useful in the long-term study of the disease if new babies were born without immunity to an S-I-R disease. For simplicity, we will assume that births equal deaths so that the size of our population remains constant.

Variables for S-I-R with Births=Deaths:

$$t \;=\; \text{time in days with } t = 0 \text{ at outbreak}$$
$$s \;=\; \text{susceptible fraction}$$
$$i \;=\; \text{infected fraction}$$
$$r \;=\; \text{removed fraction with } s + i + r = 1$$

Again, we have the parameters a and b where

$$a\,s\,i \;=\; \text{rate of spreading the infection}$$
$$b\,i \;=\; \text{rate of recovery}$$

With births and deaths we add a rate parameter k so that there are

$$k \cdot n = \text{total number of births per day} = \text{total number of deaths per day}$$

and we assume that deaths are from causes other than the disease we study, so deaths leave the compartments at rates proportional to their sizes as indicated in the figure below.

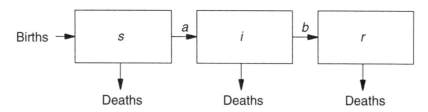

Figure 22.13: S-I-R with birth and death

The rates of deaths are

$$k \cdot s \cdot n \;=\; \text{number of dead susceptibles per day}$$
$$k \cdot i \cdot n \;=\; \text{number of dead infectives per day}$$
$$k \cdot r \cdot n \;=\; \text{number of dead removeds per day}$$

Births equals deaths because $k \cdot s \cdot n + k \cdot i \cdot n + k \cdot r \cdot n = k(s + i + r)n = k \cdot 1 \cdot n$.

The change in number of susceptibles is now

$$\text{Births} - \text{deaths of susceptibles} - \text{infections}$$

$$n\frac{ds}{dt} \;=\; k\,n - k\,s\,n - a\,s\,i\,n$$

so

$$\frac{ds}{dt} \;=\; k - k\,s - a\,s\,i$$

The sketch of the compass headings for the equations

$$\frac{ds}{dt} = k - k\,s - a\,s\,i$$
$$\frac{di}{dt} = a\,s\,i - b\,i - k\,i$$

is a little more difficult than earlier examples. No change in i, or $\frac{di}{dt} = 0$, is easy, $a\,s\,i - b\,i - k\,i = 0$ if and only if either $i = 0$ (the s-axis) or $a\,s\,i - b - k = 0$, $s = (b+k)/a$ (a vertical line).

No change in s or $\frac{ds}{dt} = 0$ occurs on a hyperbola, $k - k\,s - a\,s\,i = 0$. This might be a little confusing with all the letters. Notice that $x\,y = c$ is a simple hyperbola and a change of variables makes this equation look similar,

$$k - k\,s - a\,s\,i = 0 \quad \Leftrightarrow \quad \frac{k}{a} = s\left(i + \frac{k}{a}\right) \quad \Leftrightarrow \quad c = s\,j$$

In any case, we can use the computer to plot these curves as follows:

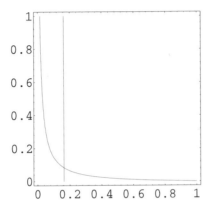

Figure 22.14: Curves of no change in s and i.

Exercise set 22.4

1. Show that the differential equation for the infectious fraction i is

$$\frac{di}{dt} = a\,s\,i - b\,i - k\,i$$

   and that $r = 1 - s - i$.

2. Sketch a compass heading chart for the S-I-R with births equations. Use the computer to help, if you wish.
   Modify the programs **AccDEsoln** and **Flow2D** to compute s and i for a model of S-I-R with births=deaths, $k = 1$ births per 1000 population, $k = 0.001$, and the parameters above for rubella, $c = 6.8$, $b = 1/11.0$, and $a = b\,c$. What does the model predict about endemic infection,

$$\lim_{t \to \infty} i[t] \neq 0?$$

   When $\frac{a}{b+k} > 1$, use the program **Flow2D** to show that

$$s \to \frac{b+k}{a} \quad \text{and} \quad i \to \frac{k}{a}\left(\frac{a}{b+k} - 1\right) \text{ as } t \to \infty$$

   Verify this in several explicit examples using **AccDEsoln**.

Prove that $(\frac{b+k}{a}, \frac{k}{a}(\frac{a}{b+k} - 1))$ is an equilibrium point. Why do we need $\frac{a}{b+k} > 1$? Are other conditions needed to make $0 < i_{equil} < 1$?

Following are two explicit solutions, but you will see that interpretation of the "phase portrait" or last frame of the flow animation reveals more about the long-term dynamics of the disease.

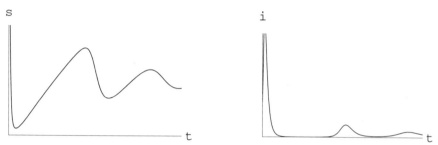

Figure 22.15: Two years of rubella infectives with births

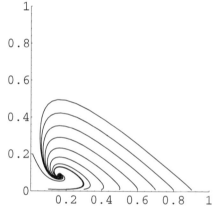

Figure 22.16: S-I-R with births = deaths and large $k = 0.01$

22.5 Projects

You are ready now to begin any of these projects and do the important first step of setting up the equations, or "speaking the language of change." You can solve these projects with approximate computer solutions of the associated differential equations, but Chapters 23 and 24 develop additional mathematical tools that can help in that analysis.

22.5.1 Herd Immunity and the S-I-S Endemic Limit

The mathematical tools that we have developed in this chapter make the epidemic projects on Herd Immunity and the S-I-S Endemic Limit into simple exercises. The questions raised in those projects are important ones like, "Why did we eradicate polio, but not measles?" The point is that calculus does give us deep insight into many scientific questions about changing quantities.

22.5.2 Mathematical Ecology

Several examples and exercises in this chapter hint at the ecological projects on competition between species. There are three extensive models in the Scientific Projects on this topic. One explores predator-prey interactions and the effect of disturbing the ecosystem. Another explores competition between species, and a third explores cooperation.

22.5.3 A Production and Exchange Economy

There is a model of production and exchange in the Scientific Projects on economics.

22.5.4 Stability of Chemical Reactions

There are projects on chemical reactions and reaction stability.

22.5.5 Low-Level Bombing

This project analyzes the affect of air resistance when aiming a bomb from a high speed airplane.

22.5.6 The Pendulum

This project studies a nonlinear pendulum. A real pendulum on a rigid arm has a period that depends on its intial position and velocity. It also has an unstable equilibrium at the top of its swing.

The Bungee-Jumping project, the air resistance project, the project on low level bombing, the study of the real and idealized pendulum all can be understood using the ideas of this chapter.

The linear differential equations in Chapter 23 have important applications in their own right, such as Lanchester's combat model and the resonance and filtering Scientific Projects. Linear equations are also important mathematically. They can be solved in terms of natural exponentials as we show in Chapter 23. These solutions then give us a new kind of "microscope." When we "look at a nonlinear equilibrium under a microscope," we see an associated linear equilibrium. Such local analysis is the subject of Chapter 24.

23

Linear Dynamical Systems

This chapter shows how to find symbolic solutions to 2-dimensional linear dynamical systems. These systems are closely related to second-order one-dimensional differential equations.

We begin this chapter by deriving a second-order differential equation for a familiar mechanical example: a simplified version of your car's suspension. Next, we show why the second-order equation is equivalent to a two-dimensional system such as the ones in the Chapter 22. Then, we show how to use exponentials $e^{r\,t}$ to solve these equations.

Solutions to linear systems are interesting in their own right but also give rise to a "dynamical system microscope" which is explained in the next chapter. This lets us apply the main idea of differential calculus to dynamical systems - namely that, under a microscope nonlinear systems appear linear. This chapter shows what the linear ones look like, so we can compute what we see in a microscopic view of the nonlinear ones.

23.1   The Shock Absorber Equation

This section sets up a differential equation describing the motion of an object of mass m suspended on a spring and moving in a fluid that acts like your car's shock absorber.

Newton's "$F = m\,A$"-law for this system becomes

$$-c\frac{dx}{dt} - s\,x = m\frac{d^2x}{dt^2}$$

where x is the displacement of the spring and t is time. The spring produces a restoring (negative) force $s\,x$ when stretched x, for the spring constant s. The fluid produces a force $-c\frac{dx}{dt}$ for the "damping" constant c. Mass times acceleration is $m\frac{d^2x}{dt^2}$ because the second derivative of position is acceleration.

You may think of this equation as the one governing the reaction of (one wheel of) your car after you go over a bump. The constants m, c, and s are a fixed part of the design of your car, at least for short periods of time. As your shock absorbers wear out, c becomes smaller. How small can c become before your car oscillates after it goes over a bump? You will be able to answer this after you complete this chapter. You know intuitively that completely worn out shocks ($c \approx 0$) make your car bounce and bounce, whereas very stiff shocks ($c >> 0$) jolt the inside passengers as they restore the level of the car.

Now, we give the details of why the mass-spring-shock absorber system satisfies the second-order differential equation above. Newton's Law says, "$F = m\,A$", the total applied force equals the mass times the acceleration that force produces. We need to use the interpretation of the first and second derivatives as velocity and acceleration (see Chapter 10 for a review).

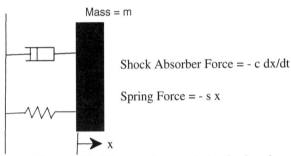

Figure 23.1: Mass, spring, and shock absorber

The position of our object of mass m is measured by x as shown in Figure 23.1, where the spring is relaxed when $x = 0$. The derivative $\frac{dx}{dt}$ represents the velocity of the object, or rate of change of position with respect to time, including sign - $(+)$ to the right and $(-)$ to the left. The second derivative $\frac{d^2x}{dt^2}$ is the acceleration of the object or rate of change of velocity (with sign, so negative acceleration is "braking.") The "$m\,A$" portion of Newton's Law becomes $m\,\frac{d^2x}{dt^2}$ and the law says,

$$m\,\frac{d^2x}{dt^2} = \text{force of the spring} + \text{force of the shock absorber}$$

The restoring force when we stretch a spring by a distance x is $-s\,x$. It is negative because it acts opposite its displacement. This formula also "pushes" when the spring is compressed (unlike the slack bungee cord). If the spring is compressed, $x < 0$, and $-s\,x > 0$, $[(-)(-) = (+)]$, so the spring pushes to the right. This formula makes "$F = m\,A$" look like:

$$\text{Force of the spring} = -s\,x$$

$$m\,\frac{d^2x}{dt^2} = -s\,x + \text{force of the shock absorber}$$

The force from the shock absorber is $-c\,\frac{dx}{dt}$. This is a perfectly linear response to movement, with the force opposite the direction of movement. (This friction force is like the project on air resistance.) The intuitive idea of this force, if not its linear nature, should be familiar. If you dip a spoon in a honey jar and move it very slowly, it resists relatively little. But if you try to move the spoon quickly, it resists a lot. We are simply making the relationship a proportion:

$$\text{Force resisting motion} \propto -\text{Speed of that motion}$$

$$\text{Force of an ideal shock absorber} = -c\,\frac{dx}{dt}$$

$$m\,\frac{d^2x}{dt^2} = -s\,x + \text{Force of the shock absorber}$$

$$m\,\frac{d^2x}{dt^2} = -s\,x - c\,\frac{dx}{dt}$$

Hence, this is the differential equation governing the response of a car's suspension or "front end" with a linear spring and shock a absorber.

Exercise set 23.1

1. *Write the equations for a heavy rider's mountain unicycle of total mass* 100 *kg suspended on a spring with* $s = 200$ *nt/m and damping* $c = 300$. *What are the units of* c?

2. *Write the equations for a unicycle of* **weight** 320 *lbs suspended on a spring with* $s = 200$ *lbs/ft and damping* $c = 300$ *lbs/(ft/sec). The units of* m *must be mass, called "slugs" in English units. The weight of* 320 *lbs is the force of gravity on a mass* m *and the gravitational constant (from Galileo's Law) is* $g = 32$ *ft/sec^2,* $F_{gravity} = m{\cdot}g = 320$.

23.2   Constant Coefficient Systems

A second-order equation $a\frac{d^2x}{dt^2} + b\frac{dx}{dt} + cx = 0$ *can be re-written as a 2-D system using the "phase" variable.*

We want to analyze the shock absorber equation in several ways. We can view it as a 2-dimensional system by introducing the "phase variable" $\frac{dx}{dt} = y$. Since $\frac{dy}{dt} = \frac{d}{dt}\left(\frac{dx}{dt}\right) = \frac{d^2x}{dt^2}$, the differential equation for the shock absorber becomes

$$m\frac{dy}{dt} = m\frac{d^2x}{dt^2} = -c\frac{dx}{dt} - s\,x = -c\,y - s\,x$$

If we put these two equations together, we have an equivalence between the system of first-order equations and the single equation with two derivatives.

Physicists call the combined position and momentum the phase of a body. The use of the vector phase $(x, \frac{dx}{dt}) = (x, y)$ is why the final flow picture is often called the phase portrait.

In general, we have

Definition 23.1. *The Phase Variable Trick*
     The phase variable form of the equation

$$a\frac{d^2x}{dt^2} + b\frac{dx}{dt} + cx = 0$$

(for constants $a \neq 0$, b, c) is given by the pair of equations

$$\frac{dx}{dt} = y$$

$$\frac{dy}{dt} = -\frac{c}{a}x - \frac{b}{a}y$$

This system is linear, a variant of

$$\frac{dx}{dt} = a_1 x + a_2 y$$

$$\frac{dy}{dt} = b_1 x + b_2 y$$

$$\Leftrightarrow \qquad \begin{bmatrix} \frac{dx}{dt} \\ \frac{dy}{dt} \end{bmatrix} = \begin{bmatrix} a_1 & a_2 \\ b_1 & b_2 \end{bmatrix} \cdot \begin{bmatrix} x \\ y \end{bmatrix} \qquad \Leftrightarrow \qquad \frac{d\mathbf{X}}{dt} = \mathbf{A}\,\mathbf{X}$$

where the constants are $a_1 = 0$, $a_2 = 1$, $b_1 = -c/a$ and $b_2 = -b/a$. In Chapter 24, we show how to convert a 2-dimensional system into a second-order equation. The two things are almost equivalent.

Complete understanding of this system will lead to a way to "localize" nonlinear problems and study equilibria with the basic idea of calculus that "nonlinear phenomena locally look linear."

Exercise set 23.2

---

1. *Write the second-order equations below as 2-D systems using the phase variable, and write your specific answers in both the classical and matrix forms:*

$$\frac{dx}{dt} = a_1 x + a_2 y$$

$$\frac{dy}{dt} = b_1 x + b_2 y$$

$$\Leftrightarrow \qquad \begin{bmatrix} \frac{dx}{dt} \\ \frac{dy}{dt} \end{bmatrix} = \begin{bmatrix} a_1 & a_2 \\ b_1 & b_2 \end{bmatrix} \cdot \begin{bmatrix} x \\ y \end{bmatrix}$$

a) $\frac{d^2x}{dt^2} + 2\frac{dx}{dt} + 3x = 0$ b) $3\frac{d^2x}{dt^2} + \frac{dx}{dt} + 2x = 0$

c) $\frac{d^2x}{dt^2} + 2\frac{dx}{dt} + x = 0$ d) $\frac{d^2x}{dt^2} + 3\frac{dx}{dt} + 2x = 0$

23.3 Symbolic Exponential Solutions

In this section, we find two "really different" solutions to our linear differential equation. The "method" of solution is simple. We will guess that the solution is exponential and figure out how to make our guess work.

Linear autonomous equations do have exponential solutions (in one form or another). To make the algebra simple, we first consider the second-order equation in the form

$$a\frac{d^2x}{dt^2} + b\frac{dx}{dt} + cx = 0$$

How many solutions can we find of the form $x[t] = e^{rt}$? We just plug in

$$x = e^{rt} \qquad\qquad cx = ce^{t}$$

$$\frac{dx}{dt} = re^{rt} \qquad\qquad b\frac{dx}{dt} = bre^{rt}$$

$$\frac{d^2x}{dt} = r^2 e^{rt} \qquad\qquad a\frac{d^2x}{dt^2} = ar^2 e^{rt}$$

so that the sum is

$$a\frac{d^2x}{dt^2} + b\frac{dx}{dt} + cx = (ar^2 + br + c)\cdot e^{rt}$$

We know that $e^{rt} \neq 0$ for any time, so the only way to make $a\frac{d^2x}{dt^2} + b\frac{dx}{dt} + cx = 0$ is to make

$$(ar^2 + br + c) = 0$$

We have reduced our calculus problem to high school algebra: Solve $(ar^2 + br + c) = 0$. This algebraic equation is called the "**characteristic equation**" of the differential equation, and its roots are called the "**characteristic roots**." The quadratic formula gives us the roots.

Procedure 23.2. Characteristic Roots & Solutions

The differential equation

$$a\frac{d^2x}{dt^2} + b\frac{dx}{dt} + cx = 0$$

has two distinct real solutions given by

$$x_1[t] = e^{r_1 t} \qquad \text{and} \qquad x_2[t] = e^{r_2 t}$$

when the roots

$$r_1, r_2 = \frac{-b \pm \sqrt{b^2 - 4ac}}{2a}$$

to the characteristic equation, $ar^2 + br + c = 0$, are distinct real numbers.

Exercise set 23.3

1. Verify Solutions
   Show that $x_1[t] = e^{2t}$, $x_2[t] = e^{-3t}$ *and* $x_3[t] = k_1 e^{2t} + k_2 e^{-3t}$ *are all solutions to the differential equation*

$$\frac{d^2x}{dt^2} + \frac{dx}{dt} - 6x = 0$$

   In each case, calculate the initial values $x_1[0]$, $x_2[0]$, $x_3[0]$, $x_1'[0]$, $x_2'[0]$, *and* $x_3'[0]$.
   Find the roots r_1 *and* r_2 *of the algebraic equation*

$$r^2 + r - 6 = 0$$

   Write the specific exponentials $e^{r_1 t}$ *and* $e^{r_2 t}$.

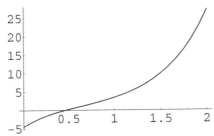

Figure 23.2: $x[t] = \frac{1}{2} e^{2t} - 5 e^{-3t}$

Usually, we can get by with high school algebra.

2. Find Solutions by Algebra
   Find two distinct exponential solutions of the equations:
   (a) $\frac{d^2x}{dt^2} - \frac{dx}{dt} - 6x = 0$
   (b) $\frac{d^2x}{dt^2} - 5\frac{dx}{dt} + 6x = 0$
   (c) $\frac{d^2x}{dt^2} + \frac{dx}{dt} - x = 0$
   Check and sketch your solutions with the computer.

3. Strange Roots
   Why does not the procedure of the last exercise work on the following equations?
   (a) $\frac{d^2x}{dt^2} - 6\frac{dx}{dt} + 9x = 0$
   (b) $\frac{d^2x}{dt^2} + \frac{dx}{dt} + x = 0$

Problem 23.1. REPEATED ROOTS ——————————————————————▼
Show that the functions

$$x_1[t] = e^{r_1 t} \qquad and \qquad x_2[t] = t e^{r_1 t}$$

are both solutions to the differential equation

$$\frac{d^2x}{dt^2} - 2r_1\frac{dx}{dt} + r_1^2 x = 0$$

The constant r_1 *satisfies* $(r - r_1)^2 = r^2 - 2r_1 r + r_1^2 = 0$.

Substitute each function into the left-hand side of the differential equation.

What are the characteristic roots of this equation? (For example, take $r_1 = -1$, so the equation is $\frac{d^2x}{dt^2} + 2\frac{dx}{dt} + x = 0$.)

Check and plot these solutions with the computer.

Show that the function $x_2[t] = e^t$ is a solution, whereas $x_3[t] = t\,e^t$ is NOT a solution to $\frac{d^2x}{dt^2} - x = 0$. What is another solution?

▲

23.4   Rotation and Euler's Formula

If $i = \sqrt{-1}$, then

$$e^{i\theta} = \mathrm{Cos}[\theta] + i\;\mathrm{Sin}[\theta]$$

Recall the following basic mathematical fact: The function $z[t] = e^{r\,t}$ is the solution of

$$z[0] = 1$$
$$\frac{dz}{dt} = r\,z$$

Also recall (or look up) the Existence and Uniqueness Theorem for Initial Value Problems, apply its hypotheses (the function $f[x] = r\,x$ is differentiable), so that the solution to this problem is unique - $z[t] = e^{r\,t}$ is the only possible solution.

We extend our "official definition" of the exponential function to complex numbers, taking $e^{i\,t} = z[t]$ to be the unique solution of the equation above when $r = i = \sqrt{-1}$. We show geometrically that the solution is $z[t] = \mathrm{Cos}[t] + i\;\mathrm{Sin}[t]$ when $r = i = \sqrt{-1}$. Together, these two things show that

The initial value problem

$$z[0] = 1$$
$$\frac{dz}{dt} = i\,z$$

(a) has the solution $z[t] = e^{i\,t}$, by the "official definition"
(b) has the solution $z[t] = \mathrm{Cos}[t] + i\;\mathrm{Sin}[t]$
(c) has only one solution; hence, $e^{i\,t} = \mathrm{Cos}[t] + i\;\mathrm{Sin}[t]$

This is a proof of Euler's important and useful formula. Euler's formula allows us to find real solutions to our differential equation when the characteristic roots are complex.

Complex numbers are vectors with a special multiplication peculiar to two dimensions. We may think of the complex number $z = x + i\,y$ as the vector

$$z = \begin{bmatrix} x \\ y \end{bmatrix}$$

We simply treat the factor of the "imaginary" number $\sqrt{-1} = i$ as the y-axis component. The complex product can be computed algebraically: $(a + i\,b)(x + i\,y) = a\,x + i\,a\,y + i\,b\,x + i^2\,b\,y = (a\,x - b\,y) + i\,(a\,y + b\,x)$.

The vector associated with the complex number $z = x + i\,y$ is perpendicular to the vector $i\,z = -y + i\,x$. To see this, compute the dot product

$$\begin{bmatrix} x \\ y \end{bmatrix} \bullet \begin{bmatrix} -y \\ x \end{bmatrix} = -x\,y + x\,y = 0$$

Vectors are perpendicular when their dot product is zero.

Now, consider the complex differential equation

$$\frac{dz}{dt} = iz \qquad \text{or} \qquad \frac{dx}{dt} + i\frac{dy}{dt} = -y + i\,x$$

Write this in vector form

$$\begin{bmatrix} \frac{dx}{dt} \\ \frac{dy}{dt} \end{bmatrix} = \begin{bmatrix} -y \\ x \end{bmatrix}$$

This vector equation says that the rate of change of the vector z has the same size as z and is a vector perpendicular (and pointing counterclockwise) to z. The solution is geometrically given by a point traveling around a circle in time 2π (independent of radius and starting point).

Example 23.1. *Geometric Solution of the Equation*

We can verify the symbolic solution $\frac{dz}{dt} = i\,z$ with $z[0] = r$ by substituting the function $z[t] = r\,\mathrm{Cos}[t] + i\,r\,\mathrm{Sin}[t]$ into the equation.

$$\frac{dz}{dt}[t] = -r\,\mathrm{Sin}[t] + i\,r\,\mathrm{Cos}[t] = i(r\,\mathrm{Cos}[t] + i\,r\,\mathrm{Sin}[t]) = i\,z[t]$$

$$r\,z[0] = r\,\mathrm{Cos}[0] + i\,r\,\mathrm{Sin}[0] = r + i\,r \times 0 = r$$

This symbolic solution is simple, but how did we guess it? Draw some circles of different radii and the same center using a pencil tied to a string. Move your pencil around the different circles and notice that you always move perpendicular to the vector pointing from the center to the tip of your pencil. The string pinned at the center and tied to your pencil is that vector.

We learned in Section 16.2 that parametric equations for a point moving around a circle at constant speed are given by the sine and cosine. When the speed of the motion is the same as the radius and we begin on the x-axis, the parametric circle is our solution.

Example 23.2. *Polar Form of Unit-Length Complex Numbers*

Unit-length complex numbers can be thought of either in terms of the sine-cosine in radian measure or in terms of the complex exponential.

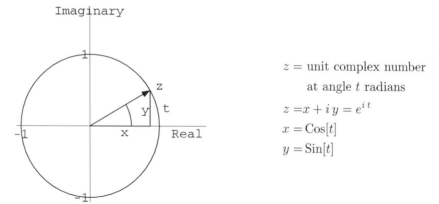

Imaginary

$z =$ unit complex number
at angle t radians

$z = x + i\,y = e^{i\,t}$

$x = \mathrm{Cos}[t]$

$y = \mathrm{Sin}[t]$

Figure 23.3: $e^{i\,t} = \mathrm{Cos}[t] + i\,\mathrm{Sin}[t]$

In general, we have

Theorem 23.3. *The natural exponential satisfies the important functional equation*

$$e^{z+w} = e^z\,e^w$$

even when z and w are complex numbers.

See Problem 23.2.

A very important special case of this identity is

$$e^{\alpha+i\beta} = e^\alpha\,e^{i\beta} = e^\alpha(\mathrm{Cos}[\beta] + i\,\mathrm{Sin}[\beta])$$

Example 23.3. *Geometric Complex Multiplication*

We can use Euler's formula to give the geometric form of complex multiplication. We write two complex numbers in "polar" form, a length (or radius) along a certain angle. The complex number $z = x + i\,y$ lies on a circle of radius $r = |z| = \sqrt{x^2 + y^2}$ and at some angle ϕ. The unit-length vector at angle $\phi = \mathrm{ArcTan}[y/x]$ is $e^{i\,\phi}$ by Euler's formula. (We know from the radian measure definition of sine and cosine that it is the vector with components $(x_1, y_1) = (\mathrm{Cos}[\phi], \mathrm{Sin}[\phi])$.) The vector stretched to length r is therefore

$$z = r\,e^{i\,\phi}$$

Similarly, we may write another complex number $w = a + i\,b$ in the form

$$w = s\,e^{i\,\psi}$$

where $\psi = \mathrm{ArcTan}[b/a]$ and $s = |w| = \sqrt{a^2 + b^2}$. The product

$$z\,w = r\,s\,e^{i\,(\phi+\psi)}$$

which means that the length of the product is the product of the lengths and the angle of the product is the sum of the angles shown in Figure 23.4

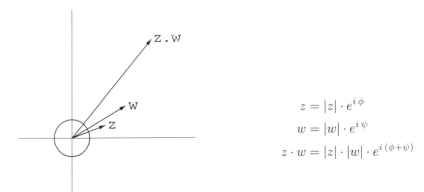

Figure 23.4: Geometric complex product

Now we return to our primary goal of solving linear differential equations by using Euler's formula in the form

$$e^{\alpha t + i\beta t} = e^{\alpha t}\, e^{i\beta t} = e^{\alpha t}(\text{Cos}[\beta\, t] + i\,\text{Sin}[\beta\, t])$$

The roots

$$r_1, r_2 = \frac{-b \pm \sqrt{b^2 - 4\,a\,c}}{2\,a} = \frac{-b}{2a} \pm i\,\frac{\sqrt{4\,a\,c - b^2}}{2a}$$

of the characteristic equation $a\, r^2 + b\, r + c = 0$ are complex exactly when the discriminant is negative,

$$b^2 - 4\,a\,c < 0$$

and, in that case, we may write $r_1 = \alpha + i\,\beta$ and $r_2 = \alpha - i\,\beta$.

Procedure 23.4. Real Solutions for Complex Roots

If the characteristic roots associated with the differential equation

$$a\frac{d^2 x}{dt^2} + b\frac{dx}{dt} + c\,x = 0$$

are complex numbers $r_1 = \alpha + i\beta$ and $r_2 = \alpha - i\beta$, then the complex functions $z_1[t] = e^{\alpha t + i\beta t}$ and $z_2[t] = e^{\alpha t - i\beta t}$ are solutions of the differential equation but so are the real functions

$$x_1[t] = e^{\alpha t}\,\text{Cos}[\beta\, t] \qquad \text{and} \qquad x_2[t] = e^{\alpha t}\,\text{Sin}[\beta\, t]$$

where

$$\alpha = \frac{-b}{2a} \qquad \text{and} \qquad \beta = \frac{\sqrt{4\,a\,c - b^2}}{2a}$$

See Exercise 23.4.5. Linear combinations of solutions are also solutions, as we show in the next section.

Exercise set 23.4

1. *Find real solutions of the form*

$$x_1[t] = e^{\alpha t} \operatorname{Cos}[\beta t] \qquad and \qquad x_2[t] = e^{\alpha t} \operatorname{Sin}[\beta t]$$

for the equations

(a) $\frac{d^2 x}{dt^2} + 4\frac{dx}{dt} + 13\, x = 0$

(b) $\frac{d^2 x}{dt^2} + \frac{dx}{dt} + x = 0$

(c) $\frac{d^2 x}{dt^2} + \omega^2\, x = 0$

Sketch your solution functions. Notice that they are sines or cosines attenuated or amplified by exponentials.

2. *Use Euler's formula to show that*

$$\operatorname{Cos}[\theta] = \frac{e^{i\theta} + e^{-i\theta}}{2} \qquad and \qquad \operatorname{Sin}[\theta] = \frac{e^{i\theta} - e^{-i\theta}}{2i}$$

Euler's formula shows that complex products have the simple geometric meaning of adding angles and multiplying lengths, but, first, here is some practice:

3. Complex Multiplication

Plot the unit vectors $\begin{bmatrix} \frac{\sqrt{3}}{2} \\ \frac{1}{2} \end{bmatrix}$ *and* $\begin{bmatrix} \frac{1}{2} \\ \frac{\sqrt{3}}{2} \end{bmatrix}$. *Compute the angle each makes with the horizontal x-axis.*

Show that the product $(\frac{\sqrt{3}}{2} + i\frac{1}{2})(\frac{1}{2} + i\frac{\sqrt{3}}{2}) = i$. *(30° + 60° = 90°)*

Show that the product $(\frac{\sqrt{3}}{2} + i\frac{1}{2})(\frac{\sqrt{3}}{2} + i\frac{1}{2}) = (\frac{1}{2} + i\frac{\sqrt{3}}{2})$. *(30° + 30° = 60°)*

Plot the unit vectors $\begin{bmatrix} \frac{\sqrt{3}}{2} \\ \frac{1}{2} \end{bmatrix}$ *and* $\begin{bmatrix} \frac{1}{\sqrt{2}} \\ \frac{1}{\sqrt{2}} \end{bmatrix}$. *Compute the angle each makes with the horizontal x-axis.*

Compute the complex product $(\frac{\sqrt{3}}{2} + i\frac{1}{2})(\frac{1}{\sqrt{2}} + i\frac{1}{\sqrt{2}}) = \frac{\sqrt{6}-\sqrt{2}}{4} + i\frac{\sqrt{6}+\sqrt{2}}{4}$

Use your calculator or the computer to plot the unit vector $\begin{bmatrix} \frac{\sqrt{6}-\sqrt{2}}{4} \\ \frac{\sqrt{6}+\sqrt{2}}{4} \end{bmatrix}$ *and show that it makes an angle of 75° with the horizontal. (30° + 45° = 75°)*

4. *Verify by direct substitution that the functions*

$$x_1[t] = e^{-t} \operatorname{Cos}[\sqrt{2}\, t] \qquad and \qquad x_2[t] = e^{-t} \operatorname{Sin}[\sqrt{2}\, t]$$

are solutions to the differential equation

$$\frac{d^2 x}{dt^2} + 2\frac{dx}{dt} + 3\, x = 0$$

Compute the derivatives of x_1 *and* x_2 *carefully using the Product Rule on the terms* e^{-t} *and* $\operatorname{Cos}[\sqrt{2}\, t]$, *and so forth. The computation is messy, but it works.*

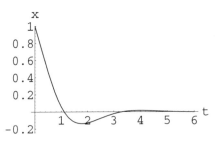

Figure 23.5: $x[t] = e^{-t} \cos[\sqrt{2}\, t]$

Also, write the solutions to this equation in the complex exponential form and use Euler's formula to find a relationship between the real and complex solutions.
Check and sketch your solutions with the computer.

5. *For the solutions z_1, z_2, x_1, and x_2, given in Procedure 23.4 above, use Euler's formula to show that*

$$x_1[t] = \frac{1}{2}[z_1[t] + z_2[t]] \qquad and \qquad x_2[t] = \frac{i}{2}[z_2[t] - z_1[t]]$$

6. Geometric Solution of $\frac{dz}{dt} = i\,z$
Suppose we start somewhere on the circle of radius r and move counterclockwise around this circle in time. The vector pointing from the origin to our position at any time is $\begin{bmatrix} x \\ y \end{bmatrix}$. Sketch this figure. What is the connection between x, y, and the radius r?
Which way will our velocity vector point for the counterclockwise motion described?
Sketch the displacement arrow $\begin{bmatrix} -y \\ x \end{bmatrix}$ with its tail at the tip of $\begin{bmatrix} x \\ y \end{bmatrix}$.

If the vector $\begin{bmatrix} -y \\ x \end{bmatrix}$ is our velocity, what is our speed? Express your answer in terms of the constant r.
How long does it take to travel around a circle of radius r at speed r? How far do you travel?

Problem 23.2. ADDITION FORMULAS ──────────────────────────────────▼
We want to use Euler's formula two ways to show the connection between the functional equation of the natural exponential and the addition formulas for sine and cosine.
  (a) *Use Euler's formula to write $e^{i\phi}$ as a sum of sine and cosine.*
  (b) *Use Euler's formula to write $e^{i\psi}$ as a sum of sine and cosine.*
  (c) *Use Euler's formula to write $e^{i(\phi+\psi)}$ as a sum of sine and cosine.*
  (d) *Multiply the complex number answer to part (1) times the complex number answer to part (2) and combine the terms in the form $A + iB$ for real expressions A and B (involving sine and cosine).*
  (e) *The answer to part (3) is $e^{i(\phi+\psi)}$ in real plus imaginary part, whereas the answer to part (4) is $e^{i\phi} \times e^{i\psi}$ in real plus imaginary part. Write the equality $e^{i(\phi+\psi)} = e^{i\phi} \times e^{i\psi}$ in real plus imaginary form,*
$$Ans.(3) = Ans.(4)$$

(f) *Show that this is equivalent to the pair of equations*

$$\text{Cos}[\phi + \psi] = \text{Cos}[\phi]\,\text{Cos}[\psi] - \text{Sin}[\phi]\,\text{Sin}[\psi]$$
$$\text{Sin}[\phi + \psi] = \text{Sin}[\phi]\,\text{Cos}[\psi] + \text{Cos}[\phi]\,\text{Sin}[\psi]$$

(g) *Use the cases shown above to prove the general case of the natural exponential equation*

$$e^z\,e^w = e^{a+ib}\,e^{x+iy} = e^{(a+x)+i(b+y)} = e^{z+w}$$

23.5 Basic Solutions

The results so far are the three cases of the result:

Theorem 23.5. *Solution of Second-Order Differential Equations*
Let $a \neq 0$, b, and c be constants. The following procedure gives a pair of real valued linearly independent basic solutions to the differential equation

$$a\frac{d^2x}{dt^2} + b\frac{dx}{dt} + cx = 0$$

Find the characteristic roots r_1 and r_2 of the algebraic characteristic equation

$$a\,r^2 + b\,r + c = 0$$

(a) *If the root is repeated, $r_1 = r_2$, then*

$$x_1[t] = e^{r_1 t} \qquad and \qquad x_2[t] = t\,e^{r_1 t}$$

(b) *If the roots are real and distinct, then*

$$x_1[t] = e^{r_1 t} \qquad and \qquad x_2[t] = e^{r_2 t}$$

(c) *If the roots are a complex conjugate pair $r_1 = \alpha + i\beta$ and $r_2 = \alpha - i\beta$, then*

$$x_1[t] = e^{\alpha t}\,\text{Cos}[\beta t] \qquad and \qquad x_2[t] = e^{\alpha t}\,\text{Sin}[\beta t]$$

Every solution to this differential equation can be written in the form

$$x[t] = k_1\,x_1[t] + k_2\,x_2[t]$$

for constants k_1 and k_2. This is the theme of the next section.

Exercise set 23.5

1. *Write the functions $x_1[t]$ and $x_2[t]$ from Theorem 23.5 in the cases:*

a) $\frac{d^2x}{dt^2} + \frac{dx}{dt} + x = 0$ b) $\frac{d^2x}{dt^2} + 2\frac{dx}{dt} + x = 0$ c) $\frac{d^2x}{dt^2} + 3\frac{dx}{dt} + x = 0$

23.6 Superposition and All Solutions

A physical system is said to satisfy the superposition principle if the response to a sum of forces is the sum of the responses to the separate forces. The damped spring system governed by

$$m\frac{d^2x}{dt^2} + c\frac{dx}{dt} + sx = \text{Applied force}$$

satisfies this principle.

The mathematical superposition principle, or linearity of the differential equation, is the key to our complete symbolic solution of $a\frac{d^2x}{dt^2} + b\frac{dx}{dt} + cx = 0$ and its associated initial value problems. So far in this chapter, we have used real or complex exponentials to find two distinct real functions $x_1[t]$ and $x_2[t]$ satisfying the equation. Mathematical superposition can be expressed as follows:

Theorem 23.6. *Superposition of Solutions*
   If real functions $x_1[t]$ and $x_2[t]$ satisfy the equation

$$a\frac{d^2x}{dt^2} + b\frac{dx}{dt} + cx = 0$$

   and k_1 and k_2 are constants, then $x[t] = k_1\,x_1[t] + k_2\,x_2[t]$ is also a solution.

PROOF OF MATHEMATICAL SUPERPOSITION:
   We show this by using the mathematical superposition property of differentiation. Basically, we just "plug in."

$c\,x_1,$	$c\,x_2,$	$c\,x = c(k_1\,x_x + k_2\,x_2)$	$= $	$k_1\,c\,x_1$	$+ \quad k_2\,c\,x_2$
$b\dfrac{dx_1}{dt},$	$b\dfrac{dx_2}{dt},$	$b\dfrac{dx}{dt} = b(k_1\dfrac{dx_1}{dt} + k_2\dfrac{dx_2}{dt})$	$=$	$k_1\,b\dfrac{dx_1}{dt}$	$+ \quad k_2\,b\dfrac{dx_2}{dt}$
$a\dfrac{dx_1^2}{dt^2},$	$a\dfrac{dx_2^2}{dt^2},$	$a\dfrac{dx^2}{dt^2} = a(k_1\dfrac{dx_1^2}{dt^2} + k_2\dfrac{dx_2^2}{dt^2})$	$=$	$k_1\,a\dfrac{dx_1^2}{dt^2}$	$+ \quad k_2\,a\dfrac{dx_2^2}{dt^2}$

Sum Vertically
   $\quad 0 \qquad\qquad 0 \qquad\qquad\qquad\qquad\qquad\qquad\qquad\qquad k_1\cdot 0 \quad + \quad k_2\cdot 0 = 0$

This general case may seem too difficult, but you should compare it to the next example, which is a specific case. The details of the specific case obscure the simple regrouping of the general computation above.

Example 23.4. *A Specific Linear Combination of Solutions*

Suppose we have a system consisting of a unit mass, $m = 1$; a unit spring constant, $s = 1$, and unit damping friction, $c = 1$. This satisfies the second-order equation $\frac{d^2x}{dt^2} + \frac{dx}{dt} + x = 0$.

It has the characteristic equation $r^2 + r + 1 = 0$ with characteristic roots $-\frac{1}{2} + i\frac{\sqrt{3}}{2}$ and $-\frac{1}{2} - i\frac{\sqrt{3}}{2}$. This makes the basic solutions

$$e^{-\frac{t}{2}} \operatorname{Cos}[\frac{\sqrt{3}}{2}t] \qquad \text{and} \qquad e^{-\frac{t}{2}} \operatorname{Sin}[\frac{\sqrt{3}}{2}t]$$

We try a solution of the form

$$x[t] = k_1\, x_1[t] + k_2\, x_2[t]$$

$$= k_1\, e^{-\frac{t}{2}} \operatorname{Cos}[\frac{\sqrt{3}}{2}t] + k_2\, e^{-\frac{t}{2}} \operatorname{Sin}[\frac{\sqrt{3}}{2}t]$$

by simply substituting it in the differential equation. Notice the use of the Product Rule for differentiation.

$$x = e^{-\frac{t}{2}}(k_1\, \operatorname{Cos}[\frac{\sqrt{3}}{2}t] + k_2\, \operatorname{Sin}[\frac{\sqrt{3}}{2}t])$$

$$\begin{aligned}
\frac{dx}{dt} &= -\frac{k_1}{2}e^{-\frac{t}{2}} \operatorname{Cos}[\frac{\sqrt{3}}{2}t] - \frac{k_1\sqrt{3}}{2}e^{-\frac{t}{2}} \operatorname{Sin}[\frac{\sqrt{3}}{2}t] \\
&\quad - \frac{k_2}{2}e^{-\frac{t}{2}} \operatorname{Sin}[\frac{\sqrt{3}}{2}t] + \frac{k_2\sqrt{3}}{2}e^{-\frac{t}{2}} \operatorname{Cos}[\frac{\sqrt{3}}{2}t] \\
&= e^{-\frac{t}{2}}(\frac{k_2\sqrt{3} - k_1}{2} \operatorname{Cos}[\frac{\sqrt{3}}{2}t] - \frac{k_1\sqrt{3} + k_2}{2} \operatorname{Sin}[\frac{\sqrt{3}}{2}t])
\end{aligned}$$

$$\begin{aligned}
\frac{d^2x}{dt^2} &= e^{-\frac{t}{2}}(\frac{k_1 - k_2\sqrt{3}}{4} \operatorname{Cos}[\frac{\sqrt{3}}{2}t] + \frac{k_1\sqrt{3} + k_2}{4} \operatorname{Sin}[\frac{\sqrt{3}}{2}t] \\
&\quad + \frac{k_1\sqrt{3} - k_2\, 3}{4} \operatorname{Sin}[\frac{\sqrt{3}}{2}t] - \frac{k_1\, 3 + k_2\sqrt{3}}{4} \operatorname{Cos}[\frac{\sqrt{3}}{2}t]) \\
&= e^{-\frac{t}{2}}(\frac{-k_1 - k_2\sqrt{3}}{2} \operatorname{Cos}[\frac{\sqrt{3}}{2}t] + \frac{k_1\sqrt{3} - k_2}{2} \operatorname{Sin}[\frac{\sqrt{3}}{2}t])
\end{aligned}$$

Adding, we obtain

$$\begin{aligned}
\frac{d^2x}{dt^2} + \frac{dx}{dt} + x &= e^{-\frac{t}{2}}[(k_1 + \frac{k_2\sqrt{3} - k_1}{2} + \frac{-k_1 - k_2\sqrt{3}}{2}) \operatorname{Cos}[\frac{\sqrt{3}}{2}t] \\
&\quad + (k_2 - \frac{k_1\sqrt{3} + k_2}{2} + \frac{k_1\sqrt{3} - k_2}{2}) \operatorname{Sin}[\frac{\sqrt{3}}{2}t]] \\
&= e^{-\frac{t}{2}}(0\, \operatorname{Cos}[\frac{\sqrt{3}}{2}t] + 0\, \operatorname{Sin}[\frac{\sqrt{3}}{2}t]) = 0
\end{aligned}$$

This verifies that this is a solution no matter which values k_1 and k_2 take.

We really never wanted to find solutions to the differential equation alone, but rather wanted to solve the associated initial value problems. Now, we have the tools to give a form for all solutions of all such problems. Two solutions $x_1[t]$ and $x_2[t]$ form a "basis" for all solutions of these initial value, if they are "really different" or linearly independent. In this case, all solutions may be written in the form

$$x[t] = k_1\, x_1[t] + k_2\, x_2[t]$$

for unknown constants k_1 and k_2. Theorem 23.5 gives a list of basic solutions depending on the roots of the characteristic equation.

The phase variable form of the equation $a\frac{d^2x}{dt^2} + b\frac{dx}{dt} + cx = 0$, $(a \neq 0)$ is given by the pair of equations

$$\frac{dx}{dt} = y$$
$$\frac{dy}{dt} = -\frac{c}{a}x - \frac{b}{a}y$$

For example, the equation $\frac{d^2x}{dt^2} + \frac{dx}{dt} + x = 0$ can be written as the two-dimensional system

$$\frac{dx}{dt} = y$$
$$\frac{dy}{dt} = -x - y$$

Initial conditions will consist of values for both $x[0]$ and $y[0]$. This is clear from the 2-D system point of view, but why is it true physically? The velocity of the mass initially certainly affects the solution. A mass initially at $x = 0$ and at rest, $y = 0$, will remain at rest, but a fast moving mass just passing through $x = 0$ at time zero does not have a constant solution.

Example 23.5. *A Specific Solution*

We solve the system above with such initial conditions and show the nontrivial motion. Here are our example conditions:

$$x[0] = 0$$
$$y[0] = 1$$

Now, we match the initial conditions by setting up equations for k_1 and k_2. The initial values are

$$x[0] = e^{-\frac{0}{2}}(k_1 \cos[\frac{\sqrt{3}}{2}0] + k_2 \sin[\frac{\sqrt{3}}{2}0])$$
$$= k_1$$
$$y[0] = \frac{dx}{dt}[0] = e^{-\frac{0}{2}}(\frac{k_2\sqrt{3} - k_1}{2} \cos[\frac{\sqrt{3}}{2}0] - \frac{k_1\sqrt{3} + k_2}{2} \sin[\frac{\sqrt{3}}{2}0])$$
$$= \frac{k_2\sqrt{3} - k_1}{2}$$

Notice the Product Rule in use when computing $\frac{dx}{dt}$.

We want

$$x[0] = 0$$

$$\frac{dx}{dt}[0] = 1$$

so we take $k_1 = 0$ and $k_2 = \frac{2}{\sqrt{3}}$, making $x[t] = \frac{2}{\sqrt{3}} e^{-\frac{t}{2}} \operatorname{Sin}[\frac{\sqrt{3}}{2}t]$ the unique solution to our initial value problem.

The solution of this initial value problem can be plotted in the phase plane as a parametric curve (or in the the the computer flow animation as a movie.) Notice that the velocity y starts high, $y[0] = 1$, whereas position x starts at $x[0] = 0$. Oscillation of the explicit solution appears as a spiral in the parametric phase plot.

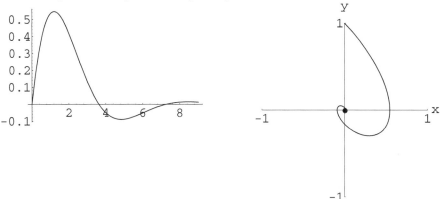

Figure 23.6: $x[t] = \frac{2}{\sqrt{3}} e^{-t/2} \operatorname{Sin}[\frac{\sqrt{3}t}{2}]$ and its phase plot

Example 23.6. *General Initial Conditions*

Any initial conditions for the equation $\frac{d^2x}{dt^2} + \frac{dx}{dt} + x = 0$ can be found by setting up these equations for unknown k_1 and k_2 in terms of given initial conditions x_0 and y_0. We know the basic solutions $x_1[t]$ and $x_2[t]$, so our equations are $x[0] = k_1 x_1[0] + k_2 x_2[0] = x_0$ and $x'[0] = k_1 x_1'[0] + k_2 x_2'[0] = y_0$. The linear equations in the example $\frac{d^2x}{dt^2} + \frac{dx}{dt} + x = 0$ are

$$x[0] = e^{-\frac{0}{2}}(k_1 \operatorname{Cos}[\frac{\sqrt{3}}{2}0] + k_2 \operatorname{Sin}[\frac{\sqrt{3}}{2}0])$$

$$= k_1$$

$$y[0] = \frac{dx}{dt}[0] = e^{-\frac{0}{2}}(\frac{k_2\sqrt{3} - k_1}{2} \operatorname{Cos}[\frac{\sqrt{3}}{2}0] - \frac{k_1\sqrt{3} + k_2}{2} \operatorname{Sin}[\frac{\sqrt{3}}{2}0])$$

$$= \frac{k_2\sqrt{3} - k_1}{2}$$

or, in matrix form

$$\begin{bmatrix} 1 & 0 \\ -\frac{1}{2} & \frac{\sqrt{3}}{2} \end{bmatrix} \begin{bmatrix} k_1 \\ k_2 \end{bmatrix} = \begin{bmatrix} x_0 \\ y_0 \end{bmatrix}$$

If we begin with basic solutions $x_1[t]$ and $x_2[t]$ of any second-order linear autonomous equation, we will always be able to solve the conditions that result from assuming that a solution has the form $x[t] = k_1 x_1[t] + k_2 x_2[t]$ and calculating k_1 and k_2 from the initial value equations $t = t_0$:

$$x_0 = x[t_0] = k_1 x_1[t_0] + k_2 x_2[t_0]$$
$$y_0 = \frac{dx}{dt}[t_0] = k_1 \frac{dx_1}{dt}[t_0] + k_2 \frac{dx_2}{dt}[t_0]$$
$$= k_1 y_1[t_0] + k_2 y_2[t_0]$$

where $y_j[t] = x_j'[t]$. Remember that x_0 and y_0 are given initial conditions and that $x_1[t]$, $y_1[t]$, $x_2[t]$, and $y_2[t]$ are computed by our algebraic procedure (plus differentiation for the y's.) If we specify the initial time t_0, the quantities $x_1[t_0]$, $y_1[t_0]$, $x_2[t_0]$, and $y_2[t_0]$ can be computed, so the equations

$$x_0 = k_1 x_1[t_0] + k_2 x_2[t_0]$$
$$y_0 = k_1 y_1[t_0] + k_2 y_2[t_0]$$

have only k_1 and k_2 as unknowns. The computations are long and tedious but so straightforward that the computer can do them. They result in a system of equations that can be written in matrix form as

$$\begin{bmatrix} x_1[t_0] & x_2[t_0] \\ y_1[t_0] & y_2[t_0] \end{bmatrix} \begin{bmatrix} k_1 \\ k_2 \end{bmatrix} = \begin{bmatrix} x_0 \\ y_0 \end{bmatrix} \qquad \Leftrightarrow \qquad \mathbf{a} \cdot \mathbf{k} = \mathbf{b}$$

where the matrices are

$$\mathbf{a} = \begin{bmatrix} x_1[t_0] & x_2[t_0] \\ y_1[t_0] & y_2[t_0] \end{bmatrix}, \qquad \mathbf{k} = \begin{bmatrix} k_1 \\ k_2 \end{bmatrix}, \qquad \mathbf{b} = \begin{bmatrix} x_0 \\ y_0 \end{bmatrix}$$

The computer can solve such equations as follows:

Example 23.7.   *Computer Solution of Linear Equations*

The computer solves the equations

$$\begin{array}{rl} k_1 \quad -k_2 &= 5 \\ -\frac{1}{2}k_1 +\frac{\sqrt{3}}{2}k_2 &= 7 \end{array} \qquad \Leftrightarrow \qquad \mathbf{a} \cdot \mathbf{k} = \mathbf{b}$$

where the matrices are $\mathbf{a} = \begin{bmatrix} 1 & -1 \\ -1/2 & \sqrt{3}/2 \end{bmatrix}, \mathbf{k} = \begin{bmatrix} k_1 \\ k_2 \end{bmatrix}, \mathbf{b} = \begin{bmatrix} 5 \\ 7 \end{bmatrix}.$

This whole section is summarized in the single result:

Theorem 23.7. *Solution of Second-Order Initial Value Problems*
   The solution of an initial value problem

$$x[t_0] = x_0$$

$$\frac{dx}{dt}[t_0] = y_0$$

$$a\frac{d^2x}{dt^2} + b\frac{dx}{dt} + cx = 0$$

is given by

$$x[t] = k_1\, x_1[t] + k_2\, x_2[t]$$

where $x_1[t]$ and $x_2[t]$ are the basic solutions given in Theorem 23.5 and the constants k_1 and k_2 for specific initial conditions $x[t_0] = x_0$ and $x'[t_0] = y_0$ are solutions of the linear algebraic equations

$$x_0 = k_1\, x_1[t_0] + k_2\, x_2[t_0]$$

$$y_0 = k_1\, x_1'[t_0] + k_2\, x_2'[t_0]$$

The next section uses Theorem 23.5 and Theorem 23.7 to make a computational procedure. You can write a symbolic computer program to implement the procedure.

Exercise set 23.6

1. *Find the solution to $\frac{d^2x}{dt^2} + \frac{dx}{dt} + x = 0$ satisfying $x[0] = 1$ and $\frac{dx}{dt}[0] = 0$.*

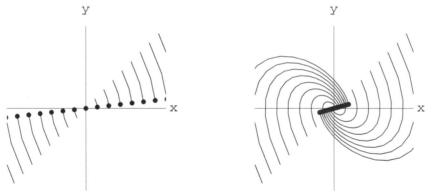

Figure 23.7: Flow of $x'' + x' + x = 0$

2. Transients
   The damped spring equation

$$m\frac{d^2x}{dt^2} = -c\frac{dx}{dt} - s\,x \qquad or \qquad m\frac{d^2x}{dt^2} + c\frac{dx}{dt} + s\,x = 0$$

has $c > 0$. Show that this makes every solution satisfy

$$\lim_{t\to\infty} x[t] = 0$$

by considering the 3 cases of Theorem 23.5.

Sketch the explicit graphs of $x_1[t]$ and $x_2[t]$ in the cases $m = 1$, $s = 1$, and

$$(a) \quad c = 1 \qquad (b) \quad c = 2 \qquad (c) \quad c = \frac{5}{2}$$

*Use the computer program **Flow2D.ma** to make flow animations of all the initial value problems associated with these three choices of parameters. (Note: Use the phase variable trick when you set **Flow2D.ma** up.)*

23.7   Solutions of Second-Order IVPs

The ingredients of the last two sections solve all linear initial value problems.

Procedure 23.8.   Solution of Second-Order IVPs

(a)   Use Theorem 23.5 to find two basic solutions,

   (1)   If $b^2 = 4\,a\,c$, the characteristic roots are repeated, so $r = -\frac{b}{2a}$ and

$$x_1[t] = e^{r\,t} \qquad \text{and} \qquad x_2[t] = t\,e^{r\,t}$$

   (2)   If $b^2 > 4\,a\,c$, the characteristic roots are $r_1 = \frac{-b+\sqrt{b^2-4\,a\,c}}{2\,a}$ and $r_1 = \frac{-b-\sqrt{b^2-4\,a\,c}}{2\,a}$ with basic solutions

$$x_1[t] = e^{r_1\,t} \qquad \text{and} \qquad x_2[t] = e^{r_2\,t}$$

   (3)   If $b^2 < 4\,a\,c$, the characteristic roots are complex, $\alpha \pm i\beta$ with $\alpha = -\frac{b}{2a}$ and $\beta = \frac{\sqrt{4\,a\,c - b^2}}{2\,a}$ with basic solutions

$$x_1[t] = e^{\alpha\,t}\,\mathrm{Cos}[\beta\,t] \qquad \text{and} \qquad x_2[t] = e^{\alpha\,t}\,\mathrm{Sin}[\beta\,t]$$

(b)   Symbolically compute the derivative of $x[t] = k_1\,x_1[t] + k_2\,x_2[t]$. (This involves the Product Rule in cases 1 and 3.)

(c)   Evaluate $x[t_0]$ and $\frac{dx}{dt}[t_0]$ numerically except for the unknowns k_1 and k_2, and write the system of linear equations in these unknowns

$$x_0 = x[t_0]$$

$$y_0 = \frac{dx}{dt}[t_0]$$

   This can be put in the matrix form $\mathbf{a} \cdot \mathbf{k} = \mathbf{b}$, $\begin{bmatrix} x_1[t_0] & x_2[t_0] \\ y_1[t_0] & y_2[t_0] \end{bmatrix} \begin{bmatrix} k_1 \\ k_2 \end{bmatrix} = \begin{bmatrix} x_0 \\ y_0 \end{bmatrix}$,

   for constants $x_1[t_0]$, $x_2[t_0]$, $y_1[t_0]$, $y_2[t_0]$, and the given values x_0, y_0.

(d)   Solve this equation for k_1 and k_2.

The specific solution $x[t] = k_1\,x_1[t] + k_2\,x_2[t]$ is the answer.

This procedure is so mechanical that we will ask you to write a computer program to implement it after you have some practice at using it.

Example 23.8. *Solution When $a = 1$, $b = 0$, $c = -1$, $x_0 = y_0 = 1$*

$$\frac{d^2x}{dt^2} - x = 0 \qquad \Leftrightarrow \qquad \begin{array}{l} \frac{dx}{dt} = y \\ \frac{dy}{dt} = x \end{array}$$

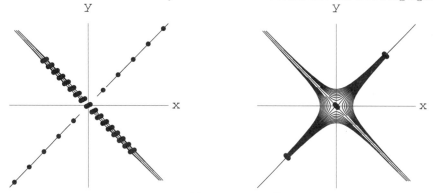

Figure 23.8: Explicit and parametric solution

The full solution flow for many initial conditions looks like the following figure.

Figure 23.9: Flow of $x'' - x = 0$

Characteristic equation:

$$r^2 - 1 = (r - 1)(r + 1) = 0$$

Roots: $r_1 = 1$, $r_2 = -1$
Basic solutions: $x[t] = k_1\, e^t + k_2\, e^{-t}$
Symbolic derivative: $x'[t] = k_1\, e^t - k_2\, e^{-t}$
Initial values at $t = 0$:

$$x[0] = k_1\, e^0 + k_2\, e^0 = k_1 + k_2$$
$$x'[0] = k_1\, e^0 - k_2\, e^0 = k_1 - k_2$$

Matrix linear equation:

$$\begin{bmatrix} 1 & 1 \\ 1 & -1 \end{bmatrix} \begin{bmatrix} k_1 \\ k_2 \end{bmatrix} = \begin{bmatrix} x_0 \\ y_0 \end{bmatrix}$$

When $(x_0, y_0) = (1.0, 1.0)$, $k_1 = 1$, and $k_2 = 0$, this is the solution $x[t] = e^t$.

Example 23.9. *Solution When $a = 1$, $b = 3$, $c = 1$, $x_0 = y_0 = 1$*

$$\frac{d^2 x}{dt^2} + 3\frac{dx}{dt} + x = 0 \qquad \Leftrightarrow \qquad \begin{array}{l} \frac{dx}{dt} = y \\ \frac{dy}{dt} = -x - 3y \end{array}$$

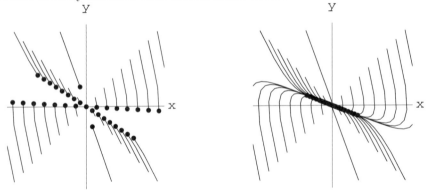

Figure 23.10: Explicit and parametric solution

A flow of many solutions with various initial conditions looks as follows:

Figure 23.11: Flow of $x'' + 3x' + x = 0$

Characteristic equation: $r^2 + 3r + 1 = 0$

Roots: $r_1 = \frac{-3+\sqrt{5}}{2}$, $r_2 = \frac{-3+\sqrt{5}}{2}$

Basic solutions: $x[t] = k_1\, e^{\frac{-3+\sqrt{5}}{2}\, t} + k_2\, e^{\frac{-3-\sqrt{5}}{2}\, t}$

Symbolic derivative: $x'[t] = k_1\, \frac{-3+\sqrt{5}}{2} e^{\frac{-3+\sqrt{5}}{2}\, t} - k_2\, \frac{-3-\sqrt{5}}{2} e^{\frac{-3-\sqrt{5}}{2}\, t}$

Initial values at $t = 0$:

$$x[0] = k_1\, e^{\frac{-3+\sqrt{5}}{2}\, 0} + k_2\, e^{\frac{-3-\sqrt{5}}{2}\, 0} = k_1 + k_2$$

$$x'[0] = k_1\, \frac{-3+\sqrt{5}}{2} e^{\frac{-3+\sqrt{5}}{2}\, 0} + k_2\, \frac{-3-\sqrt{5}}{2} e^{\frac{-3-\sqrt{5}}{2}\, 0} = k_1\, \frac{-3+\sqrt{5}}{2} + k_2\, \frac{-3-\sqrt{5}}{2}$$

Matrix linear equation:

$$\begin{bmatrix} 1 & 1 \\ \frac{-3+\sqrt{5}}{2} & \frac{-3-\sqrt{5}}{2} \end{bmatrix} \begin{bmatrix} k_1 \\ k_2 \end{bmatrix} = \begin{bmatrix} x_0 \\ y_0 \end{bmatrix}$$

When $(x_0, y_0) = (1.0, 1.0)$, $k_1 = \frac{1+\sqrt{5}}{2}$ and $k_2 = \frac{1-\sqrt{5}}{2}$.

Exercise set 23.7

1. Symbolic Solutions by Hand

Solve the initial value problem

$$x[0] = 1$$

$$y[0] = \frac{dx}{dt}[0] = 1$$

$$a\frac{d^2x}{dt^2} + b\frac{dx}{dt} + cx = 0$$

for the cases
  (a) $a = 1$, $b = 0$, $c = -4$
  (b) $a = 1$, $b = 3$, $c = 2$
  (c) $a = 1$, $b = 2$, $c = 2$
  (d) $a = 1$, $b = 2$, $c = 1$

Use your hand calculations to verify the correctness of your the computer program in the next exercise.

2. Computer Symbolic Solutions

Write a the computer program to find specific solutions to the IVP

$$x[0] = x_0$$

$$y[0] = \frac{dx}{dt}[0] = y_0$$

$$a\frac{d^2x}{dt^2} + b\frac{dx}{dt} + cx = 0$$

Include text cells in your program that explain each part of your program.
See the program ***SecondOrder***.

23.7.1   Your Car's Worn Shocks

When a car hits a bump it sets up an initial value problem

$$x[0] = 0$$

$$\frac{dx}{dt}[0] = v_0 \neq 0 \qquad \text{an initial velocity}$$

$$m\frac{d^2x}{dt^2} + c\frac{dx}{dt} + sx = 0$$

The initial position is still zero - the level position of the car - but the jolt (impulse) imparts an initial velocity to the wheel.

Problem 23.3.

What is the algebraic condition among the physical parameters m, c, and s that determines whether or not your car oscillates after it hits a bump?

*Use the computer program **SpringEQ** to solve the above initial value problem with* $m = s = v_0 = 1$ *and values of* $c = 3, 2.9, 2.8, \cdots, 2.0, \cdots, 0.0.$

▲

23.8 Projects

23.8.1 Lanchester's Combat Models

This project studies a combat model in which the size and effectiveness of the armies are compared.

23.8.2 Compartment Models for Drug Dosages

This project studies the dynamics of the blood concentration of a drug.

23.8.3 Resonance of Linear Oscillators

This project studies the response of a linear oscillator to shaking. A maximal response is called "resonance."

23.8.4 A Notch Filter

This studies a two-loop circuit that has a minimal response at a certain frequency. This can be used to filter out that frequency.

24

Equilibria of Continuous Systems

This chapter shows how to use calculus to "see" what is in a micro-scope focused at an equilibrium point of a dynamical system. The main idea of calculus is that you see the "linear approximation," which in this case is a linear dynamical system.

24.1 Equilibria in One Dimension

This section helps you discover a symbolic test for local attractors or re-pellers in 1-D. The dynamical system microscope in one dimension is so simple it is confusing, but we can compare the 1-D and 2-D cases later. In two dimensions, we really need help from calculus.

Consider the general one-dimensional dynamical system

$$\frac{dx}{dt} = f[x]$$

It is easy to find out whether an equilibrium point x_e (with $f[x_e] = 0$) attracts or repels. Plot the direction line immediately on each side of the equilibrium (before another zero of $f[x]$) and see if both arrows point toward the equilibrium (attractor), both point away (repeller), or arrows point away on one side and toward the point on the other (neither).

Example 24.1. *The Linear Case, $f[x] = a\,x$ and $f'[x] = a$*

In the linear case,

$$\frac{dx}{dt} = a\,x$$

where $f[x] = a\,x$, we know the whole story. The point $x_e = 0$ is the only equilibrium point (unless $a = 0$), and it attracts if $a < 0$ and repels if $a > 0$. You should solve this graphically by drawing direction lines in these two cases; however, the symbolic solution is

$$x[t] = x_0\, e^{a\,t}$$

and

$$\lim_{t \to \infty} x[t] = 0 \qquad \text{if} \quad a < 0$$

whereas

$$\lim_{t \to \infty} |x[t]| = \infty \qquad \text{if} \qquad a > 0$$

where the sign of $x[t]$ remains the same as x_0. This is a symbolic proof that any x_0 makes $x[t] \to 0$ when $a < 0$ and is repelled when $a > 0$.

Example 24.2. *The Affine Linear Case, $f[x] = a\,x + b$ and $f'[x] = a$*

In the affine linear case, $f[x] = a\,x + b$,

$$\frac{dx}{dt} = a\,x + b$$

which has its equilibrium when $\frac{dx}{dt} = f[x_e] = 0$ or $x_e = -b/a$. We can change variables to the difference between x and the equilibrium, $x = x_e + u$ or $u = x - x_e$. This makes $\frac{du}{dt} = \frac{dx}{dt}$ and $a\,u = a(x - \frac{-b}{a}) = a\,x + b$, so that the $a\,x + b$ equation is equivalent to

$$\frac{du}{dt} = a\,u$$

This linear equation has $u[t] \to 0$ when $a < 0$, so $x[t] = u[t] + x_e \to x_e$. Similarly, if $a > 0$, solutions are repelled from equilibrium. In either case, the coefficient a determines the type of equilibrium.

Example 24.3. *A Graphical Approach to the General Problem*

Let us draw the direction line for the function $f[x] = (x - 1)(x + 1)$. This has equilibria, $f[x_e] = 0$ at $x_e = -1$ and $x_e = +1$. The function is positive for $|x| > 1$ and negative for $|x| < 1$ as shown on Figure 24.1.

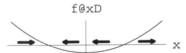

Figure 24.1: Attraction, repulsion, and $f'[x_e]$

We add direction arrows on the x-axis to indicate increasing x when $\frac{dx}{dt} = f[x] > 0$ and decreasing x when $\frac{dx}{dt} = f[x] < 0$. When the arrows point toward the equilibrium from both sides, the equilibrium is attracting. This happens when $f[x]$ changes from positive (x increasing) below the equilibrium to negative (x decreasing) above the equilibrium. What condition on $f'[x_e]$ makes $f[x]$ change from positive before x_e, (zero at x_e) to negative after x_e? This is and its repulsive counterpart are Exercise 24.1.1 below.

Example 24.4. *Linear Approximation*

We can analyze nonlinear equilibria of the equation

$$\frac{dx}{dt} = f[x]$$

with an equilibrium at x_e by linearly approximating $f[x]$ near x_e. For smooth functions $f[x]$, we know the approximation

(Incr) $$f[x] = f[u + x_e] = f[x_e] + a\,u + \varepsilon\,u$$

when $x = u + x_e$, with $\varepsilon \approx 0$ for $u \approx 0$. Because x_e is an equilibrium point, $f[x_e] = 0$, the approximation becomes

$$f[u + x_e] = a\,u + \varepsilon\,u$$

We also have $u = x - x_e$, so that $\frac{du}{dt} = \frac{d(x - x_e)}{dt} = \frac{dx}{dt}$, because the derivative of a constant, x_e, is zero. The differential equation $\frac{dx}{dt} = f[x]$ written in terms of $u = (x - x_e)$ near x_e is

$$\frac{du}{dt} = f[x_e + u] = a\,u + \varepsilon\,u$$

Dropping the term $\varepsilon\,u$, it is approximately the particular linear equation

$$\frac{dx}{dt} = f[x_e + (x - x_e)] \qquad \Leftrightarrow \qquad \frac{du}{dt} = f[x_e + u] \quad \text{``} \approx \text{''} \quad \frac{du}{dt} = a\,u$$

This derivation means that we have the approximation

$$\frac{dx}{dt} = f[x] \quad \text{near } x = x_e \qquad \text{``} \approx \text{''} \qquad \frac{du}{dt} = a\,u$$

so an equilibrium point x_e is a local attractor if $f'[x_e] < 0$ and is a local repeller if $f'[x_e] > 0$. This is stated more precisely in Theorem 24.4.

Exercise set 24.1

1. **Your Very Own Graphical Proof of the Local Equilibrium Result**
   We study the differential equation $\frac{dx}{dt} = f[x]$ where $f[x]$ is a smooth but unknown function. This exercise refers to Example 24.3. We want to find a condition on $f'[x_e]$ that tells us whether an equilibrium point is attracting or repelling. Refer to Figure 24.1 to get ideas for solving this exercise.
   (a) *If x_e is an equilibrium point of the differential equation, what is the value of $f[x_e] = ?$*
   (b) *If x_e is an attracting equilibrium, what is the sign of $f'[x_e]$?*
   (c) *If $f'[x_e]$ is positive and $f[x_e] = 0$, does x_e have to be a repelling equilibrium point?*
   HINTS:
   (d) *If $f'[x_e]$ is positive and $f[x_e] = 0$, what is the sign of $f[x]$ for $x < x_e$ but near x_e?*
   (e) *If $f'[x_e]$ is positive and $f[x_e] = 0$, what is the sign of $f[x]$ for $x > x_e$ but near x_e?*
   (f) *If $f[x]$ changes from negative to positive as x moves from just below x_e to just above x_e, what is the sign of $f'[x_e]$?*

2. **Increments and Equilibria**
   Let x_e be an equilibrium point of the one-dimensional dynamical system $\frac{dx}{dt} = f[x]$.
   (a) *When does the approximation (Incr)*

   $$f[u + x_e] = f[x_e] + a\,u + \varepsilon\,u$$

   with $\varepsilon \approx 0$ hold? What is the constant a?

(b) *Show in general that when x_e is an equilibrium point,*

$$f[u + x_e] = a\,u + \varepsilon u \qquad with \ \varepsilon \approx 0, \quad when \ u \approx 0$$

(c) *Show that the function $f[x] = x(1 - x/2)$ corresponds to the logistic growth differential equation $\frac{dx}{dt} = x(1 - x/2)$.*

(d) *Compute the general symbolic $f'[x]$ when $f[x] = x(1 - x/2)$.*

(e) *Show that the equation $\frac{dx}{dt} = x(1 - x/2)$ has equilibria only at the points $x_e = 2$ and $x_e = 0$. What is the value of $f[x_e]$ in these two cases?*

(f) *Compute $f'[x_e]$ at the two points $x_e = 2$ and $x_e = 0$.*

(g) *Sketch the whole nonlinear direction line of the differential equation $\frac{dx}{dt} = x(1 - x/2)$ and then draw magnified versions of the direction line near $x_e = 0$ and $x_e = 2$.*

(h) *Sketch direction lines of the differential equation $\frac{du}{dt} = u$ and the differential equation $\frac{du}{dt} = -u$. Why do these look like your microscopic flows? Which is which?*

3. Microscopic Dynamics

Now use the idea of the previous problem with parameters.

(a) *Sketch the direction line of*

$$\frac{dx}{dt} = f[x] = r\,x(1 - x/c)$$

assuming $r > 0$ and $c > 0$.

(b) *Compute the constant a_1 for the approximating equation at $x_{e_1} = 0$, using the method of the previous exercise.*

(c) *Sketch the direction line of $\frac{du_1}{dt} = a_1\,u_1$.*

(d) *Compute the constant a_2 for the approximating equation at $x_{e_2} = c$, using the method of the previous exercise.*

(e) *Sketch the direction line of $\frac{du_2}{dt} = a_2\,u_2$.*

Figure 24.2: Microscopic views

24.2   Linear Equilibria in 2-D

This section lists the various ways the equilibrium point at zero can "look" for a 2-D linear system. It uses the solutions of Chapter 23.

Our first task is to cast the results of Chapter 23 in terms of 2-by-2 systems so that we know what the linear systems look like. Then, we use partial derivatives to show which linear dynamical system approximates a nonlinear one near an equilibrium point.

A general linear system in two dimensions may be written

$$\frac{dx}{dt} = a_1\, x + a_2\, y$$

$$\frac{dy}{dt} = b_1\, x + b_2\, y$$

for constants a_1, a_2, b_1, b_2. This may also be written in matrix form as

$$\begin{bmatrix} \frac{dx}{dt} \\ \frac{dy}{dt} \end{bmatrix} = \begin{bmatrix} a_1 & a_2 \\ b_1 & b_2 \end{bmatrix} \begin{bmatrix} x \\ y \end{bmatrix}$$

The matrix form is a little more compact, and we want to use the four coefficients to compute the exponential solutions by a determinant related to the matrix.

We are already familiar with the linear system

$$\begin{bmatrix} \frac{dx}{dt} \\ \frac{dy}{dt} \end{bmatrix} = \begin{bmatrix} 0 & 1 \\ b_x & b_y \end{bmatrix} \begin{bmatrix} x \\ y \end{bmatrix} \qquad \longleftrightarrow \qquad \frac{d^2x}{dt^2} = b_y\, \frac{dx}{dt} + b_x\, x$$

as a geometric representation of the second-order one-dimensional equation. This is simply the "phase variable trick" $y = \frac{dx}{dt}$ from Chapter 23. Every solution of one system is also a solution of the other. In fact, if $x[t] = k_1\, x_1[t] + k_2\, x_2[t]$, then $y[t] = \frac{dx}{dt}$ is also a linear combination of $x_1[t]$ and $x_2[t]$. For example, if $x_1[t] = e^{r_1 t}$ and $x_2[t] = e^{r_2 t}$, then $y[t] = r_1\, k_1\, x_1[t] + r_2\, k_2\, x_2[t]$.

Because of this form of solutions, the zero equilibrium of both systems is attracting if and only if $\lim_{t \to \infty} x_1[t] = 0$ and $\lim_{t \to \infty} x_2[t] = 0$. It is repelling if and only if the real parts of the roots of the characteristic equation are positive. It is neither attracting nor repelling if one root is positive and the other is negative.

The various things that can happen to second-order equations $\frac{d^2x}{dt^2} = b_x\, x + b_y\, \frac{dx}{dt}$ are

(a) The roots of the characteristic equation are distinct and negative,

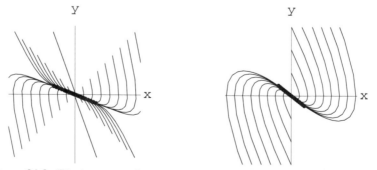

Figure 24.3: Distinct negative roots   and   repeated negative roots

so the basic solutions are $x_1[t] = e^{-r_1 t}$ and $x_2[t] = e^{-r_2 t}$, both with limit zero as $t \to \infty$.

(b) The roots of the characteristic equation are negative and "repeated," so the basic solutions are $x_1[t] = e^{-r t}$ and $x_2[t] = t\, e^{-r t}$, both with limit zero as $t \to \infty$.

(c) The roots of the characteristic equation are complex conjugates with negative real part, so the basic solutions are $x_1[t] = e^{-\alpha t}\cos[\beta t]$ and $x_2[t] = e^{-\alpha t}\sin[\beta t]$, both with limit zero as $t \to \infty$.

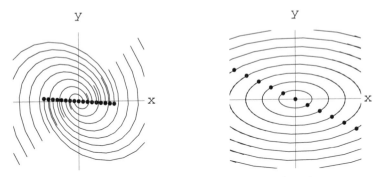

Figure 24.4: Complex negative real part   and   pure imaginary roots

(d) The roots of the characteristic equation are complex conjugates with zero real part, so the basic solutions are $x_1[t] = \cos[\beta t]$ and $x_2[t] = \sin[\beta t]$, both of absolute value one for all t. This is "neutral" stability; nearby solutions "orbit" the equilibrium forever.

(e) The roots of the characteristic equation are distinct and positive, so the basic solutions are $x_1[t] = e^{+r_1 t}$ and $x_2[t] = e^{+r_2 t}$, both with limit ∞ as $t \to \infty$.

(f) The roots of the characteristic equation are positive and "repeated," so the basic solutions are $x_1[t] = e^{r t}$ and $x_2[t] = t\,e^{r t}$, both with limit ∞ as $t \to \infty$.

(g) The roots of the characteristic equation are complex conjugates with positive real part, so the basic solutions are $x_1[t] = e^{+\alpha t}\cos[\beta t]$ and $x_2[t] = e^{+\alpha t}\sin[\beta t]$, both with the limit of the absolute value ∞ as $t \to \infty$.

(h) The roots of the characteristic equation are real of opposite sign, so the basic solutions are $x_1[t] = e^{+r_1 t}$ and $x_2[t] = e^{-r_2 t}$. One tends to ∞, and the other tends to zero as $t \to \infty$. This equilibrium is neither attracting nor repelling.

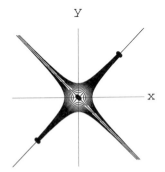

Figure 24.5: Roots of opposite sign

The computer programs in the **Linear Equilibria** folder illustrate the various attracting cases. The repelling cases can be obtained by running these animations backward (which is the solution of the equation in which t is replaced by $-t$). Figures in the text are single frames from these animations.

Theorem 24.1. *Every solution $(x[t], y[t])$ of a two-dimensional linear system*

$$\begin{bmatrix} \frac{dx}{dt} \\ \frac{dy}{dt} \end{bmatrix} = \begin{bmatrix} a_x & a_y \\ b_x & b_y \end{bmatrix} \begin{bmatrix} x \\ y \end{bmatrix}$$

with

$$\det \begin{vmatrix} a_x & a_y \\ b_x & b_y \end{vmatrix} = a_x b_y - a_y b_x \neq 0 \quad \text{and either } a_y \neq 0 \text{ or } b_x \neq 0$$

can be written as a certain specific linear combination of the basic solutions $x_1[t]$ and $x_2[t]$ of the second-order scalar equation

$$\frac{d^2 x}{dt^2} = \beta \frac{dx}{dt} + \gamma x$$

with the characteristic equation

$$\det \begin{vmatrix} a_x - r & a_y \\ b_x & b_y - r \end{vmatrix} = (a_x - r)(b_y - r) - a_y b_x$$

$$= r^2 - (a_x + b_y) r - (a_y b_x - a_x b_y)$$

$$= r^2 - \beta r - \gamma = 0$$

The basic solutions $x_1[t]$ and $x_2[t]$ of the second-order scalar equation can be found from the roots of the algebraic characteristic equation as in Theorem 23.5.

The proof of the equivalence in case $a_y \neq 0$ is to solve the dx equation for y

$$\frac{1}{a_y} \frac{dx}{dt} = \frac{a_x}{a_y} x + y$$

$$y = \frac{1}{a_y} \frac{dx}{dt} - \frac{a_x}{a_y} x$$

differentiate

$$\frac{dy}{dt} = \frac{1}{a_y} \frac{d^2 x}{dt^2} - \frac{a_x}{a_y} \frac{dx}{dt}$$

use the dy equation

$$b_y y + b_x x = \frac{1}{a_y} \frac{d^2 x}{dt^2} - \frac{a_x}{a_y} \frac{dx}{dt}$$

$$\frac{1}{a_y} \frac{d^2 x}{dt^2} - b_y y - b_x x - \frac{a_x}{a_y} \frac{dx}{dt} = 0$$

and the y equation

$$\frac{1}{a_y} \frac{d^2 x}{dt^2} - b_y \left(\frac{1}{a_y} \frac{dx}{dt} - \frac{a_x}{a_y} x \right) - b_x x - \frac{a_x}{a_y} \frac{dx}{dt} = 0$$

$$\frac{d^2 x}{dt^2} - (a_x + b_y) \frac{dx}{dt} + (a_x b_y - a_y b_x) x = 0$$

This has characteristic equation

$$r^2 + (-a_x - b_y)\, r + (a_x\, b_y - a_y\, b_x) = 0$$

which also can be computed from the determinant above.

Once we write a solution of the second-order equation $x[t] = k_1\, x_1[t] + k_2\, x_2[t]$, we can use the equation

$$y = \frac{1}{a_y} \frac{dx}{dt} - \frac{a_x}{a_y} x$$

to find a solution to the linear system.

There are two exceptional cases in which we cannot associate a second-order scalar equation with a first-order two-dimensional linear vector equation.

> **Theorem 24.2.** *The linear system*
>
> $$\begin{bmatrix} \frac{dx}{dt} \\ \frac{dy}{dt} \end{bmatrix} = \begin{bmatrix} a_x & 0 \\ 0 & b_y \end{bmatrix} \begin{bmatrix} x \\ y \end{bmatrix}$$
>
> *has basic solutions $x_1[t] = e^{a_x t}$ and $y_2[t] = e^{b_y t}$. Every other solution may be written in the form*
>
> $$\begin{bmatrix} x[t] \\ y[t] \end{bmatrix} = k_1 \begin{bmatrix} x_1[t] \\ 0 \end{bmatrix} + k_2 \begin{bmatrix} 0 \\ y_2[t] \end{bmatrix}$$

Even though we do not associate this case with a second-order equation, the basic solutions are still of the form $e^{r_1 t}$ and $e^{r_2 t}$, where r_1 and r_2 are roots of the equation

$$\det \begin{vmatrix} a_x - r & 0 \\ 0 & b_y - r \end{vmatrix} = (a_x - r)(b_y - r) = 0$$

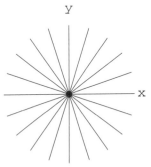
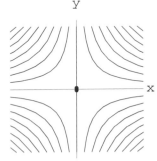

Figure 24.6: $a_x <= b_y < 0,\ a_y = b_x = 0$ and $a_x > 0$ and $b_y < 0,\ a_y = b_x = 0$

If the linear system

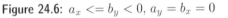

$$\begin{bmatrix} \frac{dx}{dt} \\ \frac{dy}{dt} \end{bmatrix} = \begin{bmatrix} a_x & a_y \\ b_x & b_y \end{bmatrix} \begin{bmatrix} x \\ y \end{bmatrix}$$

has

$$\det \begin{vmatrix} a_x & a_y \\ b_x & b_y \end{vmatrix} = 0$$

then the one row of the matrix of coefficients is a multiple of the other, that is, the linear system is of the form given in Exercise 24.2.3. This system is degenerate from the point of view of equilibria.

The subsection can be summed up in this way:

Theorem 24.3. *Linear Equilibria*
  All solutions of a two-dimensional linear system

$$\begin{bmatrix} \frac{dx}{dt} \\ \frac{dy}{dt} \end{bmatrix} = \begin{bmatrix} a_x & a_y \\ b_x & b_y \end{bmatrix} \begin{bmatrix} x \\ y \end{bmatrix}$$

  with

$$\det \begin{vmatrix} a_x & a_y \\ b_x & b_y \end{vmatrix} = a_x\, b_y - a_y\, b_x \neq 0$$

  may be written as a linear combination of (real or complex) exponentials e^{rt} (and $t\, e^{rt}$) for constants r satisfying the characteristic equation

$$\det \begin{vmatrix} a_x - r & a_y \\ b_x & b_y - r \end{vmatrix} = (a_x - r)(b_y - r) - a_y\, b_x = r^2 - (a_x + b_y)\, r + (a_x\, b_y - a_y\, b_x) = 0$$

Therefore, zero is an attracting equilibrium if the real parts of the roots of the characteristic equation are negative. Zero is a repeller if the real parts are positive. Zero attracts for some initial conditions and repels for others if one root is positive and the other is negative.

Exercise set 24.2

1. Checking Linear Equilibria
   *The point $(0,0)$ is always an equilibrium for a linear dynamical system. Use Theorem 24.3 to show which of the following dynamical systems have $(0,0)$ as an attractor, repeller, or neither. Modify one of the **Linear Equilibria** programs to animate each of the flows and verify your computations.*
   (a)
   $$\begin{bmatrix} du \\ dv \end{bmatrix} = \begin{bmatrix} 2 & 0 \\ 0 & 1 \end{bmatrix} \begin{bmatrix} u \\ v \end{bmatrix} dt$$
   (b)
   $$\begin{bmatrix} du \\ dv \end{bmatrix} = \begin{bmatrix} -1 & -1 \\ -\frac{2}{3} & -\frac{1}{3} \end{bmatrix} \begin{bmatrix} u \\ v \end{bmatrix} dt$$
   (c)
   $$\begin{bmatrix} du \\ dv \end{bmatrix} = \begin{bmatrix} -2 & -2 \\ 0 & -\frac{1}{3} \end{bmatrix} \begin{bmatrix} u \\ v \end{bmatrix} dt$$
   (d)
   $$\begin{bmatrix} du \\ dv \end{bmatrix} = \begin{bmatrix} -1 & 0 \\ -2 & -1 \end{bmatrix} \begin{bmatrix} u \\ v \end{bmatrix} dt$$

2. Attraction and Repulsion
   Show that the linear system
   $$\begin{bmatrix} \frac{dx}{dt} \\ \frac{dy}{dt} \end{bmatrix} = \begin{bmatrix} 0 & 1 \\ b_x & b_y \end{bmatrix} \begin{bmatrix} x \\ y \end{bmatrix}$$

with $b_x \neq 0$ has zero as its only equilibrium point. Show that the equation

$$\det \begin{vmatrix} -r & 1 \\ b_x & b_y - r \end{vmatrix} = r^2 - b_y\, r - b_x = 0$$

is the same as the characteristic equation of the associated second-order one-dimensional equation obtained from the phase variable trick.

3. Dependent Linear Systems
   Show that every point of the line $a_x \cdot x + a_y \cdot y = 0$ is an equilibrium point for the system

$$\begin{bmatrix} \frac{dx}{dt} \\ \frac{dy}{dt} \end{bmatrix} = \begin{bmatrix} a_x & a_y \\ c\,a_x & c\,a_y \end{bmatrix} \begin{bmatrix} x \\ y \end{bmatrix}$$

Also, show that the solutions travel along the lines of slope

$$\frac{dy}{dx} = c$$

so that the phase portrait consists of lines of slope c meeting a stagnant line

$$a_x \cdot x + a_y \cdot y = 0$$

*Use one of the **Linear Equilibria** computer programs to make an animation of the flow of a pair of linear equations in which the a's are multiples of the b's.*

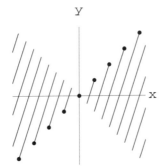

Figure 24.7: $b_x = 3\,a_x = 1$ and $b_y = 3\,a_y = -1$

4. Time Reversal
   Consider the two-dimensional linear system

$$\begin{bmatrix} \frac{dx}{dt} \\ \frac{dy}{dt} \end{bmatrix} = \begin{bmatrix} a_x & a_y \\ b_x & b_y \end{bmatrix} \begin{bmatrix} x \\ y \end{bmatrix}$$

and the related second-order scalar equation

$$\frac{d^2 x}{dt^2} = \beta\, \frac{dx}{dt} + \gamma\, x$$

with the characteristic equation

$$\det \begin{vmatrix} a_x - r & a_y \\ b_x & b_y - r \end{vmatrix} = (a_x - r)(b_y - r) - a_y b_x$$

$$= r^2 - (a_x + b_y)r - (a_y b_x - a_x b_y)$$

$$= r^2 - \beta r - \gamma = 0$$

Show that the equations obtained by time reversal have negative characteristic roots. That is, let $\tau = -t$ and write new differential equations for $x[\tau]$. Then, show that the roots of the τ-characteristic equation are the negatives of the originals.

24.3   Nonlinear Equilibria in 2-D

This section gives the "dynamical system microscope" as a precise theorem. It says in technical language that a microscopic view of a flow near a non-linear equilibrium point looks like the flow of the linear system given by the total differential at the equilibrium.

Now, we consider a nonlinear dynamical system

$$\begin{bmatrix} \frac{dx}{dt} \\ \frac{dy}{dt} \end{bmatrix} = \begin{bmatrix} f[x,y] \\ g[x,y] \end{bmatrix}$$

near an equilibrium point (x_e, y_e) where $f[x_e, y_e] = g[x_e, y_e] = 0$. We want to "look at the flow under a microscope focused at (x_e, y_e)." The total differential approximations for f and g say symbolically how the functions behave microscopically,

$$f[u + x_e, v + y_e] = \frac{\partial f}{\partial x}[x_e, y_e] \cdot u + \frac{\partial f}{\partial y}[x_e, y_e] \cdot v + \varepsilon_1 \sqrt{u^2 + v^2}$$

$$g[u + x_e, v + y_e] = \frac{\partial g}{\partial x}[x_e, y_e] \cdot u + \frac{\partial g}{\partial y}[x_e, y_e] \cdot v + \varepsilon_2 \sqrt{u^2 + v^2}$$

or

$$f[u + x_e, v + y_e] \approx a_x \cdot u + a_y \cdot v$$

$$g[u + x_e, v + y_e] \approx b_x \cdot u + b_y \cdot v$$

where $u = x - x_e$ and $v = y - y_e$ are the local variables, and the approximating constants are $a_x = \frac{\partial f}{\partial x}[x_e, y_e]$, $a_y = \frac{\partial f}{\partial y}[x_e, y_e]$, $b_x = \frac{\partial g}{\partial x}[x_e, y_e]$ and $b_y = \frac{\partial g}{\partial x}[x_e, y_e]$. Near the equilibrium, we have

$$\begin{bmatrix} \frac{dx}{dt} \\ \frac{dy}{dt} \end{bmatrix} = \begin{bmatrix} \frac{du}{dt} \\ \frac{dv}{dt} \end{bmatrix} = \begin{bmatrix} f[u + x_e, v + y_e] \\ g[u + x_e, v + y_e] \end{bmatrix} \approx \begin{bmatrix} a_x & a_y \\ b_x & b_y \end{bmatrix} \begin{bmatrix} u \\ v \end{bmatrix}$$

So, we expect that the flow of

$$\begin{bmatrix} \frac{dx}{dt} \\ \frac{dy}{dt} \end{bmatrix} = \begin{bmatrix} f[x,y] \\ g[x,y] \end{bmatrix}$$

near equilibrium microscopically looks like the flow of the linear approximation.

Theorem 24.4. *The Dynamical Equilibrium Microscope*
Let $f[x,y]$ and $g[x,y]$ be smooth functions with $f[x_e, y_e] = g[x_e, y_e] = 0$, and define coefficients by the partial derivatives at the equilibrium,

$$\begin{bmatrix} a_x & a_y \\ b_x & b_y \end{bmatrix} = \begin{bmatrix} \frac{\partial f}{\partial x} & \frac{\partial f}{\partial y} \\ \frac{\partial g}{\partial x} & \frac{\partial g}{\partial y} \end{bmatrix} [x_e, y_e]$$

Then, for a fixed amount of time and sufficient magnification at (x_e, y_e), the flow of the nonlinear system

$$\frac{dx}{dt} = f[x,y]$$

$$\frac{dy}{dt} = g[x,y]$$

appears the same as the flow of the local linear approximation given by

$$\begin{bmatrix} \frac{du}{dt} \\ \frac{dv}{dt} \end{bmatrix} = \begin{bmatrix} a_x & a_y \\ b_x & b_y \end{bmatrix} \begin{bmatrix} u \\ v \end{bmatrix}$$

We see the linearized flow in the microscope for a finite time. In the limit as t tends to infinity, the nonlinear system can "look" different. Here is another way to say this. If we magnify a lot, but not an infinite amount, then we may see a separation between the linear and nonlinear system after a very long time.

Informally, if our magnification is $1/\delta$, for $\delta \approx 0$, and our solution starts in our view,

$$(x[0] - x_e, y[0] - y_e) = \delta \cdot (a, b)$$

for finite a and b, then the solution satisfies

$$(x[t] - x_e, y[t] - y_e) = \delta \cdot (u[t], v[t]) + \delta \cdot (\varepsilon_x[t], \varepsilon_y[t])$$

where $(u[t], v[t])$ satisfies the linear equation and starts at $(u[0], v[0]) = (a, b)$ and where $(\varepsilon_x[t], \varepsilon_y[t]) \approx (0, 0)$ for all finite t. (For every screen resolution θ and every bound β on the time of observation and observed scale of initial condition, there is a magnification large enough so that if $|a| \leq \beta$ and $|b| \leq \beta$, then the error observed at that magnification is less than θ for $0 \leq t \leq \beta$.) A complete proof of this result is in the Mathematical Background material on the CD.

The point of this theorem is that it lets us compute what we would see if we looked at a nonlinear equilibrium point in a powerful microscope. Here is an example of how we "look" at equilibria:

Example 24.5. *A Sample Equilibrium Calculation*

Consider the differential equations

$$\frac{dx}{dt} = f[x,y] = y(1 - \frac{x}{7} - \frac{y}{5})$$

$$\frac{dy}{dt} = g[x,y] = x(1 - \frac{x}{8} - \frac{y}{6})$$

This system has an equilibrium point at $(x_e, y_e) = (0,5)$ because

$$f[0,5] = y(1 - \frac{x}{7} - \frac{y}{5}) = 5(1 - \frac{0}{7} - \frac{5}{5}) = 0$$
$$g[0,5] = x(1 - \frac{x}{8} - \frac{y}{6}) = 0(1 - \frac{0}{8} - \frac{5}{6}) = 0$$

The matrix of symbolic partial derivatives is

$$\begin{bmatrix} \frac{\partial f}{\partial x} & \frac{\partial f}{\partial y} \\ \frac{\partial g}{\partial x} & \frac{\partial g}{\partial y} \end{bmatrix} = \begin{bmatrix} -\frac{y}{7} & 1 - \frac{x}{7} - \frac{2y}{5} \\ 1 - \frac{x}{4} - \frac{y}{6} & -\frac{x}{6} \end{bmatrix}$$

which gives the matrix for the linear system

$$\begin{bmatrix} a_x & a_y \\ b_x & b_y \end{bmatrix} = \begin{bmatrix} \frac{\partial f}{\partial x} & \frac{\partial f}{\partial y} \\ \frac{\partial g}{\partial x} & \frac{\partial g}{\partial y} \end{bmatrix} (0,5) = \begin{bmatrix} -\frac{5}{7} & -1 \\ +\frac{1}{6} & 0 \end{bmatrix}$$

The characteristic equation is

$$\det \begin{vmatrix} a_x - r & a_y \\ b_x & b_y - r \end{vmatrix} = \det \begin{vmatrix} -\frac{5}{7} - r & -1 \\ \frac{1}{6} & -r \end{vmatrix} = (-\frac{5}{7} - r)(-r) + \frac{1}{6} = r^2 + \frac{5}{7}r + \frac{1}{6} = 0$$

with roots

$$r_1, r_2 = \frac{-\frac{5}{7} \pm \sqrt{(\frac{5}{7})^2 - 4 \cdot \frac{1}{6}}}{2} = -\frac{5}{14} \pm \frac{i}{14}\sqrt{\frac{23}{3}} \approx -0.357143 \pm i\,0.197777$$

These complex roots mean that a microscopic view of the equilibrium point at $(0,5)$ of the system spirals inward as in Figure 24.8.

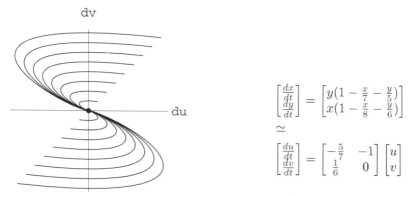

Figure 24.8: A microscopic view near $(0,5)$

Procedure 24.5. Looking at Equilibria
To "see" the equilibria of the dynamical system

$$\frac{dx}{dt} = f[x, y]$$

$$\frac{dy}{dt} = g[x, y]$$

compute the following things in the following order: (Use the computer.)

(a) Find the solutions (x_{e1}, y_{e1}), (x_{e2}, y_{e2}), (x_{e3}, y_{e3}), $\cdots$ to the equilibrium equations

$$0 = f[x, y]$$

$$0 = g[x, y]$$

(b) Use rules of differentiation (the symbolic part of the "microscope") to find the four symbolic partial derivatives:

$$\frac{\partial f}{\partial x}[x, y] \qquad\qquad \frac{\partial f}{\partial y}[x, y]$$

$$\frac{\partial g}{\partial x}[x, y] \qquad\qquad \frac{\partial g}{\partial y}[x, y]$$

(c) For each equilibrium point (x_{e1}, y_{e1}), (x_{e2}, y_{e2}), (x_{e3}, y_{e3}), $\cdots$,

(1) Evaluate the symbolic partial derivatives at the specific equilibrium to obtain a matrix of constants

$$\begin{bmatrix} \frac{\partial f}{\partial x}[x_e, y_e] & \frac{\partial f}{\partial y}[x_e, y_e] \\ \frac{\partial g}{\partial x}[x_e, y_e] & \frac{\partial g}{\partial y}[x_e, y_e] \end{bmatrix} = \begin{bmatrix} a_x & a_y \\ b_x & b_y \end{bmatrix}$$

(2) Calculate the roots of the characteristic polynomial

$$\det \begin{vmatrix} a_x - r & a_y \\ b_x & b_y - r \end{vmatrix} = r^2 - (a_x + b_y)r + (a_x b_y - a_y b_x) = 0$$

(3) Characterize the equilibrium according to the basic cases given above.

As long as the characteristic values are not zero, what you "see" in the calculus microscope is one of the five basic pictures either flowing in or out depending on the sign of the characteristic roots. (Zero characteristic values produce degenerate pictures like the sixth described below.)

The computer can help with the computations needed in using the local stability theorem, as in the general program **LocalStability**.

The continuous equilibrium microscope is the most sophisticated one we have learned. First, we need to use partial derivatives for two functions, but even after we calculate the associated linear dynamical system, we have substantial work left in sketching its flow.

It is helpful to remember the five basic cases, which could either be flowing in or out depending on the signs of the characteristic roots. These cases are: (1) Real roots of the same sign; (2) real roots of opposite sign; (3) purely imaginary roots; (4) complex roots that spiral in if the real part of the roots is negative; (5) a "repeated" real root, or non-distinct real root; (6) a degeneracy in the linear problem - for example, if the equations are just multiples, so that $y[t] = k\,x[t]$ for some constant k. Figure 24.9 shows sample flows for the six baisc cases. Except for the exact directions, these are the views in a microscope focused at an equilibrium point of a smooth dynamical system.

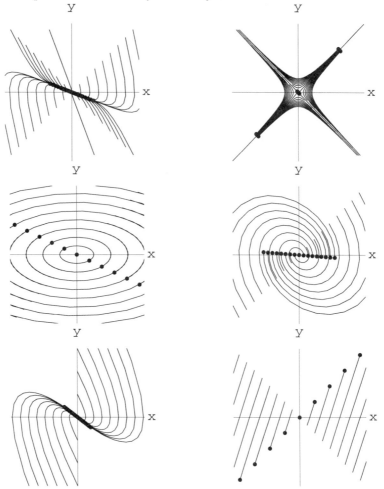

Figure 24.9: Figures for the six cases above

The local stability of an equilibrium point for a dynamical system is formulated in Theorem 24.6. Stability is an "infinite time" result, whereas the localization of Theorem 24.4 is a finite time result. In effect, the stability result says that when the real parts of the characteristic roots are negative, we can look in the microscope infinitely long and never tell the difference between the linear and nonlinear flows.

Theorem 24.6.   *Local Stability*

Let $f[x, y]$ and $g[x, y]$ be smooth functions with $f[x_e, y_e] = g[x_e, y_e] = 0$.   The coefficients given by the partial derivatives evaluated at the equilibrium

$$\begin{bmatrix} a_x & a_y \\ b_x & b_y \end{bmatrix} = \begin{bmatrix} \frac{\partial f}{\partial x} & \frac{\partial f}{\partial y} \\ \frac{\partial g}{\partial x} & \frac{\partial g}{\partial y} \end{bmatrix} [x_e, y_e]$$

define the characteristic equation of the equilibrium,

$$\det \begin{vmatrix} a_x - r & a_y \\ b_x & b_y - r \end{vmatrix} = (a_x - r)(b_y - r) - a_y\, b_x = r^2 - (a_x + b_y)\, r + (a_x\, b_y - a_y\, b_x) = 0$$

Suppose that the real parts of both of the roots of this equation are negative. Then there is a real neighborhood of (x_e, y_e) or a non-infinitesimal $\varepsilon > 0$ such that when a solution satisfies

$$\frac{dx}{dt} = f[x, y]$$

$$\frac{dy}{dt} = g[x, y]$$

with initial condition in the neighborhood, $|x(0) - x_e| < \varepsilon$ and $|y(0) - y_e| < \varepsilon$, then

$$\lim_{t \to \infty} x[t] = x_e \qquad and \qquad \lim_{t \to \infty} y[t] = y_e$$

The proof of this result is given in the Mathematical Background Chapter on Theory of Initial Value Problems. A local repeller result is also given. You may find it interesting to look at the phase diagrams of the various "exceptional" cases given there.

Exercise set 24.3

1. Nonlinear Equilibria

Show that the nonlinear system

$$\begin{bmatrix} \frac{dx}{dt} \\ \frac{dy}{dt} \end{bmatrix} = \begin{bmatrix} 2\, x(1 - \frac{x}{4} - \frac{y}{4}) \\ y(1 - \frac{x}{3} - \frac{y}{6}) \end{bmatrix}$$

has equilibria at $(x, y) = (0, 0)$, $(2, 2)$, $(4, 0)$ and $(0, 6)$.   Also, show that the linear approximation at these equilibria are as follows:
At $(0, 0)$:

$$\begin{bmatrix} du \\ dv \end{bmatrix} = \begin{bmatrix} 2 & 0 \\ 0 & 1 \end{bmatrix} \begin{bmatrix} u \\ v \end{bmatrix} dt$$

a local repeller.
At $(2, 2)$:

$$\begin{bmatrix} du \\ dv \end{bmatrix} = \begin{bmatrix} -1 & -1 \\ -\frac{2}{3} & -\frac{1}{3} \end{bmatrix} \begin{bmatrix} u \\ v \end{bmatrix} dt$$

a "saddle point," neither attracting nor repelling.

At $(4,0)$:

$$\begin{bmatrix} du \\ dv \end{bmatrix} = \begin{bmatrix} -2 & -2 \\ 0 & -\frac{1}{3} \end{bmatrix} \begin{bmatrix} u \\ v \end{bmatrix} dt$$

a local attractor.
At $(0,6)$:

$$\begin{bmatrix} du \\ dv \end{bmatrix} = \begin{bmatrix} -1 & 0 \\ -2 & -1 \end{bmatrix} \begin{bmatrix} u \\ v \end{bmatrix} dt$$

a local attractor.
Check your four local results by using the the computer program **Flow2D** to make a flow animation that includes initial points around all these equilibria.

2. **The Computer and Stability**

Suppose we wish to investigate the equilibria of the particular case of the system of equations:

$$\frac{dx}{dt} = f[x,y] = x(1 - x/7 - y/7)$$
$$\frac{dy}{dt} = g[x,y] = y(1 - x/10 - y/5)$$

See the **LocalStability** program.

(a) Use the computer to show that the equilibrium $(4,3)$ of the above system of differential equations is a local attractor near $(4,3)$ with characteristic roots -1 and $-6/35$.

(b) Use the computer to classify the other equilibria of this system at $(0,0)$, $(7,0)$ and $(0,5)$. Compare your analytical results with the geometric analysis of Example 22.2.

(c) Use the computer to show that the positive equilibrium of the system of differential equations below is neither attracting nor repelling because the characteristic roots are -1 and $+6/35$.

$$\frac{dx}{dt} = f[x,y] = x(1 - x/10 - y/5)$$
$$\frac{dy}{dt} = g[x,y] = y(1 - x/7 - y/7)$$

(d) Use the computer to classify the other equilibria of this system at $(0,0)$, $(10,0)$ and $(0,7)$. Compare your analytical results to the geometric results in Example 22.4.

3. **A Look in the Microscope**

(a) The differential equation

$$x'' + 2x' + x + 15x^3 = 0$$

is "autonomous." What does that mean? Why cannot we study flows of nonautonomous equations, but rather only their explicit solutions?

(b) *This differential equation is equivalent to the system*

$$dx = y \, dt$$
$$dy = (-x - 15x^3 - 2y) \, dt$$

Why?

(c) *If we write this system of differential equations in the form*

$$\frac{dx}{dt} = f[x, y]$$
$$\frac{dy}{dt} = g[x, y]$$

then $f[x, y] = ?$ and $g[x, y] = ??$ and the system satisfies the hypotheses of the Existence and Uniqueness Theorem. Why?

(d) *Show that the only equilibrium point of this system is $(x_e, y_e) = (0, 0)$.*

(e) *At the equilibrium, show that*

$$\frac{\partial f}{\partial x}[x_e, y_e] = 0 \qquad\qquad \frac{\partial f}{\partial y}[x_e, y_e] = 1$$

$$\frac{\partial g}{\partial x}[x_e, y_e] = -1 \qquad\qquad \frac{\partial g}{\partial y}[x_e, y_e] = -2$$

(f) *The characteristic roots of the system*

$$\frac{du}{dt} = v$$
$$\frac{dv}{dt} = -u - 2v$$

are repeated; that is, there really is only one, $r = -1$.

(g) *Which of the figures above is similar to the flow of the (u, v) system?*

(h) *What does the flow of the (x, y) system look like under a powerful microscope focused at the point $(0, 0)$?*

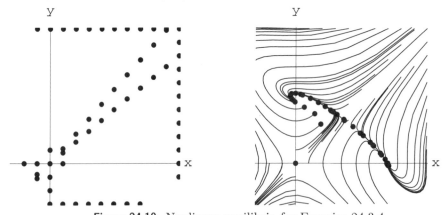

Figure 24.10: Nonlinear equilibria for Exercise 24.3.4

4. *The system of differential equations*

$$dx = y(1 - \frac{x}{10} - \frac{y}{5}) \, dt$$
$$dy = x(1 - \frac{x}{7} - \frac{y}{7}) \, dt$$

has equilibria at the points $(0,0)$, $(0,5)$, $(7,0)$ *and* $(4,3)$. *What would you see in a powerful microscope focused at each of these points?*

5. *What do you see if you look inside a microscope at a smooth flow at a non-equilibrium point?*

6. Uniqueness
A 2-D flow given by smooth differential equations has a large closed orbit, in fact, points move around the circle of radius 1,000 centered at (2,5). For example, the initial point (1002,5) travels around the circle and returns at $t = 100$. *Can any solution that starts inside this circle tend to infinity? In other words, could a solution starting inside the circle cross the circle? (HINT: See the title of this exercise.)*

Problem 24.1. S-I-R WITH BIRTHS = DEATHS ⸻▼
Show that the equilibrium point of the dynamical system

$$\frac{ds}{dt} = k - k\,s - a\,s\,i$$
$$\frac{di}{dt} = a\,s\,i - b\,i - k\,i$$

is an attractor when the parameters produce a positive equilibrium. In particular, show that the equilibrium is at

$$s_e = \frac{b+k}{a} \qquad and \qquad i_e = k\,\frac{a-b-k}{a\,(b+k)}$$

with derivative matrix at equilibrium,

$$\begin{bmatrix} -\frac{ak}{b+k} & -(b+k) \\ \frac{k(a-b-k)}{b+k} & 0 \end{bmatrix}$$

and the characteristic equation at equilibrium,

$$r^2 + \frac{ak}{b+k}\,r + k[a - (b+k)] = 0$$

In general, $a > 0$, $b > 0$, $k > 0$, *and* k *is small compared to* a *and* b. *Mathematically, we need* $a > b + k$ *in order that* $s_e > 0$ *and* $i_e > 0$. *Why are the real parts of the roots of this characteristic equation negative? Why does this make the equilibrium point an attractor?*
    For rubella with $a = 6.8/11.0$, $b = 1/11.0$ *and* $k = 0.001$, *show that this is a spiral attractor (with complex roots).*

24.4   Solutions vs. Invariants - CD

Energy is a very important example of a general mathematical idea asso-
ciated with dynamical systems, an invariant quantity. We also used an
invariant of the SIR equations in Chapter 2 to find the limit of the sus-
ceptible population in an epidemic. This section studies invariants more
carefully.

The energy

$$E(x[t], y[t]) = \frac{1}{2}m\,y^2[t] + \frac{1}{2}s\,x^2[t]$$

in an undamped spring

$$m\,\frac{d^2x}{dt^2} + s\,x = 0$$

is constant. That is, $\frac{dE[x[t],y[t]]}{dt} = 0$.
   Similarly, the quantity

$$Q(s[t], i[t]) = i[t] + s[t] - \frac{b}{a}\,\text{Log}[s[t]]$$

is constant when s and i satisfy the differential equations

$$\frac{ds}{dt} = -a\,s\,i$$

$$\frac{di}{dt} = a\,s\,i - b\,i$$

for the disease parameters a and b.
   We can show that the orbits of the Lotka-Volterra predator-prey equations are closed or permanently oscillating by computing an invariant.

Remainder of this section only on CD

24.5   Projects

24.5.1   The Second Derivative Test for Max-Min in 2-D as Stability

This project shows how to test a critical point in two variables as a max or min. The idea is to imagine being on a mountain, moving down in the steepest direction. If a place with a horizontal tangent plane is a minimum, then it is a stable equilibrium for this motion.

24.5.2   The Perfecto Skier

The gradient "motion" in the mathematical project on max-min in 2 variables is quite interesting mathematically, but it does not correspond to sliding down a mountain. If a skier slides down a slippery slope, the component of force acting tangent to the mountain produces an acceleration or second derivative. This produces different motion than the gradient flow.

24.5.3   Vectors, Bombing, The Pendulum and Jupiter

Vector geometry underlies the derivation of many differential equations in science. There are projects on low-level bombing, the pendulum, and shooting a satellite near Jupiter to "sling shot" it out of the solar system.

24.5.4   Differential Equations from Increment Geometry

Many important curves in mathematics are given by their differential equation. The tractrix, isochrone, and catenary are three such examples explored in the project book.

24.5.5   Applications to Ecology

We have seen competition and predator-prey equations in the last three chapters. The Scientific Projects on predator-prey interactions, on competition, and on cooperation study these and related examples in more detail.

24.5.6   Chemical Reactions

Two projects on nonlinear chemical reactions are given (and the drug absorption project uses a linear model of elimination of a chemical). One nonlinear model studies stability of a chemical reaction using the mathematical stability theory. The other studies approximations to the nonlinear chemical dynamics of enzyme reactions.

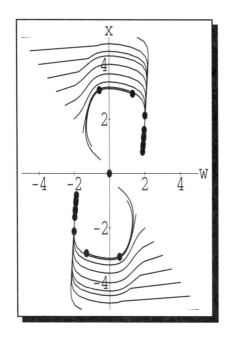

Part 6

Infinite Series

Geometric Series

This chapter begins the study of infinite series.

Series are simply discrete sums

$$u_1 + u_2 + u_3 + \cdots$$

for example,

$$\frac{1}{2} + \frac{1}{4} + \frac{1}{8} + \frac{1}{16} + \cdots$$

or

$$1 + x + \frac{x^2}{2!} + \frac{x^3}{3!} + \cdots$$

or

$$\mathrm{Cos}[x] + \frac{1}{2}\,\mathrm{Cos}[3x] + \frac{1}{2^2}\,\mathrm{Cos}[3^2 x] + \cdots$$

Finite series end at a specified point, such as

$$1 + r + r^2 + r^3 + \cdots + r^{356}$$

whereas infinite series "keep going to infinity."

Various kinds of infinite series are important as approximations to functions. Power series and Fourier series are the most important kinds of series approximations (for different reasons). In this case, "approximation" means that you can get "close enough" to the "infinite sum" by taking "sufficiently many" terms. The expression

$$e^x = 1 + x + \frac{x^2}{2!} + \frac{x^3}{3!} + \cdots$$

means that for "sufficiently large" n,

$$e^x \approx 1 + x + \frac{x^2}{2!} + \frac{x^3}{3!} + \cdots + \frac{x^n}{n!}$$

However, there are some difficulties with adding sums of functions. If we find "sufficiently many" terms to approximate at one value of x, we may not necessarily assume a good approximation at another value of x.

The goal of this chapter is to calculate geometric series and to use geometric series to estimate series approximations. The geometric series can be viewed as a function series that is dominated by a high numerical value. This estimate also applies to more complicated function series. A geometric estimate of another series is especially useful, because we can explicitly compute the error in a geometric series. This makes the geometric series the most fundamental series in mathematics.

The computer programs **BasicSeries**, **ClassicalSeries**, and **FourierSeries** visually illustrate the way various approximations depend on the value of x. You can look at the animations now and return to the electronic exercises when they arise in the text. The first five approximations associated with the power series

$$\text{Cos}[x] = 1 - \frac{x^2}{2} + \frac{x^4}{24} + \cdots + \frac{(-1)^n x^{2n}}{(2n)!} + \cdots$$

are shown in Figure 25.1. Notice that the "worst errors" or biggest differences between the cosine curve and the polynomials always occur at x values of largest absolute value. This is a general property of power series that helps to make their theory simple and powerful.

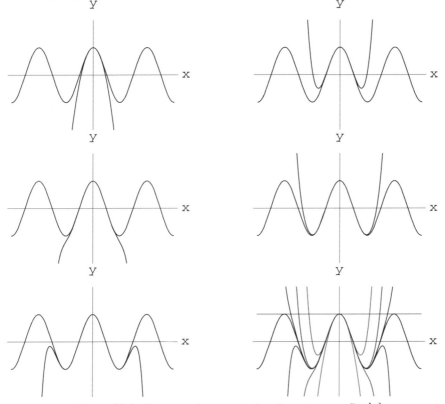

Figure 25.1: Power series approximations to $y = \text{Cos}[x]$

Approximation or convergence, of Fourier series can be more complicated than for power series. For example, the approximation

$$\frac{x}{2} = \text{Sin}[x] + \frac{-1}{2}\text{Sin}[2x] + \frac{1}{3}\text{Sin}[3x] + \cdots + \frac{(-1)^{n+1}}{n}\text{Sin}[nx] + \cdots$$

is valid for $-\pi < x < \pi$; but, because all the sine terms are 2π periodic, the limit of the series must also be periodic. When we repeat a piece of the curve $y = x$, for $-\pi < x < \pi$, periodically, we get a "sawtooth" wave with jump discontinuities at multiples of π, shown in Figure 25.2. Convergence of the infinite Fourier series is necessarily complicated near the jumps; in fact, you can see that every partial sum is a bad approximation at some x near π. However, if we stay away from the trouble spot, this remarkable combination of sines approximates a linear function. Notice in particular that, if $x = \pi/2$, then

$$\frac{\pi}{4} = 1 + 0 - \frac{1}{3} + 0 + \frac{1}{5} + 0 - \frac{1}{7} + \cdots$$

This curious numerical fact falls out of the symbolic expression once we know which values of x produce a valid approximation.

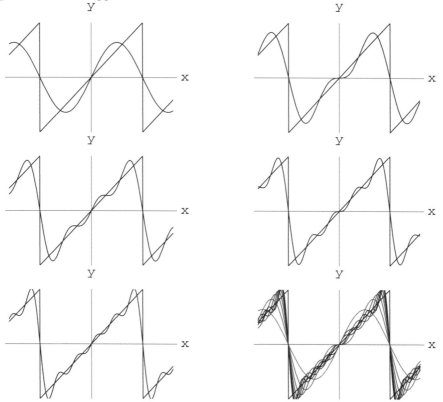

Figure 25.2: Fourier series approximations to $y = x$

Approximation is not just a question of estimating errors in one series; it also includes operations on series. In Chapter 26, we will see that we have a lot of freedom to treat power series like very long polynomials, which means that we can study transcendental functions with polynomial computations plus approximations. Fourier series do not allow us as much computational freedom as power series, but they are important for other reasons.

25.1 Geometric Series: Convergence

The geometric series

$$1 + r + r^2 + \cdots + r^n + \cdots$$

is the most important series in mathematics. It is a fundamental building block for the theory of series, and it arises in many applications.

We have the convergence formula

$$\frac{1}{1-r} = 1 + r + r^2 + \cdots + r^n + \cdots$$

$$= \sum_{k=0}^{\infty} r^k, \qquad \text{for } -1 < r < 1$$

Moreover, if $|x| \le r < 1$, then the error between $\frac{1}{1-x}$ and $1 + x + \cdots + x^n$ is less than or equal to the error between $\frac{1}{1-r}$ and $1 + r + \cdots + r^n$; that is, the worst error occurs at the largest x. You can see this on Figure 25.3 and can see it even better in the computer animation of the program **ClassicalSeries**.

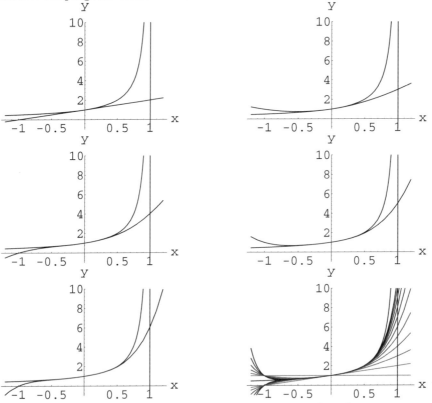

Figure 25.3: More and more terms of $1 + x + x^2 + x^3 + \cdots$

We want you to understand this convergence and its limitations. In other words, we want you to be able to estimate errors in finite approximations and to know why the formula does not work for all values of x. Once you understand this simple example thoroughly, you will be able to extend your knowledge to many rather complicated series.

First, the good news:

Theorem 25.1. *The Geometric Series Formula*

$$1 + r + r^2 + \cdots + r^n + \cdots = \sum_{k=0}^{\infty} r^k = \frac{1}{1-r} \qquad \textit{for } |r| < 1$$

This equation means that $\lim_{n \to \infty}[1 + r + r^2 + \cdots + r^n] = \frac{1}{1-r}$, *for* $|r| < 1$.
The error in summing n terms of the series is given by

$$Error_n = \frac{r^{n+1}}{1-r} = \sum_{k=n+1}^{\infty} r^k$$

If $|r| \geq 1$, *the limit* $\lim_{n \to \infty}[1 + r + r^2 + \cdots + r^n]$ *does not exist and we say*

$$\sum_{k=0}^{\infty} r^k \quad \textit{diverges for } |r| \geq 1$$

How can we see the formula? One case that is intuitively clear is given in Exercise 25.1.3. In general, we prove it as follows.

PROOF STEP 1:

We need to use the result of an exercise that is reviewed in the **GeomSeries** program.

Theorem 25.2. $\lim_{n \to \infty} r^n$

$$\lim_{n \to \infty} r^n = 0, \qquad \textit{for } |r| < 1$$
$$= 1, \qquad \textit{for } r = 1$$
$$= \infty, \qquad \textit{for } r > 1$$
$$= \textit{diverges by oscillation}, \qquad \textit{for } r \leq -1$$

---

y

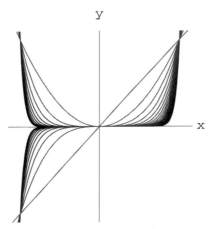

x

Figure 25.4: Limit of r^n

PROOF STEP 2: Multiply a finite partial sum by $(1 - r)$,

$$(1 + r + r^2 + \cdots + r^{n-1} + r^n)(1 - r) = 1 + r + \cdots + r^n$$
$$- r - \cdots - r^n - r^{n+1}$$
$$= 1 - r^{n+1}$$

so, if $r \neq 1$,

$$1 + r + r^2 + \cdots + r^{n-1} + r^n = \frac{1 - r^{n+1}}{1 - r}$$
$$= \frac{1}{1 - r} - \frac{r^{n+1}}{1 - r}$$

The limit is calculated using the exercise as follows:

$$\lim_{n \to \infty} [1 + r + \cdots + r^n] = \lim_{n \to \infty} [\frac{1}{1 - r} - \frac{r^{n+1}}{1 - r}]$$
$$= \frac{1}{1 - r} - \frac{1}{1 - r} \lim_{n \to \infty} r^{n+1}$$
$$= \frac{1}{1 - r}$$

when $|r| < 1$.

When $|r| < 1$, we have

$$\sum_{k=0}^{n} r^k =$$

$$1 + r + r^2 + \cdots + r^n = \frac{1}{1 - r} - \frac{r^{n+1}}{1 - r}$$

and

$$\sum_{k=0}^{\infty} r^k =$$

$$1 + r + r^2 + \cdots + r^n + \cdots = \frac{1}{1-r}$$

The error in summing only n terms of the infinite series is

$$[1 + r + r^2 + \cdots + r^n + \cdots] - [1 + r + r^2 + \cdots + r^n] = [r^{n+1} + r^{n+2} + \cdots]$$

$$\sum_{k=0}^{\infty} r^k - \sum_{k=0}^{n} r^k = \sum_{k=n+1}^{\infty} r^k$$

$$= [\frac{1}{1-r}] - [\frac{1}{1-r} - \frac{r^{n+1}}{1-r}]$$

$$= \frac{r^{n+1}}{1-r}$$

Rephrasing this observation, the error for summing n terms instead of infinitely many is

$$\text{Error}_n[r] = \frac{r^{n+1}}{1-r}$$

However, for the material ahead on more general series, it is important to realize that the error is also an infinite series,

$$\text{Error}_n[r] = r^{n+1} + r^{n+2} + \cdots$$

The error formula in this case tells us how much more there is to add from n to infinity. In more general series, we will use this basic formula to estimate the error caused by stopping at n.

The biggest pitfall in estimating the accuracy of a partial sum of a series

$$\sum_{k=1}^{\infty} u_k = u_1 + u_2 + u_3 + \cdots$$

is to sum the terms until u_n is small and then assume that the error

$$\text{Error}_n = \sum_{k=1}^{\infty} u_k - \sum_{k=1}^{n} u_k$$

is small. This estimate is almost always false. It can be infinite! The error itself is an infinite series,

$$\text{Error}_n = \sum_{k=n+1}^{\infty} u_k = u_{n+1} + u_{n+2} + \cdots$$

In other words, the error is the potential amount that remains to be added from n to infinity. The formula in the geometric series is very simple, while, but it tells exactly how much more is left to be added from n to infinity.

The proof of the formula for the geometric series gives the error as a closed formula. The error term of the proof is analogous to the question, "How far is the remaining distance to the cliff?" in the cliffhanger Exercise 25.1.3.

Example 25.1. *Distance to the Edge*

In the "cliff" Exercise 25.1.3 you step one half way to the edge of the cliff each step. The total distance is the series,

$$\frac{1}{2} + \frac{1}{4} + \frac{1}{8} + \cdots = \frac{1}{2}\left(1 + \frac{1}{2} + \frac{1}{4} + \frac{1}{8} + \cdots\right)$$

$$= \frac{1}{2}\frac{1}{1-r}, \qquad \text{with } r = \frac{1}{2}$$

$$= \frac{1}{2}\frac{1}{1-1/2} = \frac{1}{2}\frac{1}{\frac{1}{2}} = 1$$

The distance remaining after n steps is the error formula computation

$$\left(\frac{1}{2} + \frac{1}{4} + \frac{1}{8} + \cdots\right) - \left(\frac{1}{2} + \frac{1}{4} + \frac{1}{8} + \cdots + \frac{1}{2^n}\right)$$

$$= \frac{1}{2}\left[\left(1 + \frac{1}{2} + \frac{1}{4} + \frac{1}{8} + \cdots\right) - \left(1 + \frac{1}{2} + \frac{1}{4} + \frac{1}{8} + \cdots + \frac{1}{2^{n-1}}\right)\right]$$

$$= \frac{1}{2}\left(\frac{1}{2^n} + \frac{1}{2^{n+1}} + \cdots\right)$$

$$= \frac{1}{2} \cdot \frac{r^{n-1+1}}{1-r}, \qquad \text{with } r = \frac{1}{2}$$

$$= \frac{1}{2} \cdot \frac{1/2^n}{1-1/2} = \frac{1}{2^n}$$

You can show this by simple direct reasoning in Exercise 25.1.3.

Similarly, your bold cousin travels the distance,

$$\frac{2}{3} + \frac{2}{3}\left(\frac{1}{3}\right) + \frac{2}{3}\left(\frac{1}{3}\frac{1}{3}\right) + \cdots = \frac{2}{3}\left(1 + \frac{1}{3} + \frac{1}{9} + \frac{1}{27} + \cdots\right)$$

$$= \frac{2}{3} \cdot \frac{1}{1-r}, \qquad \text{with } r = \frac{1}{3}$$

$$= \frac{2}{3} \cdot \frac{1}{1-1/3} = \frac{2}{3}\frac{1}{\frac{2}{3}} = 1$$

Her distance remaining after n steps is the error formula computation

$$\left(\frac{2}{3}+\frac{2}{3}\left(\frac{1}{3}\right)+\frac{2}{3}\left(\frac{1}{9}\right)+\cdots\right)-\left(\frac{2}{3}+\frac{2}{3}\left(\frac{1}{3}\right)+\frac{2}{3}\left(\frac{1}{9}\right)+\cdots+\frac{2}{3}\frac{1}{3^{n-1}}\right)$$

$$=\frac{2}{3}\left[\left(1+\frac{1}{3}+\frac{1}{9}+\frac{1}{27}+\cdots\right)-\left(1+\frac{1}{3}+\frac{1}{9}+\frac{1}{27}+\cdots+\frac{1}{3^{n-1}}\right)\right]$$

$$=\frac{2}{3}\left(\frac{1}{3^n}+\frac{1}{3^{n+1}}+\cdots\right)$$

$$=\frac{2}{3}\cdot\frac{r^{n-1+1}}{1-r},\qquad\text{with }r=\frac{1}{3}$$

$$=\frac{2}{3}\cdot\frac{1/3^n}{1-1/3}=\frac{1}{3^n}$$

Again, this computation should only confirm your direct calculation about your cousin from California in Exercise 25.1.4.

25.1.1 The Bad and Strange News: Divergence

The geometric series diverges for $x\geq 1$; clearly

$$1+x+x^2+x^3+\cdots+x^n\geq 1+1+1+\cdots+1=n+1\to\infty$$

If $x<-1$, successive partial sums change by increasing amounts that grow in magnitude - for example,

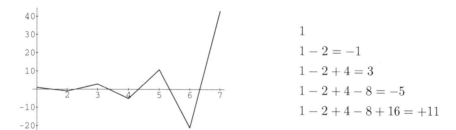

$$1$$
$$1-2=-1$$
$$1-2+4=3$$
$$1-2+4-8=-5$$
$$1-2+4-8+16=+11$$

Figure 25.5: Zigzag geometric growth

There is one strange case of divergence in our formula

$$1+x+x^2+x^3+\cdots=\frac{1}{1-x}$$

Suppose that we take the limit as x tends to -1 on both sides of our expression. There is no problem with

$$\lim_{x\to-1}\frac{1}{1-x}=\frac{1}{2}$$

See Problem 25.2 to try to make sense out of the left-hand side,

$$\lim_{x\to-1}1+x+x^2+x^3+\cdots=1-1+1-1+\cdots$$

Exercise set 25.1

Practice using the series formula:

1. Start at Zero - Somehow
Use the formula $1 + r + r^2 + \cdots + r^n + \cdots = \frac{1}{1-r}$, for $|r| < 1$ to calculate the following:

(a) $\frac{1}{2} + \frac{1}{4} + \frac{1}{8} + \cdots$ (HINT: Use $-1 + [1 + \frac{1}{2} + \frac{1}{4} + \frac{1}{8} + \cdots]$.)

(b) $\frac{1}{2} + \frac{1}{4} + \frac{1}{8} + \cdots$ (HINT: Use $\frac{1}{2}[1 + \frac{1}{2} + \frac{1}{4} + \frac{1}{8} + \cdots]$.)

(c) $\frac{1}{3} + \frac{1}{9} + \frac{1}{27} + \cdots$

(d) $1 - \frac{3}{4} + \frac{9}{16} - \frac{27}{64} + \cdots$

(e) $3 + \frac{3}{2} + \frac{3}{4} + \frac{3}{8} + \cdots = 3[1 + \frac{1}{2} + \frac{1}{4} + \frac{1}{8} + \cdots]$

(f) $a\,x + a\,x^2 + a\,x^3 + \cdots$ for a constant a and variable x. State restrictions on a or x if needed.

Practice using the error formula:

2. Geometric Errors
In each of the specific numerical series of the Exercise 25.1.1, compute the value of n so that the sum of n terms of that series is within 6 significant digits (machine precision) of the final limit. (You need a calculator or the computer to take logarithms.)
Compute r^n and $\frac{r^{n+1}}{1-r}$ in each of these cases. When is the error equal to the last term added?

3. Approaching the Cliff
You visit Yosemite National Park and are thrilled to see the magnificent view from the top of the cliff at the south overlook. However, after spending time in Iowa, you are a little afraid of heights. (Midwesterners are known for being sensible, if a little boring...) Beginning from a distance of 1 meter from the edge of a 900-meter vertical cliff, you step ahead half way toward the edge.
You have traveled $\frac{1}{2}$ of a meter and have $\frac{1}{2}$ of a meter to go.
Cautiously, you step ahead again half of the way from where you are to the edge, a distance of $\frac{1}{4}$ meter.

(a) How far have you traveled total?

(b) How far remains to the edge?

Again, you step half way, this time $\frac{1}{8}$ of a meter.

(a) How far have you traveled?

(b) How far remains to the edge?

Write a formula for the total distance traveled after n steps. Write a formula for the distance that remains after n steps.
How much is the limit

$$\frac{1}{2} + \frac{1}{4} + \frac{1}{8} + \frac{1}{16} + \cdots$$

How much is the error between a finite approximation to this limit,

$$\frac{1}{2} + \frac{1}{4} + \frac{1}{8} + \frac{1}{16} + \cdots + \frac{1}{2^n}$$

and the total limit? Write this error two ways - first as an infinite sum and, second, in terms of the distance remaining.

4. Your cousin from California is visiting the park with you. She is much bolder and, starting from the same place, steps ahead $\frac{2}{3}$ of the way to the edge. Each successive *step, she steps $\frac{2}{3}$ of the way to the cliff. Show that her distance traveled after n steps is*

$$\frac{2}{3} + \frac{2}{3}\left(\frac{1}{3}\right) + \frac{2}{3}\left(\frac{1}{3} \cdot \frac{1}{3}\right) + \cdots + \frac{2}{3}\left(\frac{1}{3}\right)^{n-1}$$

How much is the infinite sum

$$\frac{2}{3} + \frac{2}{3}\left(\frac{1}{3}\right) + \frac{2}{3}\left(\frac{1}{3} \cdot \frac{1}{3}\right) + \cdots + \frac{2}{3}\left(\frac{1}{3}\right)^{n-1} + \cdots$$

Does she step off the edge?
How much of the distance to the edge remains after your cousin takes n steps?

5. Telescoping Geometric Series
  Here is a variation on the approach to the geometric series formula:
   (a) *Look up the Telescoping Sum Theorem 12.2 and show that $\sum_{k=0}^{n}[F[k]-F[k+1]] = F[0] - F[n+1]$ for any function $F[k]$ defined at the index values, in particular,*

$$\sum_{k=0}^{n}[r^k - r^{k+1}] = 1 - r^{n+1}$$

   (b) *Use algebra to show that $\sum_{k=0}^{n}[r^k - r^{k+1}] = (1-r)\sum_{k=0}^{n}[r^k]$.*
   (c) *Use both parts above to show $\sum_{k=0}^{n}[r^k] = \frac{1-r^{n+1}}{1-r}$ provided $r \neq 1$.*

Problem 25.1.

*Run the program **GeomSeries** and answer the questions. This shows you how to prove the convergence theorem on geometric series.*

Problem 25.2.

Curious things happen as you approach the edge of convergence.
   (a) *What is the limit of the expression*

$$\lim_{x\downarrow-1}[1 + x + x^2 + x^3 + \cdots] = \lim_{x\downarrow-1}[\frac{1}{1-x}]$$

   (b) *Can we form the infinite sum*

$$\lim_{x\downarrow-1} 1 + \lim_{x\downarrow-1} x + \lim_{x\downarrow-1} x^2 + \lim_{x\downarrow-1} x^3 + \cdots$$

What are the partial sums of $1 - 1 + 1 - 1 + \cdots$?

$$1$$
$$1-1$$
$$1-1+1$$
$$1-1+1-1$$

$$\vdots$$

(c) *What sense can we make of the formula* $1-1+1-1+\cdots=\frac{1}{2}$, *if any?*

25.1.2 The bouncing ball

A nice new superball drops vertically from a height h onto a hard surface and rebounds a large fraction of the way back to its starting position. Denote the second height $h\,r$ for a value of r near 1, for example, $r=0.8$. An old mushy tennis ball also rebounds a fraction of its height, but with a value like $r=0.3$. If the bouncing is perfectly elastic and vertical (with no other energy loss or transfer to spin), the top of the second bounce is a fraction r times the height at the top of the first.

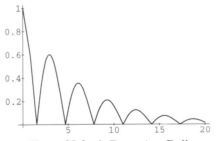

Figure 25.6: A Bouncing Ball

Problem 25.3. BOUNCING ⎯⎯⎯⎯⎯⎯⎯⎯⎯⎯⎯⎯⎯⎯⎯⎯⎯⎯⎯⎯⎯▼
Calculate the total vertical distance traveled by a perfectly elastic ball with rebound coefficient r *when dropped from an initial height* h. *Test some numerical cases such as a superball with* $r=0.8$ *dropped from a second floor window,* $h=7m$.

25.2 Convergence by Comparison

Series of functions $u_0[x]+u_1[x]+\cdots$ *can sometimes be compared to convergent series of numbers* $M_0+M_1+\cdots$ *giving us simple estimates of complicated errors.*

The easiest comparison between a series of functions, $\sum_{k=0}^{\infty} u_k[x]$ and a numerical series, $\sum_{k=0}^{\infty} M_k$ is when

$$|u_n[x]| \leq M_n, \qquad \text{for all } n \text{ and all } a \leq x \leq b.$$

This makes the maximum error in the function series no worse than the (single) error in the numerical series,

$$\text{Error}_n[x] =$$
$$|u_{n+1}[x] + u_{n+2}[x] + \cdots| \leq |u_{n+1}[x]| + |u_{n+2}| + \cdots$$
$$\leq M_{n+1} + M_{n+2} + \cdots$$
$$\leq \text{M-series Error}_n, \qquad \text{for all } a \leq x \leq b$$

This error estimate is called the Weierstrass majorization estimate. It is very simple, provided we can find the majorizing series and its error. The simplicity is that we then get a single maximum error for all x (perhaps in an interval). This produces the "strong" convergence defined as follows.

> **Definition 25.3.**   *Uniform Absolute Convergence*
> *Consider a series of functions*
>
> $$u_0[x] + u_1[x] + \cdots + u_n[x] + \cdots$$
>
> *Suppose error between the sum of n terms of* $|u_0[x]| + |u_1[x]| + \cdots + |u_n[x]| + \cdots$ *and the eventual limit at x is* $AbsError_n[x]$. *Then, the series is said to be uniformly absolutely convergent on* $[a, b]$ *if the maximum of* $AbsError_n[x]$ *for* $a \leq x \leq b$ *tends to zero as n tends to infinity.*

The general case of this criterion is difficult to state, but it means that the "worst error" tends to zero as the number of terms added increases. The Weierstrass majorization estimate always produces an absolutely uniformly convergent series.

25.2.1   Uniform Geometric Estimation

We want to view the geometric series as a sum of the functions, $1, x, x^2, \ldots$ and make a single estimate of error over a whole interval of x values. This is illustrated graphically in the the computer program **ClassicalSeries**. We really have treated the geometric series as a function series with the variable r, but now we just want the maximum error for an interval. We are being a little pedantic because we are trying to illustrate estimation in a very simple case.

Suppose $r < 1$ is a fixed value (parameter). We want to consider the maximum error between a partial sum of $1 + x + x^2 + \cdots$ and its limit $1/(1 - x)$ for $-r \leq x \leq r$. We want an estimate for

$$\text{Max Error}_n[r] =$$
$$\text{Max}|[1 + x + x^2 + \cdots + x^n + \cdots] - [1 + x + x^2 + \cdots + x^n]| =$$
$$= \text{Max}|[x^{n+1} + x^{n+2} + \cdots]|$$
$$\text{Max}|\sum_{k=0}^{\infty} x^k - \sum_{k=0}^{n} x^k| = \text{Max}|\sum_{k=n+1}^{\infty} x^k|$$

where the maximum is for $|x| \leq r$ and $0 < r < 1$ is fixed. Each term of the error series satisfies $|x^k| \leq r^k$, and we know that finite sums satisfy

$$|x^{n+1} + x^{n+2} + \cdots + x^{n+N}| \leq |x^{n+1}| + |x^{n+2}| + \cdots + |x^{n+N}|$$
$$\leq r^{n+1} + r^{n+2} + \cdots + r^{n+N}$$

and so, when $|x| \leq r$,

$$\text{Error}_n[x] = \lim_{N \to \infty} |x^{n+1} + x^{n+2} + \cdots + x^{n+N}|$$
$$\leq \lim_{N \to \infty} r^{n+1} + r^{n+2} + \cdots + r^{n+N} = \frac{r^{n+1}}{1-r}$$

This estimate does not depend on the particular x, so

$$\text{Max Error}_n \leq \frac{r^{n+1}}{1-r}$$

The estimates show that the geometric series $1 + x + x^2 + \cdots + x^n + \cdots$ is uniformly absolutely convergent on any closed interval strictly inside $-1 < x < 1$.

Theorem 25.4. *Convergence of the Geometric Series*
A geometric series $a + a\,x + a\,x^2 + \cdots + a\,x^n + \cdots$ converges uniformly absolutely to the limit $\frac{a}{1-x}$ for $|x| \leq r$ provided the constant $r < 1$. The maximum error between the sum of n terms and the limit is $a\,\frac{r^{n+1}}{1-r}$.

25.2.2 Weierstrass's Wild Wiggles

In Chapter 4, when we first studied local approximation by a linear function, we mentioned Weierstrass's famous example of a function that is continuous, but has a kink (or corner) at every point on its graph. Specifically, Weierstrass's function is nowhere differentiable. This means that if we look at the graph with microscopes of arbitrary power, we will always see wiggles - it never straightens out to look like its tangent; in other words, it never has a tangent...

Weierstrass's function is given by the series

$$W[x] = \text{Cos}[x] + \frac{\text{Cos}[3x]}{2} + \frac{\text{Cos}[9\,x]}{4} + \cdots + \frac{\text{Cos}[3^n\,x]}{2^n} + \cdots$$

This series is easy to majorize because

$$\left| \frac{\text{Cos}[3^n\,x]}{2^n} \right| \leq \frac{1}{2^n}, \qquad \text{for all } x$$

Exercise set 25.2

1. Weierstrass's Wild Wiggles

*Write a computer program to plot a piece of Weierstrass's function $W[x]$ over an arbitrary interval $a \leq x \leq b$ that is specified at the beginning of your program. Use the same **PlotRange** in the y-coordinate - for example, by including the graphics option $PlotRange -> \{W(c) - d, W(c) + d\}$ in the Plot command where $c = (a + b)/2$ and $d = (b - a)/2$.*

You can define

$$W[x] = Sum[Cos[3^k\ x]/2^k, \{k, 0, n\}]$$

but you must find an estimate of the error caused by using n instead of ∞. You can use a geometric series estimate to find the n so that the error is less than the screen resolution of about 1 in 250 (for example, suppose there are a million total pixels on a monitor, 1,000 by 1,000, but we use about one fourth of the screen for a plot). In the case that you plot in a y-range of width $2\,d$ you want the error to be less than $d/250$. Estimate with the geometric majorant and use logs to solve for n.

What is a formula for n such that the sum of n terms of Weierstrass's series has an error less than $d/250$?

Experiment with your program using small intervals (in other words, large magnification.) For example, plot the graph for $.499 < x < .501$.

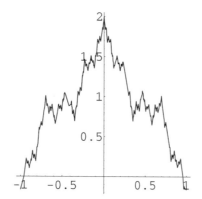

Figure 25.7: Weierstrass's wild wiggles

25.3 Compound Interest - CD

You find a snazzy red roadster, a 1994 FireBall XL, for sale at a local used car dealer. The window price is $5,000. The dealer offers financing with no down payment. The salesperson explains that the car is only $200 per month with payments for 3 years. It sounds like a great start to the summer, because you expect to have a job that will more than cover the payment. But wait, that's $200 times 36 months, $200 \times 36 = 7,200$. Whoa!

Remainder of this section only on CD

CHAPTER 26

Power Series

Power series are an important analytical tool. A convergent power series expansion for a function allows us to apply calculus and algebra to the function as if it were a "long polynomial."

Definition 26.1. *Power series are series are expressions of the form,*

$$\sum_{k=0}^{\infty}[a_k\,x^k] = a_0 + a_1\,x + a_2\,x^2 + \cdots + a_n\,x^n + \cdots$$

where the sequence of numbers a_0, a_1, a_2, $\cdots$ may depend on the index n but not on x.

For example, the geometric series is a power series,

$$\frac{1}{1-x} = 1 + x + x^2 + x^3 + \cdots + x^n + \cdots$$

where $a_n = 1$ for all n. The exponential series,

$$e^x = 1 + x + \frac{1}{2}\,x^2 + \frac{1}{6}\,x^3 + \cdots + \frac{1}{n!}\,x^n + \cdots$$

is also a power series with $a_n = \frac{1}{n!}$. (*Note:* The exclamation sign ! denotes the factorial function, $n! = n \cdot (n-1) \cdot (n-2) \cdots 3 \cdot 2$. This function is also built into the computer as !)

The series

$$|t| = 1 + \frac{1}{2}[t^2 - 1] + \frac{\frac{1}{2}(\frac{1}{2} - 1)}{2}[t^2 - 1]^2 + \frac{\frac{1}{2}(\frac{1}{2} - 1)(\frac{1}{2} - 2)}{3 \cdot 2}[t^2 - 1]^3 + \cdots$$

$$\cdots + \left(\tfrac{\frac{1}{2}}{n}\right)[t^2 - 1]^n + \cdots$$

is a convergent series for $|t| < 1$ and it is "built up from power functions," but it is not considered a power series in t. Notice that each new term of this series would produce lower powers of t after simplification, $[t^2 - 1]^{n+1} = (-1)^{n+1} + (-1)^n(n+1)t^2 + \cdots + t^{2n+2}$. Power series have the property that the partial sum $\sum_{k=0}^{n}[a_k\,x^k]$ contains all the powers lower than the nth that ever occur in the series.

26.1 Computation of Power Series

*This section has examples of calculations on power series using simple alge-
bra and calculus. The first thing you should observe is that these calculations
give us new power series expansions from known ones.*

The geometric series

$$\frac{1}{1-u} = 1 + u + u^2 + \cdots + u^n + \cdots$$

converges for $|u| < 1$. If we let $u = -x$, the resulting series will still converge for $|x| < 1$,

$$\frac{1}{1-(-x)} = 1 + (-x) + (-x)^2 + \cdots + (-x)^n + \cdots$$

$$\frac{1}{1+x} = 1 - x + x^2 - x^3 + \cdots + (-1)^n x^n + \cdots$$

We can also differentiate and integrate convergent power series term by term. At first,
it probably seems strange to suggest that this needs proof, but it does. Before we worry
about that, let us see why we would want to do these operations. Integrate both sides of
the expression for $1/(1+x)$ from 0 to u,

$$\frac{1}{1+x} = 1 - x + x^2 + \cdots + (-1)^n x^n + \cdots$$

$$\int_0^u \frac{1}{1+x} dx = \int_0^u 1 \, dx - \int_0^u x \, dx + \int_0^u x^2 \, dx + \cdots + (-1)^n \int_0^u x^n \, dx + \cdots$$

$$\mathrm{Log}[1+u] = u - \frac{1}{2}u^2 + \cdots + (-1)^n \frac{1}{n+1}u^{n+1} + \cdots$$

or written in terms of x,

$$\mathrm{Log}[1+x] = x - \frac{1}{2}x^2 + \cdots + (-1)^n \frac{1}{n+1}x^{n+1} + \cdots$$

We expect this series to converge for $|x| < 1$ (and know that there is trouble with the
logarithm when $x = -1$.)

Why do we want a theorem to justify term by term integration of power series? Such a
theorem would tell us the above expression is a valid expansion for the logarithm. Notice
that this gives the curious special case

$$\mathrm{Log}[2] = 1 - \frac{1}{2} + \frac{1}{3} - \frac{1}{4} + \cdots$$

when we take $x = 1$. We could use this to approximate the natural log of 2 with pencil-and-
paper calculations.

Example 26.1. *The Exponential Series*

$$e^x = 1 + x + \frac{1}{2}x^2 + \frac{1}{3 \cdot 2}x^3 + \frac{1}{4 \cdot 3 \cdot 2}x^4 + \cdots + \frac{1}{n!}x^n + \cdots$$

Term by term differentiation is also allowed for convergent power series. We can use this to find a power series expansion for e^x. Recall our official definition that $y = e^x$ is the unique solution to

$$y[0] = 1$$
$$\frac{dy}{dx} = y$$

Suppose we have a series

$$y = a_0 + a_1\,x + a_2\,x^2 + \cdots$$

We want to find coefficients a_0, a_1, $\cdots$, so that y is also a solution to this initial value problem. If the series converges and differentiation is valid, this y must equal e^x, by the uniqueness of the solution to the initial value problem.

Let $x = 0$ to obtain

$$y[0] = a_0 = 1$$

Differentiate y term by term to obtain

$$\frac{dy}{dx} = a_1 + 2a_2\,x + 3a_3\,x^2 + 4a_4\,x^3 + \cdots$$

In order for the series to satisfy the differential equation, we must have $y = \frac{dy}{dx}$ or

$$1 + a_1\,x + a_2\,x^2 + a_3\,x^3 + \cdots = a_1 + 2a_2\,x + 3a_3\,x^2 + 4a_4\,x^3 + \cdots$$

We can recursively solve for the coefficients. If $x = 0$, we get $a_1 = 1$,

$$1 + x + a_2\,x^2 + a_3\,x^3 + \cdots = 1 + 2a_2\,x + 3a_3\,x^2 + 4a_4\,x^3 + \cdots$$

Subtract 1 from both sides and divide by x, and we see

$$1 + a_2\,x + a_3\,x^2 + \cdots = 2a_2 + 3a_3\,x + 4a_4\,x^2 + \cdots$$

Let $x = 0$ to obtain $2a_2 = 1$ or $a_2 = \frac{1}{2}$.

The preceding algebra simply says that we can equate coefficients of like powers, obtaining

$$a_0 = 1 \qquad\qquad \text{or} \qquad a_0 = 1$$
$$a_n = (n+1)a_{n+1} \qquad a_{n+1} = \frac{1}{n+1}a_n$$

so

$$a_0 = 1$$
$$a_1 = a_0 = 1$$
$$a_2 = \frac{1}{2}a_1 = \frac{1}{2}$$
$$a_3 = \frac{1}{3}a_2 = \frac{1}{3} \cdot \frac{1}{2}$$
$$a_4 = \frac{1}{4}a_3 = \frac{1}{4} \cdot \frac{1}{3} \cdot \frac{1}{2}$$
$$\vdots$$
$$a_n = \frac{1}{n!}$$

So term by term differentiation and proof of convergence would justify the important formula

$$e^x = 1 + x + \frac{x^2}{2} + \frac{x^3}{3 \cdot 2} + \cdots + \frac{x^n}{n!} + \cdots$$

This series does converge and gives an expansion for the natural exponential at all values of x.

Exercise set 26.1

1. *Use a substitution to find a power series expansion for*

$$\frac{1}{1+x^2} = 1 + a_1 x + a_2 x^2 + \cdots + a_n x^n + \cdots$$

2. ArcTan[x] Series
   The integral of $\frac{1}{1+x^2}$ from 0 to u is ArcTan[u]. Use your series from the exercise above to show that

$$\text{ArcTan}[x] = \sum_{k=0}^{\infty} [(-1)^k \frac{x^{2k+1}}{2k+1}]$$

   and express this in the more familiar form

$$\text{ArcTan}[x] = x - ? + ? \cdots$$

   The tangent of $\pi/4$ is 1, so the arctangent of 1 is $\pi/4$. Substitute $x = 1$ in your series to obtain the special case

$$\frac{\pi}{4} = 1 - \frac{1}{3} + \frac{1}{5} - \frac{1}{7} + \cdots$$

   *Compare your work with the **ClassicalSeries** program.*

3. Sin[x] and Cos[x] Power Series

(a) *Substitute the complex variable $x = i\theta$ into the series above for e^x. Note that $(i\theta)^2 = -\theta$, $(i\theta)^3 = -i\theta^3$, $(i\theta)^4 = \theta^4$, $\cdots$. Rewrite the series in the form*

$$e^{i\theta} = \left[1 - \frac{\theta^2}{2} + \cdots + (-1)^n \frac{\theta^{2n}}{(2n)!} + \cdots\right] + i\left[\theta - \frac{\theta^3}{6} + \cdots + (-1)^n \frac{\theta^{2n+1}}{(2n+1)!} + \cdots\right]$$

and write out the first five terms of both the real and complex parts.

(b) *Use Euler's formula $e^{i\theta} = \text{Cos}[\theta] + i\,\text{Sin}[\theta]$ and the previous part of the exercise to find series for sine and cosine.*

(c) *Check your series for cosine by showing that the series satisfies*

$$y[0] = 1$$
$$y'[0] = 0$$
$$\frac{d^2y}{dx^2} = -y$$

(d) *Check your series for sine by showing that the series satisfies*

$$y[0] = 0$$
$$y'[0] = 1$$
$$\frac{d^2y}{dx^2} = -y$$

*Plot graphs of the partial sums with the **ClassicalSeries** program.*

4. **ClassicalSeries**

    *Run the the computer **ClassicalSeries** program and complete the electronic exercises of making animations of the convergence of the series for sine, log, and arctangent. The animations of cosine and the geometric series are already complete in that program, but you should check the intervals of convergence and compare the animations to the theory.*

    The purpose of the animations is to show you which parts of the graphs converge to the limit functions. Convergence that is "uniform" in x for an interval simply means that the graphs converge over that interval.

5. *Substitute the expression $x = t^2 - 1$ into the series for $\sqrt{1+x}$ (given in Theorem 26.2), obtaining a series for $\sqrt{1 + [t^2 - 1]} = \sqrt{t^2} = |t| =$???. Is this a power series? Compare your work with the **AbsSeries** program.*

6. *Find a power series for the function e^{-x^2}, and check your work with the **FormalT-Series** program.*

Between the text discussion and the exercises, we now have all the basic power series:

$$e^x = 1 + x + \frac{x^2}{2} + \frac{x^3}{3 \cdot 2} + \cdots + \frac{x^n}{n!} + \cdots$$

$$= \sum_{k=0}^{\infty} [\frac{x^k}{k!}], \qquad \text{for all } x$$

$$\text{Cos}[x] = 1 - \frac{x^2}{2} + \frac{x^4}{4 \cdot 3 \cdot 2} + \cdots + (-1)^n \frac{x^{2n}}{(2n)!} + \cdots$$

$$= \sum_{k=0}^{\infty} [(-1)^k \frac{x^{2k}}{(2k)!}], \qquad \text{for all } x$$

$$\text{Sin}[x] = x - \frac{x^3}{3 \cdot 2} + \frac{x^5}{5 \cdot 4 \cdot 3 \cdot 2} + \cdots + (-1)^n \frac{x^{2n+1}}{(2n+1)!} + \cdots$$

$$= \sum_{k=0}^{\infty} [(-1)^k \frac{x^{2k+1}}{(2k+1)!}], \qquad \text{for all } x$$

$$\frac{1}{1-x} = 1 + x + x^2 + \cdots + x^n + \cdots$$

$$= \sum_{k=0}^{\infty} [x^k], \qquad \text{for } |x| < 1$$

$$\text{Log}[1 + x] = x - \frac{x^2}{2} + \frac{x^3}{3} + \cdots + (-1)^n \frac{x^{n+1}}{n+1} + \cdots$$

$$= \sum_{k=0}^{\infty} [(-1)^k \frac{x^{k+1}}{k+1}], \qquad \text{for } |x| < 1$$

$$\text{ArcTan}[x] = x - \frac{x^3}{3} + \frac{x^5}{5} + \cdots + (-1)^n \frac{x^{2n+1}}{(2n+1)} + \cdots$$

$$= \sum_{k=0}^{\infty} [(-1)^k \frac{x^{2k+1}}{2k+1}], \qquad \text{for } |x| < 1$$

The binomial theorem from algebra says,

$$(1 + x)^n = 1 + n\,x + \cdots + \binom{n}{k} x^k + \cdots + x^n$$

$$= \sum_{k=0}^{n} [\binom{n}{k} x^k]$$

where $\binom{n}{k} = \frac{n \cdot (n-1) \cdots (n-k+1)}{k \cdot (k-1) \cdot (k-2) \cdots 1}$ are the binomial coefficients. For example,

$$(1+x)^3 = 1 + \frac{3}{1}x + \frac{3 \cdot 2}{2 \cdot 1}x^2 + \frac{3 \cdot 2 \cdot 1}{3 \cdot 2 \cdot 1}x^3 = 1 + 3x + 3x^2 + x^3$$

We can generalize this to

Theorem 26.2. *The Binomial Series*

$$(1+x)^p = \sum_{k=0}^{\infty} [\tbinom{p}{k} x^k] \qquad \text{for } |x| < 1$$

where the generalized binomial coefficients are given by

$$\tbinom{p}{k} = \frac{p \cdot (p-1) \cdots (p-k+1)}{k \cdot (k-1) \cdot (k-2) \cdots 1}$$

and p is any real power, not necessarily an integer or even positive.

Here is the way to prove this:

Problem 26.1. COMPUTE THE BINOMIAL SERIES ───────────────────────▼

(a) *Show that* $y = (1+x)^p$ *satisfies the initial value problem*

$$y[0] = 1$$

$$(1+x)\frac{dy}{dx} = p\, y$$

(b) *Show that the generalized binomial coefficients defined above have the property*

$$(k+1)\tbinom{p}{k+1} = (p-k)\tbinom{p}{k}$$

for any real p and any positive integer k. Use this to show

$$(k+1)\tbinom{p}{k+1} + k\tbinom{p}{k} = p\tbinom{p}{k}$$

(c) *Differentiate the series*

$$y = \sum_{k=0}^{\infty} [\tbinom{p}{k} x^k]$$

term by term, and show that it satisfies

$$y' = \sum_{k=1}^{\infty} [k\tbinom{p}{k} x^{k-1}] = \sum_{k=0}^{\infty} [(n+1)\tbinom{p}{n+1} x^n, \{n, 0, \infty\}]$$

(d) *Multiply* $(1 + x)$ *times* y' *to obtain*

$$(1 + x)y' = \sum_{k=0}^{\infty}[\{(k + 1)\binom{p}{k+1}\} + k\binom{p}{k}\}x^k]$$

$$= p\sum_{k=0}^{\infty}[\binom{p}{k}x^k]$$

$$= p\,y$$

(e) *Use the Uniqueness Theorem for Initial Value Problems 21.6 to prove the binomial series formula, given that the series converges for* $|x| < 1$ *and that term by term differentiation is valid.*

(f) *Write the binomial series for the case* $p = -\frac{1}{2}$ *and check the first six terms of the series for* $1/\sqrt{1 - x} = 1 + \frac{1}{2}x + \frac{3}{8}x^2 + \frac{5}{16}x^3 + \frac{35}{128}x^4 + \frac{63}{256}x^5 + \cdots.$

The computer can help with binomial series. Use the program **BinomialSeries** to see the series for $(1 + x)^p$ with $p = 1/2$:

$$1+\frac{1}{2}x - \frac{1}{8}x^2 + \frac{1}{16}x^3 - \frac{5}{128}x^4 + \frac{7}{256}x^5 - \frac{21}{1024}x^6 + \frac{33}{2048}x^7 - \frac{429}{32768}x^8 + \frac{715}{65536}x^9 - \frac{2431}{262144}x^{10}$$

Try several values of the power p.

26.2   The Ratio Test for Power Series

Theorem 26.3.   *The Ratio Test*
   Consider a power series

$$a_0 + a_1\,x + a_2\,x^2 + \cdots + a_n\,x^n + \cdots$$

Suppose that $\lim_{n\to\infty}\frac{|a_n|}{|a_{n+1}|}$ *exists or tends to* $+\infty$. *Then, the series converges uniformly and absolutely for all* $|x| \le \rho$, *for any constant* $\rho < \lim_{n\to\infty}\frac{|a_n|}{|a_{n+1}|}$.

PROOF:

We will use the Weierstrass majorization estimate with a geometric series.

There is an integer N such that for all $|x| < \rho$ and all $n \ge N$, the ratio of successive terms, $|a_{n+1}\,x^{n+1}|$ over $|a_n\,x^n|$ satisfies

$$|x|\frac{|a_{n+1}|}{|a_n|} \le \rho\frac{|a_{n+1}|}{|a_n|} \le r < 1$$

for some constant $r < 1$. This is because $\rho\lim_{n\to\infty}\left|\frac{a_{n+1}}{a_n}\right| < 1$, so eventually $\rho\frac{|a_{n+1}|}{|a_n|} \le r < 1$, for a constant $r < 1$.

This makes the error in the absolute series,

$$|a_{n+1}\, x^{n+1}| + |a_{n+2}\, x^{n+2}| + \cdots \le$$

$$\le |a_{n+1}\, \rho^{n+1}| + |a_{n+2}\, \rho^{n+2}| + \cdots$$

$$\le |a_n\, \rho^n|[r^1 + r^2 + \cdots] = |a_n\, \rho^n|\frac{r}{1-r}$$

because

$$\frac{|a_{n+1}\, \rho^{n+1}|}{|a_n\, \rho^n|} \le r \qquad \text{or} \qquad |a_{n+1}\, \rho^{n+1}| \le |a_n\, \rho^n|\, r$$

$$\frac{|a_{n+2}\, \rho^{n+2}|}{|a_{n+1}\, \rho^{n+1}|} \le r \qquad \text{or} \qquad |a_{n+2}\, \rho^{n+2}| \le |a_{n+1}\, \rho^{n+1}|\, r \le |a_n\, \rho^n|\, r^2$$

$$\frac{|a_{n+3}\, \rho^{n+3}|}{|a_{n+2}\, \rho^{n+2}|} \le r \qquad \text{or} \qquad |a_{n+3}\, \rho^{n+3}| \le |a_{n+2}\, \rho^{n+2}|\, r \le |a_n\, \rho^n|\, r^3$$

and this error tends to zero as n tends to infinity - uniformly in x because x does not appear in the expression.

The proof that $\lim_{n\to\infty} a_n\, \rho^n = 0$ is as follows. For all $n \ge N$,

$$a_n\, \rho^n \le a_N\, \rho^N r^{n-N} = \frac{a_N \rho^N}{r^N} r^n \to 0 \qquad \text{as } n \to \infty \qquad \text{(for } N \text{ fixed.)}$$

for the constant $r < 1$ above.

That is all there is to the proof. Remember that it amounts to comparison to a geometric series by estimating ratios.

The moral of the story is clear:

Procedure 26.4.

   (a)   Compute $\lim_{n\to\infty} \frac{|a_n|}{|a_{n+1}|}$.

   (b)   Let $|x| \le \rho < \lim_{n\to\infty} \left|\frac{a_n}{a_{n+1}}\right|$, for any constant ρ.
This is your interval of safe convergence. Be careful of values of x that make

$$|x| \lim_{n\to\infty} \frac{|a_{n+1}|}{|a_n|} = 1$$

The theorem makes no guarantees in that case.

In other words, you cannot take $\rho = \lim_{n\to\infty} \frac{|a_n|}{|a_{n+1}|}$ and apply the theorem. We will see why below; sometimes the series still converges and sometimes it does not.

Example 26.2.   *Radius of Convergence*

The coefficients in the exponential series are $a_n = \frac{1}{n!}$, so

$$\lim_{n\to\infty} \frac{a_n}{a_{n+1}} = \lim_{n\to\infty} \frac{(n+1)!}{n!} = \lim_{n\to\infty} n+1 = \infty$$

This means that for any positive ρ, the series

$$1 + x + \frac{x^2}{2} + \frac{x^3}{3\cdot 2} + \cdots + \frac{x^n}{n!} + \cdots$$

converges uniformly absolutely for $|x| \leq \rho < \infty$.

The coefficients of the logarithm series are $\frac{(-1)^n}{n+1}$, so

$$\lim_{n \to \infty} \left| \frac{a_n}{a_{n+1}} \right| = \lim_{n \to \infty} \frac{n+1}{n} = \lim_{n \to \infty} 1 + \frac{1}{n} = 1$$

This means that the log series converges uniformly absolutely for $|x| \leq \rho < 1$. The theorem does not apply when $x = 1$, and we know there is trouble when $x = -1$ because we have $\text{Log}[1 - 1] = \text{Log}[0] = -1 - \frac{1}{2} - \frac{1}{3} - \cdots$.

Exercise set 26.2

1. *Write each of the following series in the old-fashioned style, such as*

$$e^x = 1 + x + \frac{x^2}{2} + \frac{x^3}{3 \cdot 2} + \cdots$$

For which values of the constant ρ will the following series converge absolutely and uniformly for $|x| \leq \rho$? What is the exact symbolic value of the series?

(a) $\sum_{k=0}^{\infty} [k \, x^k]$

(b) $\sum_{k=0}^{\infty} [(-1)^k \, k \, x^k]$

(c) $\sum_{k=0}^{\infty} [\frac{x^k}{3^k}]$

(d) $\sum_{k=0}^{\infty} [\frac{x^k}{3^{k+1}}]$

(e) $\sum_{k=0}^{\infty} [(-2)^k \frac{k+2}{k+1} x^k]$

(f) $\sum_{k=0}^{\infty} [\frac{x^k}{(k+3)!}]$

The important series for sine and cosine skip terms, so $a_n = 0$ for every other coefficient in the formula of the ratio test theorem. The limit of $\frac{a_n}{a_{n+1}}$ does not exist, because it is alternately zero and undefined. However, the same basic idea of comparison with a geometric series still works.

2. Ratios for Series that Skip Terms

Suppose a power series skips every other term as in the sine or the cosine series,

$$\text{Cos}[x] = 1 + 0 \cdot x - \frac{x^2}{2} + 0 \cdot x^3 + \frac{x^4}{4 \cdot 3 \cdot 2} + \cdots$$

Successive nonvanishing terms have ratios $x^2 \frac{a_{n+2}}{a_n}$. Suppose that the $\lim_{n \to \infty} \left| \frac{a_n}{a_{n+2}} \right|$ exists or tends to $+\infty$. Show that the series is absolutely uniformly convergent for $|x| \leq \sigma$ for constants σ related to the limit.

Apply your test to the series and, if possible, give the exact symbolic sum of the series.

(a) *The sine series.*

(b) *The cosine series.*

(c) *The arctangent series.*

(d) $\sum_{k=0}^{\infty} [\frac{x^{3k}}{k!}]$

(e) $\sum_{k=0}^{\infty} [\frac{(-1)^k}{2k+1} [\frac{x}{2}]^{2k}]$

26.3   Integration of Series

Integration of uniformly convergent series is always possible term by term. You might wonder why we do not always have

$$\lim_{n \to \infty} \int_a^b f_n[x]\ dx = \int_a^b \lim_{n \to \infty} f_n[x]\ dx$$

where $f_n[x] = u_0[x] + u_1[x] + \cdots + u_n[x]$. This section shows why.

The functions

$$f_n[x] = (n+1)\, 2x \left(1 - x^2\right)^n$$

all have area 1 under their graphs between zero and one, as shown in Figure 26.1.

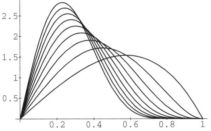

Figure 26.1: Equal areas under $f_n[x] = (n+1)\, 2x \left(1 - x^2\right)^n$

The limit of these functions is still zero, though not uniformly in x.

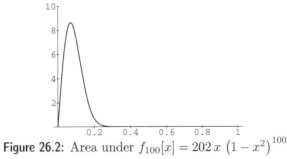

Figure 26.2: Area under $f_{100}[x] = 202\, x \left(1 - x^2\right)^{100}$

In this example, we have $\int_0^1 f_n[x]\ dx = 1$ for all n, so

$$\lim_{n \to \infty} \int_0^1 f_n[x]\ dx = 1$$

However, $\lim_{n \to \infty} f_n[x] = 0$ for every x, so

$$\int_0^1 \lim_{n \to \infty} f_n[x]\ dx = 0$$

This is because we do not have the whole graph close to its limit. When the whole graph is close, we can interchange limits.

Theorem 26.5. *Term-by-Term Integration of Series*
Suppose that the series of continuous functions

$$u_o[x] + u_1[x] + \cdots$$

converges uniformly to a "sum"

$$S[x] = \lim_{n \to \infty} u_0[x] + \cdots + u_n[x]$$

on the interval $[a, b]$; *that is, suppose*

$$M_n = Max[|S[x] - [u_0[x] + \cdots + u_n[x]]| : a \le x \le b] \to 0$$

Then the limit $S[x]$ *is continuous and*

$$\int_a^b \lim_{n \to \infty} u_0[x] + \cdots + u_n[x] \ dx = \lim_{n \to \infty} \int_a^b u_0[x] + \cdots + u_n[x] \ dx$$

PROOF

We omit the proof of continuity. It simply means that if $x_1 \approx x_2$, then $S[x_1] \approx S[x_2]$. The integral part is easy,

$$\left| \int_a^b S[x] \ dx - \left(\int_a^b u_0[x] \ dx + \cdots + \int_a^b u_n[x] \ dx \right) \right| =$$

$$\left| \int_a^b S[x] - (u_0[x] + \cdots + u_n[x]) \ dx \right| =$$

$$\le \int_a^b |S[x] - (u_0[x] + \cdots + u_n[x])| \ dx$$

$$\le \int_a^b M_n \ dx = (b - a) M_n \to 0$$

That is the proof.

Example 26.3. *Convergence of the Log Series*

Let us justify the computation of the log series. We know

$$\frac{1}{1 + x} = 1 - x + x^2 - x^3 + \cdots$$

converges uniformly absolutely on any interval $-r \le x \le r$, for $r < 1$. This means that

$$\int_0^u \frac{1}{1 + x} \ dx = \int_0^u 1 \ dx - \int_0^u x \ dx + \int_0^u x^2 \ dx - \int_0^u x^3 \ dx + \cdots$$

$$Log[u] = u - \frac{1}{2}u^2 + \frac{1}{3}u^3 - \frac{1}{4}u^4 + \cdots$$

for $|u| \le r < 1$.

Notice that this still does not justify the formula

$$Log[2] = 1 - \frac{1}{2} + \frac{1}{3} - \frac{1}{4} + \cdots$$

because we would need to integrate all the way to $u = 1$. We take this up later.

Exercise set 26.3

1. *Show that*

$$\int_0^1 f_n[x] = \int_0^1 (n+1) \, 2x \, \left[1 - x^2\right]^n \, dx = 1$$

2. (a) *Show that* $\lim_{n \to \infty} f_n[0] = \lim_{n \to \infty} f_n[1] = 0$.
   (b) *If* $0 < x < 1$, *show that* $r = [1 - x^2] < 1$, *so* $\lim_{n \to \infty} f_n[x] = 0$.
   (c) *Find* $Max[f_n[x] : 0 \le x \le 1] = M_n$ *and show that* $M_n \to \infty$. *The maximum error between* $f_n[x]$ *and its zero limit does not tend to zero.*

3. *The series for arctangent can be obtained by integrating*

$$\frac{1}{1 + x^2} = 1 - x^2 + x^4 - x^6 + \cdots$$

from zero to u and using the fact that $\mathrm{ArcTan}[0] = 0$, *as you showed in an exercise above. For which values of u is this computation valid?*

We can use term by term integration to derive more power series formulas:

4. ArcSin[x]
   Substitute $-x^2$ *in the binomial series to obtain a series for* $1/\sqrt{1 - x^2}$. *Then integrate to obtain*

$$\mathrm{ArcSin}[x] = x + \sum_{k=1}^{\infty} \left[\frac{1 \cdot 3 \cdot 5 \cdots (2k - 1)}{2 \cdot 4 \cdot 6 \cdots (2k)} \frac{x^{2k+1}}{2k + 1}\right]$$

For which values of x is this series uniformly absolutely convergent? HINT: Make the substitution $x = \mathrm{Sin}[u]$, *so*

$$\int \frac{dx}{\sqrt{1 - x^2}} = \int \frac{\mathrm{Cos}[u] \, du}{\sqrt{\mathrm{Cos}^2[u]}} = \int du$$

5. *The Dirichlet convergence theorem from Fourier series shows that for all* $0 \le x \le 2\pi$,

$$\frac{x^2}{4} - \pi\frac{x}{2} + \frac{\pi^2}{6} = \sum_{k=1}^{\infty} \left[\frac{1}{k^2} \mathrm{Cos}[k\,x]\right] = \mathrm{Cos}[x] + \frac{1}{4} \mathrm{Cos}[2\,x] + \frac{1}{9} \mathrm{Cos}[3\,x] + \cdots$$

Use this fact to deduce the following special cases:

$$\frac{\pi^2}{6} = 1 + \frac{1}{4} + \frac{1}{9} + \cdots = \sum_{k=1}^{\infty} \left[\frac{1}{k^2}\right] \qquad \text{and} \qquad \frac{\pi^3}{32} = \sum_{k=1}^{\infty} \left[\frac{(-1)^{n+1}}{(2n - 1)^3}\right]$$

*See the program **SomeSums**.*

Mathematics is not magic. If the hypotheses are not satisfied, the conclusions need not hold.

Problem 26.2. INTEGRATION ▬▬▬▬▬▬▬▬▬▬▬▬▬▬▬▬▬▼

Is the following proof that $\text{Log}[3] = \dfrac{8}{3} - \dfrac{16}{4} + \dfrac{32}{5} - \dfrac{64}{6} + \dfrac{128}{7} + \cdots$ *correct? If so, explain the steps. If not, explain the error(s). We have* $\dfrac{1}{1-x} = 1 + x + x^2 + x^3 + \cdots$ *So, letting* $x = -u$, *we get*

$$\frac{1}{1+u} = 1 - u + u^2 - u^3 + \cdots$$

Integrating both sides from 0 to x gives

$$\int_0^x \frac{1}{1+u}\, du = \int_0^x 1\, du - \int_0^x u\, du + \int_0^x u^2\, du + \cdots$$

$$\text{Log}[1+x] = x - \frac{x^2}{2} + \frac{x^3}{3} + \cdots + (-1)^{n+1}\frac{x^n}{n} + \cdots \quad \text{and letting } x = 2, \text{ we obtain,}$$

$$\text{Log}[1+2] = 2 - 2 + \frac{8}{3} + \cdots + (-1)^{n+1}\frac{2^n}{n} + \cdots$$

$$\text{Log}[3] = \frac{8}{3} - \frac{16}{4} + \frac{32}{5} - \frac{64}{6} + \frac{128}{7} + \cdots$$

Sum several terms of this with the computer or a calculator and show that it is nonsense.

▬▬▬▬▬▬▬▬▬▬▬▬▬▬▬▬▬▬▬▬▬▬▬▬▬▬▬▬▬▬▬▬▲

The next example gives a different formula for Log[3].

Example 26.4. $\text{Log}[3] = 2\left(\frac{1}{1\cdot2} + \frac{1}{3\cdot2^3} + \frac{1}{5\cdot2^5} + \cdots\right)$

We know

$$\text{Log}[1+x] = x + \frac{x^2}{2} + \frac{x^3}{3} + \cdots$$

so substituting $-x$ gives

$$-\text{Log}[1-x] = x - \frac{x^2}{2} + \frac{x^3}{3} + \cdots$$

We also know $\text{Log}[a] - \text{Log}[b] = \text{Log}[a/b]$ from high school and Chapter 28. Adding the series above and using this identity, we have

$$\text{Log}\left[\frac{1+x}{1-x}\right] = 2\left(x + \frac{x^3}{3} + \frac{x^5}{5} + \cdots\right)$$

Some algebra gives us

$$\text{Log}[3] = \text{Log}\left[\frac{1+\frac{1}{2}}{1-\frac{1}{2}}\right]$$

Together, these computations yield

$$\text{Log}[3] = 2\left(\frac{1}{2} + \frac{1}{2^3\cdot3} + \frac{1}{2^5\cdot5} + \cdots\right)$$

All of these computations are valid because the log series is uniformly absolutely convergent for $|x| \leq 1/2$. See the program **SomeSums**.

26.4   Differentiation of Power Series

Theorem 26.6.   *Differentiation of Power Series*
Suppose that a power series converges uniformly absolutely for $|x| \leq \rho$ to $S[x]$,

$$S[x] = a_0 + a_1 x + a_2 x^2 + a_3 x^3 + \cdots$$

Then, the derivative of $S[x]$ exists and the series obtained from term by term differentiation converges uniformly absolutely to it on $|x| \leq \rho$,

$$\frac{dS[x]}{dx} = a_1 + 2 a_2 x + 3 a_3 x^2 + \cdots + n a_n x^{n-1} + \cdots$$

We omit the proof, but warn you that this theorem applies only to power series. Here are some familiar examples to remind you of this limited applicability.

Weierstrass's nowhere differentiable function

$$W[x] = \mathrm{Cos}[x] + \frac{1}{2}\mathrm{Cos}[3x] + \frac{1}{2^2}\mathrm{Cos}[3^2 x] + \cdots$$

does not have a derivative at any value of x. Differentiate the series term by term and see what happens.

The identity

$$|t| = 1 + \frac{1}{2}[t^2 - 1] + \frac{\frac{1}{2}(\frac{1}{2} - 1)}{2}[t^2 - 1]^2 + \frac{\frac{1}{2}(\frac{1}{2} - 1)(\frac{1}{2} - 2)}{3 \cdot 2}[t^2 - 1]^3 + \cdots$$
$$\cdots + \left(\begin{smallmatrix}\frac{1}{2}\\n\end{smallmatrix}\right)[t^2 - 1]^n + \cdots$$

is perfectly valid for $|t| \leq 1$. However, the absolute value does not have a derivative at $t = 0$. Term by term differentiation of this series is still possible, but to what does the series converge? This series might be fun to explore with the computer. You can define it by substituting $x = t^2 - 1$ into the convergent binomial series for $\sqrt{1+x}$. Some partial sums are shown in Figure 26.3.:

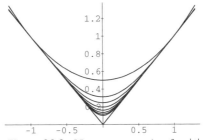

Figure 26.3: Nonpower series for $|t|$

If we have a convergent power series, its derivative is also a convergent power series. Hence, we can differentiate that and obtain another convergent power series. In other words, convergent power series have infinitely many derivatives on intervals of convergence.

Exercise set 26.4

1. *Show that the series* $y = \sum_{k=0}^{\infty} \left[\frac{x^{4k}}{(4k)!} \right]$ *converges for all* x *(uniformly and absolutely on* $|x| \leq \rho$, *for any* ρ*) and satisfies*

$$\frac{d^4 y}{dx^4} = y$$

with certain initial conditions.

What initial conditions are needed to make the solution unique? (HINT: Apply the phase variable trick enough times to make it a first-order system.)

2. Cosh[x] and Sinh[x]

The hyperbolic sine and cosine arise in some applications such as the catenary project. They can be defined as solutions to initial value problems, as series, or by identities with the natural exponential function. All of these descriptions are useful.

  (a) *Find power series solutions to the initial value problems*

$$
\begin{array}{ccc}
y = \text{Cosh}[x]: & & y = \text{Sinh}[x]: \\
y[0] = 1 & and & y[0] = 0 \\
y'[0] = 0 & & y'[0] = 1 \\
\dfrac{d^2 y}{dx^2} = y & & \dfrac{d^2 y}{dx^2} = y
\end{array}
$$

  (b) *Where do your series converge?*
  (c) *Prove the connections with the natural exponential*

$$\text{Cosh}[x] = \frac{e^x + e^{-x}}{2} \qquad\qquad \text{Sinh}[x] = \frac{e^x - e^{-x}}{2}$$

(NOTE: You can use either series or differential equations.) Notice the analogy to Euler's formula, which gives

$$\text{Cos}[x] = \frac{e^{ix} + e^{-ix}}{2} \qquad\qquad \text{Sin}[x] = \frac{e^{ix} - e^{-ix}}{2i}$$

  (d) *Show that* $\text{Cosh}[x] = \text{Cos}[ix]$ *and* $\text{Sinh}[x] = \frac{\text{Sin}[ix]}{i}$.
  (e) *Show that* $\frac{d\,\text{Cosh}[x]}{dx} = \text{Sinh}[x]$ *and* $\frac{d\,\text{Sinh}[x]}{dx} = \text{Cosh}[x]$.
  (f) *Show that* $\text{Cosh}^2[t] - \text{Sinh}^2[t] = 1$. *This implies that these functions parametrize a hyperbola. (Sine and cosine parametrize the circle* $x^2 + y^2 = 1$.) *If you let* $x = \text{Cosh}[t]$ *and* $y = \text{Sinh}[t]$, *Mathematica's ParametricPlot[.] command will let you draw this hyperbola.*

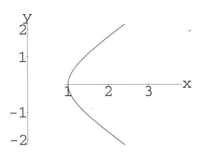

Figure 26.4: *Hyperbola*

(g) *Show that these hyperbolic functions satisfy the nonlinear first-order differential equations*

$$\frac{dy}{dx} = \sqrt{y^2 - 1} \qquad and \qquad \frac{dy}{dx} = \sqrt{y^2 + 1}$$

(HINT: You can just solve it. Separate variables and integrate with the change of variables $\text{Sinh}[u] = y$*, so* $dy = \text{Cosh}[u]\, du$ *and* $1 + y^2 = 1 + \text{Sinh}^2[x] = \text{Cosh}^2[u]$*, making*

$$\int \frac{dy}{\sqrt{1 + y^2}} = \int \frac{\text{Cosh}[u]\, du}{\sqrt{\text{Cosh}^2[u]}} = \int du$$

Can you think of an easier way?)

3. Bessel Functions

The Bessel functions $J_0[x]$ *and* $J_1[x]$ *are defined as solutions of the differential equations*

$$x^2 \frac{d^2 y}{dx^2} + x \frac{dy}{dx} + (x^2 - n^2)y = 0$$

for $n = 0$ *and* $n = 1$*, respectively. Bessel functions arise in the study of wave propagation or heat flow when there is cylindrical symmetry. (In this case, the partial differential equations describing those phenomena specialize to the differential equation.) Notice that there is a difficulty with these equations when we rewrite them as two-dimensional first-order systems,*

$$\frac{dy}{dx} = z = f(x, y, z)$$
$$\frac{dz}{dx} = -\frac{1}{x} z - \frac{x^2 - n^2}{x^2} y = g(x, y, z)$$

We cannot apply the Uniqueness Theorem 21.6 for initial conditions at zero because $g(x, y, z)$ *is discontinuous at* $x = 0$*. However,*

(a) *Show that the series*

$$J_0[x] = \sum_{k=0}^{\infty} [(-1)^k \frac{x^{2k}}{(k!)^2\, 2^{2k}}]$$

$$J_1[x] = \sum_{k=0}^{\infty} [(-1)^k \frac{x^{2k+1}}{k!\, (k+1)!\, 2^{2k+1}}]$$

converge for all x.

(b) *Show that these series satisfy the differential equations with the respective values of $n = 0$ and $n = 1$.*

(c) *Show the identities $J_0'[x] = -J_1[x]$ and $x\, J_0[x] = \frac{d(x\, J_1[x])}{dx}$.*

4. Taylor's Formula

Suppose that a function is represented by a uniformly absolutely convergent power series,

$$f[x] = a_0 + a_1\, x + a_2\, x^2 + \cdots + a_n\, x^n + \cdots$$

for $|x| \le \rho$. We know that the derivative exists and that it is given by

$$f'[x] = a_1 + 2\, a_2\, x + \cdots + n\, a_n\, x^{n-1} + \cdots$$

Let $x = 0$ in this series to show that $f'[0] = a_1$.
Differentiate again, obtaining the convergent series

$$f''[x] = 2\, a_2 + 3 \cdot 2a_3\, x + \cdots + n(n-1)\, a_n\, x^{n-2} + \cdots$$

Set $x = 0$ in this series to see that $f''[0] = 2\, a_2$.
Differentiate again, set $x = 0$, and show that $f^{(3)}[0] = 3 \cdot 2\, a_3$ or that $a_3 = \frac{1}{3 \cdot 2} f^{(3)}[0]$.
Generalize and show Taylor's formula

$$a_n = \frac{1}{n!}\, f^{(n)}[0]$$

5. *Show that the function $f[x] = \frac{\mathrm{Sin}[x]}{x}$ is actually infinitely smooth at $x = 0$ with the obvious extension of the formula. (See the program **Classical Series**.)*

CHAPTER 27

The Edge of Convergence

This chapter investigates some weakly convergent series.

Inside their interval of convergence, power series are strongly convergent. For example, the ratio test says that convergence is as strong as a geometric series. At the radius of convergence, a power series may either converge or diverge. Also, Fourier series often cannot be estimated uniformly. In these cases, the series are converging by cancellations. We begin this chapter with the simplest kind of convergence by cancellation.

27.1 Alternating Series

Alternating series have the form

$$a_0 - a_1 + a_2 - a_3 + \ldots$$

where a_k is a sequence of positive numbers and the signs in the sum alternate. If the sequence a_k decreases toward zero, the series converges.

The alternating series

$$1 - \frac{1}{2} + \frac{1}{3} - \frac{1}{4} + \cdots = \sum_{k=1}^{\infty} [(-1)^{k+1} \frac{1}{k}]$$

is called "a conditionally convergent series." It is a very sensitive kind of convergence. It converges, but the sum of its absolute value terms does not.

Definition 27.1. *Conditional Convergence*
If the series of positive terms $a_0 + a_1 + a_2 + a_3 + \ldots$ diverges, but the series $a_0 - a_1 + a_2 - a_3 + \ldots$ converges, we say the alternating series converges conditionally.

The harmonic series

$$1 + \frac{1}{2} + \frac{1}{3} + \frac{1}{4} + \cdots = \sum_{k=1}^{\infty} [\frac{1}{k}]$$

diverges to infinity; in fact, we will see that

$$1 + \frac{1}{2} + \frac{1}{3} + \frac{1}{4} + \cdots + \frac{1}{n} > \text{Log}[n] \to \infty$$

The reason that the alternating harmonic series converges is simply decreasing cancellation. First, we add 1. Then, we subtract a half. Next, we add a third. We never go back above one, nor back below a half, because the size of the oscillations decreases. At each point in this process, we never move more than the next term in the direction of the sign of that term. This is an error estimate but a very dangerous one.

Figure 27.1: Alternating decreasing moves

The fact that the partial sums of the series never move beyond one of the established positions means that the next term is an estimate of the error between the infinite sum and the partial one. This is dangerous because alternating series are about the only general case where this estimate works.

This idea of decreasing oscillations generalizes to prove:

Theorem 27.2.   *The Alternating Series Test*
If $a_0, a_1, a_2, \ldots$ is a positive and decreasing sequence, the alternating series

$$a_0 - a_1 + a_2 - a_3 + \ldots \quad converges$$

if and only if

$$\lim_{k \to \infty} a_k = 0$$

When the limit is zero, the error in summing $a_0 - a_1 + a_2 - a_3 + \ldots + a_n$ is no more than the next term, a_{n+1}.

Usually, the next term is not a good estimate of the error in an infinite series, but we always have the "necessary condition."

Theorem 27.3.
If the series $a_0 + a_1 + a_2 + a_3 + \ldots$ converges (with terms of arbitrary sign), then $\lim_{k \to \infty} a_k = 0$. However, there are divergent series $a_0 + a_1 + a_2 + a_3 + \ldots$ with $\lim_{k \to \infty} a_k = 0$.

For example, the next term of the positive harmonic series, $1/n$, can be small and the error in this case is always infinite. Moreover, floating point computer arithmetic often renders the alternating series estimate useless. (This is explored in The Big Bite of the Subtraction Bug project. Conditionally convergent series and improper integrals are explored in additional projects.)

Example 27.1. $\text{Log}[2] = 1 - \frac{1}{2} + \frac{1}{3} + \cdots$

The observation that the alternating harmonic series converges shows that the log series may be used when $x = 1$:

$$\text{Log}[1 + x] = \sum_{k=1}^{\infty} [(-1)^{k+1} \frac{x^k}{k}] = x - \frac{x^2}{2} + \frac{x^3}{3} + \cdots$$

the limit of the terms on the right hand side is a convergent series. On the left side of this identity (which we know for $|x| < 1$) we know

$$\lim_{x \to 1} \text{Log}[1 + x] = \text{Log}[2]$$

For a partial sum, we know

$$\lim_{x \to 1} [1 - \frac{x^2}{2} + \cdots + (-1)^{n+1} \frac{x^n}{n}] = 1 - \frac{1}{2} + \frac{1}{3} - \frac{1}{4} + \cdots + (-1)^n \frac{1}{n}$$

Our estimate, described above, for alternating series shows that

$$|\text{Error}_n[x]| \leq \frac{1}{n+1}$$

for all x. Hence, we do have the formula

$$\text{Log}[2] = 1 - \frac{1}{2} + \frac{1}{3} + \cdots$$

A general fact about power series is that if we can find a point of convergence, even conditional convergence, then we can use geometric comparison to prove convergence at smaller values.

Theorem 27.4. *If the power series*

$$a_0 + a_1 x + a_2 x^2 + \cdots + a_n x^n + \cdots$$

converges for a particular $x = x_1$, then the series converges uniformly and absolutely for $|x| \leq \rho < |x_1|$, for any constant ρ.

PROOF:

Because the series converges at x_1, we must have $a_n x_1^n \to 0$. If $|x| \leq \rho < |x_1|$, then

$$|a_n x^n| = |a_n x_1^n| \left| \frac{x}{x_1} \right|^n \leq \left| \frac{\rho}{x_1} \right|^n = r^n$$

is a geometric majorant for the tail of the series.

──────────── **Exercise set 27.1** ────────────

1. *Verify that*

$$\frac{\pi}{4} = 1 - \frac{1}{3} + \frac{1}{5} - \frac{1}{7} + \cdots$$

(HINT: ArcTan[1] = $\pi/4$.)

27.2 Telescoping Series - CD

A series of the form $\sum_{k=1}^{\infty}[a_{k+1} - a_k]$ collapses like a telescope.

Remainder of this section only on CD

27.3 Integrals Compared to Series - CD

A way to estimate series above or below is to compare them with integrals.

Remainder of this section only on CD

27.4 Limit Comparisons - CD

Each time we learn a new convergent or divergent series, we can use it to compare to many other series.

Theorem 27.5. *Limit Comparison of Series*
Suppose the sequences a_k and b_k satisfy $\lim_{k \to \infty} \frac{a_k}{b_k} = L \neq 0$. Then
(a) *If $\sum_{k=1}^{\infty}[a_k]$ converges, so does $\sum_{k=1}^{\infty}[b_k]$.*
(b) *If $\sum_{k=1}^{\infty}[a_k]$ diverges, so does $\sum_{k=1}^{\infty}[b_k]$.*

Remainder of this section only on CD

27.5 Fourier Series - CD

Fourier series arise in many mathematical and physical problems.

Remainder of this section only on CD

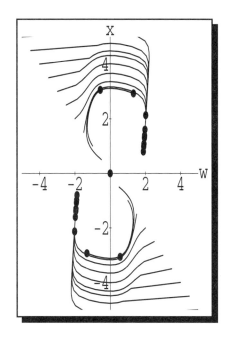

Part 7

Appendices on CD

High School Review with Computing

This chapter reviews the ideas of independent and dependent variables and parameters. We do this in the context of some down-to-earth applications. We want to help you to develop careful working habits to use in calculus. We need you to understand function notation in order to communicate ideas.

Chapter only on CD

CHAPTER 29 Complex Numbers

This Chapter only appears on CD

Index